# Advances in Polymer Composite Research

This book emphasizes the importance of experimental characterization techniques and computational modeling tools in polymer composites. The topics covered include finite element analysis, computational fluid dynamics, molecular dynamics simulations, machine learning, material informatics, multiscale modeling, advanced characterization techniques, and the emerging field of nanocomposites. Each chapter provides detailed discussions, case studies, and examples to illustrate the practical application of these techniques in polymer composite research.

Features:

- Offers a comprehensive exploration of polymer composites encompassing both experimental and computational approaches.
- Showcases most recent findings, methodologies, technologies, and applications in the field.
- Explores real-world case studies, industrial applications, and potential commercialization opportunities.
- Discusses the understanding, analysis, and design of polymer composites.
- Includes LAMMPS-, Ansys-, ABAQUS-, and Materials Studio–based simulation examples.

This book is aimed at graduate students and researchers in polymers, polymer composites, and materials science.

# Advances in Polymer Composite Research

## Integrating Experimental and Computational Approaches

Edited by Atul K. Maurya, Gaurav Manik,
and Sushanta K. Sethi

CRC Press
Taylor & Francis Group
Boca Raton London New York

CRC Press is an imprint of the
Taylor & Francis Group, an **informa** business

First edition published 2025
by CRC Press
2385 NW Executive Center Drive, Suite 320, Boca Raton FL 33431

and by CRC Press
4 Park Square, Milton Park, Abingdon, Oxon, OX14 4RN

*CRC Press is an imprint of Taylor & Francis Group, LLC*



ISBN: 9781032713946 (hbk)
ISBN: 9781032759548 (pbk)
ISBN: 9781003476337 (ebk)

DOI: 10.1201/9781003476337

Typeset in Times
by Apex CoVantage, LLC

# Contents

# Acknowledgments

**Atul K. Maurya** is highly grateful to the University of Massachusetts, Lowell, United States. The research environment and data repository of the University of Massachusetts, Lowell, United States, was instrumental in achieving the project goal of the current book submission.

**Gaurav Manik** is highly appreciative of the infrastructural and research support provided by IIT Roorkee and gratefully acknowledges that enabled smooth completion of the initiative of this book submission for the benefits of scientific fraternity.

**Sushanta K. Sethi** extends his heartfelt gratitude to IIT Dharwad for their invaluable financial support in the form of a seed grant. This assistance has been instrumental in propelling forward his research endeavors, enabling exploration and innovation in his academic pursuits.

# Preface

The need for this book on polymer composites study from both experimental and computational perspectives arises from the increasing relevance and demand for advanced composite materials in various industries. Polymer composites offer synergistic properties, such as superior strength, stiffness, and thermal and electrical characteristics from two or more distinct phases of materials simultaneously, making them highly attractive for applications in aerospace, construction, automotive, and many other sectors. However, developing and optimizing polymer composites require a comprehensive understanding of their behavior at different length scales, from the molecular level to macroscopic performance. This book aims to bridge the gap between experimental and computational approaches by providing a holistic view of the study of polymer composites. It recognizes the importance of experimental characterization techniques and computational modeling tools in advancing the knowledge and design of these materials.

The proposed book on polymer composites study on experimental and computational approaches is a comprehensive resource that addresses the critical aspects of understanding and optimizing polymer composites. It covers a wide range of topics, including experimental characterization techniques, finite element analysis, computational fluid dynamics, molecular dynamics simulations, machine learning, material informatics, multiscale modeling, advanced characterization techniques, and the emerging field of nanocomposites and filler/reinforcements treatment and use of coupling agent for interfacial interaction enhancement. Each chapter provides in-depth discussions, case studies, and examples to illustrate the practical application of the techniques in polymer composite research. By combining experimental and computational perspectives, this book aims to offer researchers, engineers, and students a comprehensive understanding of polymer composites and equip them with the necessary tools and knowledge to design and develop advanced composite materials with enhanced properties and performance.

This book on polymer composites study from both experimental and computational perspectives caters to a global audience of researchers, engineers, and students involved in developing and applying advanced composite materials to various industrial applications. With the increasing demand for lightweight and high-performance materials in industries worldwide, readers from countries such as the United States, United Kingdom, France, Germany, Japan, China, and India, among others, will find valuable insights and knowledge in this book. The diverse range of topics covered, including experimental characterization techniques, computational modeling, and advancements in nanocomposites, offers a global perspective on the latest research and advancements in the field. By addressing the needs of readers across different countries and cultures, this book aims to contribute to the ongoing progress and innovation in polymer composite research on a global scale.

# Editor Biographies

**Atul K. Maurya** is currently working as a postdoctoral research associate in the Department of Chemical Engineering and Materials Science, Michigan State University, East Lansing (MSU), USA. He has done his PhD (IIT Roorkee) in polymer composites, MTech (IIT Hyderabad) in material science and metallurgical engineering, and BTech (CIPET Lucknow) in plastics engineering. Prior to joining MSU, he served at UMass Lowell and Pittsburg State University as a postdoc fellow, at Reliance Industries Limited as a research scientist, and in academia (CIPET Ahmedabad) as a faculty. His research area encompasses the processing and optimization of polymer and polymer composites. Dr. Maurya has an impressive publication record, with around 22 research publications in internationally recognized SCI journals, international conferences, and book chapters. He has also published a patent in the field of polymer composites. He has made significant contributions to esteemed peer-reviewed journals in the field of polymer engineering and science, in cellulose polymer composites, in the *Journal of Polymers and the Environment*, among others.

**Gaurav Manik** is currently acting as a head (HOD) of the Department of Polymer and Process Engineering at IIT Roorkee. He joined IIT Roorkee in January 2013 as an assistant professor in the same department. Dr. Manik holds a PhD in chemical engineering from IIT Bombay, as well as an MTech from IIT Kanpur, and a BTech from HBTI Kanpur. Prior to his tenure at IIT Roorkee, he gained extensive experience in both industry and academia, accumulating over 18 years of professional expertise. He has worked with esteemed organizations, such as 3M, Indo Gulf Fertilizers & Chemicals Ltd., Classic Stripes Pvt. Ltd., BITS Pilani, and BIET Jhansi. Dr. Manik's industry contributions include the development and commercialization of numerous innovative products, as well as the protection of intellectual property through a substantial number of patents and record of inventions.

**Sushanta K. Sethi** is currently an assistant professor in the Department of Mechanical, Materials, and Aerospace Engineering at IIT Dharwad, since July 2023. Prior to this, Dr. Sethi worked as a postdoctoral fellow in the Department of Metallurgical Engineering and Materials Science (MEMS) at IIT Bombay. In February 2021, he successfully completed his PhD in the Department of Polymer and Process Engineering at IIT Roorkee. Dr. Sethi holds an MTech from IIT Kharagpur and a BTech in mechanical engineering from BPUT Odisha. His research interests encompass various areas, such as computational materials science, self-clean coatings, functional nanomaterials/quantum dots and their applications, multiscale molecular modeling and simulations of polymer materials, development of novel and environmentally sustainable polymer composites and nanocomposites, as well as coatings and adhesives. Dr. Sethi has an impressive publication record, with around 35 research publications in internationally recognized SCI journals, international conferences, and book chapters. He has made significant contributions to esteemed peer-reviewed journals in the field of polymer engineering and science, ACS sensors, and polymers, polymer composites, and micromachines, in *Frontiers in Bioengineering and Biotechnology*, among others. His remarkable achievements include receiving the Best Doctoral Thesis Award from the Department of Polymer and Process Engineering from IIT Roorkee and recognition from the science, technology, engineering, and management (STEM) society.

# List of Contributors

**Abhiram B. R.**
Department of Civil Engineering
Indian Institute of Science
Bangalore

**Mariappan Anandkumar**
South Ural State University
Chelyabinsk, Russian Federation

**Mukesh Bajya**
Department of Textile Technology
Dr. B. R. Ambedkar National Institute of Technology
Jalandhar, Punjab, India

**O. B. Bembalge**
General Department of Mechanical Materials, and Aerospace Engineering
Indian Institute of Technology (IIT)
Dharwad, India

**Sampat Bhati**
MM Group
Kotka, Finland

**G Bhavani**
Department of Physics
PSG Institute of Technology and Applied Research
Coimbatore, Tamilnadu, India

**Sakchi Bhushan**
Department of Polymer and Process Engineering
Indian Institute of Technology–Roorkee
Uttarakhand, India

**Gloria Biswal**
Department of Mechanical Engineering
Indian Institute of Technology–Kharagpur
West Bengal, India

**Shourav Biswas**
Department of Materials Engineering
Rajshahi University of Engineering and Technology (RUET)
Rajshahi, Bangladesh

**Sujay Chattopadhyay**
Department of Polymer and Process Engineering
Indian Institute of Technology–Roorkee
Uttarakhand, India

**Akshay K. Chaudhari**
Aavis Pharmaceuticals
Jackson Trail Road
Hoschton, GA, USA

**Arjun S. Chaudhari**
Aavis Pharmaceuticals
Hoschton, GA, USA

**Vishwa Dave**
Department of Chemical Sciences
School of Science, GSFC University
Vadodara, Gujarat, India

**Saurav Dixit**
Division of Research and Development
Lovely Professional University
Phagwara, Punjab, India

**Praveen Eliyas**
Department of Basic Engineering
Government Polytechnic College–Kottur
Theni, Tamilnadu, India

**Kirtiraj K. Gaikwad**
Indian Institute of Technology–Roorkee Saharanpur Campus
India

**Rahul Gautam**
Department of Polymer and Process Engineering
Indian Institute of Technology–Roorkee Saharanpur Campus
Paper Mill Road, Saharanpur, India

**Priya Goel**
Department of Chemical Sciences
School of Science, GSFC University
Vadodara, Gujarat, India

**Hariome Sharan Gupta**
Department of Polymer and Process Engineering
Indian Institute of Technology–Roorkee Saharanpur Campus
Saharanpur, India

**Vinay Gupta**
Department of Physics
Khalifa University
Abu Dhabi, United Arab Emirates

**Pramod M. Gurave**
Department of Textile and Fiber Engineering
Indian Institute of Technology
Delhi, India

**Arti Hansda**
Department of Life Sciences
School of Science, GSFC University
Vadodara, Gujarat, India

**Alisan Imam**
Department of Materials Engineering
Rajshahi University of Engineering and Technology (RUET)
Rajshahi, Bangladesh

**Akib Jabed**
Department of Materials Engineering
Rajshahi University of Engineering and Technology (RUET)
Rajshahi, Bangladesh

**Kannan PK**
Department of Physics
PSG Institute of Technology and Applied Research
Coimbatore, Tamilnadu, India

**Mosa. Afroja Khatun**
Department of Materials Engineering
Rajshahi University of Engineering and Technology (RUET)
Rajshahi, Bangladesh

**Sanjay Krishna**
Shiv Nadar Institution of Eminence
Greater Noida, India

**Pankaj Kumar**
School of Nanoscience and Technology
Indian Institute of Technology–Kharagpur
West Bengal, India

**Sachin Kumar**
Department of Polymer and Process Engineering
Indian Institute of Technology–Roorkee
Uttarakhand, India

**Rajesh Mahadeva**
Department of Computer Science and Engineering
Manipal Institute of Technology Manipal Academy of Higher Education
Manipal, India
and
Department of Physics, Khalifa University
Abu Dhabi, United Arab Emirates

**Subrata Kumar Maiti**
Department of Polymer and Process Engineering
Indian Institute of Technology–Roorkee
Uttarakhand, India

**Priyabrata Mandal**
Research and Technical Division,
Permionics Membranes Pvt. Ltd.
Vadodara, Gujarat, India

**Tushar Kanti Maiti**
Department of Polymer and Process Engineering
Indian Institute of Technology–Roorkee
Uttarakhand, India
and
Department of Chemical Engineering and Centre for Bioengineering and Biomedical Technologies (CBio)
University of Bath
Bath, United Kingdom

**Gaurav Manik**
Department of Polymer and Process Engineering
Indian Institute of Technology (IIT)–Roorkee
Uttarakhand, India

**Yesha Master**
Department of Life Sciences
School of Science, GSFC University
Vadodara, Gujarat, India

**Atul K. Maurya**
Department of Chemical Engineering and Materials Science
Composite Materials and Structures Center
Michigan State University
East Lansing, MI, USA

**Ganesh Sahadeo Meshram**
Department of Mechanical Engineering
Indian Institute of Technology–Kharagpur
West Bengal, India

**Jolly Mistry**
Department of Chemical Sciences
School of Science, GSFC University
Vadodara, Gujarat, India

**Swati Mohapatra**
Department of Life Sciences
School of Science, GSFC University
Vadodara, Gujarat, India

**Debajyoti Palai**
Department of Metallurgical and Materials Engineering
Indian Institute of Technology–Kharagpur
Kharagpur, India

**Akanksha Pandey**
Department of Chemical Engineering
Indian Institute of Technology–Bombay (IIT-B)
India

**Vishwa Patel**
Department of Chemical Sciences
School of Science, GSFC University
Vadodara, Gujarat, India

**Zarana Patel**
Department of Chemical Sciences
School of Science, GSFC University
Vadodara, Gujarat, India

**Sharad Raghuvanshi**
Department of Paper Technology
Indian Institute of Technology–Roorkee Saharanpur Campus
Paper Mill Road, Saharanpur, India

**Harsh Satiya**
Indian Institute of Technology–Roorkee Saharanpur Campus
India

**Sushanta K. Sethi**
Mechanical Materials and Aerospace Engineering
Indian Institute of Technology (IIT)–Dharwad
Karnataka, India

**Annu Sharma**
Indian Institute of Technology
Chennai, India

**Abhinav K. Singh**
Department of Polymer and Process Engineering
Indian Institute of Technology–Roorkee
Uttarakhand, India

**Manjinder Singh**
Department of Polymer and Process Engineering
Indian Institute of Technology–Roorkee
Uttarakhand, India
and
School of Chemical Engineering
University of Birmingham
UK

**Nisha Tamta**
Department of Polymer and Process Engineering
Indian Institute of Technology–Roorkee DPPE IIT-R SRE Campus
Saharanpur, India

**Md Mahamud Hasan Tusher**
Department of Materials Engineering
Rajshahi University of Engineering and Technology (RUET)
Rajshahi, Bangladesh

**Akarsh Verma**
Department of Mechanical Engineering
University of Petroleum and Energy Studies
Dehradun, India
and
Department of Mechanical Science and Bioengineering
Osaka University, Osaka, Japan

**Anilkumar Yadav**
Department of Textile Technology
Dr. B. R. Ambedkar National Institute of Technology
Jalandhar, Punjab, India

# List of Figures

# List of Tables

# 1 Introduction to Polymer Composites

*Hariome Sharan Gupta, Rahul Gautam, Sharad Raghuvanshi, Sushanta K. Sethi, and O. B. Bembalge*

## 1.1 INTRODUCTION

The term "polymer composite" refers to an assembled material of two or more components which generates better characteristics than individual ingredient. In the polymer composites (PCs), reinforcers (fillers/fibers) are embedded in a polymer matrix. In general, one or more discontinuous phases (fillers/fibers) dispersed across a continuous phase (polymer matrix) make PCs. The phase that is discontinuous is referred to as the reinforcing material, while the phase that is continuous is called the polymer matrix. The idea behind composite materials is that mixing various materials together frequently results in a new material that performs better than any of its component parts alone [1, 2]. The properties of the PCs are subject to the nature of the matrix and the filler/fiber reinforcer. Some factors of polymer matrix can affect the PCs, namely, the type of polymer matrix, the shape and proportion, the interface properties, and also the method of manufacture or compounding. The PCs depend on the properties of the internal structure of the polymer matrix, including their density, electrical resistivity, mechanical tensile strength, and mechanical tensile modulus. Similarly, the reinforcer fillers/fibers' properties and other characteristics, namely, size, maturity, and the techniques used in the fillers/fibers extraction process, also have an influence on PCs. The core and chemical structures of the fillers/fibers are related to properties including density, electrical resistivity, and mechanical properties. Due to these characteristics, PCs are highly used in different industries, like energy, industrial, construction, packaging, agricultural, sports, marine, automobile, aircraft, aerospace, medical, civil, electronics, communications, kitchen, and others involving home items applications [1, 3–5].

Research interest has increased in the preceding decades on the use of natural fillers/fibers as a reinforcement for PCs because of their low cost, light weight, affordability, environmental friendliness, easy availability, and biodegradable characteristics. Some natural fillers/fibers are mostly used for PCs, namely, cotton, sisal, jute, banana, coir, oil palm, flax, hemp, abaca, alfa, bamboo, bagasse, kenaf ramie, pineapple, etc. [4, 6, 7]. The compounding and processing of natural fillers/fibers with polymer matrices are very easy, which have processes at low temperatures. But handling and safety problems have been found during the use of natural filler/fiber reinforcer in PCs. The most significant issue with natural fillers/fibers is their hydrophilic nature, which makes them expand and eventually smell or rot due to fungal attack or invasion. Because they are made of lignocellulose, which has highly polarized hydroxyl groups, natural fillers/fibers are polar and hydrophilic in nature. Therefore, these fillers/fibers are fundamentally incompatible with nonpolar thermoplastics that are hydrophobic in nature, such as polyolefins polyethylene (PE), polyolefins polypropylene (PP), polystyrene (PS), polyvinyl chloride (PVC), polyamides (PAs), etc. The main drawback of utilizing fibers in the matrices is their inability to form a strong interfacial bond between nonpolar hydrophobic polymer matrix and the polar hydrophilic fillers/fibers. In addition, inadequate matrix fiber wetting causes difficulty in mixing and compounding, which contributes to composites with weak interfaces [7–9]. To improve the adhesion between reinforcing fillers/fibers and the polymer matrix, there is a need for fillers/fibers to undergo surface coatings, physical, chemical, and biological treatments which decrease their absorption of moisture and their hydrophilicity. Physical treatment includes plasma, ozone, and gamma ray treatment; chemical treatments, meanwhile, include alkaline, silane, acetylation, permanganate,

DOI: 10.1201/9781003476337-1

peroxide, benzoylation, acrylation, acrylonitrile, isocyanate, stearic acid, and malleated treatment; and biological treatments come in the form of fungal and enzyme treatments. Also one of the methods used to reduce the moisture absorption and hydrophilicity of natural fillers/fibers is to treat them hydrothermally. This can make cellulose more crystallin and thus help lower the amount of moisture the fillers/fibers absorb. Furthermore, a portion of the hemicellulose is removed during the hydrothermal process, which lowers moisture absorption and hydrophilicity [9–15].

In natural filler/fiber–reinforced PCs, water absorption is a big problem, so to reduce the problems, treatment of synthetic fillers/fibers is needed so they can act as a reinforcement in PCs. The need for synthetic fillers/fibers is growing worldwide as it is an essential material for filler/fiber-reinforced composites. Due to their extraordinary properties, synthetic fillers/fibers are becoming more and more necessary as the market for distinctive composite materials expands. Synthetic fillers/fibers and natural fillers/fibers are strong competitors in the market for PCs. However, there has been interest between synthetic fillers/fibers and natural fillers/fibers in the last 20 years. Synthetic fillers/fibers, such as ceramic textiles, carbon, glass, aramid, and polymeric fibers, are frequently utilized as reinforcement fillers/fibers in PCs since they have higher mechanical properties (tensile strength and tensile modulus). However, synthetic fillers/fibers have the drawback of potentially irritating the human skin if not handled properly, which causes a health risk. Synthetic fillers/fibers clearly have higher desirable tensile strength and tensile modulus than natural fillers/fibers. Synthetic fillers/fibers are frequently used to make sophisticated PCs in the automotive industry, and in aerospace and aircraft applications, due to their higher mechanical, thermal, physical, chemical, and inertia properties [16–18].

Now the time has come that researchers are focusing on polymer nanocomposites, where nanofillers/fibers are used as a reinforcement in PCs. Some nanofiller/fiber reinforcers include clays and carbon nanotubes, and other nanofillers/fibers are used for PCs. Nanofiller/fiber reinforcers are reinforced in PCs in the nanoscale (size of nanofillers/fibers is 1–100 nm). The dispersion of the nanofiller/fiber reinforcer with the polymer matrix in the nanoscale provides macroscopic homogeneity in PCs. The nanocomposites provide better mechanical, thermal, and physical properties due to the dispersion of the nanofillers/fibers' macroscopic homogeneity level in the PCs. Consequently, nanocomposites have gained an enormous popularity, and scientists are always working to enhance the processes used in their manufacturing to provide them useful features [19–21].

This view aims to investigate the current state of PCs. The numerous types of reinforcing fillers/fibers and polymer matrices relevant to the PCs are provided, considering the significance and interaction between fillers/fibers and polymer matrices. PCs are categorized separately in this study, and their uses are provided in detail.

## 1.2 REQUIRED MATERIALS FOR POLYMER COMPOSITES

For the preparation of PCs, mainly two materials are need, that is, polymer matrices and reinforcing fillers/fibers. The properties of the PCs depend on the properties and nature of the reinforcing fillers/fibers and polymer matrices.

### 1.2.1 Polymer Matrices

Polymer matrices play an important role in PCs which are used as a base material in the composites. Mainly, the polymer matrices are used to hold or bind the reinforcing fillers/fibers in the PCs. Considerably, the two types of polymer matrix utilized in PCs are thermosets and thermoplastics. It is difficult to choose the right polymer matrix for the PCs based on their usage, because the polymer matrix has created a direct impact on the final composite properties. The polymer matrix employed in the composite determines its transverse tensile properties, shear strength, compressive strength, and resistance to heat and environmental factors, while the filler/fiber reinforcement utilized in the composite determines its longitudinal tensile properties. The polymer matrices protect the reinforcing fillers/fibers in PCs from external factors, including air, humidity, heat, moisture, friction, abrasions, chemicals, environmental factors, etc. Polymer matrices are also used to protect reinforcing fillers/fibers from the external applied load, wherein the PCs' load is transferred from matrix to reinforcing fillers/fibers and also aligns and stabilizes the fillers/fibers [22–24].

**Figure 1.1** shows the different types of polymer matrices [23, 25] which have been used for PCs. Polymer matrices play an important role in PCs, where the properties of the PCs depend on the nature of the polymer matrices. The polymer matrix helps to significantly improve the mechanical properties of PCs, making it a suitable composite material for different applications. **Table 1.1** shows the mechanical properties of the polymer matrices which have been used for PCs.

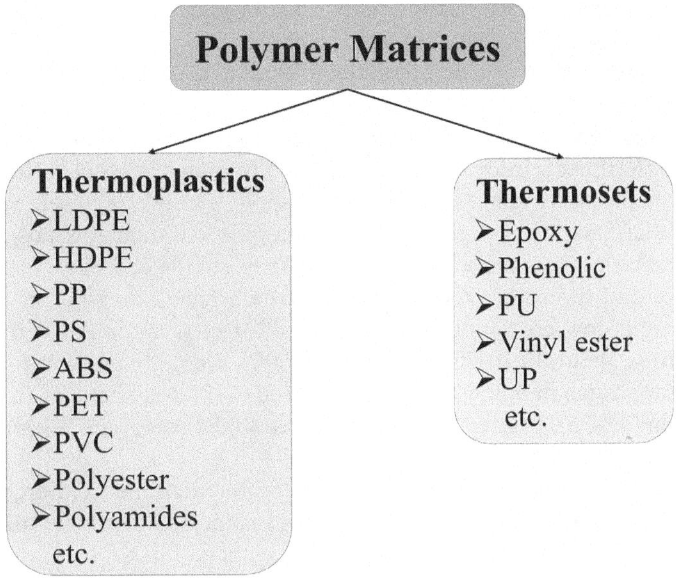


**FIGURE 1.1** Types of polymer matrices.

**TABLE 1.1**
**Mechanical Properties of the Polymer Matrices**

| Polymer Matrices | Density (gm/cm$^3$) | Tensile Modulus (GPa) | Tensile Strength (MPa) | Elongation at Break (%) | References |
|---|---|---|---|---|---|
| LDPE | 0.91–0.925 | 0.055–0.38 | 40–78 | 90–800 | [26] |
| HDPE | 0.94–0.96 | 0.4–1.5 | 14.5–38 | 2.0–130 | [26] |
| PP | 0.899–0.92 | 0.95–1.77 | 26–41.4 | 15–700 | [26] |
| PS | 1.04–1.05 | 1.2–2.6 | 35.9–56.6 | 1.2–3.6 | [27] |
| ABS | 1–1.2 | 1.1–2.9 | 27.6–55.2 | 1.5–100 | [27, 28] |
| PET | 1.29–1.4 | 2.76–4.14 | 48.3–72.4 | 30–300 | [27, 28] |
| PVC | 1.38 | 3 | 53 | – | [23] |
| Polyester | – | 4.03 | 50 | – | [28] |
| Polyamides | 1.12–1.14 | 2.59–3.2 | 71–165 | 30–100 | [27, 28] |
| Epoxy | 1.1–1.4 | 0.5–6 | 26–90 | 1.9–3.5 | [28, 29] |
| Phenolic | 1.3 | 2.7–4.1 | 50–55 | – | [29] |
| Vinyl ester | 1.4 | 3.8 | 86 | – | [23] |
| Polyurethane (PU) | 1.12 | 1.31 | 40 | – | [23] |
| Unsaturated polyester (UP) | 1.2–1.5 | 1–4 | 61–63 | 2.5–4.7 | [29] |

### 1.2.2 Fillers/Fibers

In PCs, reinforcing fillers/fibers play a very important role, where the reinforcing fillers/fibers provide the strength and stiffness of the composites. Reinforcing fillers/fibers used in PCs mainly come in either a natural or synthetic form. In case of synthetic reinforcing filler/fiber composites, synthetic fillers/fibers are formed by humans, and they are categorized as either organic or inorganic, which depends on the elements they include. They are made by chemical synthesis. In PCs, some synthetic filler/fiber reinforcers are used the most, including carbon, aramid, glass, graphene, and Kevlar. The synthetic filler/fiber reinforcers in a composite sustain load because of their strength and stiffness, which are often substantially higher than those of polymer matrices. Synthetic filler/fiber composites provide greater strength to the composites compared with natural filler/fiber composites. But the availability of synthetic fillers/fibers is more difficult compare with natural fillers/fibers composites [30–32].

The abundantly available natural fillers/fibers are a relatively simple material to obtain. In PCs, some natural filler/fiber reinforcers are used the most, namely, abaca, alfa, bagasse, bamboo, banana, coir, flax, hemp, jute, kenaf, oil palm, pineapple, ramie, and sisal. Some of the exceptional advantages that have been found in these materials are their low cost, renewability and biodegradable materials, and high strength, particularly stiffness. Over synthetic fillers/fibers, PCs are used as natural filler/fiber reinforcements which have a number of unique properties, including lower weight, low cost, biodegradability, no toxicity, no pollution to the environment, and easy recyclability. Natural filler/fiber–reinforced PCs are more popular than synthetic filler/fiber–reinforced composites in modern applications due to their advantages in both the economy and the environment [33–35]. **Figure 1.2** shows the classifications of the fillers/fibers used in PCs as a reinforcement.

The properties of the PCs depend on the properties of the filler/fiber reinforcer used in the composites, where reinforcers are used in PCs as a form of natural or synthetic filler/fiber. **Table 1.2** shows the mechanical properties of some natural and synthetic fillers/fibers which have been used in PCs as reinforcement.

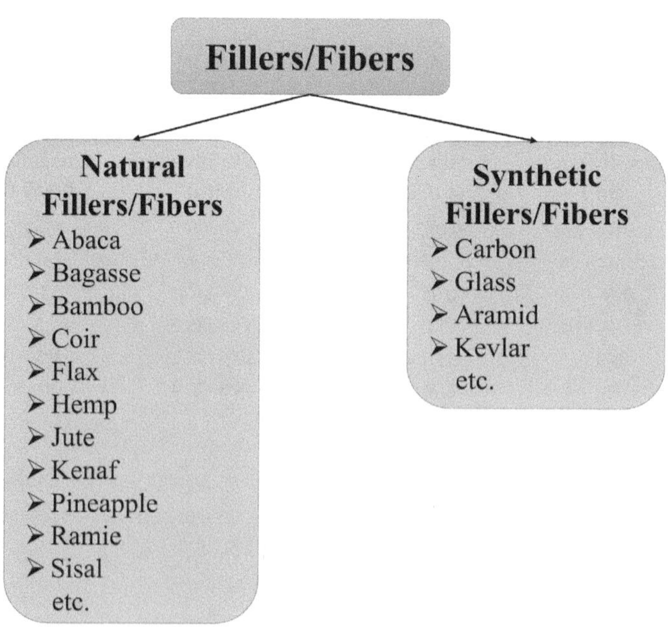


**FIGURE 1.2** Classifications of fillers/fibers.

**TABLE 1.2**
**Mechanical Properties of the Fillers/Fibers**

| Fiber | Density (gm/cm³) | Tensile Modulus (GPa) | Tensile Strength (MPa) | Elongation at Break (%) | References |
|---|---|---|---|---|---|
| Abaca | 1.5 | 12 | 400 | 3–10 | [36] |
| Bagasse | 0.55–1.25 | 2.7–17 | 20–290 | 3–5 | [37, 38] |
| Bamboo | 0.8 | 48–89 | 391–1,000 | 3.5 | [38, 39] |
| Coir | 1.2 | 4–6 | 131–593 | 15–40 | [38, 40] |
| Flax | 1.5 | 27.6 | 345–1,100 | 2.7–3.2 | [36, 38] |
| Hemp | 1.47 | 30–70 | 690 | 1.6–4 | [36, 38] |
| Jute | 1.3–1.4 | 13–26.5 | 393–773 | 1.2–1.5 | [36, 41] |
| Kenaf | 1.45 | 53 | 930 | 1.6 | [36, 38] |
| Pineapple | 1.52–1.56 | 34.5–82.5 | 413–1,627 | 1.6 | [36, 38] |
| Ramie | 1.5 | 61.4–128 | 400–938 | 1.2–3.8 | [36] |
| Sisal | 1.4 | 9.4–22 | 468–640 | 3–7 | [36] |
| Carbon | 1.4–1.74 | 227–240 | 3,800–4,000 | 1.4–1.8 | [29, 42, 43] |
| E-glass | 2.5 | 70 | 2,000–3,500 | 0.5–3 | [2, 26, 29] |
| S-glass | 2.5 | 86 | 4,570 | 2.8 | [2, 26, 29] |
| C-glass | 2.49 | 69 | 3,300 | – | [29] |
| Aramid | 1.4 | 63–67 | 3,000–3,150 | 3.3–3.7 | [2, 26] |
| Kevlar | 1.44–1.47 | 55–143 | 2,300–3,800 | 1.5–4.5 | [44] |

## 1.3 INTERACTION BETWEEN POLYMER MATRICES AND FILLERS/FIBERS

Interactions between polymer matrices and reinforcing fillers/fibers affect the properties (i.e., mechanical, thermal, chemical, physical, environmental effects, life cycle, etc.) of PCs. In the PCs, the interface between the reinforcing fillers/fibers (strong component) and the polymer matrices (weak component) is common. They share a common boundary that allows load transfer from the polymer matrices to the reinforcing fillers/fibers in accordance with strain compatibility. The load cannot be transferred from the polymer matrices to the filler/fiber reinforcer if the interface fails between the reinforcing fillers/fibers and the polymer matrices, because the deformation of the polymer matrices and the fillers/fibers is incompatible. If the interactions fail between fillers/fibers and polymer matrices, the reinforcing effect of fillers/fibers on the polymer matrices cannot be achieved in the PCs and required properties will not be found. The properties of the PCs mainly depend on the interactions and interface between the polymer matrix and the reinforcing fillers/fibers. PCs are influenced by a number of factors, including the composition, size, and form of their constituents, as well as the interactions that occur between the polymer matrix and dispersion phase at the interface. The adhesion between the fillers/fibers and polymer matrix in the PCs depends on the polymer matrices and fillers/fibers aspect ratio, and other fillers/fibers features can be alternated to achieve further advancements in the properties. For several features of the composite, including its mechanical, physical, electrical, and thermal properties, the fillers/fibers aspect ratio plays an important role, because these fillers/fibers enhance the characteristics of the PCs, and also in low filler/fiber volume fractions [19, 45, 46].

## 1.4 COMPOUNDING/PROCESSING OF POLYMER COMPOSITES

The PCs' compounding/processing uses fillers/fibers as reinforcements and polymer as matrix. The reinforcing fillers/fibers may be synthetic or natural, and the polymer matrix may be thermoplastic or thermoset. In this case, the process of reinforcing PCs with synthetic fillers/fibers make such products which have higher mechanical properties, namely, tensile, flexural, and impact. Synthetic filler/fiber–reinforced PCs are mostly used in the automotive industry and in aerospace and aircraft applications because of the higher mechanical, thermal, and physical properties of synthetic fillers/fibers. Synthetic filler/fiber–reinforced PCs have higher and more desirable mechanical and thermal properties than natural filler/fiber–reinforced PCs [47–49]. The compounding/processing of natural filler/fiber–reinforced PCs is primarily influenced by some key factors, such as temperature, fiber type, fiber volume fraction, and moisture content. The moisture content of both the fillers/fibers and the polymer matrices is controlled before processing with some necessary adjustments, that is, filler/fiber modifications, pre-heat treatments of fillers/fibers, and also for polymer matrices. The PCs' performance, sustainability, and processing are significantly impacted by the polymer matrices, filler/fiber length, aspect ratio, and chemical compositions and properties. Additionally, the processing temperature of the natural filler/fiber–reinforced PCs is crucial, and they are processed at a low temperature. Most of the natural filler/fiber–reinforced PCs are processed below 200°C for 20 min. If the processing temperature used for the natural filler/fiber–reinforced PCs is above 200°C, the results may change, that is, degradation, shrinkage, and low performance of the composites. Because of dehydration, depolymerization, decarboxylation, oxidation, hydrolysis, and recrystallization after the use of a high processing temperature, natural fillers/fibers undergo changes in their chemical, physical, and mechanical properties. Natural filler/fiber–reinforced PCs are mostly used in automobile, construction, agriculture, home alliance, sports, etc. [50, 51].

Generally, the selection of the compounding/processing methods of the PCs depends on the kind of, and procedure used for, materials, the quality of the parts, the parts' design, the production capacity, and also the cost requirement. For the processing of natural filler/fiber–reinforced PCs, some methods are used, like extrusion, pultrusion, compression, injection, resin transfer molding (RTM), hand layup, vacuum bag, autoclave, etc. [52]. The extrusion process offers good functional adaptability and better compounding or mixing capabilities of reinforcing fillers/fibers and polymer matrices. In the extruder process, mostly thermoplastic matrices are used as matrix in the PCs. Similarly, the pultrusion process works like extrusion, although pultrusion involves pulling the materials through a die rather than pushing it through a die. The die cross section determines the final shape of the required products. In the pultrusion process, the continuous fillers/fibers are used as reinforcing materials, and thermoset polymer matrices are used as matrix in the PCs. From both the extrusion and pultrusion processes, approximately similar products, such as a bar or a rod, may be made, depending on the used die shape. The injection molding process has the advantages of reduced cost, increased time efficiency, and enhanced processability in comparison to various composite manufacturing methods. Furthermore, one of the PCs' manufacturing processes that can be utilized for mass production is the injection molding process [53]. The autoclave process is primarily used in high-quality parts of the aerospace industry, but autoclave procedures have undergone high manufacturing costs and a labor-intensive manufacturing process. Resin transfer molding (RTM) is being utilized more and more to make large-scale composite structures, like wind turbine blades, boats, and aircraft frames [54, 55].

## 1.5 POLYMER COMPOSITES

Polymer–matrix composites (PMCs) have garnered noteworthy attention primarily attributed to their cost-effectiveness and superior specific strength and stiffness compared to traditional metal alloys [56, 57]. In addition, PMCs provide increased design flexibility and demonstrate superior resistance to corrosion and fatigue. The utilization of PMCs has been well-established for many years within the aerospace and automotive sectors, and its application is continually growing into diverse fields, including the biomedical, marine, energy, and infrastructure industries [57]. Polymer–matrix composites are comprised of two primary elements: (1) the matrix, the continuous component, which usually has a greater ductility and functions as an anchor for the dispersed phase, and (2) the reinforcement, the secondary or scattered component. The combined properties of the matrix and the reinforcement make the PMCs stiffer, more rigid, and less dense, with noteworthy advantages, such as corrosion resistance and thermal insulation. In addition, their production costs are more cost-effective than those of their metallic counterparts. The extensive range of available polymer matrices, combinable with various reinforcements, enables the creation of diverse PMC compositions and combinations [58, 59].

An emerging trend of polymer composites involves the adoption of natural fibers and particulates extracted from plants, fruits, and seeds as key reinforcing materials [60–62]. This shift toward bio-based reinforcing agents is driven by a growing inclination to lessen dependence on traditional plastic and to upsurge the awareness of advanced sustainable materials. Polymers with natural fibers and particles garnered extensive attention in different applications, like automotive, physical components, cosmetics, and household equipment [63–65]. The automotive industry showed a notable interest in these materials because of their lightweight characteristics and the potential to lessen the environmental footprint of vehicles [63]. The use of these composites aligns with the industry's pursuit of more sustainable practices and offers a promising avenue for developing lighter, eco-friendly components in various engineering domains.

Over the past few decades, significant progress has been made in the field of polymer–matrix composites through the integration of diverse reinforcing agents, including fibers and particles [58, 59]. These reinforcing agents, sourced from both natural and synthetic materials, play a vital part in enhancing the properties of polymer–matrix composites. Notably, there has been a growing emphasis on utilizing reinforcing materials derived from plants, vegetables, fruits, and seeds as an eco-friendly alternative to condense the dependence on traditional plastics [66]. This shift toward incorporating natural fibers and particles has witnessed a notable surge in the application of polymer composites across various engineering domains, including automotive, structural engineering, cosmetics, and household appliances [64]. This trend underlines the increasing recognition of sustainable and environmentally friendly solutions in particulate and fiber-reinforced design of polymer composites. The subsequent segments describe the polymer composite design based on particulate and fiber reinforcement.

### 1.5.1 Particulate Polymer Composites

The application of particle-reinforced polymer composites extends across different sectors, benefiting from their impressive strength-to-weight ratio. There is a growing mandate for materials like PMCs that possess unique combinations of properties, surpassing what conventional materials like ceramics, metals, and polymers can individually offer. This imperative has directed the focus of material engineers and scientists toward the development of sustainable PMC materials [67]. Industries such as cosmetics and household appliances value these composites for their aesthetic appeal and environmentally friendly attributes. As the demand for sustainable materials rises, incorporating natural fibers and particulates into polymer composites plays a pivotal role in addressing these needs across an extensive variety of industrial applications [68]. The flexibility and usefulness of these composites make them a valuable solution in meeting both performance and sustainability requirements in various industries.

A number of natural cellulose fillers created from the by-products of agricultural fields exist in the form of biowaste [69]. Some of these fillers are effectively used for energy generation, and a significant portion is unfortunately treated as waste, contributing to environmental pollution. Biowaste elements are nowadays gaining recognition as a viable alternative to traditional elements, such as alumina, glass, and aramid [69, 70]. This shift is attributed to their numerous advantages, including being plenty in nature, cost-effectiveness, high hardness, favorable strength-to-weight ratio, decent mechanical properties, biodegradability, nonabrasiveness properties, and eco-friendliness [67, 71]. These biowaste particles serve as reinforcing elements in PMCs, used in reducing ductility and increasing modulus. The utilization of biowaste particles in polymer composites offers a cost-effective alternative to fiber-reinforced materials, resulting in reduced production costs [72, 73].

This highlights the potential of biowaste particles as a valuable component in enhancing the mechanical properties of PMCs. Various biowaste elements, including, but not limited to, groundnut shell, coconut shell, oil palm shell, coir pith, wood dust, rice husk, bamboo fiber, and cashew nut shell particles, have found applications as reinforcing elements in diverse PMCs [74]. Polypropylene composites reinforced with rice husk particles [75] have been studied for their mechanical characteristics. This suggests a potential for expanding the utilization of recycled polymers in the production of biowaste particle–reinforced polymer composites. Andrzej and Abdullah [76] studied a barley husk/coconut shell particle–reinforced thermoplastic polymer composite. Bledzki et al. [77] investigated the characteristics, such as thermal fragmentation structures, water absorption, and solubility indicators, of rye and wheat husk elements. Raju and Kumarappa et al. [78] investigated the mechanical possessions of chemically altered groundnut shell–built composites. Husseinsyah and Mostapha [79] investigated the influence of the coconut shell element content on water absorption, and the morphological properties of polyester composites with coconut shell particles. Ojha et al. [80] conducted an assessment and comparison of the mechanical and tribological characteristics of PMC with particulates from wood apple and coconut shells.

### 1.5.2 Fiber Polymer Composites

Synthetic and natural fibers are generally used in the development of polymer–matrix composites. Among the synthetic fibers, carbon, Kevlar, and glass fiber are currently used to develop PMCs. In addition to these synthetic options, natural fiber reinforcements are also incorporated into polymers. These natural fibers possess a light weight and contribute to linking the polymer matrix with significant strength and stiffness. The use of both synthetic and natural fibers provides a comprehensive approach to reinforcing polymer matrices. It also combines the desirable characteristics of each to enhance the overall performance of the composites. The emphasis on lightweight natural fiber reinforcements highlights their importance in achieving strength and stiffness without compromising on the weight of the composite materials.

Glass fibers are one of the synthetic fibers widely used as reinforcement in a polymer matrix. There are different types of glass fibers, like the E-glass (electrical-grade glass fiber), which is the oldest one; the A-glass and C-glass (both chemical-resistant); and the strong R-glass, also known as S-glass. E-glass, being electrical-grade, has a long history of use in various applications. The S-glass is known for its high strength. In a laboratory testing, glass fibers can withstand tensile pressures of up to 7,000 N/mm$^2$, while commercially, they can withstand loads of up to 2,800 N/mm$^2$ [81]. In today's world, these fiber-reinforced polymer composites are manufactured by 3D printing. Wang et al. [82] initiated the design of the 3D printing of composites using acrylonitrile butadiene styrene (ABS) as the base material. Maqsood et al. [83] revealed significant improvements in the tensile strength of hybrid composites made from polylactic acid (PLA) reinforced with short carbon fiber (SCF) and continuous carbon fiber (CCF). The findings showed the potential benefits of hybridization in improving the mechanical properties of polymer composites for various applications. The work done by Wang et al. [84] showed that incorporating both short and continuous carbon fibers as reinforcements in polyamide (PA)–based composites could lead to enhanced mechanical properties. Pande et al. [85] gave an information that the addition of functionalized carbon nanotubes (CNTs) to the PMMA (polymethyl methacrylate) matrix resulted in an increase in the mechanical properties of the nanocomposites. Ma et al. [86] proposed that achieving a homogeneous distribution of carbon nanotubes (CNTs) in polymer–matrix composites (PMCs) depends on several critical parameters, including volume fraction of CNTs, CNT length and entanglement, matrix viscosity, and intertube attraction, alongside various dispersion methods. Xie et al. [87] reported that well-aligned and homogeneously dispersed CNTs in polymer matrices significantly improve various properties of polymer nanocomposites (PNCs). Wang et al. [88] highlighted the critical role of the functionalization process in achieving a well-distributed gel of ultra-high-molecular-weight polyethylene (UHMWPE)/multiwalled carbon nanotubes (MWCNTs). Cadek et al. [89] studied the mechanical and microstructure characteristics of polymer nanocomposites (PNCs) with polyvinyl alcohol (PVA)/multiwalled carbon nanotube (MWCNT) and polyvinyl carbazole (PVK)/MWCNT.

Qian et al. [90] mentioned that nanocomposites with MWCNTs exhibit higher mechanical properties compared to the unreinforced PS matrix. Safadi et al. [91] produced a PS/MWCNT composite using the spin casting and film casting methods. Ferreira et al. [92] developed a nanocomposite consisting of CNTs and high-density polyethylene (HDPE) through the melt mixing process. The use of natural fibers in the composites substitutes the traditional options for synthetic fibers [93, 94]. Senthilkumar et al. [95] conducted alkaline treatment of sisal fiber–reinforced PLA composites.

Nowadays, there is a trend toward the utilization of natural fibers [93, 94] for the hybridization approach; this shows the potential benefits of combining different natural fibers and using different treatment methods to optimize the overall performance of composite materials. The experimental effort on the hybridization of banana and sisal fibers with a polyester matrix was done by Idicula et al. [96]. The influence of chemical treatments on the binding and tensile strength of sisal fiber composite was studied by Sood et al. [97]. Jacob et al. [98] developed a composite with hybrid natural rubber and studied its dynamic properties. Nimanpure et al. [99] developed a hybrid kenaf- and sisal fiber–reinforced composite and studied its thermal behavior. Atiqah et al. [100] conducted research aimed at hybridizing natural and synthetic fibers, specifically sugar palm and glass fibers, in a polyurethane matrix. Pappu et al. [101] provided a detailed study of hybrid reinforced sisal/hemp fibers in poly(lactic acid) composites with a fixed composition of sisal and hemp fiber. Joseph et al. [102] hybridized more than two natural and synthetic fibers and developed the composites. With newly developed materials [103–108], technologies have strongly extended the potential of fiber composites and hybrid composites. Fiber composites and hybrid composites are used to develop the fundamental mechanisms and for the use in industry.

## 1.6 CONCLUSION

In summary, the reviewed articles offer valuable insights into the diverse sides of particulate and fiber reinforcements and their effects, with a primary emphasis on their processing conditions and on environmental factors, on the mechanical properties of polymer composites. This emphasis on rigorous assessment is crucial for gaining a nuanced understanding of how these materials respond to different conditions over extended periods.

## ACKNOWLEDGMENTS

First author wishes to thank the Indian Institute of Technology–Roorkee for providing an excellent research environment.

## REFERENCES

[1] R. Hsissou, R. Seghiri, Z. Benzekri, M. Hilali, M. Rafik, A. Elharfi, Polymer composite materials: A comprehensive review, Compos Struct 262 (2021) 113640. https://doi.org/10.1016/J.COMPSTRUCT.2021.113640.

[2] D. My, A.H.A.M. Laila, M. Nurazzi, A. Khalina, S.M. Sapuan, D. Laila, M. Rahmah, Z. Hanafee, Science & technology a review: Fibres, polymer matrices and composites, Pertanika J Sci Technol 25 (2017) 1085–1102.

[3] R.-M. Wang, S.-R. Zheng, Y.-P. Zheng, Introduction to polymer matrix composites, Polym Matr Compos Technol (2011) 1–548. https://doi.org/10.1533/9780857092229.1.

[4] I.O. Oladele, T.F. Omotosho, A.A. Adediran, Polymer-based composites: An indispensable material for present and future applications, Int J Polym Sci 2020 (2020). https://doi.org/10.1155/2020/8834518.

[5] S. Raghuvanshi, H. Khan, V. Saroha, H. Sharma, H.S. Gupta, A. Kadam, D. Dutt, Recent advances in biomacromolecule-based nanocomposite films for intelligent food packaging: A review, Int J Biol Macromol 253 (2023). https://doi.org/10.1016/J.IJBIOMAC.2023.127420.

[6] S. Das Lala, A.B. Deoghare, S. Chatterjee, Effect of reinforcements on polymer matrix bio-composites: An overview, IEEE J Sel Top Quantum Electron 25 (2018) 1039–1058. https://doi.org/10.1515/SECM-2017–0281/ASSET/GRAPHIC/J_SECM-2017–0281_FIG_001.JPG.

[7] S. Das Lala, P. Deb, E. Barua, A. Deoghare, C. Bhowmik, Investigation on Effect of reinforcing fillers and matrix on polymer bio-composites: A review, Lecture Notes in Intelligent Transportation and Infrastructure Part F 1361 (2022) 339–348. https://doi.org/10.1007/978-981-16-7660-4_30/COVER.

[8] T. Väisänen, O. Das, L. Tomppo, A review on new bio-based constituents for natural fiber-polymer composites, J Clean Prod 149 (2017) 582–596. https://doi.org/10.1016/J.JCLEPRO.2017.02.132.

[9] S. Kalia, B.S. Kaith, I. Kaur, Pretreatments of natural fibers and their application as reinforcing material in polymer composites: A review, Polym Eng Sci 49 (2009) 1253–1272. https://doi.org/10.1002/PEN.21328.

[10] X. Li, L.G. Tabil, S. Panigrahi, Chemical treatments of natural fiber for use in natural fiber-reinforced composites: A review, J Polym Environ 15 (2007) 25–33. https://doi.org/10.1007/S10924-006-0042-3.

[11] J.J. Kenned, K. Sankaranarayanasamy, C.S. Kumar, Chemical, biological, and nanoclay treatments for natural plant fiber-reinforced polymer composites: A review, Polym Polym Compos 29 (2021) 1011–1038. https://doi.org/10.1177/0967391120942419/ASSET/IMAGES/LARGE/10.1177_0967391120942419-FIG14.JPEG.

[12] P. Jagadeesh, M. Puttegowda, S. Mavinkere Rangappa, S. Siengchin, A review on extraction, chemical treatment, characterization of natural fibers and its composites for potential applications, Polym Compos 42 (2021) 6239–6264. https://doi.org/10.1002/PC.26312.

[13] H. Sharma, A. Kumar, S. Rana, N.G. Sahoo, M. Jamil, R. Kumar, S. Sharma, C. Li, A. Kumar, S.M. Eldin, M. Abbas, Critical review on advancements on the fiber-reinforced composites: Role of fiber/matrix modification on the performance of the fibrous composites, J Mater Res Technol 26 (2023) 2975–3002. https://doi.org/10.1016/J.JMRT.2023.08.036.

[14] H.S. Gupta, S.K. Sethi, A. Verma, Applications of coating materials: A critical overview, Coat Mater 1086 (2023) 81–109. https://doi.org/10.1007/978-981-99-3549-9_4.

[15] H.S. Gupta, U. Shankar, A. Verma, R. Gogoi, S.K. Sethi, Computational aspects: Self-clean coatings, plastics and polymers in coatings, In Coating Materials: Computational Aspects, Applications and Challenges. Singapore: Springer Nature Singapore 1086 (2023) 167–184. https://doi.org/10.1007/978-981-99-3549-9_8.

[16] D.K. Rajak, P.H. Wagh, E. Linul, A review on synthetic fibers for polymer matrix composites: Performance, failure modes and applications, Materials 15 (2022) 4790. https://doi.org/10.3390/MA15144790.

[17] M.M. Camargo, E.A. Taye, J.A. Roether, D.T. Redda, A.R. Boccaccini, A review on natural fiber-reinforced geopolymer and cement-based composites, Materials 13 (2020) 4603. https://doi.org/10.3390/MA13204603.

[18] S. Begum, S. Fawzia, M.S.J. Hashmi, Polymer matrix composite with natural and synthetic fibres, Adv Mater Process Technol 6 (2020) 547–564. https://doi.org/10.1080/2374068X.2020.1728645.

[19] S. Puggal, N. Dhall, N. Singh, M.S. Litt, A review on polymer nanocomposites: Synthesis, characterization and mechanical prop. Ind J Sci Technol 9 (2016) 10. https://doi.org/10.17485/ijst/2016/v9i4/81100.

[20] E. Omanović-Mikličanin, A. Badnjević, A. Kazlagić, M. Hajlovac, Nanocomposites: A brief review, Health Technol (Berl) 10 (2020) 51–59. https://doi.org/10.1007/S12553-019-00380-X/FIGURES/3.

[21] S.K. Kumar, R. Krishnamoorti, Nanocomposites: Structure, phase behavior, and properties, Annu Rev Chem Biomol Eng 1 (2010) 37–58. https://doi.org/10.1146/ANNUREV-CHEMBIOENG-073009–100856.

[22] I.Y. Chang, J.K. Lees, Recent development in thermoplastic composites: A review of matrix systems and processing methods. J Thermoplast Compos Mater 1 (1988) 277–296. https://doi.org/10.1177/089270578800100305.

[23] V. Mahesh, S. Joladarashi, S.M. Kulkarni, A comprehensive review on material selection for polymer matrix composites subjected to impact load, Defence Technol 17 (2021) 257–277. https://doi.org/10.1016/J.DT.2020.04.002.

[24] M.R. Kessler, Polymer matrix composites: A perspective for a special issue of polymer reviews, Polym Rev 52 (2012) 229–233. https://doi.org/10.1080/15583724.2012.708004.

[25] R. Hsissou, R. Seghiri, Z. Benzekri, M. Hilali, M. Rafik, A. Elharfi, Polymer composite materials: A comprehensive review, Compos Struct 262 (2021) 113640. https://doi.org/10.1016/J.COMPSTRUCT.2021.113640.

[26] H. Ku, H. Wang, N. Pattarachaiyakoop, M. Trada, A review on the tensile properties of natural fiber reinforced polymer composites, Compos B Eng 42 (2011) 856–873. https://doi.org/10.1016/J.COMPOSITESB.2011.01.010.
[27] V. Chauhan, T. Kärki, J. Varis, Review of natural fiber-reinforced engineering plastic composites, their applications in the transportation sector and processing techniques, J Thermoplast Compos Mater 35 (2022) 1169–1209. https://doi.org/10.1177/0892705719889095/ASSET/IMAGES/LARGE/10.1177_0892705719889095-FIG7.JPEG.
[28] S. Kangishwar, N. Radhika, A.A. Sheik, A. Chavali, S. Hariharan, A comprehensive review on polymer matrix composites: Material selection, fabrication, and application, Polym Bull 80 (2022) 47–87. https://doi.org/10.1007/S00289-022-04087-4.
[29] S. Kumar, K.K. Singh, Tribological behaviour of fibre-reinforced thermoset polymer composites: A review, Proc Inst Mech Eng Part L J Mater Des Appl 234 (2020) 1439–1449. https://doi.org/10.1177/1464420720941554/ASSET/IMAGES/LARGE/10.1177_1464420720941554-FIG5.JPEG.
[30] R. Rahman, S.Z.F.S. Putra, Tensile properties of natural and synthetic fiber-reinforced polymer composites, In Mechanical and Physical Testing of Biocomposites, Fibre-Reinforced Composites and Hybrid Composites (2019) 81–102. https://doi.org/10.1016/B978-0-08-102292-4.00005-9.
[31] D.K. Rajak, D.D. Pagar, R. Kumar, C.I. Pruncu, Recent progress of reinforcement materials: A comprehensive overview of composite materials, J Mater Res Technol 8 (2019) 6354–6374. https://doi.org/10.1016/J.JMRT.2019.09.068.
[32] M.A. Ghalia, A. Abdelrasoul, Compressive and fracture toughness of natural and synthetic fiber-reinforced polymer, In Mechanical and Physical Testing of Biocomposites, Fibre-Reinforced Composites and Hybrid Composites (2019) 123–140. https://doi.org/10.1016/B978-0-08-102292-4.00007-2.
[33] E. Omrani, P.L. Menezes, P.K. Rohatgi, State of the art on tribological behavior of polymer matrix composites reinforced with natural fibers in the green materials world, Eng Sci Technol Int J 19 (2016) 717–736. https://doi.org/10.1016/J.JESTCH.2015.10.007.
[34] W. Ouarhim, N. Zari, R. Bouhfid, A.E.K. Qaiss, Mechanical performance of natural fibers-based thermosetting composites, In Mechanical and Physical Testing of Biocomposites, Fibre-Reinforced Composites and Hybrid Composites (2019) 43–60. https://doi.org/10.1016/B978-0-08-102292-4.00003-5.
[35] H. Jariwala, P. Jain, A review on mechanical behavior of natural fiber reinforced polymer composites and its applications, J Reinf Plast Compos 38 (2019) 441–453. https://doi.org/10.1177/0731684419828524/ASSET/IMAGES/LARGE/10.1177_0731684419828524-FIG20.JPEG.
[36] S.K. Ramamoorthy, M. Skrifvars, A. Persson, A review of natural fibers used in biocomposites: Plant, animal and regenerated cellulose fibers, Polym Rev 55 (2015) 107–162. https://doi.org/10.1080/15583724.2014.971124.
[37] W.G. Trindade, W. Hoareau, J.D. Megiatto, I.A.T. Razera, A. Castellan, E. Frollini, Thermoset phenolic matrices reinforced with unmodified and surface-grafted furfuryl alcohol sugar cane Bagasse and Curaua fibers: Properties of fibers and composites, Biomacromolecules (2005). https://doi.org/10.1021/bm058006.
[38] A.R. Pai, R.N. Jagtap, Surface morphology & mechanical properties of some unique natural fiber reinforced polymer composites: A review, J Mater Environ Sci 6 (2015) 902–917.
[39] N. Tien, P. Graduate, T. Fujii, B. Chuong, K. Okubo, Study on how to effectively extract bamboo fibers from raw bamboo and wastewater treatment, J Mater Sci Res 1 (2012). https://doi.org/10.5539/jmsr.v1n1p144.
[40] S.K. Ramamoorthy, M. Skrifvars, A. Persson, A review of natural fibers used in biocomposites: Plant, animal and regenerated cellulose fibers, Polym Rev 55 (2015) 107–162. https://doi.org/10.1080/15583724.2014.971124.
[41] M.R. Sanjay, P. Madhu, M. Jawaid, P. Senthamaraikannan, S. Senthil, S. Pradeep, Characterization and properties of natural fiber polymer composites: A comprehensive review, J Clean Prod 172 (2018) 566–581. https://doi.org/10.1016/J.JCLEPRO.2017.10.101.
[42] M.H. Al-Saleh, U. Sundararaj, Review of the mechanical properties of carbon nanofiber/polymer composites, Compos Part A Appl Sci Manuf 42 (2011) 2126–2142. https://doi.org/10.1016/J.COMPOSITESA.2011.08.005.
[43] A. Sayam, A.N.M.M. Rahman, M.S. Rahman, S.A. Smriti, F. Ahmed, M.F. Rabbi, M. Hossain, M.O. Faruque, A review on carbon fiber-reinforced hierarchical composites: Mechanical performance, manufacturing process, structural applications and allied challenges, Carbon Lett 32 (2022) 1173–1205. https://doi.org/10.1007/S42823-022-00358-2.
[44] Ş. Ursache, C. Cerbu, A. Hadăr, Characteristics of carbon and Kevlar fibres, their composites and structural applications in civil engineering: A review, Polymers 16 (2023) 127. https://doi.org/10.3390/POLYM16010127.

[45] S. Huang, Q. Fu, L. Yan, B. Kasal, Characterization of interfacial properties between fibre and polymer matrix in composite materials: A critical review, J Mater Res Technol 13 (2021) 1441–1484. https://doi.org/10.1016/J.JMRT.2021.05.076.

[46] J. Karger-Kocsis, H. Mahmood, A. Pegoretti, Recent advances in fiber/matrix interphase engineering for polymer composites, Prog Mater Sci 73 (2015) 1–43. https://doi.org/10.1016/J.PMATSCI.2015.02.003.

[47] S. Begum, S. Fawzia, M.S.J. Hashmi, Polymer matrix composite with natural and synthetic fibres, Adv Mater Process Technol 6 (2020) 547–564. https://doi.org/10.1080/2374068X.2020.1728645.

[48] J.S. Chohan, K.S. Boparai, R. Singh, M.S.J. Hashmi, Manufacturing techniques and applications of polymer matrix composites: A brief review, Adv Mater Process Technol 8 (2022) 884–894. https://doi.org/10.1080/2374068X.2020.1835012.

[49] H.S. Gupta, P. Kumar, A.K. Sethi, A. Verma, S.K. Sethi, Software-based simulations of the creep recovery model of polymer composites, In Dynamic Mechanical and Creep-Recovery Behavior of Polymer-Based Composites (2024) 271–289. https://doi.org/10.1016/B978-0-443-19009-4.00016-3.

[50] A. Gholampour, T. Ozbakkaloglu, A review of natural fiber composites: Properties, modification and processing techniques, characterization, applications, J Mater Sci 55 (2020) 829–892. https://doi.org/10.1007/S10853-019-03990-Y/TABLES/23.

[51] P. Jagadeesh, M. Puttegowda, P. Boonyasopon, S.M. Rangappa, A. Khan, S. Siengchin, Recent developments and challenges in natural fiber composites: A review, Polym Compos 43 (2022) 2545–2561. https://doi.org/10.1002/PC.26619.

[52] J.S. Chohan, K.S. Boparai, R. Singh, M.S.J. Hashmi, Manufacturing techniques and applications of polymer matrix composites: A brief review, Adv Mater Process Technol 8 (2022) 884–894. https://doi.org/10.1080/2374068X.2020.1835012.

[53] G.S. Mann, L.P. Singh, P. Kumar, S. Singh, Green composites: A review of processing technologies and recent applications, J Thermoplast Compos Mater 33 (2020) 1145–1171. https://doi.org/10.1177/0892705718816354/FORMAT/EPUB.

[54] J. Lee, X. Ni, F. Daso, X. Xiao, D. King, J.S. Gómez, T.B. Varela, S.S. Kessler, B.L. Wardle, Advanced carbon fiber composite out-of-autoclave laminate manufacture via nanostructured out-of-oven conductive curing, Compos Sci Technol 166 (2018) 150–159. https://doi.org/10.1016/J.COMPSCITECH.2018.02.031.

[55] T. Centea, L.K. Grunenfelder, S.R. Nutt, A review of out-of-autoclave prepregs: Material properties, process phenomena, and manufacturing considerations, Compos Part A Appl Sci Manuf 70 (2015) 132–154. https://doi.org/10.1016/J.COMPOSITESA.2014.09.029.

[56] S.M. Harle, Durability and long-term performance of fiber reinforced polymer (FRP) composites: A review, Structures 60 (2024) 105881. https://doi.org/10.1016/j.istruc.2024.105881.

[57] T. Mishra, P. Mandal, A.K. Rout, D. Sahoo, A state-of-the-art review on potential applications of natural fiber-reinforced polymer composite filled with inorganic nanoparticle, Compos Part C Open Access 9 (2022). https://doi.org/10.1016/j.jcomc.2022.100298.

[58] O. Huseynov, S. Hasanov, I. Fidan, Influence of the matrix material on the thermal properties of the short carbon fiber reinforced polymer composites manufactured by material extrusion, J Manuf Process 92 (2023) 521–533. https://doi.org/10.1016/j.jmapro.2023.02.055.

[59] Akhyar, A. Gani, M. Ibrahim, F. Ulmi, A. Farhan, The influence of different fiber sizes on the flexural strength of natural fiber-reinforced polymer composites, Res Mater 21 (2024) 100534. https://doi.org/10.1016/j.rinma.2024.100534.

[60] F. Qin, F. Lu, K. Chen, Y. Hou, C. Zhang, L. Huang, Numerical simulation and experimental validation of ratchetting deformation of short fiber-reinforced polymer composites, Compos B Eng 266 (2023). https://doi.org/10.1016/j.compositesb.2023.110974.

[61] S. Li, P. Cheng, S. Ahzi, Y. Peng, K. Wang, F. Chinesta, J.P.M. Correia, Advances in hybrid fibers reinforced polymer-based composites prepared by FDM: A review on mechanical properties and prospects, Compos Commun 40 (2023). https://doi.org/10.1016/j.coco.2023.101592.

[62] N.I.N. Haris, M.Z. Hassan, R.A. Ilyas, M.A. Suhot, S.M. Sapuan, R. Dolah, R. Mohammad, M.R.M. Asyraf, Dynamic mechanical properties of natural fiber reinforced hybrid polymer composites: A review, J Mater Res Technol 19 (2022) 167–182. https://doi.org/10.1016/j.jmrt.2022.04.155.

[63] K.M.F. Hasan, K.N. Al Hasan, T. Ahmed, S.T. György, M.N. Pervez, L. Bejó, B. Sándor, T. Alpár, Sustainable bamboo fiber reinforced polymeric composites for structural applications: A mini review of recent advances and future prospects, Case Stud Chem Environ Eng 8 (2023). https://doi.org/10.1016/j.cscee.2023.100362.

[64] M.K. Singh, R. Tewari, S. Zafar, S.M. Rangappa, S. Siengchin, A comprehensive review of various factors for application feasibility of natural fiber-reinforced polymer composites, Res Mater 17 (2023). https://doi.org/10.1016/j.rinma.2022.100355.
[65] X. Ao, A. Vázquez-López, D. Mocerino, C. González, D.Y. Wang, Flame retardancy and fire mechanical properties for natural fiber/polymer composite: A review, Compos B Eng 268 (2024). https://doi.org/10.1016/j.compositesb.2023.111069.
[66] R. Nanoth, K. Jayanarayanan, P. Sarath Kumar, M. Balachandran, A. Pegoretti, Static and dynamic mechanical properties of hybrid polymer composites: A comprehensive review of experimental, micromechanical and simulation approaches, Compos Part A Appl Sci Manuf 174 (2023). https://doi.org/10.1016/j.compositesa.2023.107741.
[67] A. Sharma, T. Mukhopadhyay, V. Kushvaha, Experimental data-driven uncertainty quantification for the dynamic fracture toughness of particulate polymer composites, Eng Fract Mech 273 (2022). https://doi.org/10.1016/j.engfracmech.2022.108724.
[68] D. Karimi, A.S. Milani, SRVE modeling of particulate polymer matrix composites with irregularly shaped inclusions: Application to a green stone composite, Compos Struct 228 (2019). https://doi.org/10.1016/j.compstruct.2019.111331.
[69] A.S. Jose, A. Athijayamani, S.P. Jani, A review on the mechanical properties of bio waste particulate reinforced polymer composites, Mater Today Proc, Elsevier Ltd (2020) 1757–1760. https://doi.org/10.1016/j.matpr.2020.07.360.
[70] T. Raja, V. Mohanavel, S. Suresh Kumar, S. Rajkumar, M. Ravichandran, R. Subbiah, Evaluation of mechanical properties on kenaf fiber reinforced granite nano filler particulates hybrid polymer composite, Mater Today Proc 59 (2022) 1345–1348. https://doi.org/10.1016/j.matpr.2021.11.548.
[71] J.Z. Liang, Reinforcement and quantitative description of inorganic particulate-filled polymer composites, Compos B Eng 51 (2013) 224–232. https://doi.org/10.1016/j.compositesb.2013.03.019.
[72] S.Y. Fu, X.Q. Feng, B. Lauke, Y.W. Mai, Effects of particle size, particle/matrix interface adhesion and particle loading on mechanical properties of particulate-polymer composites, Compos B Eng 39 (2008) 933–961. https://doi.org/10.1016/j.compositesb.2008.01.002.
[73] Y. Li, W. Li, X. Lin, M. Yang, Z. Zhao, X. Zhang, P. Dong, N. Xu, Q. Sun, Y. Dai, X. Zhang, L. Chen, Theoretical modeling of the temperature dependent tensile strength for particulate-polymer composites, Compos Sci Technol 184 (2019). https://doi.org/10.1016/j.compscitech.2019.107881.
[74] A. Catapano, J. Jumel, A numerical approach for determining the effective elastic symmetries of particulate-polymer composites, Compos B Eng 78 (2015) 227–243. https://doi.org/10.1016/j.compositesb.2015.03.085.
[75] S.K. Najafi, E. Hamidinia, M. Tajvidi, Mechanical properties of composites from sawdust and recycled plastics, J Appl Polym Sci 100 (2006) 3641–3645. https://doi.org/10.1002/app.23159.
[76] A.K. Bledzki, A.A. Mamun, J. Volk, Barley husk and coconut shell reinforced polypropylene composites: The effect of fibre physical, chemical and surface properties, Compos Sci Technol 70 (2010) 840–846. https://doi.org/10.1016/j.compscitech.2010.01.022.
[77] A.K. Bledzki, A.A. Mamun, J. Volk, Physical, chemical and surface properties of wheat husk, rye husk and soft wood and their polypropylene composites, Compos Part A Appl Sci Manuf 41 (2010) 480–488. https://doi.org/10.1016/j.compositesa.2009.12.004.
[78] G.U. Raju, S. Kumarappa, Experimental study on mechanical properties of groundnut shell particle-reinforced epoxy composites, J Reinf Plast Compos 30 (2011) 1029–1037. https://doi.org/10.1177/0731684411410761.
[79] C.I. Idumah, A. Hassan, Emerging trends in flame retardancy of biofibers, biopolymers, biocomposites, and bionanocomposites, Rev Chem Eng 32 (2016) 115–148. https://doi.org/10.1515/revce-2015–0017.
[80] O. Shakuntala, G. Raghavendra, A. Samir Kumar, Effect of filler loading on mechanical and tribological properties of wood apple shell reinforced epoxy composite, Adv Mater Sci Eng 2014 (2014). https://doi.org/10.1155/2014/538651.
[81] N. Saha, A.N. Banerjee, B.C. Mitra, Dynamic mechanical study on unidirectional polyethylene fibers-PMMA and glass fibers-PMMA composite laminates, J Appl Polym Sci 60 (1996) 657–662. https://doi.org/10.1002/(sici)1097–4628(19960502)60:5<657::aid-app2>3.0.co;2-p.
[82] K. Wang, S. Li, Y. Rao, Y. Wu, Y. Peng, S. Yao, H. Zhang, S. Ahzi, Flexure behaviors of ABS-based composites containing carbon and Kevlar fibers by material extrusion 3D printing, Polymers (Basel) 11 (2019). https://doi.org/10.3390/polym11111878.

[83] N. Maqsood, M. Rimašauskas, Characterization of carbon fiber reinforced PLA composites manufactured by fused deposition modeling, Compos Part C Open Access 4 (2021). https://doi.org/10.1016/j.jcomc.2021.100112.
[84] F. Wang, Z. Zhang, F. Ning, G. Wang, C. Dong, A mechanistic model for tensile property of continuous carbon fiber reinforced plastic composites built by fused filament fabrication, Addit Manuf 32 (2020). https://doi.org/10.1016/j.addma.2020.101102.
[85] S. Pande, R.B. Mathur, B.P. Singh, T.L. Dhami, Synthesis and characterization of multiwalled carbon nanotubes-polymethyl methacrylate composites prepared by in situ polymerization method, Polym Compos 30 (2009) 1312–1317. https://doi.org/10.1002/pc.20696.
[86] P.C. Ma, N.A. Siddiqui, G. Marom, J.K. Kim, Dispersion and functionalization of carbon nanotubes for polymer-based nanocomposites: A review, Compos Part A Appl Sci Manuf 41 (2010) 1345–1367. https://doi.org/10.1016/j.compositesa.2010.07.003.
[87] X.L. Xie, Y.W. Mai, X.P. Zhou, Dispersion and alignment of carbon nanotubes in polymer matrix: A review, Mater Sci Eng R Rep 49 (2005) 89–112. https://doi.org/10.1016/j.mser.2005.04.002.
[88] Y. Wang, R. Cheng, L. Liang, Y. Wang, Study on the preparation and characterization of ultra-high molecular weight polyethylene-carbon nanotubes composite fiber, Compos Sci Technol 65 (2005) 793–797. https://doi.org/10.1016/j.compscitech.2004.10.012.
[89] M. Cadek, J.N. Coleman, V. Barron, K. Hedicke, W.J. Blau, Morphological and mechanical properties of carbon-nanotube-reinforced semicrystalline and amorphous polymer composites, Appl Phys Lett 81 (2002) 5123–5125. https://doi.org/10.1063/1.1533118.
[90] D. Qian, E.C. Dickey, R. Andrews, T. Rantell, Load transfer and deformation mechanisms in carbon nanotube-polystyrene composites, Appl Phys Lett 76 (2000) 2868–2870. https://doi.org/10.1063/1.126500.
[91] B. Safadi, R. Andrews, E.A. Grulke, Multiwalled carbon nanotube polymer composites: Synthesis and characterization of thin films, J Appl Polym Sci 84 (2002) 2660–2669. https://doi.org/10.1002/app.10436.
[92] F.V. Ferreira, B.R.C. Menezes, W. Franceschi, E.V. Ferreira, K. Lozano, L.S. Cividanes, A.R. Coutinho, G.P. Thim, Influence of carbon nanotube concentration and sonication temperature on mechanical properties of HDPE/CNT nanocomposites, Fuller Nanotub Carbon Nanostruct 25 (2017) 531–539. https://doi.org/10.1080/1536383X.2017.1359553.
[93] M. Ramesh, K. Palanikumar, K.H. Reddy, Mechanical property evaluation of sisal-jute-glass fiber reinforced polyester composites, Compos B Eng 48 (2013) 1–9. https://doi.org/10.1016/j.compositesb.2012.12.004.
[94] M.R. Sanjay, G.R. Arpitha, B. Yogesha, Study on mechanical properties of natural—Glass fibre reinforced polymer hybrid composites: A review, Mater Today Proc, Elsevier Ltd (2015) 2959–2967. https://doi.org/10.1016/j.matpr.2015.07.264.
[95] K. Senthilkumar, N. Saba, N. Rajini, M. Chandrasekar, M. Jawaid, S. Siengchin, O.Y. Alotman, Mechanical properties evaluation of sisal fibre reinforced polymer composites: A review, Constr Build Mater 174 (2018) 713–729. https://doi.org/10.1016/j.conbuildmat.2018.04.143.
[96] M. Idicula, K. Joseph, S. Thomas, Mechanical performance of short banana/sisal hybrid fiber reinforced polyester composites, J Reinf Plast Compos 29 (2010) 12–29. https://doi.org/10.1177/0731684408095033.
[97] M. Sood, D. Dharmpal, V.K. Gupta, Effect of fiber chemical treatment on mechanical properties of sisal fiber/recycled HDPE composite, Mater Today Proc, Elsevier Ltd (2015) 3149–3155. https://doi.org/10.1016/j.matpr.2015.07.103.
[98] M. Jacob, S. Thomas, K.T. Varughese, Mechanical properties of sisal/oil palm hybrid fiber reinforced natural rubber composites, Compos Sci Technol 64 (2004) 955–965. https://doi.org/10.1016/S0266–3538(03)00261–6.
[99] S. Nimanpure, S.A.R. Hashmi, R. Kumar, H.N. Bhargaw, R. Kumar, P. Nair, A. Naik, Mechanical, electrical, and thermal analysis of sisal fibril/kenaf fiber hybrid polyester composites, Polym Compos 40 (2019) 664–676. https://doi.org/10.1002/pc.24706.
[100] A. Afzaluddin, M. Jawaid, M.S. Salit, M.R. Ishak, Physical and mechanical properties of sugar palm/glass fiber reinforced thermoplastic polyurethane hybrid composites, J Mater Res Technol 8 (2019) 950–959. https://doi.org/10.1016/j.jmrt.2018.04.024.
[101] A. Pappu, K.L. Pickering, V.K. Thakur, Manufacturing and characterization of sustainable hybrid composites using sisal and hemp fibres as reinforcement of poly (lactic acid) via injection moulding, Ind Crops Prod 137 (2019) 260–269. https://doi.org/10.1016/j.indcrop.2019.05.040.

[102] J. Joseph, P.R. Munda, M. Kumar, A.M. Sidpara, J. Paul, Sustainable conducting polymer composites: Study of mechanical and tribological properties of natural fiber reinforced PVA composites with carbon nanofillers, Polym Plast Technol Mater 59 (2020) 1088–1099. https://doi.org/10.1080/25740881.2020.1719144.

[103] A. Verma, H.S. Gupta, S.K. Sethi, Hybrid Composite Materials: Experimental and Theoretical Analysis, Springer (2024). https://doi.org/10.1007/978-981-97-2104-7.

[104] S.K. Sethi, H.S. Gupta, A. Verma, Polymer Composites: From Computational to Experimental Aspects, Springer (2024). https://doi.org/10.1007/978-981-97-0888-8.

[105] D. Verma, A. Verma, H.S. Gupta, Applications of deep learning for composites materials, In Hybrid Composite Materials: Experimental and Theoretical Analysis (2024) 195–213. https://doi.org/10.1007/978-981-97-2104-7_7.

[106] R. Gautam, H.S. Gupta, A. Chauhan, A.K. Jaiswal, A. Verma, Composite materials for bio-energy. In Hybrid Composite Materials: Experimental and Theoretical Analysis (2024) 91–104. https://doi.org/10.1007/978-981-97-2104-7_4.

[107] J. Kumar, A. Kumar, A.K. Maurya, H.S. Gupta, S.P. Singh, C. Sharma, Utilization of Ananas comosus crown residue husk as a sustainable strength additive for EPR/LDPE blend composites, ACS Omega 9 (2024) 2740–2751. https://doi.org/10.1021/acsomega.3c07697.

[108] J. Kumar, A.K. Maurya, H.S. Gupta, S.P. Singh, C. Sharma, Development of eco-friendly bio-composite by reinforcing pineapple fruit waste crown fiber to ethylene-propylene rubber modified polyethylene, Polym Compos 43 (2022) 8259–8273. https://doi.org/10.1002/pc.26997.

# 2 Experimental Characterization of Polymer Composites

*Akanksha Pandey*

## 2.1 INTRODUCTION

Nature has always blended organic and inorganic elements to create intelligent materials with amazing features and capabilities (hydrophobicity, density, permeability, mechanics, color, etc.). By observing nature, scientists have created new synthesis techniques that result in materials with several simultaneous uses. This interest originated because it is possible to create new materials with specific qualities by adding filler, usually fibers, flakes, spheres, or small particles. These substances are part of a novel class of material known as composites. A *composite* is a structural material that macroscopically blends two or more insoluble components. The matrix is one of the components, and an additional component, known as the reinforcing phase, will be embedded, improving the thermal and mechanical properties of the matrix (Datsyuk et al., 2020; Tarhini et al., 2019; Pallab and Pankaj, 2019; Rachid et al., 2018a, 2021; Vaggar and Kamate, 2020).

The inability to find a single material with all the qualities needed for a given purpose led to the development of composite materials. The usage of adobe blocks in construction is a historical example. Clay, water, and an organic material like straw or bamboo shoots were used to make these blocks. The lower production costs to attain high-performance materials, weight, and qualities (strength, thermal, stiffness, and surface topography) of composites nowadays make them a preferable material.

## 2.2 CHARACTERIZATION TECHNIQUES FOR POLYMER COMPOSITE

Characterization techniques are a vital tool to analyze a material comprehensively. Various characterization techniques have been developed to understand the physical, mechanical, chemical, and thermal properties of a composite. Sections 2.2.1 to 2.2.6 are a few experimental characterization techniques used during the analysis.

### 2.2.1 Physical Characterization

*Water absorption.* For a polymer composite, transport phenomena such as heat, mass, or fluid flow is an important subject of study. Natural fibers are susceptible to water absorption, could cause erosion in dimensional stability or mechanical properties (Sanjeevi et al., 2021). Affinity to water absorption limits the applicability of natural fiber–reinforced composites in water-based applications (Vigneshwaran et al., 2021). Water absorption occurs through capillary flow, diffusion, or water molecule transportation (Ahmad et al., 2011; Dhakal et al., 2007). Capillary flow occurs in gaps present between the reinforcement and the matrix due to the incomplete impregnation of the reinforcement with the matrix. Diffusion mechanism can be observed inside the microgaps among polymer chains (MuNoz and Garcia, 2015). Water molecule transport in the matrix over the microcrack is due to the swelling of fibers. The three types of diffusion behavior identified by the water molecule transport consist of an anomalous, or non-Fickian, Fickian diffusion model (Collings, 1994; Shen and Springer, 1976).

 DOI: 10.1201/9781003476337-2

*Procedure for measurement.* ASTM D5229 or D570 is used to measure the water absorption behavior of a specimen. Here, the procedure for analysis is explained using ASTM D 520 standard.

1. *Conditioning.* The specimen whose water absorption is strongly influenced by temperature should be dried at 50 ± 3°C in an oven for 24 hr before immersion in distilled water, cooled, and weighted, with tolerance of 0.001 g. If the specimen is unaffected by temperature, should be dried at 105–110°C for 1 hr.
2. *Immersion.* The specimen should immediately be immersed in distilled water kept at 23 ± 1°C for 24 hr after dry weight measurement. After the specified time duration, the sample will be removed from the water bath, dried using tissue paper, and weighted again. The procedure will be repeated for several times, up to the saturation of water absorption percentage. The following equation is used to calculate the water absorption percentage.

$$Change\,in\,weight(\%) = \frac{Wet\,weight - Conditioned\,weight}{Conditioned\,weight} \times 100$$

The thickness swelling is calculated using the following formula:

$$Thickness\,swelling(\%) = \frac{Specimen\,thickness\,at\,time\,t - Initial\,thickness\,of\,specimen}{Initial\,thickness\,of\,specimen} \times 100$$

### 2.2.2 Mechanical Characterization

Mechanical properties include tensile strength, modulus, flexural resistance, impact strength, toughness, resilience, hardness, etc. Explanations related to respective properties are provided in what follows.

#### 2.2.2.1 Tensile Properties

Tensile properties are most commonly quantified parameters and are used as an indicator of strength for a composite material. It analyzes the extent of deformation or applied force a material can withstand before failure. The selection of a material for a specific application, tensile parameter, such as modulus, elongation on stretching, and strength is required.

##### *2.2.2.1.1 Tensile Strength*

The maximum tensile stress a specimen can sustain at the time of tensile measurement before breaking is known as tensile strength. Nath et al. measured the tensile strength of PPFA composite according to ASTM D638 and reported the decrease in the stated parameter with the increment in fly ash content (Dilip et al., 2009).

##### *2.2.2.1.2 Tensile Modulus*

A mechanical parameter that quantifies the relative stiffness of a material is called tensile modulus. It is defined as a ratio of tensile stress to the corresponding strain within the proportionality limit. Hajnalka et al. measured the modulus of a HEMP fiber–reinforced polypropylene composite and concluded the increment in modulus value with the increase in fiber dosage (Hajnalka et al., 2008).

$$Tensile\,modulus = \frac{Tensile\,stress}{Corresponding\,tensile\,strain}$$

#### 2.2.2.1.3 *Elongation on Stretching*

Elongation is calculated by applying a tensile force to the specimen, and change in length is measured. Elongation is measured at two points, that is, at yield and break point in stress vs. strain curve. Elongation at yield is the ratio of increase in length at yield point to the initial length, whereas elongation at break is the maximum extension in the specimen before it breaks. Ahmed M. El-hadi measured the elongation at break of polylactic acid (PLLA) composites reinforced with cellulose nanocrystal, plasticizer, and poly (3-hydroxy butyrate) (PHB) (Hadi et al., 2017). A study reports that elongation at break of composite approaches 250% while pristine PLLA exhibited brittle behavior.

$$Elongation\ on\ stretching = \frac{Final\ length - Initial\ lenght}{Initial\ length\ of\ specimen} \times 100$$

*Procedure for measurement.* Various processing techniques, namely, injection molding, compression molding, or die cutting, are used to prepare the sample for tensile testing. The specimen shape used for the testing is the dumbbell form. In order to create a specimen, five categories are available based on material type and thickness. For rigid or semirigid plastic, generally, type I specimen dimension is preferred when the thickness is 7 mm or less. When specimen does not fracture in the narrow space, using type I rather than type II specimen is suggested. Types III and V specimen are used for a material having a thickness of 4 mm or less and thickness higher than 7 mm, respectively. Type IV specimen is used for nonrigid material with a thickness of 4 mm or lesser. In case of reinforced composite polymer, type I specimen is used during the analysis.

ASTM D 638 is used to analyze the tensile properties of a polymeric material (Srinivasan et al., 2014; Sims, 2003). All the dimensions of the specimen are measured, with the tolerance of 0.025 mm in several places, before the analysis. A gauge mark on the specimen should be placed using an ink that should not influence the material to be tested. Mount the specimen between the grips of universal testing machine and ensure that alignment of specimen long axis and the grip' imaginary line connecting the points of attachment of grip with the machine. Fasten the extensometer to the sample to measure the extension parameter. To remove any bending in the sample, a preload is applied at a crosshead speed of 0.1 mm/min. Crosshead speed can vary depending on the type of material being analyzed. If crosshead speed is not specified, then selection is based on the trial method that the specimen fractures within five minutes. Tensile modulus, tensile strength, and elongation percentage are calculated based on formulas explained earlier.

### 2.2.2.2 Flexural Properties

Flexural measurement, despite being simple in theory and to perform, has its own set of difficulties (Sims, 2003). Flexural loads comprise of shear, compression, and tensile loads. During loading, the upper surface faces compression, the center section experiences shear, and the lower surface faces tension. The variation in stress happens from maximum compressive stress at one layer and maximum tensile stress on another surface. In the case of isotropic material, if compressive and tensile modulus are the same, then the top layer will be in compression and the bottom face will be in equal tension. For several applications, flexural properties are highly quantified.

#### 2.2.2.2.1 *Flexural Strength*

*Flexural strength* is a mechanical constraint of a specimen to resist deformation due to applied force perpendicular to the specimen's longitudinal axis. The calculation is based on the maximum stress and strain that arise at the top and bottom surface of a sample. Asma Benkhelladi et al. measured the flexural strength of sisal-, flex-, and jute-reinforced polymer composites (Asma et al., 2020). The authors concluded that alteration in the flexural strength of the fiber-reinforced epoxy composite is strongly governed by the surface treatment of fiber. The $NaHCO_3$- and NaOH-treated fiber-filled epoxy composite showed higher flexural strength than the untreated fiber-filled composite.

$$Flexural\ strenght = \frac{3 \times applied\ load \times disstance\ of\ supports}{2 \times Widht\ of\ specimen \times thickness\ of\ specimen}$$

#### 2.2.2.2.2 Flexural Modulus

*Flexural modulus* is the ratio of stress to the strain experienced by a specimen due to bending. Tension and compression phenomenon will occur simultaneously. Yahaya et al. analyzed the influence of layer sequence and chemical treatment on the flexural modulus of an aramid–kenaf hybrid composite (Yahaya et al., 2015). The study concluded that, compared to three-layered Kevlar–kenaf, four-layered hybrid laminated composite exhibited improved flexural properties.

$$Flexural\ modulus = \frac{Applied\ load\,(F) \times distance\ of\ span\ support^3\,(L^3)}{4 \times Widht\ of\ specimen\,(b) \times thickness\ of\ specimen^3\,(d^3) \times diflection\ due\ to\ applied\ load\,(Y)}$$

*Procedure for measurement.* Two distinct techniques are used to examine the flexural properties of a sample, namely, three-point and four-point bending loading. In the three-point bending procedure, the specimen rests on two supports, and load is applied in between specified supports. In the four-point bending method, load is applied in two places equally spaced from the support, and the remaining is the same. The ASTM D790 standard is used to examine the flexural properties of a material (Hamed et al., 2013; Yan et al., 2016). The dimensions specified for the sample are 127 mm long and 12.7 mm wide, and the thickness should be 3.2 mm. The conditioning of the sample is carried out at 25°C temperature with 50 ± 5% relative humidity (RH).

Measure all three dimensions of the sample, namely, length, width, and thickness, with an accuracy of 0.03 mm. Adjust the support span to sample thickness ratio of 16:1. The stated ratio can be varied depending on some tensile and shear parameters of material. Ensure that the loading nose is placed halfway between the support and that the cylindrical surface axes are parallel. Position the sample, with the specimen's longitudinal axis perpendicular to the loading nose. At a particular crosshead rate, apply force to the specimen and collect the load-deflection data instantaneously. Stop the measurement when the maximum strain in the top layer reaches 0.05 mm/mm or a break occurs before reaching the maximum strain.

### 2.2.2.3 Compressive Properties

Compressive properties explain the material response when it is squeezed, compressed, crushed, etc. In order to quantify the performance, compressive parameters are necessary to be examined. These parameters include yield stress, modulus of elasticity, compressive strength, deformation at yield point, compressive strain percentage, etc.

The compression test for a composite material is the reverse of the tension test in terms of loading direction, though more difficult to understand (Ramamurty, 2001; Wisnom, 2000). Failure during compressive loading contains buckling of whole sample or microbuckling of particular reinforcement, such as fiber (Ramamurty, 2001). End-loaded sample have the tendency to spilt from the end; therefore, indirect loading through shearing is frequently employed.

#### 2.2.2.3.1 Compressive Strength

In compression measurement, the maximum stress borne by the specimen is called the compressive strength. It is a ratio of the maximum compressive load borne by the sample to the initial cross-sectional area of the sample. Subramaniyan et al. analyzed the effect of nanoclay on the compressive strength of glass fiber–filled vinyl ester matrix (Arun and Sun, 2005). A substantial increment in compressive strength of the composite is reported due to the addition of montmorillonite clay.

$$Compressive\ strenght = \frac{Applied\ load\ on\ the\ specimen}{Cross\ sectional\ area\ of\ specimen}$$

#### 2.2.2.3.2 *Modulus of Elasticity*

The modulus of elasticity is the ratio of stress to strain within the proportionality limit. Zakaria et al. studied the compressive properties of an epoxy composite filled with alumina-carbon nanotube ($Al_2O_3$-CNT) by varying the $Al_2O_3$-CNT synthesis techniques, that is, physical mixing, or hybrid, via chemical vapor deposition (CVD) (Muhammad et al., 2016). The study reported that hybrid $Al_2O_3$-CNT enhances the elastic modulus more than the physically mixed $Al_2O_3$-CNT-filled epoxy composite.

*Procedure for measurement.* To analyze the compressive properties of a polymeric material, ASTM D695 is used for the measurement (Gaurav et al., 2013). The first step includes the careful measurement of specimen dimensions with a precision of 0.025 mm at different places. In the next step, fix the specimen between the compression tool, and ensure the alignment of the long axis of the specimen with the plunger center line. Along with that, the end surface of the specimen and the compression plunger should parallel each other. Adjust the crosshead such that it hardly touches the upper surface of the compression tool plunger. Start the measurement at a speed of 1.3 mm/min, and record the maximum load supported by the sample at the time of failure.

### 2.2.2.4 Impact Properties

The impact properties of a material are quantitative demonstrations of the toughness of the material. The area under the stress-vs.-strain curve obtained in flexural or tensile testing provides quantitative information about the toughness. Toughness is an alternative representation of energy absorbed by a material during loading. The higher the toughness, the more the impact resistance will be for a material. There are three groups of material, namely, tough without modification, modified using another polymeric material, and modified using some reinforcement. Materials such as polypropylene or polyethylene are tough polymers without any modification, while styrene-acrylonitrile resin (SAN) requires modification to improve the impact properties (Liao et al., 2018). Molecular flexibility is an important entity to decide the toughness or brittleness of macromolecules. The large segment of molecules in flexible polymer disentangles effortlessly and responds to applied stress, whereas in stiff polymers, the disentanglement phenomenon is not easy to occur, which results in brittle failure due to applied load. In order to measure the stated parameters, various testing methods, such as pendulum impact test, drop weight impact, or instrumented impact test, are employed for the measurement (Ankur et al., 2022).

#### 2.2.2.4.1 *Impact Strength*

*Impact strength* is the quantitative evaluation of a material's ability to resist fracture due to shock loading or resist fracture under applied stress at high speed, commonly applicable for a polymer that endures brittle fracture. A ductile material might show brittle fractures below the ambient temperature. Hom et al. examined the effect of basalt fiber hybridization on the impact behavior of a hemp–polyester composite (Hom Nath et al., 2015). The study raises concern for the fiber laminate with low-velocity impact resistance. Noteworthy damage because of cracking or fiber fracture is observed in impact loading.

$$Impact\,strenght\,by\,izod = \frac{Impact\,value\,obtained\,from\,the\,scale}{Thickness\,of\,specimen}$$

$$Impact\ strenght\ by\ Charpy = \frac{Impact\,value\,obtained\,from\,the\,scale}{Area \times Thickness\,of\,specimen}$$

*Procedure for measurement.* In order to determine the impact strength or impact resistance, ASTM D256, D1822, D5420, D1709, and D2444 standards were proposed based on application requirements or material types. Here, only ASTM D256 procedure has been explained elaborately, to minimize the data presentation. The main principle for this test is to let a known mass pendulum descend to a predefined height and hit the specimen at the swing's lowest point, then record the height to which the pendulum swings further.

*Izod/Charpy test.* These measurements are performed on notched samples. The reason to cut the notch is to concentrate the stress point in one particular region and develop brittle failure rather than ductile. The notch is cut using a notch cutter at an angle of 45°, 0.25 mm radius, and 2.54 mm notch depth. Condition the specimen at 23 ± 2°C temperature and 50 ± 5% relative humidity for 40 hr. Measure the specimen width after the notch with a tolerance of 0.025 mm. Calibrate the pendulum hammer by releasing it from its defined position. Fix the specimen vertically such that the notch faces the striking end of a pendulum. Release the pendulum impact meter, and hit the specimen freely to cause failure. In the case of the Charpy method, the specimen is placed horizontally, and the notch should face the opposite side of the pendulum hammer.

### 2.2.2.5 Hardness

Hardness is one of the very popular parameters used for quality control and an indicator of durability. It is useful for grading as well as for comparative assessment of a material. Hardness is the quantification of resistance to localized plastic deformation for plastic, rubber, composite, etc. Depending on the type of material, hardness parameters vary; hence, a single type of hardness analysis will not be able to cover the entire polymeric material. Therefore, to measure the hardness of material, various techniques are available, and a few of them have been explained comprehensively in this chapter.

#### *2.2.2.5.1 Durometer Hardness*

Durometer hardness is determined by the amount an indenter can penetrate onto the surface of a specimen. It is an indirect way to present the elastic modulus or the viscoelastic behavior of a material. Durometers are divided into two categories, namely, shore A and shore D. Shore A is employed when the soft material will be analyzed, that is, rubber and elastomers. Whereas for hard materials, such as polyethylene, polystyrene, polypropylene, etc., shore D is used for the measurement.

The measurement is performed as per ASTM D2240 for soft materials. Sample with a minimum of 3 mm thickness is hand-pressed using an indenter against a plain hard surface. The hardness value is directly noted from the dial indicator.

#### *2.2.2.5.2 Rockwell Hardness*

Rockwell hardness is measured using a dial meter, depending on the number of times the dial needle passes from zero. ASTM D785 standard is followed to measure the hardness of hard material. The sample thickness should be more than 6 mm with a 25 mm$^2$ area. Place the sample on the anvil, and apply a minor load using an indenter. Minor load confirms the contact between the indenter and the specimen. Increase the load within 10 sec after ensuring that the dial of the apparatus is set to zero. The applied major load will be removed after 15 sec, and the specimen is allowed to recover for 15 sec. Count the number of times the needle passes from zero within 15 sec.

#### *2.2.2.5.3 Barcol Hardness*

The hardness of rigid plastics, whether reinforced or pristine, is measured using the Barcol hardness technique. Similar to previous techniques, a penetration indenter is used to evaluate the hardness of the sample. The indenter has a flat surface tip and frustum of 26°. Hand pressure is applied using an indenter on the 1.5 mm thick specimen.

### 2.2.2.6 Creep Properties

In polymer composite, a temporary change in free energy occurs when the polymer matrix is under stress. Momentarily, an increase in free energy changes when the segmental reorientation of the polymer occurs to attain a lower-energy state. The transition of molecules depends on stress, temperature, and neighboring molecules (Matsuoka, 1992). If the service temperature is lower than the glass transition temperature, measuring the time-dependent deterioration is time-intensive. Hence, extrapolation is performed to reduce the analysis time and obtain the required data. Creep failure is a collective incident, such as shear yielding, void formation, chain slippage, viscoelastic deformation, interfacial strength of polymer/reinforcement, etc. (Spathis and Kontou, 2012). The creep analysis can be performed in various modes, that is, tensile, shear, and compressive, and named as tensile, flexural, and compression creep, respectively.

When a constant load is applied on the material, it deforms instantly, and then continuation deformation happens with time. The process of deformation due to applied load as a function of time is known as creep. Yanjun Xu et al. examined the creep behavior of pristine and recycled high-density polyethylene (HDPE) and polyvinyl chloride (PVC) filled with bagasse fiber (Yanjun et al., 2010). At lower temperature, the bagasse fiber/PVC composite exhibited higher creep than the bagasse fiber/HDPE. All composites showed temperature-dependent creep behavior.

*Procedure for measurement.* The specimen dimension varies depending on the type of creep test selected. ASTM D638 standard is used to prepare the specimen for tensile creep, whereas D695 is recommended for compressive creep. Before the analysis, condition the specimen at 25°C temperature and 50 ± 5% RH for more than 40 hr. For tensile creep, the extension in the specimen as a function of time is measured using a tensile creep setup and strain gauge. Flexural creep is measured by recording the deflection as a function of time by applying a fixed load.

### 2.2.3 Chemical Characterization

The chemical analysis of a material is one of the important aspects to check its applicability in various industrial applications, such as fuel tanks and tubing for automotives, storage tanks for chemicals, rings used in joints, etc. The interaction of fillers, polymer matrix, additives, etc. can occur in different ways that may lead to erosion in properties at different rates and degrees. The characteristics of the material that govern the chemical behavior include chemical bond strength, polarity, branching, and crystallinity.

#### 2.2.3.1 Fourier Transform Infrared (FTIR) Spectroscopy

Fourier transform infrared (FTIR) spectroscopy is one of the prevailing methods to classify the type of bonds present in any organic or inorganic material, using infrared radiation. To comprehend the chemical nature and examine the surface chemistry of a material, Fourier transform infrared spectroscopy has been considered as one of the effective techniques. The analysis is carried out by exposing the specimen to infrared waves using an infrared spectrometer. When a material is exposed to IR waves, the chemical bonds present in it vibrate at different frequencies due to absorption, that is, a given bond can vibrate at different frequencies. These frequencies are related to the lowest energy state, that is, ground state and excited energy state, as per quantum mechanics. By allowing the bond to absorb light energy, one can excite the bond and raise molecular vibration frequency. Since IR radiation ranging from 2 to 16 μm is not energetic excitation, it is able to cause only bending or stretching vibrations. The Beer–Lambert law is used to analyze the IR absorption of molecular vibration, which represents FTIR, as a quantitative technique (Peak, 2005).

$$A = \epsilon bC$$

where $A$, $b$, $C$, and $\epsilon$ represent absorbance, thickness of sample, molecule concentration in a sample, and molar absorptivity coefficient, respectively. Peak height in the FTIR spectra represents absorbance ($A$), and molar absorptivity coefficient ($\epsilon$) is specific for a particular vibration.

Based on the state of a substance, different types of modes are used to analyze the FTIR spectra, such as attenuated total reflectance (ATR) for suspension and aqueous solution, diffuse reflectance with powder material, and transmission mode for pressed disks. ATR mode is a recent addition to FTIR techniques which allows effective evaluation with nominal sample preparation and higher sensitivity. The technique is based on measuring the entirely internally reflected IR spectra. The sample is pressed against a highly refractive index crystal, and IR waves are directed to the crystal at a critical angle. The evanescent waves, produced because of the interaction of IR waves with the crystal surface, are used to generate the FTIR spectra for a sample (Peak, 2005). It is a commonly used mode in FTIR techniques, since it is quick to operate and different types of samples, namely, solid, liquid, and powder, can be analyzed. Diffuse reflectance mode was proposed to address the issue of the pressed disk method. Powdered samples scatter light in a diffuse manner in all directions. This technique separates the diffusely scattered IR light from the specularly reflected or transmitted and collects it as diffuse reflectance spectra. The transmission mode for pressed disks is the most common mode of FTIR techniques. The solid material is mixed with 1–2 wt.% of KBr salt and pressed into a thin disk. Along with that, the sample should be conditioned before mixing with KBr, to remove the moisture as much as possible. Fakhrul et al. analyzed the biodegradability of polypropylene and wood sawdust or wheat flour composite using FTIR technique (Fakhrul and Islam, 2013). The FTIR spectra showed the disappearance of peaks at 1,376.6, 1,646.9, and 1,725 $cm^{-1}$, which correspond to a methyl group, C = C bond, and C = O bond (carbonyl bond), respectively. The disappearance of peaks exhibits the degradation in polypropylene composite.

## 2.2.4 Morphological Characterization

The morphology includes crystallinity, arrangement of molecules, branching, and molecular shape of a polymeric material. The molecular structure and orientation, that is, configuration and conformation, have a major impact on the macroscopic properties of a polymeric material. The arrangement of unit segments (i.e., monomer) of the polymer provides different molecular structures, such as linear, network, and branched shapes. For example, styrene butadiene styrene is a copolymer of polystyrene and polybutadiene available in linear or branched molecular structures.

In the case of orientation, a configuration is an ordering of monomeric units or atoms of monomeric units by the chemical bonds. Therefore, configuration orientation cannot be altered until chemical bonds are reformed. Contrary, conformation arrangement results from the rotation of the monomeric segment around a single bond. Here, rotation happens without reforming or breaking of a chemical bond, for a polymeric chain with specific configuration might have several conformations over a time period, depending on applied stress, solvent interaction, thermal energy, or filler types. Polymeric materials are mostly semicrystalline. The packing of molecules in polymer consists of crystalline and amorphous regions mixed in a random fashion. The addition of filler, fiber, or additives may alter the crystalline or amorphous phases. Therefore, various morphological techniques are used to analyze the polymeric or composite material.

### 2.2.4.1 Scanning Electron Microscopy

To examine the microstructural characteristics of a material, scanning electron microscopy is one of the versatile techniques for analysis. This technique is an inexpensive, relatively rapid, and basically nondestructive method. It offers direct observation of the sample surface with good resolution and depth of field. It covers a wide range of applicability, from metallurgy, polymer, composite, and geology to biology.

The scanning electron microscope (SEM) utilizes an electron beam to obtain an image of a specimen with high magnification. An electron gun, enclosed with a vacuum column, emits a beam of electrons that travels through a number of magnetic lenses and focuses on a fine spot. The interaction of the electron beam with the specimen surface results in the emission of secondary electrons along with backscattered. The detector counts the secondary electrons (SE), and the image is constructed depending on the received number of SE. Due to high abundance, energy less than 50 eV, and defined rather than arbitrary property, secondary electrons are selected for image analysis (Pavan and Andrew, 2012), whereas backscattered electrons (BSE) are high-energy electrons, quick to leave the sample, lower yield, and vary depending on the sample atomic number or sample interaction volume. In order to perform image analysis using the SEM technique, it is necessary to etch the sample. For instance, to perform image analysis of a nonconducting sample, coating of the conductive material is required to avoid charging up. To capture a high-resolution image, extremely focused electron beam is required. This may lead to the accumulation of high energy density on the sample surface, which causes damage to the sample. Most polymeric materials are good insulators; hence, coating of the conductive material is required. These materials can be various metals, namely, platinum (Pt), gold (Au), and titanium (Ti) (Molnar et al., 2021).

Sathishkumar et al. performed SEM analysis on snake grass–reinforced isophthalic polyester composites to measure the tensile and flexural performance (Sathishkumar et al., 2012). The SEM images, shown in Figures 2.1, 2.2, and 2.3, exhibited fiber failure when flexural and tensile were performed on composite material containing 10 and 25% fibers. Similarly, Mohanty et al. performed the analysis on a sisal fiber–reinforced polypropylene composite (Mohanty et al., 2004). The SEM images are provided in Figures 2.3 and 2.4. The high lignin content in sisal fibers causes a higher

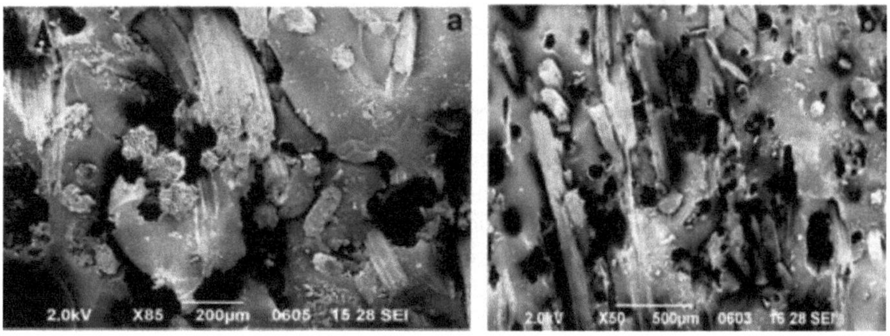


**FIGURE 2.1** SEM images of composite with (a) 10% and (b) 25% volume fraction of snake grass plant fiber in polyester resin after tensile testing.

**Source: Sathishkumar et al. (2012).**

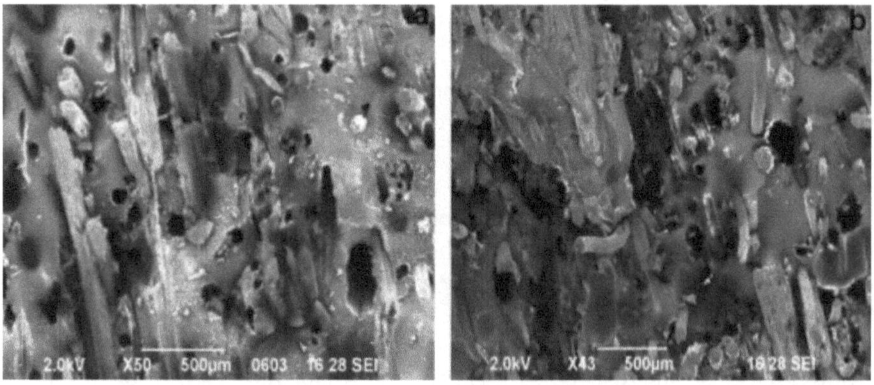


**FIGURE 2.2** SEM images of composite with (a) 10% and (b) 25% volume fraction of snake grass plant fiber in polyester resin after flexural testing.

**Source: Sathishkumar et al. (2012).**

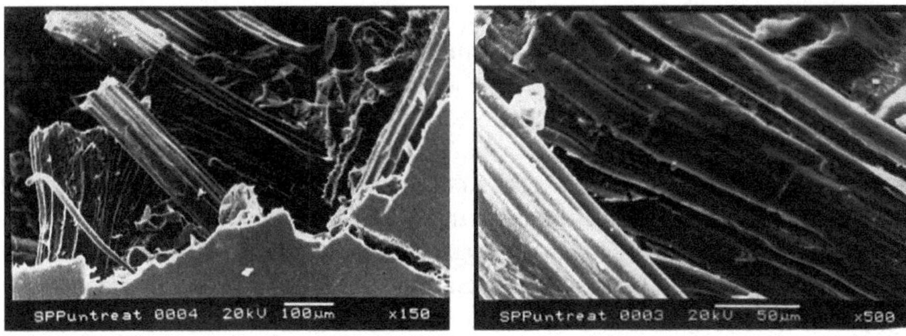


(A) (B)

**FIGURE 2.3** SEM images of sisal fiber and polypropylene composite.

**Source: Mohanty et al. (2004).**

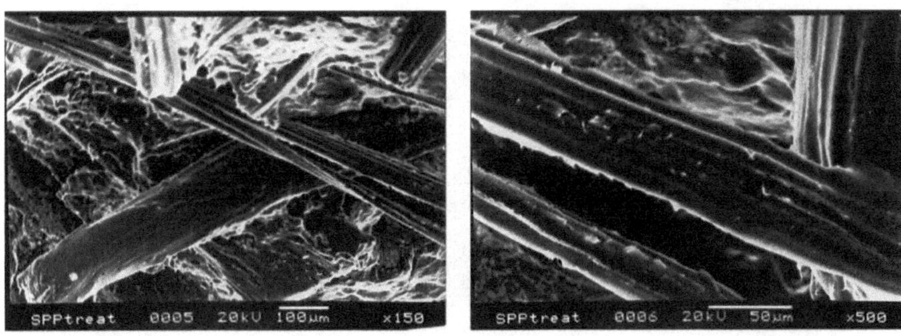


(A) (B)

**FIGURE 2.4** SEM images of sisal fiber and polypropylene composite, maleic anhydride–grafted PP (MAPP) as a coupling agent.

**Source: Mohanty et al. (2004).**

affinity toward water, which results in poor adhesion with the polypropylene matrix. The image provided in Figure 2.4 exhibits the definite holes due to the fiber coming out of the matrix during fracture analysis. Weak interfacial adhesion causes the pullout of fiber from the matrix, whereas the addition of a coupling agent, that is, maleic anhydride–grafted PP, enhances the adhesion, as shown in Figure 2.4. The presence of coupling agent fibers is well-infused by the propylene coating, which increases the adhesion between fiber and polymer and minimizes fiber removal.

### 2.2.4.2 Transmission Electron Microscopy (TEM)

TEM analysis has a vital role whenever interfaces or defects control the microscopic properties of a polymeric material, such as the second phase's intricate microstructure in the evolution of advanced structural materials or electronic materials, which severely depend on the control of interfaces. The distinctive characteristic of transmission electron microscopy (TEM) is to provide images of atomic arrangement at localized regions, distinctions in structure among the regions, and their interface.

The construction of an image is based on emerging electron intensity from the samples. An electron gun produces a monochromatic beam of electrons which is focused into a thin small coherent beam through the condenser lenses. These lenses are controlled by the spot knob and brightness or intensity knob. A condenser aperture is selected to remove the high-angle electrons that are off from the optic axis. A beam of electrons hits the sample, and a portion of it is transmitted. The objective lens focuses this transmitter part into an image. The obtained image consists of lighter and darker regions. The lighter region represents a thinner area where a beam of electrons is allowed to transmit. In contrast, denser areas restricted the transmission of electrons beam. S. Pegel et al. applied statistical analysis on TEM images to measure the distribution, alignment, and dispersion of carbon nanotubes in the polycarbonate matrix (Pegel et al., 2009). By a stereological method of TEM image processing, the study reported the macroscopic distribution and dispersion of CNT in the composite material. Dan Ciprari et al. examined polymer composite interphase and its effect on mechanical performance using FTIR, thermogravimetric analysis (TGA), dynamic mechanical analysis (DMA), etc., along with morphological analysis (Dan et al., 2006). In order to measure the average nanoparticle size, TEM technique was employed, the image captured from TEM is presented in Figure 2.5. The study reported that the size of the particles measured by TEM was two to three times higher than the size specified by the manufacturer. The difference in particle size shows the flocculation in the composite.

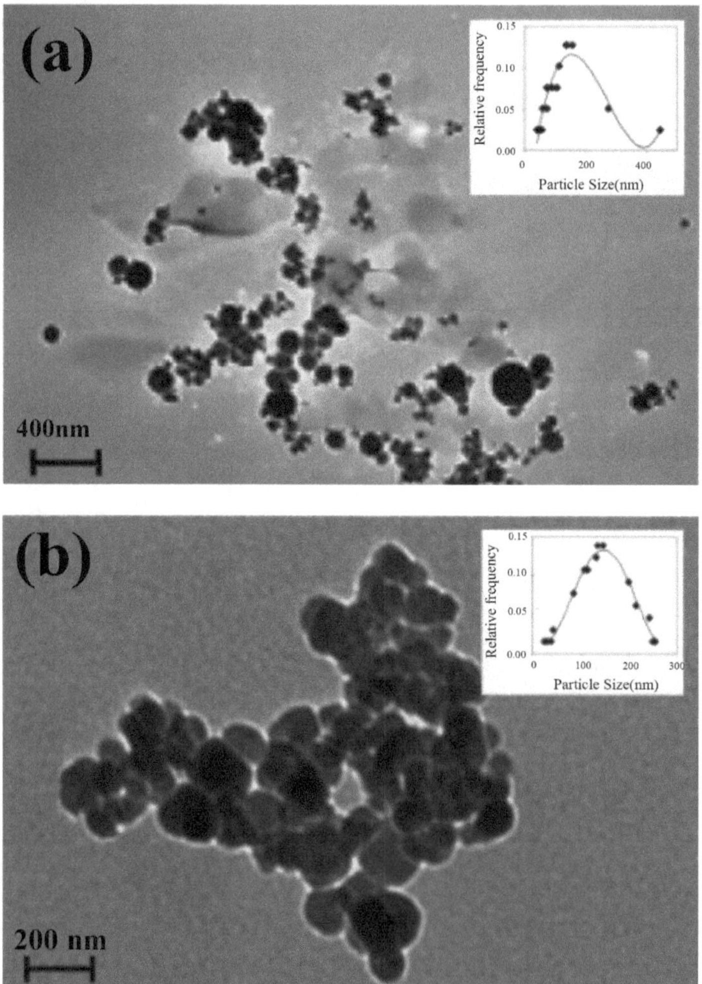


**FIGURE 2.5** TEM micrograph of a solution of reacted oxide nanoparticle and polymer (a) $Al_2O_3$/PS, (b) $Fe_3O_4$/PS. Image analysis is also performed to quantify the particle size distribution.

Source: Dan et al. (2006).

#### 2.2.4.3 X-Ray Diffraction

The X-ray diffraction (XRD) technique is a nondestructive procedure that provides comprehensive information about crystallinity, identification of crystalline phases, and the crystallographic structure of a polymer. The main concept is based on interference of X-rays with the sample having crystalline and amorphous phases. X-rays are electromagnetic radiation with shorter wavelengths produced when the deceleration of electrically charged particles with adequate energy occurs. The produced X-rays strike the sample and get diffracted. The diffracted rays are detected, counted, and processed. A unique diffraction pattern will be achieved because of the atomic arrangement and chemistry of the material.

XRD analysis is divided into two categories, namely, small-angle X-ray scattering (SAXS) and wide-angle X-ray scattering (WAXD), based on diffracted rays' angular region. SAXS relies on the elastic scattering of X-rays through the sample and records the scattering at an angle from approximately 0 to 10° (Abhilash et al., 2016). The stated characteristics exhibit that various parameters, such as pore size, shape, and size of monodispersed polymer; characteristic distance of ordered material; etc., can be measured using the SAXS technique. One of the important uses of SAXS is to measure lamellar thickness. Irrespective of the origin or nature of the microstructure, the existence of two regions with different electron densities can be detected using the SAXS technique. WAXD technique focused on analyzing the Bragg peak scattering to a wide angle, indicating that nanoscale structures are responsible for such scattering. Crystallite size and chemical composition can be determined by the generated diffractogram. Debye–Scherrer equation is used to calculate the crystallite size:

$$L = \frac{K\lambda}{\beta(2\theta)} cos\theta$$

where $K$, $\lambda$, and $\beta(2\theta)$ are shape factor, wavelength of X-ray, and full width at half-maximum, respectively.

As per the technique, the specimen is scanned in a goniometer, and scattering intensity is analyzed by varying the $2\theta$ angle. The quantity of electrons found in imaginary space is directly proportional to the intensity of d-spaced patterns. The reason to use X-ray is that the wavelength is similar to the spacing between atoms. Hence, the diffraction angle will be affected by the d-spacing only. Each crystalline solid has a discrete d-spacing pattern that serves as its fingerprint. So depending on d-spaced patterns, a sample with the same chemical composition but a different crystalline or amorphous region can be identified. The rays strike the sample and diffract at some angle $\theta$ because of the presence of an atomic layer in the mid of the incident wave path. The beam with the same wavelength resulted in a constructive interference with a higher amplitude. Bragg's law ($d = \frac{\lambda}{2sin\theta}$) and angle of diffraction are used to calculate the d-spacing between atomic planes, where the entities $d$, $\lambda$, and $\theta$ represent distance between atomic planes, wavelength, and angle of diffraction, respectively.

The major use for XRD in polymer composite is to analyze the reinforcements that will be used for the preparation of the polymer composite, an observation made from the literature review. Yoldas Seki et al. have used XRD to examine the crystallite size and crystallinity index of *Ferula communis* fiber using Scherrer's formula and the Segal empirical method, respectively (Yoldas et al., 2013). The diffraction peak at $2\theta$ = 22.2°, 16.8°, and 15.1° corresponds to the (002), (110), and ($1\bar{1}0$) planes of cellulose I structure. P. Narayanasamy et al. examined a cellulose fiber extracted from *Calotropis gigantea* fruit to utilize in composite preparation (Narayanasamy et al., 2020). Both alkali and untreated fibers exhibited peaks at $2\theta$ = 16.9°, 16.6°, and 23°, 22.7° that correspond to amorphous and crystalline phases. The crystallinity index is 39.8% and 36% for alkali-treated and untreated fibers, respectively. The increment in crystallinity index shows the elimination of noncellulosic components, that is, hemicellulose, lignin, and pectin.

### 2.2.5 Rheological Characterization

Since polymeric materials are many-body systems, therefore, experiments should be able to guide theoretical advancement. *Rheology* is an experimental science and is able to provide information at the segmental level of polymer. It is more convenient if we divide rheometry into categories, namely, drag flow and pressure-driven rheometry. The fixtures belonging to shear rheometry–drag flow are sliding plates, cone and plates, concentric cylinders, and parallel plates. Capillary flow, axial annulus flow, and slit flow come under shear rheometry–pressure-driven rheometry. In the case of polymer composite, drag flow and pressure-driven shear rheometers are used for the analysis. In this chapter, a few of them have been explained.

The flow behavior and deformation may not belong to these two straightforward categories, namely, purely viscous liquid and purely elastic solid material, known from Maxwell's area. Several materials are neither liquid nor elastic solid, which means the mechanical performance is time-dependent. A larger timescale shows liquid-like behavior, while at a shorter timescale, solid-like behavior can be noticed. These materials belong to another class, termed as non-Newtonian material. These non-Newtonian materials will have both viscous and elastic behavior, depending on analysis timescale. The viscoelastic behavior comes from the structural change in the material because of applied stress/strain that gives a relaxation spectrum. The structural variation can be gels, micelles, foam, emulsion, polymers, colloids, and many more.

The viscoelastic behavior because of applied stress/strain or temperature is examined under rheological analysis. The characterization of rheological parameters is one way to study molecular structure, branching, morphology, information about the processibility of a material, and temperature-dependent properties. In order to understand the spectra of flow behavior, namely, non-Newtonian, Newtonian, thixotropic, Bingham plastic, etc., parameters such as stress/strain, modulus, shear rate, elasticity, frequency, temperature, and segmental relaxation are needed to be examined. A single parameter, that is, viscosity, cannot provide all the information required during melt processing and solution casting. Therefore, various measurement techniques have been proposed to generate spectra of performance for a material, depending on the application requirement. *Capillary rheometer*, *torque rheometer*, and *dynamic shear rheometer* are the instruments used to analyze the rheological behavior of a polymeric material.

The *capillary rheometer* is based on examining the flow properties of a melt by measuring the force required to extrude the material via a capillary tube. It consists of a capillary tube with a particular length to measure the flow rate and pressure difference. The material is pushed by a plunger running at a definite speed, extruded through the die of a specific specification and pressure, and recorded at a set volumetric flow rate. The force applied by the piston on the molten material gives shear stress.

So from the Navier–Stokes equation, shear stress is:

$$\sigma(r) = -\frac{rdP}{2dZ}$$

Shear stress on the wall:

$$\sigma_w = \sigma(R) = \frac{R}{2} \times \left(-\frac{dP}{dZ}\right)$$

Apparent wall shear rate:

$$\dot{\gamma}_A = \frac{4Q}{\pi R^3}$$

Rabinowitsch correction:

$$\dot{\gamma}_w = \frac{3n-1}{4n} \times \dot{\gamma}_A$$

$$\frac{dlog\left(r\Delta P/2L\right)}{dlog\,(4Q/\pi R^{3}}n=$$

Viscosity:

$$\eta=\frac{\sigma}{\gamma^{\cdot}=\frac{\sigma_{w}}{\gamma_{w}}}$$

Sara P. et al. have used a double-capillary rheometer to examine the various rheological parameters, that is, apparent shear rate, shear stress, volumetric flow rate, Rabinowitsch–Weissenberg correction, and melt viscosity, of cork–polypropylene composites (Sara et al., 2016). Compared to unfilled polypropylene, cork–polypropylene composites exhibited increases in viscosity value. In contrast, the incorporation of polypropylene grafted with maleic anhydride (PPgMA) resulted in a reduction in viscosity value. Krzysztof L. et al. have used the same technique to examine the wood–polypropylene composite material (Krzysztof et al., 2016). The study reported the shear rate dependence of rheological parameters up to some extent as a function of filler dosage, while reduction in rheological properties is observed at the higher shear rate and the significance of wood content decreases.

To evaluate the processing behavior of polymer or composite, *torque rheometry* established itself as one of the helpful techniques. The viscosity-related torque generated because of the material's resistance to the shearing force is examined in torque rheometry. The analysis covers the shear sensitivity, thermal sensitivity, moisture, and effect of additives. Therefore, it is an excellent technique to examine the process-stable formulation for a process along with melt flow, degradation time, and thermal stability of a polymeric material. A batch mixer with two co- or counter-rotating mixing elements filled with the material is operated at a defined processing temperature. The torque required to run the mixing elements is analyzed as a function of time. The torque generated due to the resistance to flow by the material is another way to look into shear stress, while the speed of the rotor indicates the shear rate. So by using correlation, depending on instrument dimension, rotor speed, and torque, these can be converted to shear rate and viscosity by making some assumptions. The stated assumption could be that the apparent shear rate is linearly related to the frequency, torque is linearly related to apparent shear stress, and the mixing chamber can be considered as a two-coaxial cylinder having irregular dimensions, etc. (Ari et al., 2010; Bousmina et al., 1999). These assumptions are used to convert the data attained from torque rheometry to rheological data. Goodrich et al. have explained the process for the conversion of torque rheometry parameters to rheological parameters (Goodrich and Roger, 1967). The equations are as follows:

$$\sigma=\tau\times K_{1}\ and\ \gamma=K_{2}\times S$$

$$S=\left(2\pi/60\right)\times R$$

where $\sigma$, $\tau$, $\gamma$, $S$, $R$ denote stress, torque, shear rate, angular speed of rotor, and rotor speed, respectively, and $K_1$ and $K_2$ are constant and depend on torque rheometer.

$$K_{1}=\frac{1}{2\pi R_{m}{}^{2}h}\ \text{and}\ K_{2}=\frac{2}{nR_{m}{}^{\frac{2}{n}}\left(R_{i}{}^{\frac{-2}{n}}-R_{e}{}^{\frac{-2}{n}}\right)}$$

where $n$, $Re$, $Ri$, $h$, and $R_m$ represent power law index, outer and inner radius, cylinder length, and average radius, respectively. The dimension calibration for the cylinder is performed using Newtonian material. In order to examine the rheological parameters for non-Newtonian material, the same value might be useful with some empirical correlations (Ari and Aydin, 2010).

Ari, G. A., et al. examined the influence of particle size of calcium carbonate on the rheological response and fusion behavior of a rigid poly(vinyl chloride) composite using torque rheometry (Ari and Aydin, 2010). The torque value of the samples is measured by operating a Haake torque rheometer. The increment in the population of the particle by reducing the particle size exhibited a rise in frictional forces. The increase in frictional forces led to an increment in fusion torque and a drop in temperature. Similarly, Abdoos, H., et al. have used the same technique to select a desired formulation of polymer–powder mixture (Abdoos et al., 2014). Polypropylene (PP), aluminum-based powder, stearic acid, and paraffin wax have been used for the analysis.

The dynamic shear rheometer is the most advanced version compared to the techniques stated earlier. It measures any parameter related to rheology, namely, rotational viscosity ($\eta$), complex viscosity ($\eta^*$), loss modulus ($G''$), storage modulus ($G'$), phase angle ($\theta$), shear rate ($\dot{\gamma}$), stress ($\sigma$)/strain ($\gamma$), etc. Cone and plates, concentric cylinders, or parallel plates—any type of geometry—can be used to analyze the specified parameters. Based on the observation made from literature, for polymer composite, parallel plate geometry has been utilized more frequently for rheological measurements. The structure consists of two circular disks with parallel contact surface, identified as parallel plate geometry. Parallel plate geometry may not be suitable to examine the nonlinear response, though it might be a helpful method to create shear rate gradient in the sample (Wang, 2018). Mooney was the first person to suggest the parallel plate fixture. If we made a few assumptions, that is, $v_\theta(r,z)$ only and $v_z = v_r = 0$, laminar, steady, and isothermal flow, cylindrical edge, and negligible force by a system (Christopher, 1994).

Shear strain:

$$\gamma = \frac{r\theta}{h}$$

Shear rate at $r = R$:

$$\dot{\gamma} = \frac{R\tau}{h}$$

Then torque:

$$\tau = 2\pi \int_0^{\dot{\gamma}_R} \left(\frac{R}{\dot{\gamma}_R}\right)^3 \dot{\gamma}^2 \sigma \, d\dot{\gamma}$$

Petra Potschke et al. measured the rheological parameters of a carbon nanotube–reinforced polycarbonate composite material (Petra et al., 2002). Complex viscosity, storage modulus, and loss modulus are measured as a function of angular frequency and dosage of carbon nanotubes. The increase in viscosity was reported with the increment in carbon nanotube concentration. As the aspect ratio of the filler changes, a more pronounced effect on the rheological parameter is observed at a lower frequency. Lim Y. et al. correlated the rheological behavior of a polymer-layered silicate composite to the phase morphology (Lim and Park, 2001). In order to measure the rheological response of a composite material, storage and loss modulus are measured by varying the frequency. The analysis is performed in oscillatory mode using a 25 mm parallel plate fixture at 220°C. An increment in storage and loss modulus value is reported, and dependence on the applied frequency reduces with the dosage of silicate. The improvement in moduli value is due to the presence of randomly oriented anisotropic stacks of silicate layers.

### 2.2.6 Thermal Characterization

An increasing demand of composite materials highlights the significance of analyzing thermal properties. An elevated temperature results in sensitivity to the mechanical stress and increases vulnerability to external stimuli, while a lower temperature increases brittle failure. Therefore, it is necessary to understand the thermal history of a material that facilitates the designing of a system, that is, processing, assembling, finishing, and end use. Various experimental processes have been implemented to analyze the thermal behavior of the polymeric material.

#### 2.2.6.1 Differential Scanning Calorimetry (DSC)

Differential scanning calorimetry is a technique where quantitative measurement of heat flow is carried out to examine the thermoresponse of a material. During the analysis, the reference and sample are maintained at approximately similar temperatures, and the amount of heat required to raise the temperature of the sample and reference is measured by varying the temperature. The basic concept of DSC is that when a material undergoes a phase transition, depending on transition, heat will flow to it, compared to a reference to keep both systems at a similar temperature. The flow of heat depends on whether the process is endothermic or exothermic. In the endothermic process, heat flows to the sample, whereas heat flow happens out of the sample in the exothermic process. The DSC technique is used to collect information related to glass transition temperature ($T_g$), melting temperature ($T_m$), heat of melting, and crystallization temperature ($T_c$), to study phase transformation and crystallization kinetics. The glass transition temperature is the temperature above which the polymer will be in a rubbery state, while below it is in a glassy state. It is also known as a second-order transition, since the material undergoes a rise in heat capacity, whereas in first-order, transient heat exchange is between the system and the surroundings, and an abrupt volume change occurs. Melting and crystallization temperatures belong to first-order transition.

ASTM D3418 is used to measure the transition temperature or phase transition behavior of a polymer using the DSC technique. A premeasured sample quantity (<5 mg) with an accuracy of 10 μg is placed into the sample pan. Ensure uniform heat transfer from the bottom of the pan to the sample by crimping the lid against the pan, and the bottom of the pan should be flat. Perform a thermal cycle of heating and then cooling at a required rate under a nitrogen atmosphere, and record the thermograph. Glass transition temperature and melting temperature are obtained from the heating cycle. In contrast, crystallization temperature is achieved from the cooling cycle. Geum-Hyun Doh et al. analyzed the thermal behavior of liquid wood–polymer composite using the DSC technique (Geum et al., 2005). The change in melting temperature ($T_m$) is observed for HDPE, LDPE/HDPE, and PP because of adding liquid wood. The study states that up to 20% dosage of liquid wood decrease in melting was observed, while increases for 30% dosage. Along with that, a reduction in $T_m$ is noticed for liquid wood and LDPE/HDPE–PP composite also. Yunsheng Xu et al. have used the DSC technique to analyze a single-wall carbon nanotube and poly(vinylidene fluoride) composite (Yunsheng et al., 2006). The addition of the nanotube resulted in an additional peak in the thermograph, while no significant change was noticed in the overall $T_m$. The additional peak might be due to the development of transcrystallinity around the nanotubes. The change in the degree of crystallinity also increases the enthalpy of fusion.

#### 2.2.6.2 Thermogravimetric Analysis

Thermogravimetric analysis (TGA) is based on measuring the change in weight as a function of temperature or time. The TGA instrument consists of a precision balance, with a sample holder positioned in the core of the furnace. The specimen is heated at a controlled rate in a required environment (nitrogen, argon, air, $CO_2$, etc.), and change in weight is recorded by varying the temperature at different time intervals. The alteration in material weight may be due to decomposition, oxidation, desorption, reduction, etc. Gases or volatile matter lost during the thermal changes can also be measured using the TGA technique. These constituents are integral parts of polymer, fiber, films, and composite (Shabnam and Amit, 2023). A TGA instrument can be operated in three modes, namely, isothermal or static, quasistatic, and dynamic mode. As the name suggests, in isothermal TGA, temperature is kept constant, and specimen weight is measured as a function of time. The quasistatic process works on heating the sample to constant weight at each series of increasing temperatures. In dynamic TGA, a commonly used method, the specimen is heated in a particular environment, and the temperature change occurs at a constant rate.

Asim, M., et al. have summarized the thermal stability of natural fiber used as a reinforcer for polymer composite using the TGA technique (Asim et al., 2020). The thermal performance of natural fiber strongly depends on the chemical compositions, that is, cellulose, lignin, and hemicellulose.

Weight loss between 50 and 100°C is because of the removal of moisture from the fiber surface (Asim et al., 2020; Sergio et al., 2012). The degradation between 300 and 450°C presents the thermal degradation of cellulose and lignin, while hemicellulose degradation is observed in the range of 200 to 300°C. Beyond 450°C, the residual could be designated as char or other element from decomposition. Lee et al. (2011) showed that the increment in magnesium chloride in magnesium chloride–polyacrylamide composite reduces thermal stability (Lee et al., 2011), while a magnesium hydroxide–polyacrylamide composite exhibited improvement in thermal stability with the increment of magnesium hydroxide dosage. Possible reason could be hydrogen bonding among magnesium hydroxide and nitrogen of the polyacrylamide polymer.

### 2.2.6.3 Dynamic Mechanical Analysis

Dynamic mechanical analysis (DMA) is used to examine the viscoelastic response as a function of time, temperature, or frequency of deformation. The analysis is based on evaluating the solid-state rheological behavior of a sample by varying the temperature of frequency. A sinusoidal force is applied to a specimen, and the resulting strain is measured. The reproducibility or wave shape of deformation depends on the quantity of elastic and viscous phases a material has. For any point on the curve, applied stress $(\sigma) = \sigma_0 sin\omega t$, and rate of stress $\left(\frac{d\sigma}{dt}\right) = \omega\sigma_0 cos\omega t$. The extreme response of the material, that is, elastic and viscous, delivers limits that will provide a strain wave. As per Hookean limit, the material will behave elastically, and response (strain) will be

$$\varepsilon(t) = E\sigma_0 sin\omega t$$

In the linear viscoelastic region, stress and strain are linearly related by *E*.

So:

$$\varepsilon(t) = \varepsilon_0 sin\omega t$$

The viscous response is expressed as

$$\varepsilon(t) = \eta \frac{d\sigma}{dt = \eta\omega\sigma_0 cos\omega t} \text{ or } \varepsilon(t) = \eta\omega\sigma_0 sin\left(\omega t + \frac{\pi}{2}\right)$$

$$\varepsilon(t) = \omega\sigma_0 cos\omega t$$

$$= \omega\sigma_0 sin\left(\omega t + \frac{\pi}{2}\right)$$

or

$$\varepsilon(t) = \varepsilon_0 sin(\omega t + \delta) = \varepsilon_0 [sin\ sin\,\omega t\, cos\,\delta + sin\ sin\,\delta\, cos\,\omega t]$$

$$\text{So } \varepsilon' = \varepsilon_0 sin\ sin\,\delta \text{ So and } \varepsilon'' = \varepsilon_0 cos\ cos\,\delta$$

The vector sum of these entities is $\varepsilon^* = \varepsilon' + i\varepsilon''$, often denoted as *E'* and *E"*.

A modulus (*E*), loss modulus (*E"*), and storage modulus (*E'*) values obtained from DMA analysis for an applied sinusoidal stress allow to sweep across frequency or temperature window. Apart from the stated parameters, thermal transition, including glass transition temperature ($T_g$), can also be measured. Saba et al. have provided a literature review on the dynamic mechanical properties of various fiber-reinforced polymer composites (Saba et al., 2016). Lahelin et al. analyzed a Ni–Mn–Ga alloy filled silicon and epoxy composite using DMA technique (Lahelin et al.,

2009). Compared to the unfilled matrix, a soft epoxy matrix reinforced with Ni–Mn–Ga showed improved damping value.

## 2.3 CONCLUSIONS

The prime requirement of the composite industries is the endeavor to provide economically attractive composite materials. To achieve that, various characterization techniques have been developed and implemented. The characterization techniques include mechanical analysis, along with chemical, morphological, rheological, thermal, etc. The stated methods correlate the material properties with their field performance. Based on these techniques, researchers have focused on process optimization, optimum material dosage, impact of reinforcement dispersion, interaction between phases of composite materials, size distribution of nanomaterial in the composite, thermal response, rheological behavior, and friction and wear behavior of a polymer composite. A brief about the implications of respective techniques, from the literature, has also been discussed. The chapter concludes that characterization techniques provide insight into material response and explain their actual performance.

## ACKNOWLEDGMENTS

The author thanks the editors of the book titled *Advances in Polymer Composite Research: Integrating Experimental and Computational Approaches* for the invitation to contribute a chapter. A special acknowledgment to the Department of Chemical Engineering, IIT Bombay.

## REFERENCES

Abdoos, H., Khorsand, H., Yousefi, A.A., (2014) Torque rheometry and rheological analysis of powder–polymer mixture for aluminum powder injection molding. *Iranian Polymer Journal*, 23:745–755. https://doi.org/10.1007/s13726-014-0268-1.

Abhilash, V., Rajender, N., Suresh K., (2016) *Spectroscopy of Polymer Nanocomposites*, William Andrew Publishing, 410–451, ISBN 9780323401838. https://doi.org/10.1016/B978-0-323-40183-8.00014-8.

Ahmad, S.H., Bonnia, N.N., Zainol, I., (2011) Polyester-Kenaf composites: Effects of alkali fiber treatment and toughening of matrix using liquid natural rubber. *Journal of Composite Materials*, 45:203–217.

Ankur, C., Rahul, S.M., and Avinash, P., (2022) Polymer-based nanocomposites for impact loading: A review. *Mechanics of Advanced Materials and Structures*, 29:2581–2606. https://doi.org/10.1080/15376494.2021.1871688.

Ari, G.A., Aydin, I., (2010) Rheological and fusion behaviors of PVC micro- and nano-composites evaluated from torque rheometer data. *Journal of Vinyl and Additive Technology*, 16:223–228. https://doi.org/10.1002/vnl.20241.

Arun, K.S., Sun, C.T., (2006) Enhancing compressive strength of unidirectional polymeric composites using nanoclay. *Composites Part A: Applied Science and Manufacturing*, 37:2257–2268. https://doi.org/10.1016/j.compositesa.2005.12.027.

Asim, M., Paridah, M.T., Chandrasekar, M., Shahroze, R.M., Jawaid, M., Nasir, M., Siakeng, R., (2020) Thermal stability of natural fibers and their polymer composites. *Iranian Polymer Journal*, 29:625–648. https://doi.org/10.1007/s13726-020-00824-6.

Asma, B., Hamdi, L., Ali, B., (2020) Tensile and flexural properties of polymer composites reinforced by flax, jute and sisal fibres. *The International Journal of Advanced Manufacturing Technology*, 108:895–916.

Bousmina, M., Ait-Kadi, A., Faisant, J.B., (1999) Determination of shear rate and viscosity from batch mixer data. *Journal of Rheology*, 43:415–433. https://doi.org/10.1122/1.551044.

Christopher, W.M., (1994) *Rheology, Principles, Measurements, and, Applications*, Wiley-VCH, Inc.

Collings, T.A., (1994) Moisture absorption–Fickian diffusion kinetics and moisture profiles. In *Handbook of Polymer Fibre Composites*, Longman Scientific and Technical, Essex, UK, 366–371.

Dan, C., Karl, J., Rina, T., (2006) Characterization of polymer nanocomposite interphase and its impact on mechanical properties. *Macromolecules*, 39:6565–6573. https://doi.org/10.1021/ma0602270.

Datsyuk, V., Trotsenko, S., Trakakis, G., Boden, A., Vyzas-Asimakopoulos, K., Parthenios, J., Galiotis, C., Reich, S., Papagelis, K., (2020) Thermal properties enhancement of epoxy resins by incorporating polybenzimidazole nanofibers filled with graphene and carbon nanotubes as reinforcing material. *Polymer Testing*, 82:106317.

Dhakal, H.N., Zhang, Z.Y., Richardson, M.O.W., (2007) Effect of water absorption on the mechanical properties of hemp fibre reinforced unsaturated polyester composites. *Composites Science and Technology*, 67:1674–1683.

Dilip, C.D.N., Bandyopadhyay, S., Aibing, Y., Qinghua, Z., Tapas, D., Darryl, B., Chris, W., (2009) Structure-property interface correlation of fly Ash-isotactic polypropylene composites. *Journal of Materials Science*, 44:6078–6089. https://doi.org/10.1007/s10853-009-3839-3.

Fakhrul, T., Islam, M., (2013) Degradation behavior of natural fiber reinforced polymer matrix composites. *Procedia Engineering*, 56:795–800.

Gaurav, L., Allan, M.H., Behzad, F., Liangjun, L., Kurtis, K., Yi-Xian, Q., Antonios, G.M., Balaji, S., (2013) Two-dimensional nanostructure-reinforced biodegradable polymeric nanocomposites for bone tissue engineering. *Biomacromolecules*, 14:900–909. https://doi.org/10.1021/bm301995s.

Geum, H., Doh, S.Y., Lee, I.-A.K., Young, T.K., (2005) Thermal behavior of liquefied wood polymer composites (LWPC). *Composite Structures*, 68:103–108. https://doi.org/10.1016/j.compstruct.2004.03.004.

Goodrich, J.E., Roger, S.P., (1967) A rheological interpretation of torque-rheometer data. *Polymer Engineering & Science*, 7:45–51.

Hadi, E.A., (2017) Increase the elongation at break of poly (lactic acid) composites for use in food packaging films. *Scientific Reports*, 7, 46767. https://doi.org/10.1038/srep46767.

Hajnalka, H., Racz, I., Anandjiwala, R.D., (2008) Development of HEMP fibre reinforced polypropylene composites. *Journal of Thermoplastic Composite Materials*, 21:165–174.

Hamed, Y.K., Mohammad, F., Zahra, R., (2013) Physical, mechanical and morphological properties of polymer composites manufactured from carbon nanotubes and wood flour. *Composites Part B: Engineering*, 44:750–755. https://doi.org/10.1016/j.compositesb.2012.04.023.

Hom, N.D., Fabrizio, S., Carlo, S., Jacopo, T., Zhongyi, Z., Vellayaraj, A., (2015) Effect of basalt fibre hybridisation on post-impact mechanical behaviour of hemp fibre reinforced composites. *Composites Part A: Applied Science and Manufacturing*, 75:54–67. https://doi.org/10.1016/j.compositesa.2015.04.020.

Krzysztof, L., Kazimierz, P., Stanisław, Z., Jacek, M., (2016) Rheological properties of wood polymer composites at high shear rates. *Polymer Testing*, 51:58–62. https://doi.org/10.1016/j.polymertesting.2016.02.004.

Lahelin, M., Aaltio, I., Heczko, O., Soderberg, O., Ge, Y., Lofgren, B., Hannula, S.P., Seppala, J., (2009) DMA testing of Ni–Mn–Ga/polymer composites. *Composites Part A: Applied Science and Manufacturing*, 40:125–129. https://doi.org/10.1016/j.compositesa.2008.10.011.

Lee, K.E., Khan, I., Morad, N., Teng, T.T., Poh, B.T., (2011) Thermal behavior and morphological properties of novel magnesium salt–polyacrylamide composite polymers. *Polymer Composites*, 10:1515–1522.

Liao, Y.J., Wu, X.L., Zhu, L., Yi, T., (2018) Synthesis and properties of novel styrene acrylonitrile/polypropylene blends with enhanced toughness. *Chemistry Central Journal*, 12(1):78. https://doi.org/10.1186/s13065-018-0447-9.

Lim, Y.T., Park, O.O., (2001) Phase morphology and rheological behavior of polymer/layered silicate nanocomposites. *Rheological Acta*, 40:220–229.

Matsuoka, S., (1992) *Relaxation Phenomena in Polymers*, Oxford University Press, New York.

Mohanty, S.V., Nayak, S.K., Tripathy, S.S., (2004) Influence of fiber treatment on the performance of sisal–polypropylene composites. *Journal of Applied Polymer Science*, 94:1336–1345. https://doi.org/10.1002/app.21161.

Molnar, J., Zuba, Z., Sepsi, O., Ujhelyi, F., Erdei, G., Lenk, S., Menyhard, A., (2021) Structural investigation of semicrystalline polymers. *Polymer Testing*, 95:107098. https://doi.org/10.1016/j.polymertesting.2021.107098.

Muhammad, R.Z., Akil, H.M., Muhammad, H.A.K., (2016) Muhammad Bisyrul Hafi Othman, compressive properties and thermal stability of hybrid carbon nanotube-alumina filled epoxy nanocomposites. *Composites Part B: Engineering*, 91:235–242. https://doi.org/10.1016/j.compositesb.2016.01.013.

MuNoz, E., Garcia, M.J.A., (2015) Water absorption behaviour and its effect on the mechanical properties of flax fibre reinforced bioepoxy composites. *International Journal of Polymer Science*. https://doi.org/10.1155/2015/390275.

Narayanasamy, P., Balasundar, P., Senthil, S., Sanjay, M.R., Suchart, S., Anish, K., Abdullah, M.A., (2020) Characterization of a novel natural cellulosic fiber from Calotropis gigantea fruit bunch for ecofriendly polymer composites. *International Journal of Biological Macromolecules*, 150:793–801. https://doi.org/10.1016/j.ijbiomac.2020.02.134.

Pallab, D., Pankaj, T., (2019) Thermal degradation study of waste polyethylene terephthalate (PET) under inert and oxidative environments. *Thermochimica Acta*, 679:178340.

Pavan, M.V.R., Andrew, R.B., (2012) *SEM and Its Applications for Polymer Science*, Rice University.

Peak, D., (2005) Fourier transform infrared spectroscopy. In *Encyclopedia of Soils in the Environment*, Elsevier, 80–85. https://doi.org/10.1016/B0-12-348530-4/00174-0.

Pegel, S., Potschke, P., Villmow, T., Stoyan, D., Heinrich, G., (2009) Spatial statistics of carbon nanotube polymer composites. *Polymer*, 50:2123–2132. https://doi.org/10.1016/j.polymer.2009.02.030.

Petra, P., Fornes, T.D., Paul, D.R., (2002) Rheological behavior of multiwalled carbon nanotube/polycarbonate composites. *Polymer*, 43:3247–3255. https://doi.org/10.1016/S0032-3861(02)00151-9.

Rachid, H., Bekhta, A., Harfi, A.E., (2018a) Synthesis and characterization of a new epoxy resin homologous of DGEBA: Diglycidyl bis disulfide carbon ether of bisphenol A. *Journal of Chemical Technology and Metallurgy*, 53:414–421.

Rachid, H., Rajaa, S., Zakaria, B., Miloudi, H., Mohamed, R., Ahmed E., (2021) Polymer composite materials: A comprehensive review. *Composite Structures*, 262:113640. https://doi.org/10.1016/j.compstruct.2021.113640.

Ramamurty, U., (2001) Mechanical testing methods of fibers and composites. In *Encyclopedia of Materials: Science and Technology*, Elsevier.

Saba, N., Jawaid, M., Othman, Y.A., Paridah, M.T., (2016) A review on dynamic mechanical properties of natural fibre reinforced polymer composites. *Construction and Building Materials*, 106:49–159. https://doi.org/10.1016/j.conbuildmat.2015.12.075.

Sanjeevi, S., Shanmugam, V., Kumar, S., Ganesan, V., Sas, G., Johnson, D.J., Shanmugam, M., Ayyanar, A., Naresh, K., Neisiany, R.E., Das, O., (2021) Effects of water absorption on the mechanical properties of hybrid natural fibre/phenol formaldehyde composites. *Scientific Reports*, 11:13385. https://doi.org/10.1038/s41598-021-92457-9\.

Sara, P., Magalhaes, D., Pulo S.L., Jose, M.O., (2016) Rheological behaviour of cork-polymer composites for injection moulding. *Composites Part B: Engineering*, 90:172–178. https://doi.org/10.1016/j.compositesb.2015.12.015.

Sathishkumar, T.P., Navaneethakrishnan, P., Shankar, S., (2012) Tensile and flexural properties of snake grass natural fiber reinforced isophthallic polyester composites. *Composites Science and Technology*, 72:1183–1190. https://doi.org/10.1016/j.compscitech.2012.04.001.

Sergio, N.M., Veronica, C., Ruben, J.S.R., Frederico, M.M., (2012) Thermogravimetric behavior of natural fibers reinforced polymer composites: An overview. *Materials Science and Engineering: A*, 557:17–28. https://doi.org/10.1016/j.msea.2012.05.109.

Shabnam, D., Amit, C., (2023) *Novel Materials for Environmental Remediation Applications*, Elsevier, 537–553. https://doi.org/10.1016/B978-0-323-91894-7.00001-3.

Shen, C.H., Springer, G.S., (1976) Moisture absorption and desorption of composite materials. *Journal of Composite Materials*, 10:2–20.

Sims, G.D., (2003) Fatigue test methods, problems and standards. In *Fatigue in Composites: Science and Technology of the Fatigue Response of Fibre-Reinforced Plastics*. https://doi.org/10.1016/B978-1-85573-608-5.50007-3.

Spathis, G., Kontou, E., (2012) Creep failure time prediction of polymers and polymer composites. *Composites Science and Technology*, 72:959–964. https://doi.org/10.1016/j.compscitech.2012.03.018.

Srinivasan, V.S., Rajendra, B.S., Sangeetha, D., Vijaya, R.B., (2014) Evaluation of mechanical and thermal properties of banana–flax based natural fibre composite. *Materials & Design*, 60:620–627. https://doi.org/10.1016/j.matdes.2014.03.014.

Tarhini, A.A., Tehrani-Bagha, A.R., (2019) Graphene-based polymer composite films with enhanced mechanical properties and ultra-high in-plane thermal conductivity. *Composites Science and Technology*, 184:107797.

Vaggar, G.B., Kamate, S.C., (2020) Thermal property characterization for enhancement of thermal conductivity of hybrid polymer composites. *Material Today: Proceeding*, 20:208–212. https://doi.org/10.1016/j.matpr.2019.11.055.

Vigneshwaran, S., Uthayakumar, M., Arumugaprabu, V., Sundarakannan, R., (2021) Numerical and experimental assessment of water absorption of red mud: An industrial waste reinforced sisal/polyester hybrid polymer composite. In *Structural Health Monitoring System for Synthetic, Hybrid and Natural Fiber Composites*, 217–229. https://doi.org/10.1007/978-981-15-8840-2_15.

Wang, S.Q., (2018) *Nonlinear Polymer Rheology: Macroscopic Phenomenology and Molecular Foundation*, John Wiley & Sons, Inc. https://doi.org/10.1002/9781119029038.

Wisnom, M.R., (2000) Test methods, nondestructive evaluation, and smart materials. In *Comprehensive Composite Materials,* Vol. 5. ScienceDirect, Pergamon.

Yahaya, R., Sapuan, S.M., Jawaid, M., Leman, Z., Zainudin, E.S., (2015) Effect of layering sequence and chemical treatment on the mechanical properties of woven kenaf–aramid hybrid laminated composites. *Materials & Design*, 67:173–179. https://doi.org/10.1016/j.matdes.2014.11.024.

Yan, L., Nawawi, C., Liang H., Bohumil K., (2016) Effect of alkali treatment on microstructure and mechanical properties of coir fibres, coir fibre reinforced-polymer composites and reinforced-cementitious composites. *Construction and Building Materials*, 112:168–182. https://doi.org/10.1016/j.conbuildmat.2016.02.182.

Yanjun, X., Qinglin, W., Yong, L., Fei, Y., (2010) Creep behavior of bagasse fiber reinforced polymer composites. *Bioresource Technology*, 101:3280–3286. https://doi.org/10.1016/j.biortech.2009.12.072.

Yoldas, S., Mehmet, S., Kutlay, S., Cenk, D., (2013) Extraction and properties of Ferula communis (chakshir) fibers as novel reinforcement for composites materials. *Composites Part B: Engineering*, 44:517–523. https://doi.org/10.1016/j.compositesb.2012.03.013.

Yunsheng, X., Gunawidjaja, R., Beckry, A.M., (2006) Thermal behavior of single-walled carbon nanotube polymer–matrix composites. *Composites Part A: Applied Science and Manufacturing*, 37:114–121. https://doi.org/10.1016/j.compositesa.2005.04.009.

# 3 Finite Element Analysis (FEA) in Polymer Composite Study

*Atul K. Maurya, Akshay K. Chaudhari, Arjun S. Chaudhari, Akarsh Verma, and Sushanta K. Sethi*

## 3.1 INTRODUCTION TO POLYMER COMPOSITES

The demand for the use of polymer composite materials in sectors such as structural, construction, biomedical, automobile, aeronautical, etc. is increasing at an accelerated pace [1–5]. Composite materials consist of two or more than two phases to take advantage of both phases synergistically. Continuous phases in the composite are termed *matrix*, while the other, secondary phases are *reinforcements*. Various types of reinforcements, such as particulate, fibrous, or a combination of them, can be used for the fabrication of polymeric composites. Various types of reinforcements, such as glass fibers, carbon fibers, natural fibers, silica, and fly ash, have already been employed in the composite industry. These reinforcements provide excellent strength, toughness, flexibility, and impact strength to the fabricated composites and, thereafter, can be used for various industrial purposes. The most common type of polymeric matrices used with these materials are epoxy, polyester, polyolefins, or a blend of rubber and polymer, depending on their end-use applications [2]. Fibers can be used either as short fibers or long fibers. Short fibers are randomly oriented inside the polymeric matrices, whereas long fibers can be either unidirectional or bidirectional [6, 7]. Interestingly, long fibers show optimum properties along the fiber direction and the least strength toward the transverse direction, exhibiting anisotropy. Unidirectional fibers can be impregnated in resin for composite laminates and may be aligned on a thin plate and may later be used for stacking and designing composite laminates.

The bidirectional orientation of the fiber can be achieved by various kinds of weaving. These anisotropic materials show different properties in different planes and axes and do not show symmetry in properties with each direction [6, 7]. Furthermore, *orthotropic materials* can be defined as a part of anisotropic materials. Where planes parallel to the fiber, direction may exhibit superior strength compared to the properties measured perpendicular to the planes. The most common example of orthotropic material is wood, where properties in the radial and tangential directions are lower than that of the properties measured toward the fiber direction [6, 7].

It has already been mentioned that the properties demonstrated by the composite materials cannot be performed by their individual constituents [6, 7]. Hence, the selection of a particular reinforcement and matrix may give superior tensile and impact strength, stiffness, corrosion resistance, and high temperature. These property-tailoring qualities make composite material of specific interest and attractive at the same time. Currently, composite materials are successfully employed in biomedical, sports, automotive, aeronautical, civil construction, and similar applications. However, the presence of heterogeneity makes them costly, and it is very difficult to characterize them.

Various researchers assessed this difficulty, which opened a new window toward computational simulations. For instance, the damage caused in composite material (matrix, fiber, or interface) during a static, dynamic loading or due to the application of temperature and pressure, among others, can be understood by using finite element analysis (FEA).

The failure phenomena of the composite materials cannot always be addressed from the analytical models. Different failure mechanics, such as detachment of the reinforcement from the interface, lead to the initiation and propagation of the voids in the matrix.

DOI: 10.1201/9781003476337-3

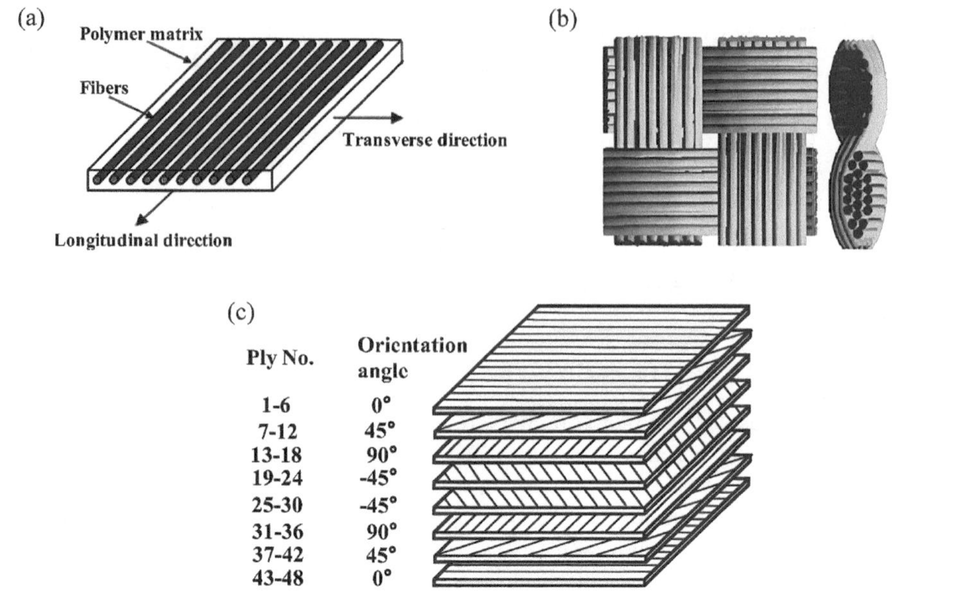


**FIGURE 3.1** Long fiber–reinforced polymer composites in (a) unidirectional, (b) bidirectional, and (c) multimerization laminates.

**Source: Used with permission from ref. [8].**

Layer delamination, fiber rupture, fiber debonding from the matrix, and transverse matrix fracture are the four major types of failure mechanism that occur during load application. Interestingly, composite failure starts with the transverse matrix fracture due to the lower stress failure value of the matrix material compared to the laminates. In the next step, fiber debonding from the interface facilitates the progress of the transverse matrix fracture in the failure mechanism further. Transverse failures occur in the fiber direction and hence lead to a decrease in stiffness without breaking the longitudinal fiber. Hence, in this case, damage criteria need the onset of the damage orientation and failure.

Application of FEM to the composites can be achieved by applying constitutive law, modeling, and failure criteria of the composite materials and the type of elements used to model the objects.

## 3.2 BASICS OF FINITE ELEMENT ANALYSIS (FEA)

Analytical models have already been employed for the prediction of polymeric composites. For instance, the inputs (volume fraction, modulus of the reinforcements and the matrix, Poisson's ratio, aspect ratio, density, viscoelastic behavior, fiber orientation, orthographic properties, etc.) of both the reinforcements and the matrix help in predicting the various properties of the composites. Likewise, the rule of the mixture and the inverse rule of the mixture are the simplest models to predict the properties of continuous and unidirectional fiber-reinforced composites. Many researchers already use the Halpin Tsai model for the prediction of short fiber–reinforced polymer composites. Interestingly, this model uses the aspect ratio of the randomly oriented fiber into the polymer matrices. Another famous analytical model is the use of an ancient model, such as Cox's model, to predict the modulus and strength of the short fiber–reinforced polymer composites. The determination of elastic modulus of short sisal fiber–, basalt fiber–, and bamboo fiber–reinforced polypropylene composite from various models was successfully implemented by Maurya and Gogoi et al. [4, 9, 10].

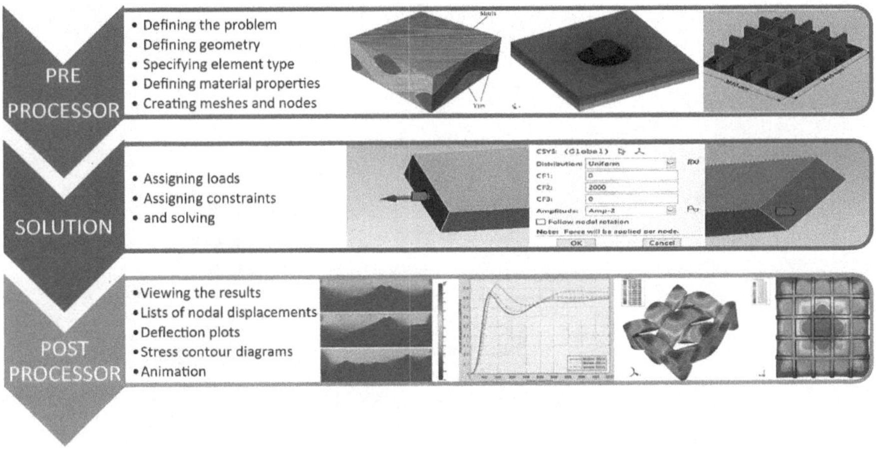


**FIGURE 3.2** Overview of the FEA.

**Source: Adopted with permission from [11].**

The FEA is an M&S tool to solve complex structures, material models, or boundary conditions both in industry and in academia. Technically, a virtual experimental condition is provided for the graph and data generation from the FEA. The product development time and lifespan of the product can be optimized by running several iterations to the model developed based on the boundary conditions. Figure 3.2 shows the various steps involved during the FEA analysis.

## 3.3 MATERIAL MODELING IN POLYMER COMPOSITE ANALYSIS

Three basic approaches have been reported for the simulation of composite materials: the micromechanics approach, the equivalent homogeneous material (EHM) approach, and the combination of both.

To report fiber debonding from the matrix, local defect, or complex deformation mechanism, the micromechanics approach is practical, since it describes the material's behavior locally [12]. However, this approach is time-consuming due to the use of a fine mesh size compared to other models. On the contrary, the EHM model shortens the simulation time, but it is not able to predict the local effects, such as fiber debonding, etc. [12–14]. However, these two models might be advantageous in determining shear force and subsurface damage [12].

This methodology was applied by Rao et al. [12], where the authors used unidirectional glass fiber–reinforced polymer composites to study the machining of the chip formation mechanism. They used micromechanics for the material adjacent to the cutting tool, whereas the portion that was away from the toll was studied using EHM modeling (Figure 3.3).

The cutting zone was modeled with the specification of the composites, whereas the region around the tool and cutting zone was modeled as per EHM, since a single result was not so important for this region.

According to Jones et al. [16], the stiffness and strength of the targeted composites, or the behavior of the composite materials, can be categorized on micro- and macroscales:

Microscale: At this level, the interaction, such as interfacial adhesion, is examined as defined by the heterogeneity of the materials.

Macroscale: The inspection/detection of the properties of the composites as homogeneous material and constituent materials is only done by the mean apparent properties of the composites.

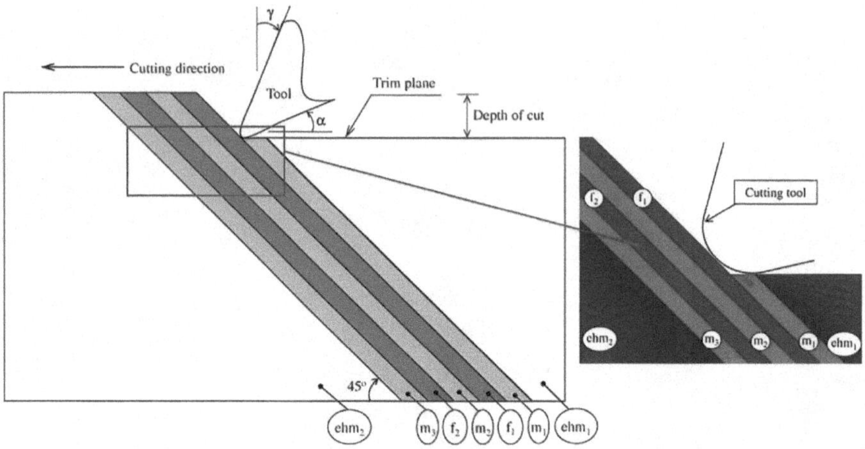


**FIGURE 3.3** Schematic of various components (fiber, matrix, and equivalent homogeneous materials [EHM]) used in FEA analysis for 45°C unidirectional fiber–reinforced composites.

**Source: Used with permission from ref. [15].**

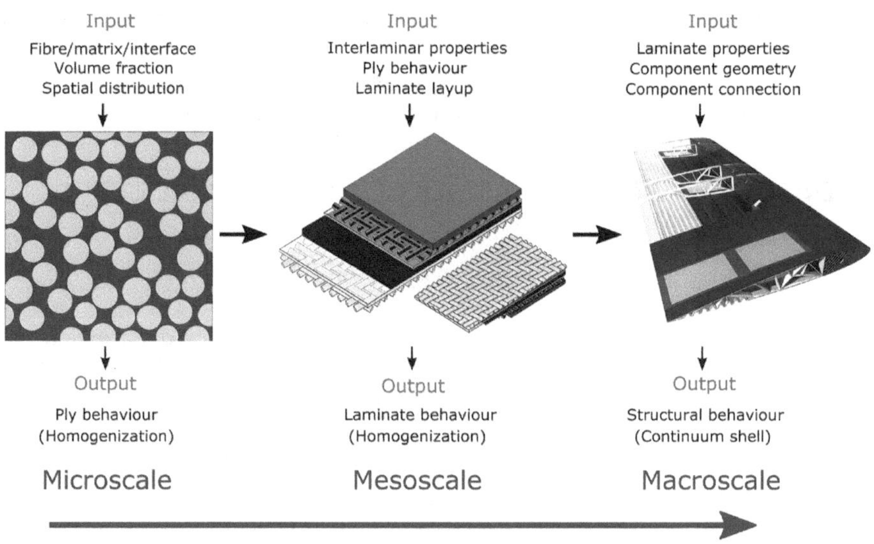


**FIGURE 3.4** Micro-, meso-, and macroscale approaches.

**Source: Used with permission from ref. [18].**

Tenek and Argyris reported that sheet properties could be defined using the microscale procedure, and these can be applied for global analysis using the macroscale procedure [17].

Furthermore, due to the uncertainties at the microscale, stochastic or statistical models are required. The goal of this approach is to define the composite modules from all the participating heterogeneous materials, or the strength of the composites, in terms of their phases. Hence, some basic approaches comprise the mechanism of materials and the theory of elasticity based on repetitive representative volumes, presuming excellent interaction at the interface, which is not true. Hence, as a result, the micromechanical approach can be validated only from experimental results. Structural evolution from ply to global composite structure is presented in Figure 3.4.

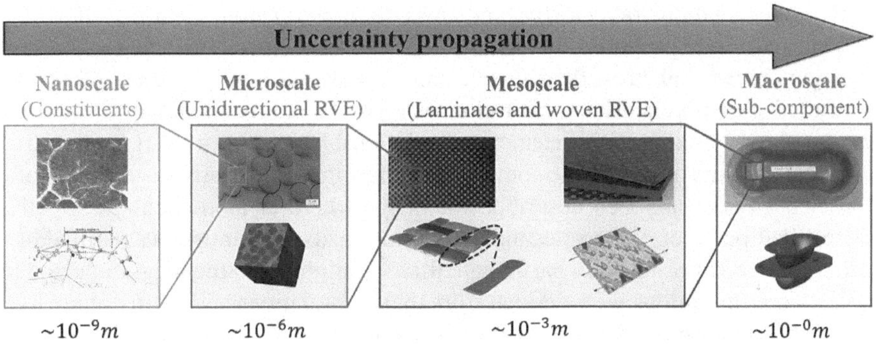


**FIGURE 3.5** Illustration of a four-scale woven fiber composite with a polymer matrix. In computational modeling, each integration point reflects a structure at a finer scale. Materials at fine scales may introduce more uncertainty in lower-scale representative volume elements (RVEs) than in higher scales. Identifying uncertainty sources at lower scales is essential for quantifying uncertainty in a macroscopic quantity of interest through propagation.

**Source: Used with permission from ref. [28].**

The simulation study of the composites facilitates the evaluation of the properties of the composite from the nanoscale to the macroscale. Hence, multiscale techniques can be applied by providing a simulation of the behavior of the composite as a consequence of multiple timescales and/or length scales [19–21]. Simulations of the microstructural and mechanical properties of the nanocomposites and other materials have already been reported by using this technique [22–26]. An ample number of studies have already been reported based on the micro-to-macroscale model approach. However, the inclusion of nanomaterials in the composite matrix leads to the development of a mesoscale approach. The mesoscale approach provides a better understanding of the damage progression inside a nanocomposite (Figure 3.5). Overall, the multiscale approach improved the classical models, and hence, information from the smallest scale to the highest level can be implemented in an archical process, through the homogenization technique [27].

## 3.4 APPLICATION OF FEA IN POLYMER COMPOSITE ANALYSIS

Researchers use various length scales of simulations, among which continuum mechanics using FEA is one. The application of FEA in polymer composite analysis ranges across diverse domains, playing a pivotal role in unraveling the complex behavior of these materials. Structural analysis stands out as a primary application, employing FEA to assess the mechanical performance and integrity of polymer composites under various loading conditions. This includes understanding stress distributions, identifying deformation patterns, and predicting potential failure modes, which are crucial for designing robust and reliable composite structures.

Mascia et al. [29] utilized FEA to create a mathematical model assessing the influence of interlayers on stress distribution around cylindrical fibers in transversely deforming fiber/polymer composites. The Galerkin finite element technique, coupled with a penalty method, was employed for its effectiveness in handling nonlinear deformations. The model was then extended to analyze composites with coated fibers of varying coating properties. Validation was demonstrated by comparing results with established solutions and experimental data. The study highlights FEA as a valuable tool for comprehending and optimizing stress distributions in complex fiber/polymer composite scenarios. Similarly, Hsu et al. [30] estimated the stress distribution along fibers in model polymer composites using FEA. They measured both thermal residual (compressive) and tensile stress distributions along reinforcement fibers with high sensitivity and spatial resolution. For blunt-ended fibers, the obtained stress distribution aligned well with expectations from an idealized shear–lag model. Tapered fibers demonstrated a constant stress distribution along their length, without a critical length identified. They proposed that the FEA predictions closely matched experimental data, providing insights into radial stress dependence and relative strain in both the fiber and the matrix. The study offered a comprehensive exploration of possible failure modes in discontinuous fiber composites, primarily based on finite-element calculations. Naseehi et al. [31] presented a potential micromechanical analysis of polymeric composites, offering a quantitative assessment of their bulk behavior. A robust FEM has been described for microstructural modeling, specifically focusing on particulate-filled polymer composites under external loads. The model was utilized to simulate stress distribution in composites with particulate fillers. Insights into the magnitude and location of maximum stress concentrations were derived from these simulations, aiding in predicting dominant failure and crack growth mechanisms. The authors compared the model predictions with experimental data and with values from other reported methods in the literature, demonstrating the developed model's validity and predictive potential within a defined range.

Moving to thermal analysis, FEA proves instrumental in evaluating the response of polymer composites to temperature variations. This is particularly significant in applications where thermal stability and resistance are highly desired, such as in aerospace or automotive components. The ability to simulate and analyze heat transfer, thermal stresses, and potential hot spots enables engineers to optimize the composite's thermal performance.

Vaidyanathan et al. [32] utilized FEA to develop and validate a novel method predicting the macroscopic effective thermal conductivity of filled composites. This FEA method incorporated microstructural characteristics, including filler aspect ratio, interfacial thermal resistance, volume fraction, and filler and fiber dispersion, for circular and rectangular fillers. The developed model considered filler interactions and chain formation effects. To streamline the process, an algorithm the authors created for automated model development allows for the generation of multiple microstructures. The method's superiority over existing models was evident in its ability to handle interactive effects, offer detailed heat flux patterns, and model high filler volume fractions, elucidating its potential for accurate predictions in complex composite scenarios. In the same line, Hillier et al. [33] employed FEM analysis with ABAQUS to address residual stresses in composite materials due to their heterogeneous nature during manufacturing. They estimated the mechanical and thermal properties through microstructure analysis. Various element types were compared for prediction accuracy and discussed the technique of FEM application. The analysis focused on temperature-induced stresses and deformations in thick cylindrical parts with arbitrary geometry. Predictions of spring-in values for 90° angles with varying curvature at the corner showed reasonably good agreement with measurements, demonstrating the effectiveness of the FEM approach in handling complex scenarios of residual stresses and deformations in thick composite components. Polymers, valued for their low specific gravity and corrosion resistance, often lack certain mechanical, thermal, and electrical properties. To address this, Mertiny et al. [34] studied on enhancing polymer properties using particulate fillers. They introduced a numerical modeling framework employing stochastic FEA to predict the effective properties, particularly thermal conductivity, of particulate-modified polymers. The model accommodated various filler shapes and composite morphologies and was validated against experimental data for composites with randomly distributed spherical particles. This approach efficiently guides the design of particulate-modified polymer composites, providing a fundamental understanding and reducing the need for labor-intensive experimental analyses. Chandrasekar et al. [35] discussed the widespread use of natural fibers and their reinforced polymeric products in recent years for non-load-bearing structures. The limitations of experimental evaluations, including human error and testing machine inaccuracies, were addressed by employing FEA to provide a more realistic and accurate assessment of mechanical, thermal, and other properties. The representative volume element (RVE) method, a homogenization-based, multiscale, finite element model, was introduced as an efficient tool to analyze the relationship between micro- and macrostructural behavior in natural fiber–reinforced polymer composites. The authors presented an overview of the modeling and simulation procedure, highlighting FEA's role in overcoming the challenges associated with experimental evaluations of these materials in their chapter.

Fatigue and durability analysis represents another crucial facet of FEA applications. By subjecting polymer composites to repeated loading cycles, FEA helps predict and mitigate fatigue-related failures. This is vital in applications where components are subjected to cyclic loading over their lifespan, ensuring that the composite materials can withstand prolonged usage without compromising structural integrity. Satyanarayana et al. [36] explored glass fiber–reinforced polymer composites as promising materials, emphasizing their diverse applications in electronics, aviation, and automobiles. The study focused on fatigue behavior in complex structures, utilizing FEA to determine strength properties. Their research aimed to compare analytical and experimental values from fatigue testing, providing insights into the durability of polymer composites. Similarly, Arifin et al. [37] investigated the durability of composite structures in marine environments through FEA. Using ANSYS Workbench, the composites with chopped strand mat (CSM) and woven roving (WR) fabric were tested for Charpy impact and three-point bending. Results showed an impact energy of 0.282 J and a peak load of 738 N. Bending test simulation indicated a bending stress of 633.43 MPa. Madhu et al. studied about replacing conventional SAE 5160 steel (chromium steel) with a carbon-reinforced polymer composite (CFRP) in automobile leaf springs to enhance strength, reduce weight, and improve fuel efficiency. The research, employing FEA, assesses flexural fatigue life and damage for both materials, considering various ply orientations for optimal results.

Vibrational analysis, yet another dimension of FEA application, is essential for understanding how polymer composites respond to dynamic forces and vibrations. This is particularly relevant in fields like aerospace and automotive engineering, where minimizing vibrations and ensuring structural stability are crucial. FEA allows for the simulation and analysis of natural frequencies, mode shapes, and damping characteristics, aiding in the design of composite structures that can effectively dampen vibrations and maintain stability. Suresha et al. [38] utilized ANSYS for FEA to determine the natural frequencies and mode shapes of cantilever glass and CFRP composites. It included eight-ply cantilevered plates with varying fiber volume fractions. Results were validated against literature, demonstrating excellent agreement. The research investigated the impact of matrix material change, hybridization, and laminate stacking sequence on natural frequencies and mode shapes. The findings emphasize that hybridization and outermost layer orientation significantly affect natural frequencies compared to fiber volume fraction and matrix material change.

The collective application of FEA across these domains provides a comprehensive understanding of the performance and limitations of polymer composites. It empowers engineers to make relevant decisions during the design and optimization phases, ensuring that the composite materials meet the specific requirements of their intended applications. As technology advances, the integration of FEA in polymer composite analysis continues to evolve, offering a robust and efficient approach to addressing the multifaceted challenges posed by these versatile materials. The insights gained from FEA applications not only enhance the current state of polymer composite engineering but also pave the way for innovations and advancements in material design and optimization.

## 3.5 CHALLENGES AND FUTURE DIRECTIONS

FEA has become essential for studying polymer composites, offering insights into their mechanical behavior and performance. However, challenges persist in accurately capturing the complex, nonlinear, and anisotropic nature of these materials. Developing, and validating, material models that can predict mechanical response under various conditions is a primary challenge.

The multiscale nature of polymer composites poses another significant challenge. With distinct length scales in matrix, reinforcement, and interphase regions, bridging the gap between macroscopic FEA simulations and microscopic features is a challenging endeavor. Researchers are working on addressing this challenge to enhance the accuracy of predictions. Durability and long-term performance are crucial considerations for polymer composites. FEA models need to account for environmental degradation effects, such as temperature, humidity, and exposure to chemicals. Understanding complex degradation mechanisms and their impact on material properties over time is an ongoing challenge.

The future directions for FEA in polymer composite studies involve overcoming these challenges and exploring new avenues. Advances in material modeling, including more accurate constitutive models, are critical. Integration with other simulation methods, like MD, can offer a more comprehensive understanding of mechanical behavior. The incorporation of uncertainty quantification techniques in FEA models will be increasingly important for robust predictions. Understanding the uncertainties related to material properties, manufacturing processes, and environmental conditions enhances the reliability of FEA predictions.

Despite the fact that FEA has significantly advanced the understanding on polymer composites, challenges persist in accurately representing their complex behavior. The future involves overcoming these challenges through improved material modeling, multiscale simulations, and the incorporation of uncertainty quantification techniques. FEA will continue to be at the forefront, providing valuable insights for designing and optimizing polymer composites across diverse applications.

## 3.6 CONCLUSION

In conclusion, the chapter provides a comprehensive and insightful exploration into the application of FEA for understanding and optimizing the behavior of polymer composites. The challenges inherent in polymer composite analysis emphasize the critical role of computational tools, and FEA emerges as a powerful methodology to address these complexities. By elucidating the basics of FEA, material modeling intricacies, and practical applications in structural, thermal, fatigue, durability, and vibrational analyses, the chapter equips researchers and engineers with valuable insights. As we navigate through the challenges of the present, the chapter not only serves as a guide to current methodologies but also points toward future research directions. The identified challenges discussed the need for ongoing innovation, while the proposed future directions pave the way for advancements in the study of polymer composites. In essence, this chapter serves as fundamental for those seeking a profound understanding of FEA's role in unraveling the mysteries and optimizing the performance of polymer composites.

## ACKNOWLEDGMENTS

All the authors wish to thank the University of Massachusetts–Lowell and the Indian Institute of Technology–Dharwad for providing excellent research environment.

## REFERENCES

[1] A.K. Maurya, F.M. de Souza, T. Dawsey, R.K. Gupta, Biodegradable polymers and composites: Recent development and challenges, Polym Compos (2023). https://doi.org/10.1002/PC.28023.

[2] A.K. Maurya, G. Manik, Advances towards development of industrially relevant short natural fiber reinforced and hybridized polypropylene composites for various industrial applications: A review, J Polym Res 30 (2023) 1–20. https://doi.org/10.1007/S10965-022-03413-8.

[3] A. K. Maurya, S. Kumar, M. Singh, G. Manik, Polyamide fiber reinforced polymeric composites: A short review, Mater Today Proc 80 (2023) 98–103. https://doi.org/10.1016/J.MATPR.2022.10.171.

[4] A.K. Maurya, R. Mahadeva, G. Manik, S.P. Patole, An investigation into performance properties of sustainable polypropylene composites reinforced with basalt fiber and fly ash, Polym Compos 44 (2023) 5104–5120. https://doi.org/10.1002/PC.27476.

[5] A.K. Maurya, *Synthesis and Mechanical Studies of Mg-Hydroxyapatite Composite for Bio materials Applications* (2015) [Doctoral dissertation, Indian Institute of Technology Hyderabad]. CRC Press. Taylor & Francis Group.

[6] M. Alhijazi, Q. Zeeshan, Z. Qin, B. Safaei, M. Asmael, Finite element analysis of natural fibers composites: A review, Nanotechnol Rev 9 (2020) 853–875. https://doi.org/10.1515/NTREV-2020-0069/ASSET/GRAPHIC/J_NTREV-2020-0069_FIG_005.JPG.
[7] S.D. Müzel, E.P. Bonhin, N.M. Guimarães, E.S. Guidi, Application of the finite element method in the analysis of composite materials: A review, Polymers 12 (2020) 818. https://doi.org/10.3390/POLYM12040818.
[8] J. Xu, A. Mkaddem, M. El Mansori, Recent advances in drilling hybrid FRP/Ti composite: A state-of-the-art review, Compos Struct 135 (2016) 316–338. https://doi.org/10.1016/J.COMPSTRUCT.2015.09.028.
[9] R. Gogoi, G. Manik, B. Arun, High specific strength hybrid polypropylene composites using carbon fibre and hollow glass microspheres: Development, characterization and comparison with empirical models, Compos B Eng 173 (2019) 106875. https://doi.org/10.1016/J.COMPOSITESB.2019.05.086.
[10] A.K. Maurya, R. Gogoi, G. Manik, Mechano-chemically activated fly-ash and sisal fiber reinforced PP hybrid composite with enhanced mechanical properties, Cellulose 28 (2021) 8493–8508. https://doi.org/10.1007/S10570-021-03995-4/TABLES/5.
[11] J. Naveen, M. Jawaid, A. Vasanthanathan, M. Chandrasekar, Finite element analysis of natural fiber-reinforced polymer composites, In *Modelling of Damage Processes in Biocomposites, Fibre-Reinforced Composites and Hybrid Composites* (2019) 153–170. https://doi.org/10.1016/B978-0-08-102289-4.00009-6.
[12] C.R. Dandekar, Y.C. Shin, Modeling of machining of composite materials: A review, Int J Mach Tools Manuf 57 (2012) 102–121. https://doi.org/10.1016/J.IJMACHTOOLS.2012.01.006.
[13] L. Yang, Y. Yan, Y. Liu, Z. Ran, Microscopic failure mechanisms of fiber-reinforced polymer composites under transverse tension and compression, Compos Sci Technol 72 (2012) 1818–1825. https://doi.org/10.1016/J.COMPSCITECH.2012.08.001.
[14] A. Mkaddem, I. Demirci, M. El Mansori, A micro–macro combined approach using FEM for modelling of machining of FRP composites: Cutting forces analysis, Compos Sci Technol 68 (2008) 3123–3127. https://doi.org/10.1016/J.COMPSCITECH.2008.07.009.
[15] G. Venu Gopala Rao, P. Mahajan, N. Bhatnagar, Machining of UD-GFRP composites chip formation mechanism, Compos Sci Technol 67 (2007) 2271–2281. https://doi.org/10.1016/J.COMPSCITECH.2007.01.025.
[16] R.M. Jones, *Mechanics of Composite Materials*, Second Edition (2018) 1–519. CRC Press. Taylor & Francis Group. SSRN/Elsevier. https://doi.org/10.1201/9781498711067/MECHANICS-COMPOSITE-MATERIALS-ROBERT-JONES.
[17] L. Tenek, J. Argyris, Finite element analysis for composite structures, 2013. https://books.google.com/books?hl=en&lr=&id=xpbpCAAAQBAJ&oi=fnd&pg=PR9&ots=VZs2Rb-9Y6&sig=XwMbDzrIibbQjNkxCDOu9Br2StE (accessed February 23, 2024).
[18] W. Tan, F. Naya, L. Yang, T. Chang, B.G. Falzon, L. Zhan, J.M. Molina-Aldareguía, C. González, J. Llorca, The role of interfacial properties on the intralaminar and interlaminar damage behaviour of unidirectional composite laminates: Experimental characterization and multiscale modelling, Compos B Eng 138 (2018) 206–221. https://doi.org/10.1016/J.COMPOSITESB.2017.11.043.
[19] D. Li, Z. Wang, C. Zhang, A multi-level and multi-site mesh refinement method for the 2D problems with microstructures, Mech Adv Mater Struct 28 (2021) 1462–1479. https://doi.org/10.1080/15376494.2019.1681036.
[20] J. Ma, W. Du, W. Gao, P. Wriggers, X. Xue, Multiscale finite element analysis of uncertain-but-bounded heterogeneous materials at finite deformation, Finite Elem Anal Des 149 (2018) 15–31. https://doi.org/10.1016/J.FINEL.2018.06.001.
[21] J. Aboudi, S. Arnold, B. Bednarcyk, Micromechanics of composite materials: A generalized multiscale analysis approach, 2013. https://books.google.com/books?hl=en&lr=&id=jKJ2VqSwu08C&oi=fnd&pg=PP2&ots=NUFfvVDAvq&sig=c-0YT7mRkQC20uEhoJJFhLBas-Y (accessed February 23, 2024).
[22] Z.X. Wang, H.S. Shen, Nonlinear vibration of nanotube-reinforced composite plates in thermal environments, Comput Mater Sci 50 (2011) 2319–2330. https://doi.org/10.1016/J.COMMATSCI.2011.03.005.
[23] Z.X. Wang, H.S. Shen, Nonlinear vibration and bending of sandwich plates with nanotube-reinforced composite face sheets, Compos B Eng 43 (2012) 411–421. https://doi.org/10.1016/J.COMPOSITESB.2011.04.040.
[24] S.S.R. Koloor, S.M. Rahimian-Koloor, A. Karimzadeh, M. Hamdi, M. Petrů, P. Petrů, M.N. Tamin, Nano-level damage characterization of graphene/polymer cohesive interface under tensile separation, Polymers 11 (2019). https://doi.org/10.3390/polym11091435.

[25] J.A. Elliott, Novel approaches to multiscale modelling in materials science, Int Mater Rev 56 (2011) 207–225. https://doi.org/10.1179/1743280410Y.0000000002.
[26] J. Ma, W. Du, W. Gao, P. Wriggers, X. Xue, Multiscale finite element analysis of uncertain-but-bounded heterogeneous materials at finite deformation, Finite Elem Anal Des 149 (2018) 15–31. https://doi.org/10.1016/J.FINEL.2018.06.001.
[27] B. Stier, J.W. Simon, S. Reese, Numerical and experimental investigation of the structural behavior of a carbon fiber reinforced ankle-foot orthosis, Med Eng Phys 37 (2015) 505–511. https://doi.org/10.1016/J.MEDENGPHY.2015.02.002.
[28] R. Bostanabad, B. Liang, J. Gao, W.K. Liu, J. Cao, D. Zeng, X. Su, H. Xu, Y. Li, W. Chen, Uncertainty quantification in multiscale simulation of woven fiber composites, Comput Methods Appl Mech Eng 338 (2018) 506–532. https://doi.org/10.1016/J.CMA.2018.04.024.
[29] V. Nassehi, M. Kinsella, L. Mascia, Finite element modelling of the stress distribution in polymer composites with coated fibre interlayers, J Compos Mater 27 (1993) 195–214. https://doi.org/10.1177/002199839302700205.
[30] C.F. Fan, S.L. Hsu, A study of stress distribution in model composites by using finite-element analysis. I. End effects, J Polym Sci B Polym Phys 30 (1992) 603–618. https://doi.org/10.1002/POLB.1992.090300612.
[31] E. Ghassemieh, V. Naseehi, Prediction of failure and fracture mechanisms of polymeric composites using finite element analysis. Part 1: Particulate filled composites, Polym Compos 22 (2001) 528–541. https://doi.org/10.1002/PC.10557.
[32] K. Ramani, A. Vaidyanathan, Finite element analysis of effective thermal conductivity of filled polymeric composites, J Compos Mater 29 (1995) 1725–1740. https://doi.org/10.1177/002199839502901304.
[33] J. Wang, D. Kelly, W. Hillier, Finite element analysis of temperature induced stresses and deformations of polymer composite components, J Compos Mater 34 (2000) 1456–1471. https://doi.org/10.1106/76G7-X9QM-C5JF-E4D5.
[34] H. Ahmadi Moghaddam, P. Mertiny, Stochastic finite element analysis framework for modelling thermal conductivity of particulate modified polymer composites, Res Phys 11 (2018) 905–914. https://doi.org/10.1016/J.RINP.2018.10.045.
[35] J. Naveen, M. Jawaid, A. Vasanthanathan, M. Chandrasekar, Finite element analysis of natural fiber-reinforced polymer composites, In *Modelling of Damage Processes in Biocomposites, Fibre-Reinforced Composites and Hybrid Composites* (2019) 153–170. https://doi.org/10.1016/B978-0-08-102289-4.00009-6.
[36] A.P. Rao, A. Satyanarayana, An investigation on fatigue behaviour of bi-directional fibre reinforced polymer nano composite using FEA solver, Mater Today Proc 49 (2022) 1737–1744. https://doi.org/10.1016/J.MATPR.2021.08.006.
[37] M. Firdaus Zainudin, A. Mubarak, T. Arifin, Study on durability of composite marine structure using FEA, Res Prog Mech Manuf Eng 4 (2023) 371–378. https://doi.org/10.30880/rpmme.2023.04.02.045.
[38] T.G. Loganathan, K. Vinoth Kumar, S. Madhu, Flexural and fatigue of a composite leaf spring using finite element analysis, Mater Today Proc 22 (2020) 1014–1019. https://doi.org/10.1016/J.MATPR.2019.11.265.

# 4 Computational Fluid Dynamics (CFD) for Polymer Composite Analysis

*Ganesh Sahadeo Meshram, Gloria Biswal, and Pankaj Kumar*

## 4.1 INTRODUCTION TO POLYMER COMPOSITES

*Polymer composites* are a type of innovative materials that combine the distinct qualities of polymers with reinforcing materials to attain improved performance characteristics. The materials consist of a polymer matrix, which acts as the main component and is strengthened by including fibers, particles, or other elements (Bergbreiter & Martin, n.d.). This results in a mix of qualities that work together beneficially. The objective is to exploit the advantages of both constituents, yielding a composite material that exceeds the individual characteristics of either the polymer or the reinforcing material in isolation (Ashish Chauhan et al., 2012).

A significant benefit of polymer composites is their adaptability, which enables engineers and material scientists to customize the material's characteristics to fulfill precise application demands (Kumar, 2023). The choice of polymer matrix and reinforcing material relies on the required mechanical, thermal, electrical, or other performance qualities. For example, fiber-reinforced polymer composites, including carbon fiber–reinforced polymers (CFRPs) and glass fiber–reinforced polymers (GFRPs), have outstanding strength-to-weight ratios (Kumar et al., 2024; Prasad et al., 2024). This makes them well-suited for situations where lightweight, high-strength materials are essential in the aerospace and automotive sectors (Frank Ko & Ko, 2008).

In addition, polymer composites frequently demonstrate enhanced resistance to corrosion, increased durability, and greater design adaptability in comparison to conventional materials. These attributes render them very suitable for a diverse array of uses, such as in construction materials, sporting goods, medical gadgets, and consumer products. As research and development in materials science continue to grow, the design and manufacture of polymer composites will undoubtedly play an increasingly vital role in fulfilling the evolving demands of diverse industries, stimulating innovation and pushing the boundaries of material performance (Zheng et al., 2023).

## 4.2 TYPES OF POLYMER COMPOSITES

Polymer *composites* are materials comprising a polymer matrix that is strengthened by fibers, particles, or other reinforcing elements. These composite materials provide enhanced mechanical, thermal, and electrical characteristics in comparison to the separate constituents. There exist various categories of polymer composites that are formed by incorporating diverse reinforcing materials.

### 4.2.1 Fiber-Reinforced Polymer Composites

**Glass fiber–reinforced polymer (GFRP).** *Glass fiber–reinforced polymer* (GFRP) is a composite material where glass fibers are utilized as a reinforcement, providing exceptional strength and rigidity (*Glass Fiber Reinforced Plastics (Gfrp) Composites Market*, n.d.).

DOI: 10.1201/9781003476337-4

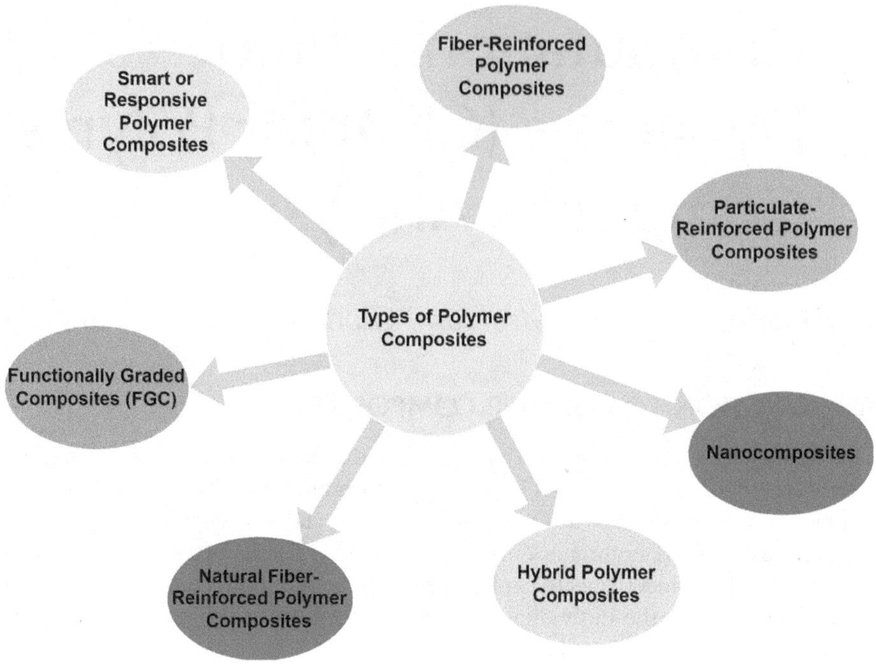


**FIGURE 4.1** Types of polymer composites.

**Carbon Fiber–Reinforced Polymer (CFRP).** *Carbon fiber–reinforced polymer* (CFRP) is a material that combines carbon fibers with a polymer. CFRP is particularly suitable for high-performance uses in industries like aerospace and sports equipment due to its exceptional strength, rigidity, and lightweight properties (*CFRP or Carbon Fiber Reinforced Plastics (Carbon FRP)*, n.d.).

**Aramid Fiber–Reinforced Polymer (AFRP).** Aramid fibers, such as Kevlar, enhance the strength and impact resistance of polymer composites (*Great Lake Supply*, n.d.).

### 4.2.2 Particulate-Reinforced Polymer Composites

**Metal Matrix Composites (MMC).** *Metal matrix composites* (MMC) refer to the incorporation of metal particles, such as aluminum or titanium, into a polymer matrix in order to improve thermal conductivity and mechanical characteristics (*Recent Studies and Advances in Metal Matrix Composites*, n.d.).

**Ceramic Particle–Reinforced Polymer Composites.** Ceramic particle–reinforced polymer composites enhance wear resistance, hardness, and thermal stability by including ceramic particles, such as alumina or silica.

### 4.2.3 Nanocomposites

**Polymer Nanocomposites.** Polymer nanocomposites involve the dispersion of nanoparticles, such as clay, carbon nanotubes, or graphene, within a polymer matrix (Talinungsang et al., 2019). This is done to improve the mechanical, thermal, and barrier properties of the material (Makevale, n.d.).

### 4.2.4 Hybrid Polymer Composites

**Hybrid Fiber Composites.** *Hybrid fiber composites* refer to the amalgamation of various fiber types, such as glass fibers and carbon fibers, in order to attain a harmonious blend of characteristics.

**Hybrid Particle Composites.** *Hybrid particle composites* refer to the integration of several types of particles, such as metal and ceramic particles, to achieve a combined enhancement in characteristics through synergy.

### 4.2.5 Natural Fiber–Reinforced Polymer Composites

**Wood Polymer Composites (WPC).** *Wood–polymer composites* (WPC) are materials that are made by combining natural fibers produced from wood, such as cellulose, with polymers. This combination results in materials that have enhanced strength and dimensional stability (*WOOD-POLYMER COMPOSITE*, n.d.).

**Flax-, Hemp-, or Jute-Reinforced Composites.** Natural plant fibers such as flax, hemp, or jute are utilized as reinforcements for polymers, offering a lightweight and environmentally beneficial substitute (Upadhaya, Kumar, et al., 2019a).

### 4.2.6 Functionally Graded Composites (FGC)

**Materials with Graded Reinforcements.** Composite materials using graded reinforcements are engineered to exhibit customized mechanical properties by strategically varying the kind or concentration of reinforcements throughout the material.

### 4.2.7 Smart or Responsive Polymer Composites

**Shape Memory Polymers (SMP).** *Shape memory polymers* (SMP) are a type of polymers that have the ability to retain their initial shape and revert back to it when activated by external factors, such as heat.

**Conductive Polymer Composites.** *Conductive polymer composites* involve the integration of conductive substances to produce composites that possess electrical or thermal conductivity (Upadhaya, Talinungsang, et al., 2019b). These composites are commonly utilized in the development of sensors and electronic devices.

These diverse polymer composites are used in numerous industries, such as aerospace, automotive, construction, electronics, and sports equipment manufacturing, among others. The selection of reinforcing and matrix materials is contingent upon the particular demands of the application.

## 4.3 INTRODUCTION TO COMPUTATIONAL FLUID DYNAMICS

A potent subfield of numerical analysis known as computational fluid dynamics (CFD) makes use of computer methods to simulate and analyze fluid flow, heat transfer, and associated physical processes. Throughout engineering, physics, and other scientific fields, CFD is widely used to provide insights into intricate fluid dynamics problems that may be difficult or impossible to answer analytically (Gopala et al., 2017; Zhang et al., 2007). Engineers and researchers are able to analyze and observe fluid behavior in a variety of scenarios, from aerodynamics and combustion to heat exchangers (Biswal & Dash, 2023a) and biomedical applications (Ray et al., 2009; Cuneyt Sert et al., 2007).

The fundamental set of equations controlling fluid flow, the Navier–Stokes equations, explains the conservation of mass and momentum (Alfredo Iranzo, 2019). These equations can be represented as partial differential equations (PDEs) in the setting of incompressible flow (Norouzi et al., 2016). Depending on the particular application, additional equations for species transport and energy conservation are frequently incorporated. Through the use of numerical techniques, such as finite difference, finite volume, and finite element methods, these continuous equations are discretized (van der Weide et al., 2015; Moukalled et al., 2015; Nikishkov, 2001).

The type of problem, the shape of the domain, and the required level of simulation accuracy all influence the discretization method selection. To solve the discretized forms of the governing equations in CFD, numerical techniques are necessary. By discretizing spatial derivatives using finite differences, the finite difference method (FDM) transforms PDEs into algebraic equations. The finite volume method (FVM) focuses on the conservation of mass and energy within control volumes, making it particularly ideal for problems involving flow through porous media or heat exchangers (Biswal & Dash, 2023b). The fluid domain is viewed as a collection of elements in the FEM which make it possible to simulate intricate geometries with asymmetric boundaries (Biswal & Meshram, 2024). The particulars of the problem and the available computer power determine which numerical approach is best.

In CFD simulations, boundary conditions are extremely important because they define the interactions that take place between the fluid and its surroundings. A vital step in producing accurate simulations is the process of meshing, which involves partitioning the computational domain into distinct parts or cells. Accuracy and computing efficiency are both affected by the type of mesh used, which can be either structured or unstructured, as well as the refinement of the mesh in close proximity to complicated geometry. In more advanced meshing approaches, such as adaptive mesh refinement, it is possible to make dynamic adjustments to the mesh based on the behavior of the solution in the local environment.

The CFD technique has numerous applications across a wide variety of industries, such as aerospace, automotive, environmental engineering, and biomedical research. Aerodynamics, combustion processes, heat transfer, and fluid–structure interactions are some of the topics that are investigated using this type of tool. Recent developments in CFD have involved the use of artificial intelligence and machine learning techniques. These techniques aim to improve the accuracy of simulations while simultaneously reducing the costs of computation. The simulation of larger and more complicated issues is made possible by high-performance computers and parallel processing. In addition, CFD is rapidly being utilized for the optimization of designs and processes, which contributes to the development of innovative and efficient practice across a wide range of engineering disciplines. As processing power continues to increase, CFD is poised to play an even more major role in the resolution of complex solutions to fluid dynamics problems.

## 4.4 GOVERNING EQUATIONS USED IN COMPUTATIONAL FLUID DYNAMICS

The governing equations employed in CFD elucidate the fundamental principles underlying fluid movement and heat transport (Yaqoob et al., 2020). The Navier–Stokes equations are the most frequently used equations for describing the flow of incompressible fluids (Tu, 2008). These equations, when coupled with other equations for energy and species transport, constitute a comprehensive system. The following are the fundamental governing equations utilized in CFD (Zhang et al., 2013).

### 4.4.1 Incompressible Flow

#### Continuity Equation

$$\nabla \cdot u = 0$$

The continuity equation represents the conservation of mass, where $u$ is the velocity vector.

**Navier–Stokes Equations**

$$\frac{\partial u}{\partial t}+(u+\nabla)u=-\frac{1}{\rho}\nabla p+\nu\nabla^2 u$$

These equations describe the conservation of momentum, where $u$ is the velocity vector, $p$ is the pressure, $\rho$ is the density, and $\nu$ is the kinematic viscosity.

**Energy Equation**

$$\rho c_p\left(\frac{\partial T}{\partial t}+u\cdot\nabla T\right)=\nabla\cdot(k\nabla T)+\Phi$$

The energy equation represents the conservation of energy, where $T$ is the temperature, $c_p$ is the specific heat at constant pressure, $k$ is the thermal conductivity, and $\Phi$ represents any internal heat sources or sinks.

### 4.4.2 Turbulent Flows

**Reynolds-Averaged Navier–Stokes (RANS) Equations**

$$\frac{\partial p^{-}}{\partial t}+\nabla\cdot(\rho^{-}u)=0\,\frac{\partial(\rho^{-}u)}{\partial t}+\nabla\cdot(\rho^{-}u.\,u)$$
$$=-\nabla p^{-}+\nabla\cdot\tau-\nabla\cdot\underline{u'u'}\,\frac{\partial\partial E}{\partial t}+\nabla\cdot\left[\left(\underline{\rho E}+p^{-}\right)u\right]$$
$$=\nabla\cdot(\lambda\nabla T)+\Phi-\nabla\cdot q$$

These equations are time-averaged versions of the compressible flow equations, where $^{-}$ denotes the Reynolds averaging operation, and $u'$ represents turbulent fluctuations.

## 4.5 TYPES OF CFD SCHEMES USED FOR THE ANALYSES

*CFD* refers to a range of techniques and methodologies used to simulate and analyze fluid flow and associated physical processes using numerical calculations. Following are several prevalent CFD methodologies, shown in Figure 4.2.

### 4.5.1 Finite Difference Method (FDM)

FDM is a numerical approach employed in CFD to discretize and solve partial differential equations that govern the movement of fluids. The FDM method approximates the spatial and temporal derivatives in governing equations by employing finite differences. It is a commonly used strategy and especially effective for issues with uncomplicated shapes and uniform grids.

### 4.5.2 Finite Volume Method (FVM)

The FVM is a prevalent numerical approach employed in CFD to solve partial differential equations that describe fluid flow and heat transport. The FVM partitions the computational domain into control volumes and emphasizes the preservation of mass, momentum, and energy within these volumes.

Both FDM and FVM are numerical methods employed to solve partial differential equations numerically, but they differ in their approach to discretizing the domain and approximating derivatives.

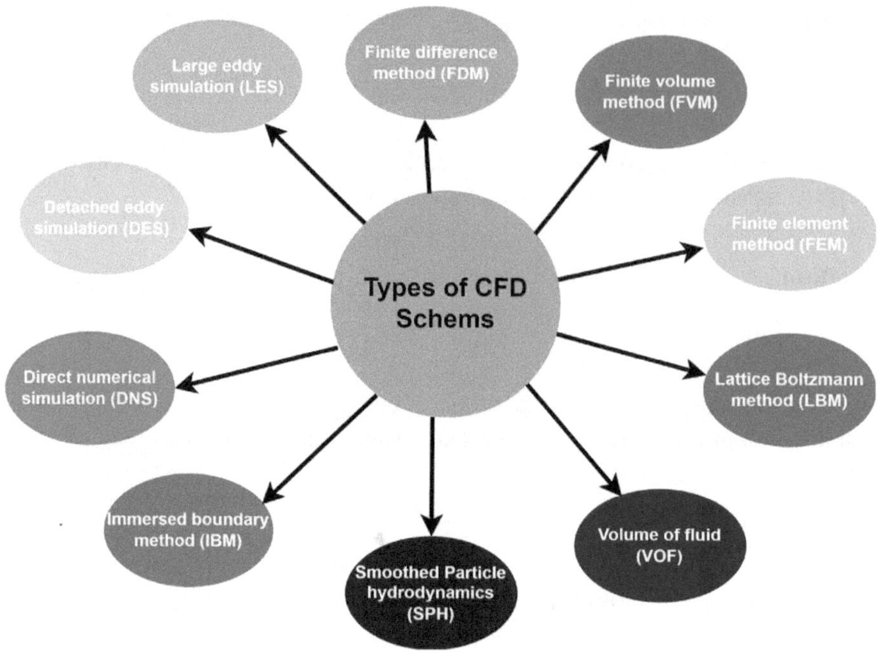


**FIGURE 4.2** Images of polymer composites.

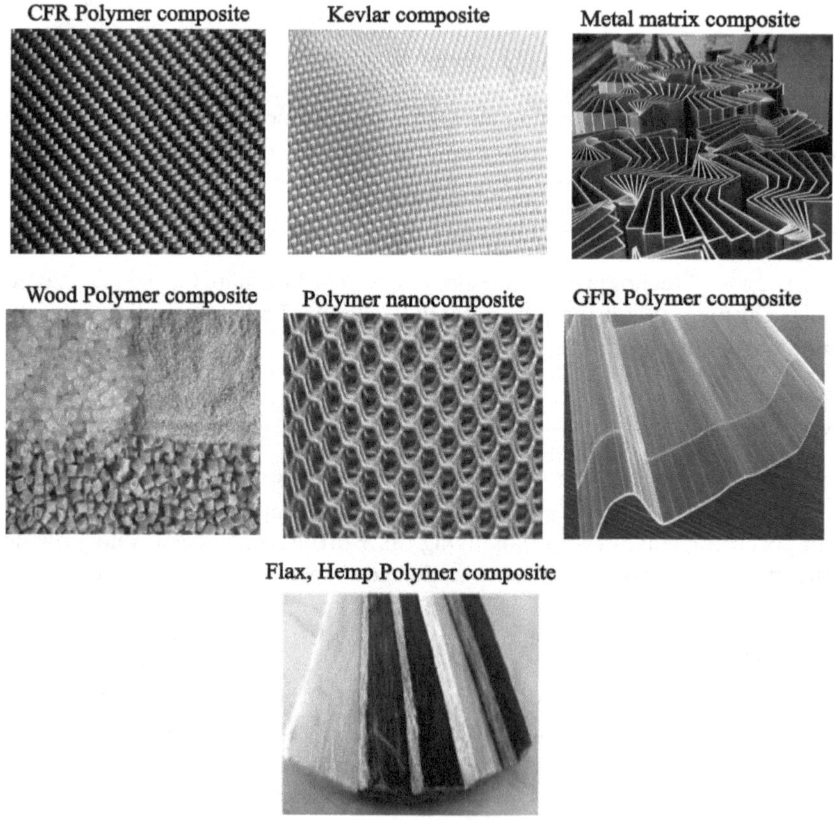


**FIGURE 4.3** Types of CFD schemes.

**TABLE 4.1**
**Difference between FDM and FVM**

| Aspect | Finite Difference Method (FDM) | Finite Volume Method (FVM) |
|---|---|---|
| Basic concept | Discretizes derivatives using the difference quotient to approximate PDEs at discrete points in the domain. | Divides the domain into control volumes, and balances fluxes across the boundaries of these volumes. |
| Spatial discretization | Grid-based approach where the domain is discretized into a grid and derivatives are approximated using finite differences. | Control volume–based approach where the domain is divided into control volumes and fluxes are balanced over these volumes. |
| Representation of unknowns | Typically uses nodal values (values at grid points) to represent the solution. | Often uses cell-centered values (averaged values within control volumes) to represent the solution. |
| Conservation | May require additional effort to enforce conservation properties. | Natural conservation laws are inherently satisfied through the flux balance across control volumes. |
| Numerical stability | Stability depends on the choice of discretization scheme (explicit, implicit, etc.) and grid resolution. | Generally more stable due to the inherent conservation principles, but stability also depends on discretization schemes. |
| Applications | Commonly used for problems with simple geometries and regular grids. | Suitable for complex geometries and unstructured grids, often used in CFD. |

### 4.5.3 Finite Element Method (FEM)

The FEM is a numerical technique used for solving engineering and mathematical problems by dividing a complex structure or domain into smaller, simpler elements. These elements are interconnected at discrete points called nodes, forming a mesh that represents the overall geometry of the system. FEM is widely employed in various fields, including structural engineering, heat transfer, fluid dynamics, electromagnetics, and more.

Here is a brief overview of the key steps involved in the FEM:

- *Discretization.* The first step is to divide the entire domain into a finite number of smaller, simple shapes or elements. These elements can be triangles, quadrilaterals (in 2D problems), or tetrahedra, hexahedra (in 3D problems). This process is known as meshing.
- *Selection of element type.* Depending on the nature of the problem and the geometry, different types of elements may be chosen. Common choices include linear elements, quadratic elements, etc.
- *Interpolation and shape function.* Within each element, the variation of the field variables (e.g., displacement, temperature) is approximated using interpolation functions or shape functions. These functions define how the variable changes within an element based on the values at the nodes.
- *Derivation of element equations.* The physical behavior of each element is described by a set of equations based on the governing principles (e.g., equilibrium, compatibility). These equations are derived by applying variational methods or energy principles.
- *Assembly of global system equations.* The individual element equations are then assembled to form a global system of equations that represents the entire problem. This system accounts for the interactions between adjacent elements through shared nodes.
- *Boundary conditions.* The next step involves applying the prescribed constraints or boundary conditions to the global system of equations. These conditions represent the known values or constraints in the problem.

- *Solution of the system of equations.* The global system of equations is solved to obtain the unknown field variables. Various numerical techniques, such as direct solvers or iterative methods, can be employed for this purpose.
- *Post-processing.* After obtaining the solution, post-processing involves extracting useful information, such as stress distribution, strain, or other relevant quantities. Visualization tools are often used to interpret and present the results.

The benefits of FEM include the capability to model intricate geometries and non-uniform boundaries. It offers versatility in managing issues related to fluid–structure interactions. The flexible nature of FEM meshes allows for easy adjustment to different resolutions. The FEM is particularly suitable for addressing problems that include fluid dynamics with time-dependent or transient behavior.

### 4.5.4 Lattice Boltzmann Method (LBM)

The lattice Boltzmann method (LBM) is a computational technique used to model fluid motion at the mesoscopic scale. It is highly suitable for simulating intricate fluid behaviors and is renowned for its adaptability in simulating various flow circumstances, such as flows around intricate shapes and flows involving many phases (Zhang et al., 2014). Following is a summary of the fundamental elements involved in applying the LBM in CFD:

- *Distribution of particles at the mesoscopic scale.* LBM models the fluid as a discrete ensemble of particles that traverse a lattice structure (Zhang et al., 2014; Chen et al., 1998). The particles exhibit discrete velocity vectors, and their distribution across lattice nodes undergoes temporal evolution. The Boltzmann equation, which characterizes the statistical dynamics of these particles, is discretized in order to simulate collisions and streaming (Meshram & Kondaraju, 2021).
- *Arrangement of a lattice.* The selection of lattice configuration is pivotal in LBM. Typical lattice structures used are the D2Q9 lattice for two-dimensional problems and D3Q27 or D3Q19 for three-dimensional problems. Every lattice point represents a node where the distribution functions of particles are monitored, and each discrete velocity vector corresponds to a collection of weights and velocities.
- *Steps for collision and streaming.* The process of LBM has two main stages: collision and streaming. During the collision step, the particle distribution functions are modified according to local equilibrium distributions and relaxation times. In the streaming phase, the particle distributions are propagated to adjacent nodes based on their discrete velocities.
- *Variables at the macroscopic level.* The particle distribution functions are used to derive the macroscopic fluid variables, such as density and velocity. These variables are utilized to calculate the local equilibrium distributions during the collision step.
- *Practical uses of LBM.* LBM is utilized in several applications, encompassing fluid dynamics, heat transfer, and multiphase flows. It is frequently employed to simulate fluid flows in porous media, microfluidics, and intricate geometries, where conventional approaches may encounter difficulties (Meshram & Kondaraju, 2021).

### 4.5.5 Volume of Fluid (VOF)

The volume-of-fluid (VOF) method is a numerical technique used in CFD to simulate the behavior of fluids, especially in multiphase flows, where multiple immiscible fluids are present (Paz et al., 2019). The method is particularly suitable for problems involving free surfaces and fluid interfaces.

Here are the key principles and features of the volume-of-fluid method:

- *Representation of fluid phases.* The VOF method tracks the volume fractions of different fluids within each computational cell. A scalar field, known as the volume fraction field, is used to represent the presence and distribution of each fluid phase in a given cell.

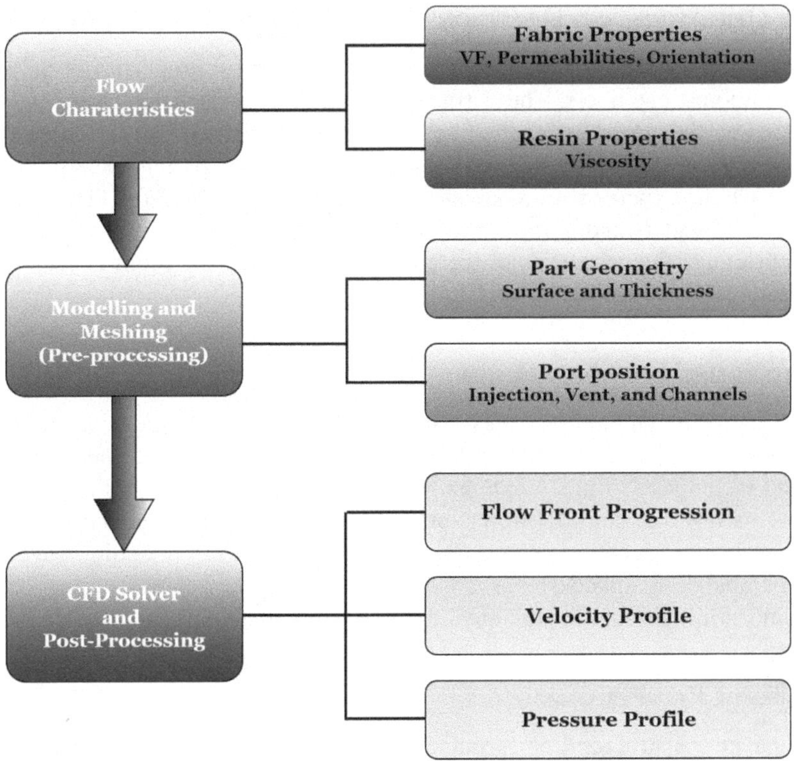


**FIGURE 4.4** Resin flow analysis using CFD.

- *Conservation of mass.* The VOF method ensures the conservation of mass for each fluid phase. The sum of volume fractions within a cell is always equal to 1, representing the entire volume of the cell.
- *Advection and diffusion.* Advection terms are used to model the movement of fluid phases through the computational domain. Diffusion terms account for the spreading or smearing of fluid interfaces over multiple grid points.

## 4.6 APPLICATIONS OF CFD IN POLYMER COMPOSITES

The application of CFD is of critical importance in the process of analyzing and optimizing a variety of processes that are associated with polymer composites. CFD has a number of applications in the field of polymer composites, including the following.

### 4.6.1 Resin Infusion and Cure Processes

When it comes to the manufacturing of composites, CFD can be used to model the flow of resin during the infusion process. Both the impregnation of fibers and the distribution of resin throughout the mold can be predicted with the assistance of this tool. The curing process has a considerable impact on the properties that are ultimately exhibited by polymer composites (de Araújo Mota et al., 2020). Simulations using CFD are helpful in forecasting the distribution of temperatures, changes in resin viscosity, and heat transmission within the composite while it is curing. Because of this, manufacturers are able to optimize cure cycles, so reducing thermal stresses, ensuring that curing is uniform, and improving the composite's overall structural integrity (Pico et al., 2023).

### 4.6.2 Fiber Orientation and Dispersion

One of the most important factors that determine the mechanical properties of polymer composites is the orientation of the fibers. The stiffness, strength, and anisotropy of the composite are all greatly impacted by the alignment of the fibers along the primary stress directions (Oumer et al., 2013). Several different processing methods, including injection molding, extrusion, and compression molding, each have their own unique effects on the orientation of the fibers themselves. It is vital to have a solid understanding of the mechanisms that determine fiber alignment in order to successfully customize the properties of the composite to match the requirements of a particular application (Letchumanan et al., 2023).

### 4.6.3 Mold Filling and Mold Design

Injection molding for polymer composites can be modeled using CFD (Oumer et al., 2015). CFD in predicting how the polymer fills the mold cavity, identifying potential issues, such as air traps, weld lines, and incomplete filling. CFD can be utilized to predict the warpage tendencies of the molded part. This involves considering factors such as temperature gradients, cooling rates, and residual stresses during the cooling phase (Bergamasco et al., 2014). In addition to assisting in the optimization of mold design, it also helps in forecasting flow patterns and ensuring that the polymer is distributed uniformly throughout the mold cavity (Oumer et al., 2015).

### 4.6.4 Composite Part Performance

In order to simulate the structural performance of composite parts under a variety of loading circumstances, CFD can be utilized (Bergamasco et al., 2014). The behavior of composite structures may be predicted with its assistance, and the design of these structures can be optimized for particular applications. Failure processes in polymer composites, such as delamination, debonding, and other problems associated with fluid-induced stresses, can be predicted and analyzed with greater accuracy.

## 4.7 COMPUTATIONAL AND EXPERIMENTAL ANALYSIS OF COMPOSITES

Experimental and computational approaches are often considered as competing resources. In order to fully comprehend the fundamental science that underlies complicated physical phenomena, both approaches must be used in their evaluation. The sole method for verifying a design or testing a theory that involves physically evaluating a part is through computational analysis; on the other hand, mechanical testing is used to examine the properties of materials and is carried out using a particular machine based on the requirements. Table 4.3 provides an overview of the research study conducted on composites using both computational and experimental methods.

## 4.8 CHALLENGES AND ADVANCES OF CFD IN POLYMER COMPOSITES

The manufacturing processes of these composites involve intricate fluid flow phenomena, making CFD an invaluable tool for predicting and optimizing process parameters. Nevertheless, there are still a number of obstacles to overcome in order to get an accurate simulation of fluid dynamics in polymer composite manufacturing. CFD presents a number of challenges for polymer composites:

- **Non-Newtonian fluid behavior.** Polymer melts are characterized by shear-thinning and viscoelastic processes, which are examples of non-Newtonian fluid behavior. In CFD simulations, accurately capturing these rheological properties continues to be a difficulty, which has an impact on the prediction of flow patterns and shear rates (Saldi & Wen, 2017).

**TABLE 4.2**
**Application of CFD for Various Polymer Composites**

| Research Study | Methodology | | Materials | Result |
|---|---|---|---|---|
| | Computational | Experimental | | |
| Validation of UAV wing structural model | Analysis of a composite wing structure using Abaqus/CAE | The actual wing was loaded using sandbags to stimulate the aerodynamic loads on the wing. | Carbon/epoxy; carbon unitape/epoxy; Kevlar/epoxy; honeycomb, glass fiber/epoxy | The differences between deflection for bending test and FE analysis results range between 0.35% and 16.4%. |
| Composite material in aerospace applications | Analysis of mechanical properties and mechanical response using Abaqus | Tensile test and thermal stress test were undertaken to the rectangular-shaped specimen. | E-glass | Composite materials have a high fatigue, corrosion resistance, and also high strength-to-weight ratio. |
| Analysis of ballistic impact on composite laminate armor | Deformation, impact, and ballistic using ANSYS to the fiber cement, Kevlar, and steel | Soft-core bullet projectile was shot toward the fiber cement layer. | Fiber cement, Kevlar, steel | Simulation can show the ballistic behavior with the limits of the design. |
| Nonlinear flexural behavior of glass/epoxy-laminated composite panel | Analysis of the nonlinear static deflections from the laminated composite using ANSYS | Bending test using universal testing machine UTM-INSTRON 5967 and tensile test via UTM-INSTRON 1195. | Glass/epoxy | Significant influence of side-to-thickness ratio, the aspect ratio, the curvature ratio, and the end conditions. |
| Factor analysis of machining parameters of fiber-reinforced polymer composites | Orthogonal cutting simulation using Abaqus software | Milling experiment via DMG Ultrasonic 20 linear high-speed machining. | Carbon/epoxy | The factors affect the cutting force, and surface roughness is the depth of cut and cutting speed. |
| Analysis of composite ankle–foot orthosis | Mechanical behavior analysis using Abaqus | Force displacement using INSTRON 5866 electromechanical materials testing system | Carbon fiber | FEA will improve carbon fiber ankle–foot orthosis without wasting resources on trial-and-error fabrication. |

- **Multiphase flows.** It is common for many polymer composite processes to incorporate multiphase flows. Some examples of these processes include resin injection and fiber impregnation (Wang et al., 2019). In CFD, one of the most difficult tasks is modeling the interaction between distinct phases, considering phase shifts, and precisely capturing interfaces.
- **Heat transfer and curing.** It is exceptionally important to accurately simulate the heat transfer that occurs throughout the curing process in order to accurately forecast the final composite properties (Duan et al., 2022). An additional layer of complexity is added to the

simulation by the fact that heat transmission is coupled with chemical reactions and phase transitions (Fu et al., 2022).

- **Computational cost.** The simulation of polymer composite processes at a high spatial and temporal resolution can be computationally costly. This is referred to as the computational cost. A fundamental challenge that continues to exist, particularly for large-scale industrial processes, is striking a balance between precision and computing cost.

## 4.9 CONCLUSION

The utilization of CFD in the analysis of polymer composites is an innovative approach that significantly enhances our understanding and optimization of these versatile materials. We have utilized CFD to investigate the intricate fluid dynamics, heat transport, and significant manufacturing procedures associated with polymer composite production.

The FVM is a widely recognized technique in CFD that is known for its ability to handle complex geometries and conserve properties. This makes it particularly well-suited for studying polymer composites. The analysis of multiple applications in this chapter demonstrates the substantial impact of CFD on various stages of polymer composite production.

CFD is essential for optimizing manufacturing operations through precise simulation of injection molding processes. It offers valuable information about filling patterns, temperature distributions, and pressure contours. Furthermore, CFD aids in comprehending the complex process of curing in thermosetting polymer composites. The analysis of resin flow is an essential determinant that showcases the ability of CFD predictions to produce a uniform distribution of resin, minimize defects, and ensure the superior quality of the end composite.

Furthermore, the chapter highlights the importance of CFD in predicting the spatial arrangement of fiber orientation within the polymer matrix. This research provides useful insights on the orientation of reinforcing fibers and its direct influence on the mechanical properties of the composite material. CFD is employed to examine fluid–structure interaction situations, yielding significant insights into the dynamic interplay between the fluid phase and the solid structure in industrial processes.

To summarize, CFD has become a crucial instrument in the realm of polymer composite analysis. This platform provides academics, engineers, and practitioners with a robust tool for virtual prototyping, design optimization, and a deeper understanding of the basic processes that drive these complex materials. The integration of CFD not only accelerates innovation in polymer composite manufacturing but also facilitates the development of materials with enhanced performance characteristics in many industries. The shift from simulation to application signifies a significant progress, presenting the possibility of a future where CFD continues to be a crucial catalyst in pushing the limits of polymer composite science and technology.

## ACKNOWLEDGMENTS

I would like to thank Dr. Sushanta K. Sethi for providing me an opportunity to contribute to this book chapter. I would like to express my heartfelt gratitude to my coauthors for their invaluable guidance and mentorship throughout the writing process of this chapter. Their expertise and support were instrumental in shaping the content and direction of the work.

## REFERENCES

AZO. (n.d.). *Recent studies and advances in metal matrix composites.* https://www.azom.com/article.aspx?ArticleID=21800

Bergamasco, L. (2014). *Multi-scale modeling of complex fluids and deformable fibrous media for liquid composite molding* [Doctoral Dissertation, Universidad de Zaragoza].

Bergbreiter, D., & Martin, C. (n.d.). Functional polymers. *Polymers and Polymeric Composites*. https://doi.org/10.1007/978-3-319-95987-0

Biswal, G., & Dash, S. K. (2023a). An isothermal constant surface area horizontal helical coil under natural convection and radiative heat loss: A CFD study. *International Communications in Heat and Mass Transfer, 145*, 106846. https://doi.org/10.1016/j.icheatmasstransfer.2023.106846

Biswal, G., & Dash, S. K. (2023b, August). *Combined natural convection and radiative heat transfer from a horizontal helical coil placed on the ground and in air: A comparative study* [Conference session]. The 9th World Congress on Mechanical, Chemical, and Material Engineering. https://doi.org/10.11159/htff23.164

Biswal, G., & Meshram, G. S. (2024). Computational investigation of aspect ratio and inclination effects on natural convection heat transfer in nanofluid-filled rough wall cavities. *Numerical Heat Transfer, Part A: Applications*, 1–16. https://doi.org/10.1080/10407782.2024.2302077

CFRP or Carbon Fiber Reinforced Plastics (Carbon FRP). (n.d.). https://www.alformet.com/composite/frp/cfrp

Chauhan, A., & Chauhan, P. (2012). Advanced polymer composites. *Journal of Analytical and Bioanalytical Techniques, 2012*(6), 1. https://doi.org/10.4172/2155-9872.1000e108

Chen, S., & Doolen, G. D. (1998). Lattice Boltzmann method for fluid flows. *Annual Review of Fluid Mechanics, 30*(1), 329–364. https://doi.org/10.1146/annurev.fluid.30.1.329

de Araújo Mota, C. A., Porto, T. R. N., de Lima, A. G. B., Figueirôa, J. A., de Almeida, M. D. A., de Mélo Vieira Neto, J., & de Souza, L. M. C. (2020). Resin injection process in the manufacture of a polymer composite reinforced with NiTi ribbons: A CFD analysis. *Diffusion Foundations, 27*, 73–89. https://doi.org/10.4028/www.scientific.net/df.27.73

Duan, S., Li, B., & Rong, W. (2022). Study on gas-solid heat transfer and decomposition reaction of calcination process in an annular shaft kiln based on the finite volume method. *Processes, 10*(4), 648. https://doi.org/10.3390/pr10040648

Fu, J., Chen, S., & Zhou, X. (2022). Effect of heterogeneity on interphase heat transfer for gas–solid flow: A particle-resolved direct numerical simulation. *Physics of Fluids, 34*(12), 123317. https://doi.org/10.1063/5.0130850

Glass Fiber Reinforced Plastics (GFRP) Composites Market. (n.d.). https://www.alliedmarketresearch.com/glass-fiber-reinforced-plastics-composites-market

Gopala, V. R., Singh, A., & Appanaboyina, S. (2017, December). *HiPC 2017 workshop 2: Second workshop on computational fluid dynamics (CFD)* [Conference session]. 2017 IEEE 24th International Conference on High Performance Computing Workshops (HiPCW) (p. 40). IEEE. https://doi.org/10.1109/hipcw.2017.00015

Great Lake Supply. (n.d.). https://greatlakesupplyco.com/blogs/articles/difference-between-aramid-kevlar-carbon-fiber

Iranzo, A. (2019). CFD applications in energy engineering research and simulation: An introduction to published reviews. *Processes, 7*(12), Article Number 883. https://doi.org/10.3390/pr7120883

Ko, F. (2008). Book review: Polymer nanocomposites—Processing, characterization, and applications. *Journal of Engineered Fibers and Fabrics, 3*(3), 155892500800300. https://doi.org/10.1177/155892500800300306

Kumar, P. (2023). Recent progress in computational techniques in various coating materials. In A. Verma, S. K. Sethi, & S. Ogata (Eds.), *Coating materials: Computational aspects, applications and challenges* (pp. 217–228). Springer Nature. https://doi.org/10.1007/978-981-99-3549-9_10

Kumar, P., Prasad, P. S., & Guerrero, J. D. M. (2024). Different curing methods used for polymer composites. In S. K. Sethi, H. S. Gupta, & A. Verma (Eds.), *Polymer composites: From computational to experimental aspects* (pp. 43–53). Springer Nature Singapore. https://doi.org/10.1007/978-981-97-0888-8_2

Letchumanan, S. M., Arifin, A. M. T., Taib, I., & Salim, N. A. N. (2023). Computational fluid dynamic analysis on carbon fibre reinforced polymer wrapped on defected oil and gas piping system using solidwork flow simulation. *Journal of Advanced Research in Fluid Mechanics and Thermal Sciences, 101*(2), 197–210. https://doi.org/10.37934/arfmts.101.2.197210

Makevale. (n.d.). *Polymer nanocomposites*. https://www.makevale.com/applications/polymer-nanocomposites/

Meshram, G., & Kondaraju, S. (2021). *Numerical investigation of wettability and its effects on flow through textured micro-channels using Lattice Boltzmann Method* [Conference session]. Proceedings of the 26th National and 4th International ISHMT-ASTFE Heat and Mass Transfer Conference December 17–20, 2021, IIT Madras, Chennai, Tamil Nadu, India. https://doi.org/10.1615/IHMTC-2021.1850

Moukalled, F., Mangani, L., Darwish, M., Moukalled, F., Mangani, L., & Darwish, M. (2016). *The finite volume method* (pp. 103–135). Springer International Publishing. https://doi.org/10.1007/978-3-319-16874-6

Nikishkov, G. (2001). Introduction to the finite element method. *Finite Element Analysis*. https://doi.org/10.1016/b978-075063266-9/50010-x

Norouzi, H. R., Zarghami, R., Sotudeh-Gharebagh, R., & Mostoufi, N. (2016). *Coupled CFD-DEM modeling: Formulation, implementation and application to multiphase flows*. John Wiley & Sons. https://doi.org/10.1002/9781119005315

Oumer, A. N., Ali, A. M., & Mamat, O. B. (2009). Numerical simulation of fibre orientation in simple injection molding processes. *International Journal of Mechanical and Mechatronics Engineering*, *9*, 361–367.

Oumer, A. N., Hamidi, N. M., & Sahat, I. M. (2015). Numerical prediction of flow induced fibers orientation in injection molded polymer composites. In *IOP conference series: Materials science and engineering* (Vol. 100, No. 1, p. 012066). IOP Publishing. https://doi.org/10.1088/1757-899x/100/1/012066

P&W. (n.d.). *Wood-polymer composite*. https://pwpolymerwood.ua/en/wpc-article-en

Paz, C., Suárez, E., Vence, J., & Cabarcos, A. (2019). Analysis of the volume of fluid (VOF) method for the simulation of the mucus clearance process with CFD. *Computer Methods in Biomechanics and Biomedical Engineering*, *22*(5), 547–566. https://doi.org/10.1080/10255842.2019.1569637

Pico, D., Machado, S., Meza, J., & Unfried-Silgado, J. (2023). Resin flow analysis during fabrication of coconut mesocarp fiber-reinforced composites using vartm process. *International Journal of Modern Manufacturing Technologies (IJMMT)*, *15*(1). https://doi.org/10.54684/ijmmt.2023.15.1.51

Prasad, P. S., Guerrero, J. D. M., & Kumar, P. (2024). Natural and synthetic fiber-filled polymer composites used as anticorrosive materials. In S. K. Sethi, H. S. Gupta, & A. Verma (Eds.), *Polymer composites: From computational to experimental aspects* (pp. 181–202). Springer Nature Singapore. https://doi.org/10.1007/978-981-97-0888-8_9

Ray, B., Bhaskaran, R., & Collins, L. R. (2012). *Introduction to CFD basics*. Cornell Online Course.

Saldi, Z. S., & Wen, J. X. (2017). Modeling thermal response of polymer composite hydrogen cylinders subjected to external fires. *International Journal of Hydrogen Energy*, *42*(11), 7513–7520. https://doi.org/10.1016/j.ijhydene.2016.06.108

Sert, C., & Nakiboglu, G. (2007, June). *Use of computational fluid dynamics (CFD) in teaching fluid mechanics* [Conference session]. 2007 Annual Conference & Exposition (pp. 12–1527).

Talinungsang, U. D., Kumar, P., & Purkayastha, D. D. (2019). Superhydrophilicity of photocatalytic $ZnO/SnO_2$ heterostructure for self-cleaning applications. *Journal of Sol-Gel Science and Technology*, *92*, 575–584. https://doi.org/10.1007/s10971-019-05127-8

Tu, J. (2008). Chapter 3: Governing equations for CFD fundamentals. https://doi.org/10.1016/b978-075068563-4.50005-7.

Upadhaya, D., Kumar, P., & Dhar Purkayastha, D. (2019a). Superhydrophobic $ZnO/TiO_2$ heterostructure with significantly enhanced photocatalytic activity. *Journal of Materials Science: Materials in Electronics*, *30*(11), 10399–10407. https://doi.org/10.1007/s10854-019-01381-2

Upadhaya, D., Kumar, P., & Purkayastha, D. D. (2019b). Tuning the wettability and photocatalytic efficiency of heterostructure $ZnO\text{-}SnO_2$ composite films with annealing temperature. *Materials Science in Semiconductor Processing*, *95*, 28–34. https://doi.org/10.1016/j.mssp.2019.02.009

Wang, H., Yuan, X., Liang, H., Chai, Z., & Shi, B. (2019). A brief review of the phase-field-based lattice Boltzmann method for multiphase flows. *Capillarity*, *2*(3), 33–52. https://doi.org/10.26804/capi.2019.03.01

Weide, E. V. D., Giangaspero, G., & Svärd, M. (2015). Efficiency benchmarking of an energy stable high-order finite difference discretization. *AIAA Journal*, *53*(7), 1845–1860. https://doi.org/10.2514/1.j053500

Yaqoob, T., Ahsan, M., Hussain, A., & Ahmad, I. (2020). Computational fluid dynamics (CFD) modeling and simulation of flow regulatory mechanism in artificial kidney using finite element method. *Membranes*, *10*(7), 139. https://doi.org/10.3390/membranes10070139

Zhang, J., & Kwok, D. Y. (2014). Lattice Boltzmann Method (LBM). In *An introduction to the Lattice Boltzmann Method*. https://doi.org/10.1007/springerreference_67034

Zhang, L., Tu, J., Yeoh, G. H., & Liu, C. (2007). *Computational fluid dynamics: A practical approach*. Elsevier.

Zhang, L., Tu, J., Yeoh, G. H., & Liu, C. (2013). Governing equations for CFD: Fundamentals. In *Computational fluid dynamics* (pp. 61–121). https://doi.org/10.1016/b978-0-08-098243-4.00003-2

Zheng, S., Wang, Y., Zhu, Y., & Zheng, C. (2024). Recent advances in structural design of conductive polymer composites for electromagnetic interference shielding. *Polymer Composites*, *45*(1), 43–76. https://doi.org/10.1002/pc.27773

# 5 Molecular Dynamics
## *Simulation of Polymer Composites*

*Abhiram B. R.*

## 5.1 INTRODUCTION

Computational methods such as the finite element method (FEM) and MD are often used for the analysis of polymer composites. While FEM provides a macroscopic perspective, MD simulations offer a molecular-level analysis of atomic interactions among the constituents of a composite. They are generally used for studying the mechanical properties, thermal behavior, and interaction between polymers and nanofillers. This chapter discusses the application of MD simulation for studying polymer composites at the nanometer scale. The major focus is to provide a preliminary understanding on the modeling and simulation of a polymer composite, with emphasis on extracting fundamental properties, such as elastic modulus and fracture toughness. The chapter is organized as follows.

The fundamental concepts on MD simulation are presented in Section 5.2. A detailed explanation on modeling and simulating a polymer composite is given in Section 5.3. The computation of the mechanical properties of polymer composites using MD simulation is highlighted in Section 5.4. Finally, the challenges and future prospects of MD simulation are presented in Sections 5.5 and 5.6.

## 5.2 BASICS OF MD SIMULATION

The objective of a molecular dynamics (MD) simulation is to generate the trajectory of every atom in a system. A successful MD run should produce the path followed by each atom under specified boundary conditions. To illustrate, Figure 5.1 depicts the positional changes of atoms over a time step ($\Delta t$). The simulation assumes that atoms adhere to Newton's laws of mechanics. In accordance with these laws, the dynamics of the system, particularly for each atom $i$, are described by the following equations:

$$F_i = m_i a_i \tag{5.1}$$

$$F_i = m_i \frac{d^2 r_i}{dt^2} \tag{5.2}$$

where *mi*, *ri*, *ai*, and *Fi* represent the mass, position, acceleration, and force acting on atom $i$, respectively. Equation (5.1) corresponds to a system of coupled second-order nonlinear differential equations for which no exact solution exists when more than two atoms are present. However, the equations can be solved by using numerical integration techniques, such as Verlet central difference, leapfrog, and velocity–Verlet algorithm. For more details on numerical integration in MD, readers are referred to [1, 2].

### 5.2.1 Force Field or Interatomic Potential

The *force field* (potential) in an MD simulation defines the interaction between atoms in a system and provides an estimate of the energy landscape. The choice of potential depends on the atom types, temperature, boundary condition, and type of simulation. The selection of potential is the

DOI: 10.1201/9781003476337-5

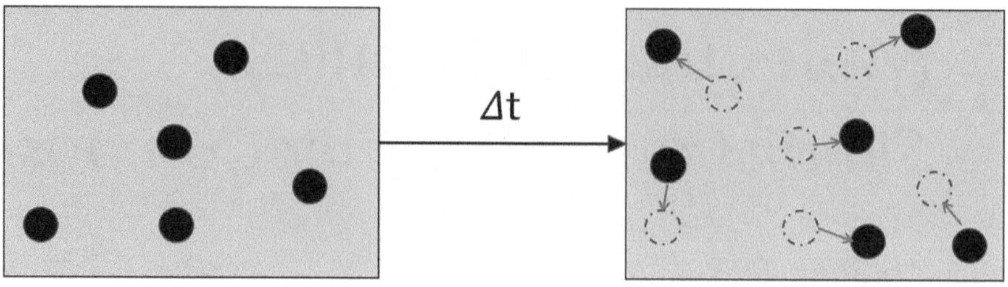


**FIGURE 5.1** Change in atom position in $\Delta t$ time step.

most crucial factor for accurate predictions from an MD simulation. Generally, the energy form of a potential can be expressed as the sum of contributions from covalent, Coulombic, van der Waals, and some other terms, as:

$$U_{total} = U_{covalent} + U_{Coulombic} + U_{vdW} + U_{bond} + \cdots \tag{5.3}$$

The terms in equation (5.3) vary with respect to the type of potential. A few commonly used potentials are Lennard–Jones, Morse, embedded atom model, COMPASS, and ReaxFF. The parameters of a potential are derived from ab initio simulations or experimental data.

### 5.2.2 Periodic Boundary Condition

In a simulation cell, the exposed atoms at the boundaries result in high energies. The concept of periodic boundary condition (PBC) is used to overcome surface effects and simulate a bulk system. The simulation cell is replicated in all directions to induce periodicity. A two-dimensional representation of PBC is shown in Figure 5.2. The simulation box 5 in Figure 5.2 is surrounded by its replicas in both directions. When an atom moves from box 5 to 2, one of its image enters through the opposite side, as shown in Figure 5.2. The number density in the central box remains conserved. Generally, attention is shifted to the image entering when one atom crosses the simulation boundary. Ideally, there are no walls at the boundary of the central box—hence the absence of surface atoms. PBC condition is used in most MD simulations, especially while simulating PNC.

### 5.2.3 Statistical Mechanics in MD

As mentioned in Section 5.2, an MD simulation provides the atomic trajectory over a period of time. However, to extract meaningful results from these simulations, it has to be related to macroscopic thermodynamics properties. In such cases, statistical mechanics help analyze MD simulations and interpret results from these simulations. It provides a direct link between an ensemble of microscopic states and the corresponding macroscopic thermodynamic properties, such as pressure, stress, and strain tensor. The instantaneous atomic positions are not valid; thus, it needs to be averaged over multiple configurations. For this purpose, the ergodic hypothesis—one of the most crucial theorems in the practical application of MD simulation—is used. The ergodic hypothesis states that the ensemble average of a property *A* equals the time average and is given by

$$A_{Ensemble} = A_{Time} \tag{5.4}$$

where the symbol *AEnsemble* denotes an average of a variable. Equation (5.4) enables the calculation of thermodynamical properties by averaging over sufficiently long time trajectories, thus providing an appropriate measurement of ensemble properties. The time average from an MD simulation can be calculated by

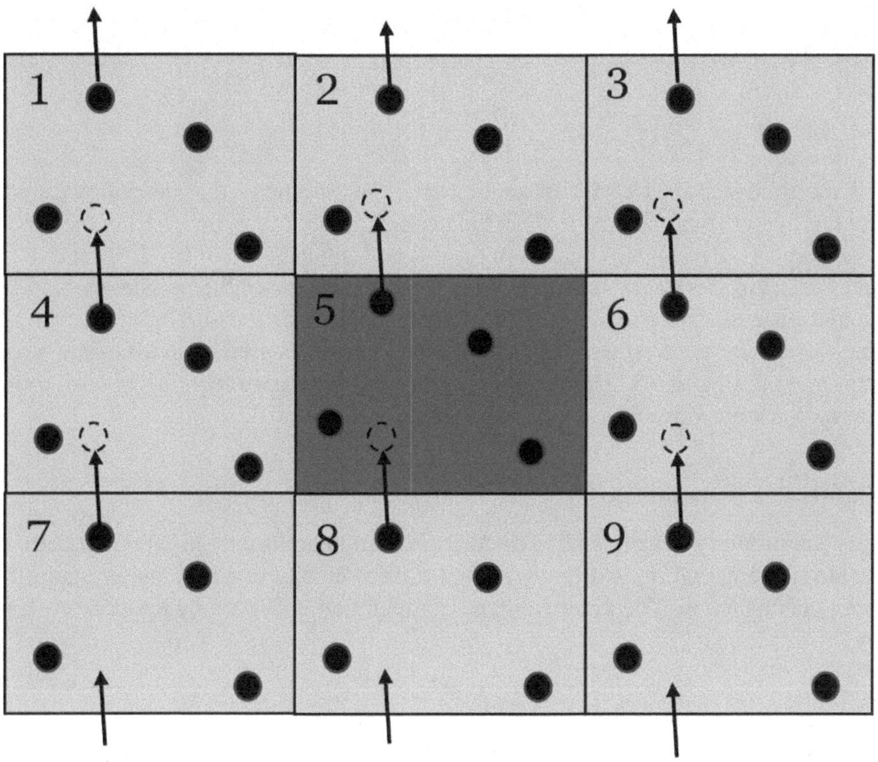


**FIGURE 5.2** Schematic representation of periodic boundary condition.

$$A_{Time} = \frac{1}{M}\sum\nolimits_{i-1}^{M} A(p,r) \tag{5.5}$$

Here, $M$ is the number of measurements taken, and $p$, $r$ represent the linear momentum and the position of atoms, respectively.

### 5.2.4 Thermodynamic Ensembles

A collection of all possible atomic configurations corresponding to a certain fixed thermodynamic quantity is known as an *ensemble*. When the equations of velocity–Verlet algorithm is integrated, it represents a microcanonical ensemble (NVE), that is, the number of atoms ($N$), the volume of the system ($V$), and the total energy of the system ($E$) remain constant throughout the simulation. Other thermodynamic ensembles, such as canonical ensemble (NVT), isobaric-isothermal ensemble (NPT), and grand canonical ensemble ($\mu$VT), can be attained by appropriately modifying the equations of motion. This modification is performed by using thermostats and barostats.

**Thermostats and barostats.** Thermostats and barostats are used to maintain temperature and pressure as a constant in an MD simulation. As mentioned earlier, thermostats and barostats are introduced to achieve various thermodynamic ensembles by modifying the equations of motion. For instance, to enable an NVT ensemble, a thermostat is used. The velocities of atoms are altered so that temperature reaches a desired value. A rescaling parameter, $\lambda$, is used to scale the velocity of each atom $i$, given by

$$v_{new,i} = \lambda v_i \tag{5.6}$$

where *vnew,i* and *vi* are the new and old velocities of atom *i*, respectively. The parameter λ can be calculated from

$$\lambda = \sqrt{1 + \frac{\Delta t}{\tau}\left(\frac{T'}{T(t) - 1}\right)} \tag{5.7}$$

where $\Delta t$ is the time step in the MD simulation, $\tau$ is the parameter that describes the strength of coupling to the hypothetical heat bath, $T'$ is the desired temperature, and $T(t)$ is the instantaneous temperature, respectively. Some other common thermostats used in an NVT ensemble are Nosé–Hoover and Langevin thermostats. Similar to NVT, an NPT ensemble can be enabled by using a thermostat and a barostat simultaneously. The barostat adjusts the pressure of the system by altering the cell size. For a better understanding of the various ensembles used in an MD simulation, a schematic is presented in Figure 5.3. The piston, weights, and insulation techniques used to maintain a specific thermodynamic variable are also represented in Figure 5.3.

### 5.2.5 Property Computation

A few important property calculations in an MD simulation are briefly highlighted here.

**Temperature.** Temperature is directly related to the velocity of atoms in a system through the kinetic energy contribution. The kinetic energy computation, followed by temperature calculation, is given by

$$\left\langle \frac{1}{2} m_i v_i^2 \right\rangle = \langle K.E \rangle \tag{5.8}$$

$$T = \frac{2}{3} \frac{\langle K.E \rangle}{N \kappa_B} \tag{5.9}$$

where $\kappa_B$ is the Boltzmann constant.

**Pressure.** The pressure $P$ of a system can be obtained from

$$P = N\kappa_B T - \frac{1}{3}\frac{1}{V}\sum_{i=1}^{n}\sum_{j=1, j<1}^{n}\left\langle r_{ij}\frac{d\phi}{dr_{ij}}\right\rangle \tag{5.10}$$

It is to be noted that the first and second terms in equation (5.10) represent the kinetic contribution from particles hitting the walls of the container and the contribution from interatomic forces.

**Virial stress.** Due to the discrete nature of the atomic microstructure, continuum-based stress cannot be applied to an MD simulation. In an MD simulation, virial stress is used to measure atomic-based stress. Virial stress is given by [3, 4]

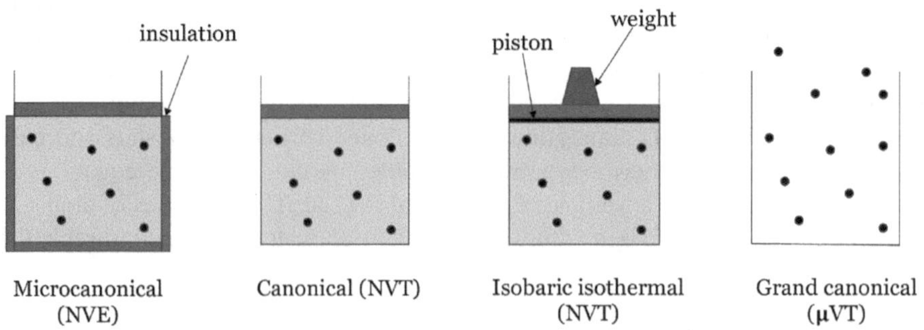


**FIGURE 5.3** Representation of various ensembles in an MD simulation.

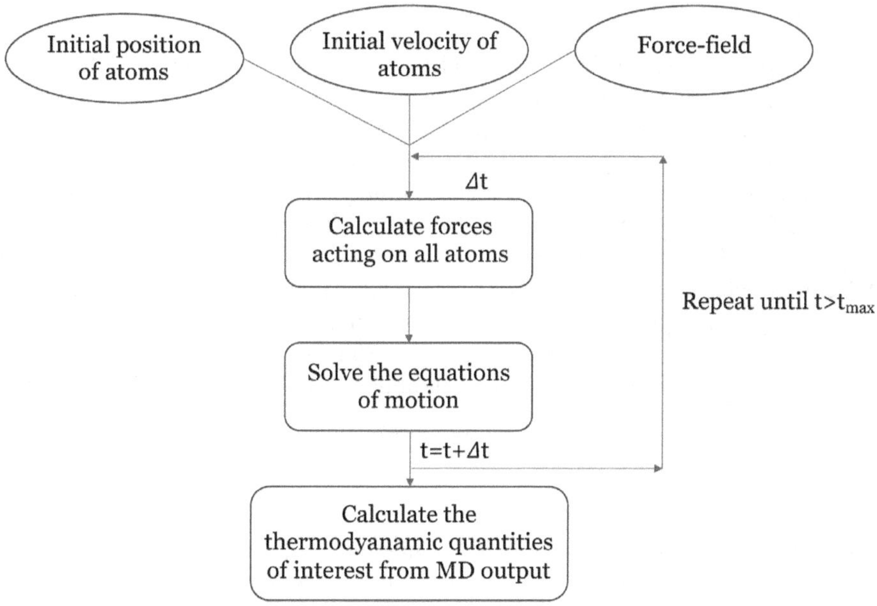


**FIGURE 5.4** Framework of a basic MD algorithm.

$$\sigma_{ij}^{v} = \frac{1}{V}\sum\nolimits_{\alpha}\left[\frac{1}{2}\sum\nolimits_{\beta=1}^{N}\left(R_i^{\beta} - R_i^{\alpha}\right)F_j^{\alpha\beta} - m^{\alpha}v_i^{\alpha}v_j^{\alpha}\right] \tag{5.11}$$

where $(i, j)$ takes values of $x$, $y$, and $z$ directions; $\beta$ takes values 1 to $N$ neighbors of atom α; $R\alpha$ is the position of atom α along the direction $i$; $F\alpha\beta$ is the force $ij$ on atom α due to atom $\beta$; $V$ is the total volume of the system; $m\alpha$ is the mass of atom α; and $v\alpha$ is the velocity of atom α. The first and second terms in equation (5.11) represent the force and kinetic contribution, respectively. The main condition for using the kinetic term in equation (5.11) is that mean velocity must be removed from the velocity to calculate stress. Otherwise, a rigid body motion would lead to stress. Also, for the equivalence of virial stress to Cauchy stress, proper spatial and temporal averages have to be taken [3, 5].

### 5.2.6 The MD Algorithm

To summarize, a basic workflow of an MD simulation is given in Figure 5.4. To initiate an MD simulation, the atomic position, velocity, and force field should be provided as input. Given these inputs, the position of atoms at the next time step $\Delta t$ can be predicted. This step involves computing the forces acting on all atoms, followed by velocity and position calculation, as shown in Figure 5.4. The step is repeated until the maximum number of steps provided is achieved. Finally, the macroscopic properties are estimated from the output of the MD simulation.

## 5.3 MD Modeling and Simulation

Here, modeling of a polymer nanocomposite (PNC) using Materials Studio 6.0 [6] and simulation using Large-Scale Atomic/Molecular Massively Parallel Simulator (LAMMPS) [7] is discussed in detail. There are various other software used for modeling and simulation of PNCs. A list of commonly used software and their features is shown in Table 5.1.

### 5.3.1 Modeling

Modeling is performed by specifying the initial atomic coordinates and velocities. For polymer composites, modeling provides the arrangement of all the polymer chains and reinforcements. In this chapter, a single-walled carbon nanotube (CNT) is used as reinforcement, and polymethyl methacrylate (PMMA) is used as polymer. The representative volume element (RVE) concept—the portion of the simulated material or system that accurately represents its overall behavior—is used while modeling a PNC. The RVE consists of CNT surrounded by polymer chains. Initially, a CNT of desired chirality and length and a single chain of polymer is generated. The atomic representation of CNT and PMMA is shown in Figures 5.5a and 5.5b. Figure 5.5a shows functionalized CNT, that is, CNT with chemical bonds attached to the carbon atoms. The CNT is placed at the center of RVE, and polymer chains are packed around it. The final configuration of atoms in the RVE is shown in Figure 5.5c. The modeled system is then imported to LAMMPS using msi2lmp package for performing MD simulations.

### 5.3.2 Simulation

In this chapter, the MD simulation procedure is explained with the help of LAMMPS script file. LAMMPS provides a robust framework for integrating the equations of motion. It is suitable for modeling atomic, polymeric, coarse-grained, or macroscopic systems using a variety of interatomic potentials and boundary conditions. Furthermore, it can output the atomic trajectories in various formats, making it compatible with several software. As mentioned in Section 5.2, a successful MD simulation provides the trajectory of atoms from which desired macroscopic properties can be extracted. For this purpose, initial atomic coordinates and interatomic potential need to be provided as an input. The modeling procedure explained in Section 5.3.1 provides the atomic coordinates, whereas interatomic potential is chosen depending on the material system and type of simulation. Here, to model the interaction in CNT-PMMA nanocomposite, a reactive potential known as ReaxFF is used [8]. ReaxFF computes bond orders for a pair of atoms based on their interatomic distances. Bond orders are updated on the fly, allowing for bond creation and dissociation. The interatomic forces are calculated from the respective bond orders. The total potential energy, $EReaxFF$, in the ReaxFF force field is calculated as the sum of various terms [9]:

$$R_{ReaxFF} = E_{bond} + E_{over} + E_{angle} + E_{tors} + E_{V\,dWaals} + E_{Coulomb} + E_{specific} \tag{5.12}$$

where $Ebond$, $Eover$, $Eangle$, $Etors$, $EV\,dW\,aals$, and $ECoulomb$ represent energy contributions corresponding to bond, overcoordination, valance angle, torsion, van der Waals, and Coulombic interactions, respectively. $Especific$ represents system-specific energy terms, such as *lone–pair*, *conjugation*, and *hydrogen binding*, that are not generally included unless specified. The parameter

**TABLE 5.1**
**Various Modeling Software and Their Features**

| Software | Type | Features |
|---|---|---|
| Materials Studio | Commercial | Suitable for modeling any PNC. Best suited for users with limited experience. |
| Moltemplate | Open-source | Suitable for modeling any PNC. However, users will have to tackle a steep learning curve. |
| VMD | Open-source | Used for modeling of nanotubes, graphenes. Not suitable for modeling complex PNCs. |
| Packmol | Open-source | Preferred for modeling basic PNCs. |

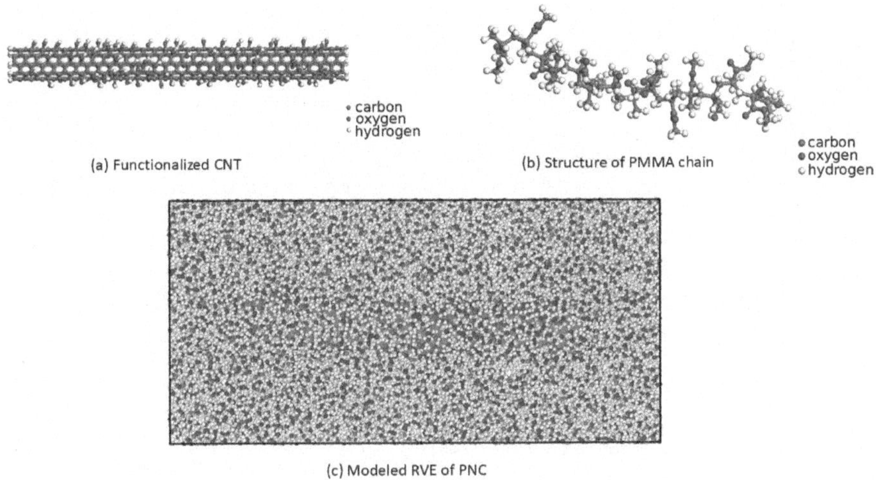


**FIGURE 5.5** (a) Molecular structures of single-walled functionalized CNT, (b) structure of PMMA chain, and (c) a modeled RVE of PNC.

values used in this study are available in [10] and downloadable from an open-source link [11]. It is to be noted that the atom types used in this example are closely related to that used in [10]. Furthermore, the same parameter values were successfully used to simulate a graphene–PMMA system previously in [12]. Next, a sample LAMMPS input script file is shown to explain the process of equilibration in an MD simulation.

**LAMMPS file.** The code shown in what follows is used to equilibrate a PNC. The system is initialized by providing the atom style, periodicity information, and input data, which consist of initial atomic coordinates. Next, the interatomic interactions and the path to the parameter file are provided. These two steps are preliminary to initiate any LAMMPS simulation. To reduce the potential energy of the system, a conjugate–gradient minimization is performed. Finally, an NPT ensemble is used to equilibrate the temperature and pressure of PNC. The atomic coordinates are dumped to a separate file for post-processing.

```
#LAMMPS code to e q u l ib r a t e a CNT-PMMA system
#I n i t i a l i z a t i o n
u n i ts     r e a l
ato m s tyle charge
boundary     p p p
read data    pmma pnc funct . data
#Define the p o t e n t i a l and i t s parameters
p a i r s t y l e    reax / c c o n t r o l f i l e lgvdw yes
p a i r c o e f f    * * f f i e l d . reax . lg C C H O H C C C O O C C
neighbor     4 bin
neigh modify every 10 delay 0 check no
#Define the output s t y l e s thermo 100
the rm o s tyle custom s tep atoms temp p re s s pe ke e t o t a l vol d
e n s i ty lx ly l z
#Conjugate gra d ie n t minimization
f i x std a l l qeq / reax 1 0 . 0 1 0 . 0 1 e-6 reax / c m in s tyle
cg
minimize     1 . 0 e-6 1 . 0 e-8 10000 100000
#Set o f e q u i l i b r a t i o n s to reduce the energy o f the system
r e s e t t i m e s t e p 0
```

```
t imestep 0. 25
#NPT ensemble
f i x npt e q ui a l l npt temp 300. 0 300. 0 2 5. 0 i s o 1. 0 1. 0 250.
0 f i x bd a l l reax / c/ bonds 500 bonds. reaxc eqbn
#Output the t r a j e c t o r y o f atoms
dump npt dump a l l custom 500 dump. npt id type x y z run 200000
w r i t e r e s t a r t   r e s t a r t . npt e q ui unf ix npt e q ui
unf ix bd
undump npt dump
```

After equilibration, the temperature and pressure need to be checked to ensure the stability of the system. Following the process, PNC can be subjected to any boundary condition, and the desired macroscopic properties can be extracted.

## 5.4 MECHANICAL PROPERTIES OF PNC

MD simulations provide insights into various mechanical properties, such as elastic modulus, fracture properties, and interfacial strength. To estimate properties, suitable boundary conditions need to be applied to the PNC. This is facilitated by adopting appropriate thermodynamical ensembles. In this section, the calculation of elastic modulus and fracture properties is provided in detail. PNCs with varying degrees of hydroxyl (-OH) functionalization are considered for analysis. Functionalization is varied by changing the number of -OH groups attached to the carbon atoms.

To compute the elastic and fracture properties, a uniaxial strain is applied on PNC with and without a pre-crack. The crack acts as a defect for initiating fracture. A uniaxial strain of $10^{10}\ s^{-1}$ is coupled with an NPT ensemble at 300 *K*. The coupling enables a realistic deformation of PNC. The schematic and atomic representation of pre-cracked PNC with applied boundary condition are shown in Figure 5.6. To compute properties, the stress–strain values are extracted from the MD simulations. The stress is measured in terms of virial stress, as discussed in Section 5.2.5.

### 5.4.1 Elastic Modulus

The elastic modulus is calculated by using a linear fit in the elastic regime of the stress–strain curve. Here, modulus is calculated for PNCs with varying degrees of functionalization. The key idea is to show the validity of MD simulations in predicting the elastic properties of PNC. The values obtained from the simulation are shown in Table 5.2. It is observed from the table that elastic modulus improves with the introduction of functionalization. This observation is consistent with the experimental studies. Furthermore, the values reported in the table are in good accordance with the published experimental results [13]. Thus, MD simulations are a capable method to estimate the elastic modulus of PNC. It is to be noted that the abbreviation *Fn* in Table 5.2 represents functionalization.

### 5.4.2 Fracture Properties

Fracture simulations on PNCs are marginally studied using MD simulations. This is due to the difficulty in modeling and the unavailability of potentials that capture fracture process. However, recently, fracture simulations are performed using the help of reactive potentials [12, 14, 15].

The fracture properties are estimated by conducting uniaxial tension on functionalized PNCs with an initial crack. The corresponding stress–strain curves from the simulation are shown in Figure 5.7. The fluctuations in these plots are due to the temperature—300 *K*— used in the simulation. The stress–strain curves are obtained by using moving averages over 100 points. From the figure, two key quantities are calculated: J-integral and ultimate strain to failure. The value of *J* is obtained from the load–displacement curve, which, in turn, is obtained from the stress–strain

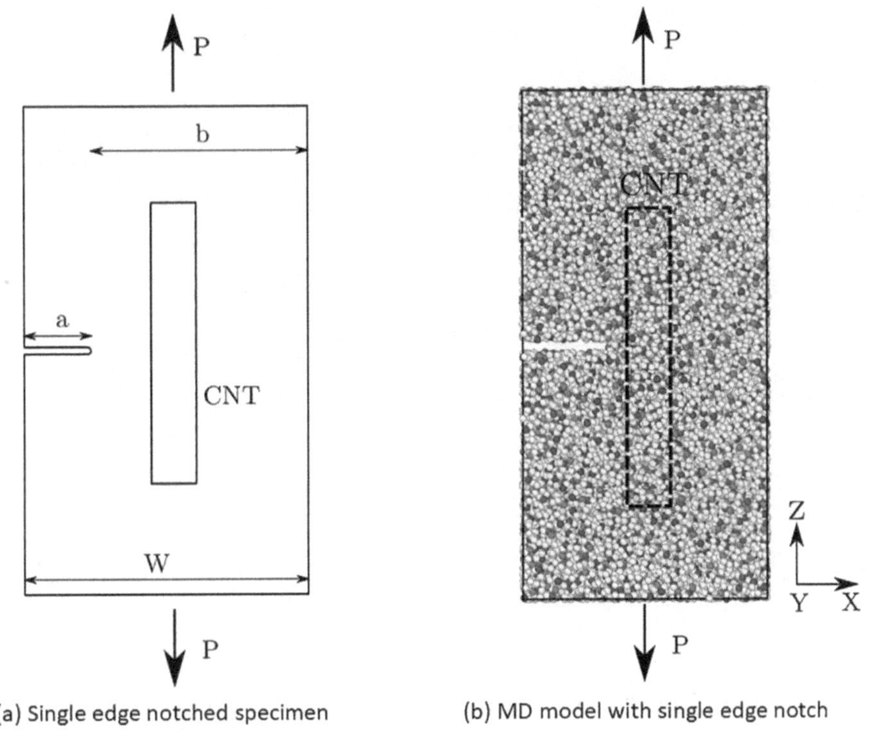


**FIGURE 5.6** Schematic: (a) single-edge notch specimen and (b) MD model with single-edge notch.

**TABLE 5.2**
**Elastic Modulus Calculated from MD Simulation**

| Material | Elastic Modulus (GPa) | Improvement (%) |
|---|---|---|
| PMMA | 2.9 | Baseline |
| PMMA + CNT | 3.4 | 17 |
| PMMA + 3.5% FnCNT | 3.7 | 28 |
| PMMA + 10% FnCNT | 4.2 | 45 |
| PMMA + 15% FnCNT | 4.4 | 52 |
| PMMA + 20% FnCNT | 4.3 | 48 |

curve. More details on the calculation of $J$ and its validity are presented in [14]. The ultimate strain corresponds to the strain at which the PNC undergoes complete failure. This can be taken as the point in Figure 5.7 where the curve intersects the $X$ axis, that is, zero strain.

The fracture quantities obtained from the curve are shown in Table 5.3. From the table it is observed that the J-integral value increases for functionalized PNC. Since $J$ is a measure of resistance to crack propagation [16], this implies functionalization enhances resistance to crack propagation. This improvement of $J$ upon functionalization is well-studied in experiments [17]. It is also observed from Table 5.3 that functionalization improves the ultimate strain to failure. This indicates that the ductility of PNC increases due to functionalization. However, J-integral and ultimate strain to failure reduces beyond a particular degree of functionalization. This is due to the stiffening of interface. A detailed discussion on this behavior is provided in [14].

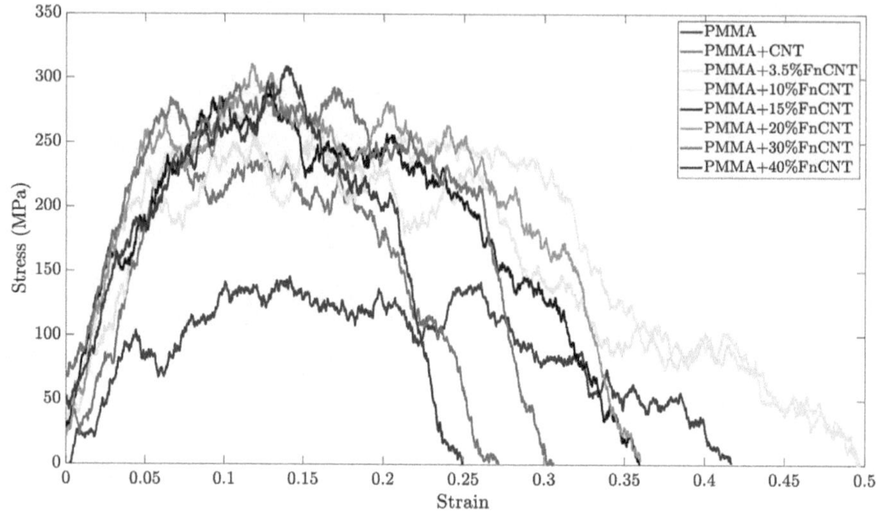


**FIGURE 5.7** Stress–strain plots of PNC with initial crack for different levels of functionalization.

**TABLE 5.3**
**Variation in Ultimate Strain and J-Integral with Respect to Degree of Functionalization**

| Material | Ultimate Strain (%) | J-Integral (N/m) | Improvement in J (%) |
|---|---|---|---|
| PMMA + CNT | 27 | 0.58 | Baseline |
| PMMA + 3.5% FnCNT | 50 | 1.02 | 76 |
| PMMA + 10% FnCNT | 50 | 1.20 | 107 |
| PMMA + 15% FnCNT | 36 | 0.92 | 59 |
| PMMA + 20% FnCNT | 36 | 0.98 | 69 |
| PMMA + 30% FnCNT | 30 | 0.87 | 53 |
| PMMA + 40% FnCNT | 25 | 0.64 | 10 |

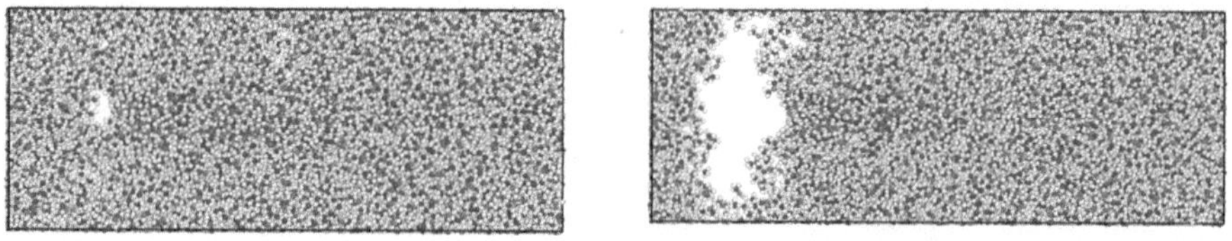

(a) Strain 20% (b) Strain 30%

**FIGURE 5.8** Crack propagation in 40% functionalized PNC: (a) strain 20% and (b) strain 30%.

The major advantage of using the MD simulation is to gain deeper insights into the behavior of PNC. To highlight this aspect, a few snapshots of crack propagation in PNC for 40% functionalization are shown in Figure 5.8. Figures 5.8a and 5.8b show crack initiation and final failure pattern in PNC. The propagation of crack at the end of CNTs has been previously reported in several studies [18]. Thus, MD simulations are crucial in understanding the crack propagation and failure pattern in PNCs. It is to be noted that the fracture quantities obtained from MD simulations are not comparable to experiments. However, they provide valuable insights into the behavior of PNCs.

## 5.5 CHALLENGES

Even though MD simulations are widely used for studying polymer composites, they bring in several challenges. Some of the major challenges in using MD simulation are as follows:

1. *Computational expense.* MD simulations are computationally expensive to conduct. This limits its applicability of scaling simulations to higher length scales.
2. *Lack of adequate potentials.* Although potentials are transferable, most of them are developed for studying a particular class of material. Appropriate potentials are not available for simulating several material properties.
3. *Limited scalability.* The MD simulations presented in this chapter are at the nanometer scale. However, actual material behavior needs to be studied at a macroscopic scale. Furthermore, MD simulations are often limited by the accessible timescale. The computational expense makes it impossible to scale these simulations.

These challenges need to be considered before performing any MD simulation.

## 5.6 FUTURE PROSPECTS

Several research works are presently conducted to reduce the computational expense of MD simulations. One of the recent successes in this field has been achieved by coupling MD simulations with machine learning (ML). It has been found that MD-informed ML models are capable of scaling the simulations to higher length scales [19, 20]. This is achieved by training ML models using the data obtained from MD simulations. A recent study has shown the potential of introducing large language models into the area of MD simulations, thereby reducing the computational cost significantly [21]. With the growth of ML models, detailed patterns can be extracted from the dataset provided by MD simulations. Nevertheless, these research domains are still in their early stages of development.

## 5.7 CONCLUSION

This chapter presents the role of MD simulation on studying the behavior of polymer composites. The modeling and simulations discussed in this chapter reveal that MD is a suitable tool for the analysis of polymer composites. It can accurately estimate the elastic properties and fracture behavior. The properties showed good agreement with the experimental studies. Furthermore, several valuable insights can be obtained through these simulations. However, the computational expense and scalability remain a challenge in this area.

## REFERENCES

[1] Buehler, M.J.: Atomistic modeling of materials failure. Springer (2008).
[2] Allen, M.P., Tildesley, D.J.: Computer simulation of liquids. Oxford University Press (2017).
[3] Subramaniyan, A.K., Sun, C.: Continuum interpretation of virial stress in molecular simulations. International Journal of Solids and Structures **45**(14–15), 4340–4346 (2008).
[4] Tsai, D.: The virial theorem and stress calculation in molecular dynamics. The Journal of Chemical Physics **70**(3), 1375–1382 (1979).
[5] Admal, N.C., Tadmor, E.B.: A unified interpretation of stress in molecular systems. Journal of Elasticity **100**, 63–143 (2010).
[6] BIOVIA Materials Studio 6.0. https://www.3ds.com/products-services/biovia/products/molecular-modeling-simulation/biovia-materials-studio/.
[7] Thompson, A.P., Aktulga, H.M., Berger, R., Bolintineanu, D.S., Brown, W.M., Crozier, P.S., In't Veld, P.J., Kohlmeyer, A., Moore, S.G., Nguyen, T.D., Shan, R.: LAMMPS: A flexible simulation tool for particle-based materials modeling at the atomic, meso, and continuum scales. Computer Physics Communications **271**, 108171 (2022).

[8] Van Duin, A.C., Dasgupta, S., Lorant, F., Goddard, W.A.: Reaxff: A reactive force field for hydrocarbons. The Journal of Physical Chemistry A **105**(41), 9396–9409 (2001).
[9] Senftle, T.P., Hong, S., Islam, M.M., Kylasa, S.B., Zheng, Y., Shin, Y.K., Junkermeier, C., Engel-Herbert, R., Janik, M.J., Aktulga, H.M., Verstraelen, T.: The Reaxff reactive force-field: Development, applications and future directions. NPJ Computational Materials **2**(1), 1–14 (2016).
[10] Liu, L., Liu, Y., Zybin, S.V., Sun, H., Goddard, W.A., III: Reaxff-lg: Correction of the Reaxff reactive force field for London dispersion, with applications to the equations of state for energetic materials. The Journal of Physical Chemistry A **115**(40), 11016–11022 (2011).
[11] LAMMPS ffield.reax.lg. https://github.com/lammps/lammps/blob/develop/potentials/ffield.reax.lg.
[12] Najafi, F., Wang, G., Mukherjee, S., Cui, T., Filleter, T., Singh, C.V.: Toughening of graphene-based polymer nanocomposites via tuning chemical functionalization. Composites Science and Technology **194**, 108140 (2020).
[13] Banks-Sills, L., Shiber, D.G., Fourman, V., Eliasi, R., Shlayer, A.: Experimental determination of mechanical properties of PMMA reinforced with functionalized CNTs. Composites Part B: Engineering **95**, 335–345 (2016).
[14] Abhiram, B.R., Ghosh, D.: Atomic investigation on optimal interfacial bonding for enhanced fracture properties in polymer nanocomposites. Engineering Fracture Mechanics 109078 (2023).
[15] Abhiram, B.R., Ghosh, D.: Influence of nanofiller agglomeration on fracture properties of polymer nanocomposite: Insights from atomistic simulation. Engineering Fracture Mechanics 109503 (2023).
[16] Anderson, T.L.: Fracture mechanics: Fundamentals and applications. CRC Press (2017).
[17] Cha, J., Jun, G.H., Park, J.K., Kim, J.C., Ryu, H.J., Hong, S.H.: Improvement of modulus, strength and fracture toughness of CNT/epoxy nanocomposites through the functionalization of carbon nanotubes. Composites Part B: Engineering **129**, 169–179 (2017).
[18] Opelt, C.V., Becker, D., Lepienski, C.M., Coelho, L.A.: Reinforcement and toughening mechanisms in polymer nanocomposites–carbon nanotubes and aluminum oxide. Composites Part B: Engineering **75**, 119–126 (2015).
[19] Lal, H.P., Abhiram, B., Ghosh, D.: Prediction of nonlocal elasticity parameters using high-throughput molecular dynamics simulations and machine learning. European Journal of Mechanics-A/Solids **103**, 105175 (2024).
[20] Yang, Z., Yu, C.H., Guo, K., Buehler, M.J.: End-to-end deep learning method to predict complete strain and stress tensors for complex hierarchical composite microstructures. Journal of the Mechanics and Physics of Solids **154**, 104506 (2021).
[21] Buehler, M.J.: Melm, a generative pretrained language modeling framework that solves forward and inverse mechanics problems. Journal of the Mechanics and Physics of Solids **181**, 105454 (2023).

# 6 Machine Learning and Artificial Intelligence in Polymer Composites

*Rajesh Mahadeva, Sushanta K. Sethi, Atul K. Maurya, Saurav Dixit, Vinay Gupta, and Gaurav Manik*

## 6.1 INTRODUCTION

### 6.1.1 Overview of Polymer Composites

Materials made of a polymer matrix reinforced with fibers or particles are known as polymer composites [1–19]. Compared to the constituent parts, these composites perform better mechanically, thermally, and chemically because they integrate the advantageous qualities of the reinforcing elements and the polymer matrix [20–23]. The following are the main features of polymer composites:

- *Polymer matrix.* Usually, a polymer resin, like vinyl ester, epoxy, or polyester, serves as the matrix. Polymers offer corrosion resistance, flexibility, and ease of processing.
- *Reinforcement.* To improve the mechanical properties of the polymer matrix, fibers or particles are added. Common reinforcing materials include fiberglass, carbon fiber, aramid (such as Kevlar), and natural fibers.
- *Types of polymer composites.* (1) Fibers are inserted in a polymer matrix to create fiber-reinforced polymer (FRP) composites. (2) The material known as carbon fiber–reinforced polymer, or CFRP, has a high strength-to-weight ratio. (3) Glass fiber–reinforced polymers, or GFRP, are frequently utilized due to their affordability and ability to withstand corrosion. (4) Aramid fiber–reinforced polymers (AFRP) are renowned for their extreme toughness and resilience to impact. (5) Particulate composites consist of polymer matrices with scattered particles, such as clay or nanoparticles, to improve qualities, like stiffness and thermal stability.
- *Benefits.* (1) Strength and stiffness: the composite gains enhanced strength and stiffness from the reinforcing elements. (2) Weight reduction is crucial, especially in the aircraft and automobile industries, where lightweight materials are preferred. (3) Corrosion resistance:—composites are appropriate for hostile settings because polymer matrices frequently resist corrosion.
- *Applications.* (1) Aerospace: used to lighten airplane parts without sacrificing strength. (2) Automotive: parts such as interiors, engine parts, and body panels. (3) Construction: structures such as buildings, pipelines, and bridges. (4) Sports and recreation: bicycles, tennis rackets, and golf clubs are examples of sporting items.
- *Methods of processing.* (1) Molding: resin transfer molding, injection molding, and compression molding are standard methods. (2) Layering: this process involves stacking polymer and reinforcing material layers to create laminates.

Because polymer composites are versatile and can be tailored to fulfill specific performance needs, they are essential to many sectors. They remain a topic of active investigation and advancement.

DOI: 10.1201/9781003476337-6

### 6.1.2 Role of ML and AI in Enhancing Polymer Composite Properties

Nowadays, ML and AI play an essential role in various engineering, medical, and industrial applications to save time, money, and energy [24–38]. This research highlights the use of ML and AI in the field of polymer composite properties. ML and AI can significantly improve the characteristics of polymer composites, which offer sophisticated tools for material design, optimization, and production processes [39–41]. An overview of their role is provided here. *Material Forecasting and Design*: ML and AI systems can analyze large datasets on the characteristics of polymers, reinforcement materials, and manufacturing conditions. They can forecast how modifications in composition and processing settings will affect the final attributes of the composite by finding patterns and correlations within these datasets. These predictive capabilities accelerate the material design process. *Property Optimization*: ML models can optimize polymer composites' properties by recommending the best polymers and reinforcing materials mix. These models assist engineers in determining the ideal balance for a given application by considering several variables, including strength, stiffness, weight, and cost. *Process Optimization*: ML algorithms can improve manufacturing procedures by examining data from different phases of the composite production process. These cover mixing ratios, pressure, temperature, and curing time. Reducing production costs and improving material qualities can result from optimizing these factors.

*Quality Control and Defect Detection*: To find flaws or irregularities in the composite material, quality control procedures can use ML and AI. By utilizing computer vision, automated inspection systems may detect flaws and guarantee that the finished product satisfies quality requirements. *Sensitivity Analysis*: ML models can analyze sensitivity to determine which variables affect the composite's characteristics most. This knowledge is helpful when concentrating efforts on optimizing the most significant parameters. *Materials Informatics*: Materials informatics analyzes enormous datasets about polymer composites to find new patterns, correlations, and hidden insights. ML and AI make this possible. This may result in the identification of new materials with improved characteristics. *Customization for Particular Uses*: ML techniques can help customize polymer composites for uses. By considering those needs, ML models can suggest the best material and processing condition combinations to meet specific industry performance criteria.

*Real-Time Process Monitoring*: AI systems can continuously monitor production processes and modify settings to guarantee constant quality. This real-time feedback loop lowers the probability of faults and increases the reproducibility of composite qualities. *Life Cycle Assessment and Sustainability*: By evaluating the environmental effects of various material selection and production techniques, ML and AI may help ensure the sustainability of polymer composites. It makes creating composites with better lifespan assessments and lower carbon footprints possible. *Continuous Improvement*: Performance data from goods in use can be analyzed using ML models to provide insights into long-term behavior and possible areas for improvement. It makes refining manufacturing procedures and material formulations possible through an ongoing feedback loop.

It should be noted that ML and AI are practical tools in polymer composites, with the potential to completely transform quality control, production procedures, and material design. By combining these technologies, new materials with customized qualities, better performance, and increased sustainability are developed.

### 6.1.3 Objectives and Contributions

**Objectives:** (1) *Exploration of ML and AI applications.* The first objective of this research is to explore and catalog how ML and AI are used in polymer composites. Predictive modeling, quality assurance, process optimization, and material design may be examples. (2) *Performance enhancement.* The second objective of this research is to evaluate how polymer composites' characteristics and performance can be improved using ML and AI techniques. This could entail forecasting long-term behavior based on real-world situations, optimizing production procedures, or material composition optimization.

**Contributions:** (1) *Survey based on Scopus database, Elsevier.* Provide an in-depth analysis of the state of ML and AI applications in the polymer composites industry. Provide an overview of the literature and point out any gaps or areas that require more research. (2) *Implementation guidelines.* Give instructions or suggestions on how to apply ML and AI to create and improve polymer composites. It could cover data needs, best practices, and things to remember at various points during the material design and manufacturing process. (3) *Prospective routes.* Analyze possible future paths for research and development at the nexus of polymer composites, AI, and ML. Determine new patterns and areas that, with more research, could result in significant breakthroughs.

## 6.2 LITERATURE REVIEW

### 6.2.1 Based on the Scopus Database, Elsevier

This study's literature review was conducted using the Scopus database, Elsevier. This survey covered the keywords "machine learning," "artificial intelligence," and "polymer composites" and was exported on February 2, 2024, obtaining 2,039 research studies. Research shows that after the year 2017, ML and AI have been involved more than in previous years (Figure 6.1). The research also highlights that, in 2023, research was more than 862, and in 2022, was more than 500, the maximum number of research applied to analyze polymer composite properties. It shows the recent trends of ML and AI in this polymer composite fields.

Further, the research highlighted that after the year 2017, research articles had been published in high-impact, peer-reviewed journals and publications such as *Advanced Materials*, *Composite Part B Engineering*, *Nano Energy*, *Advanced Functional Materials*, *Composite Structure*, *Sensors*, *Materials*, *Polymers*, *Applied Science Switzerland*, etc. (Figure 6.2). These journals have the topmost publications in this fields. Furthermore, Boopathi, S.; Chen, J.; Lee, C.; Buehler, M. J.; Gupta, V.; Ali, H. E.; Datta, S.; Paturi, U. M. R.; Wang, Z. L.; and Amin, M. N., are the top ten researchers/authors who have published research articles in this field (Figure 6.3). Boopathi, S., has published the highest number of research articles (15) in this area. In addition, Chen, J., and Lee, C., have published 11 research articles each, while Buehler, M. J., and Gupta, V., have published 10 each in this area.

The research survey shows that the affiliations of the Ministry of Education of the People's Republic of China are more involved in this field and have published 78 articles with this affiliation. In addition, the Chinese Academy of Sciences (49 articles), National University of Singapore

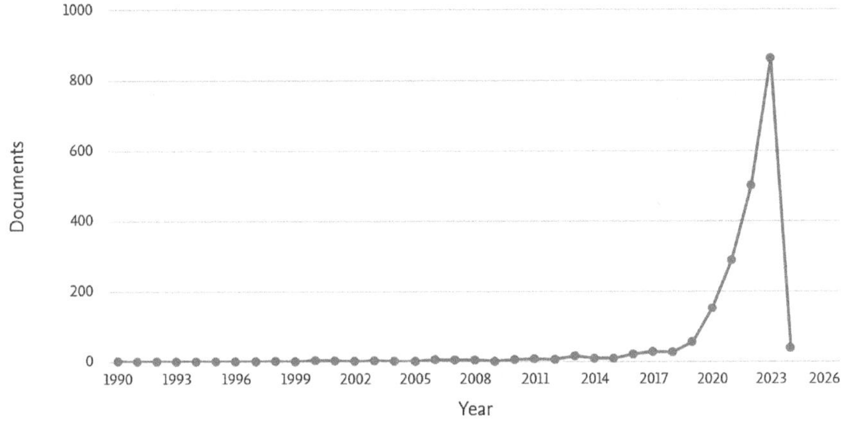


**FIGURE 6.1** Documents by year.

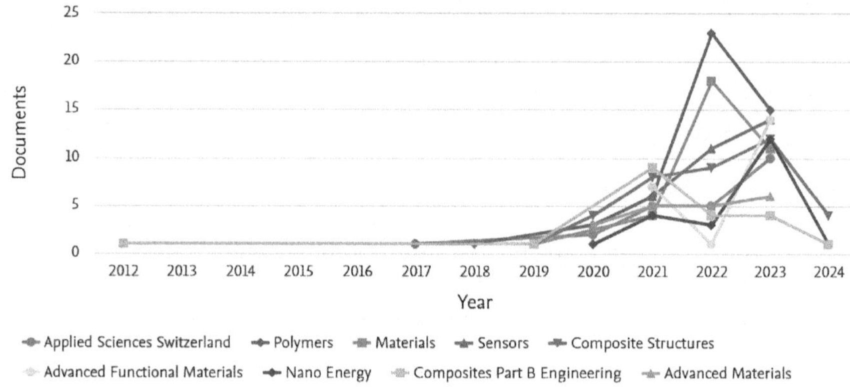


**FIGURE 6.2** Documents per year by source.

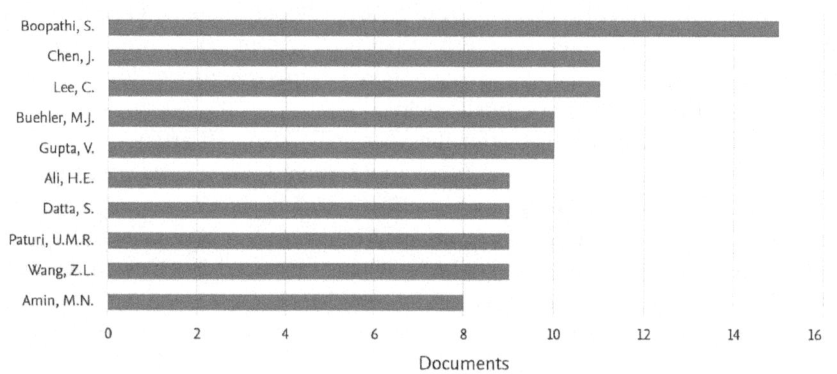


**FIGURE 6.3** Documents by author.

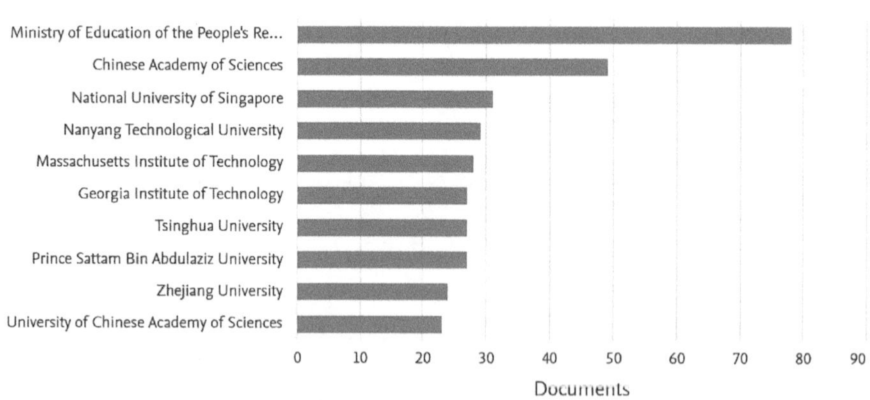


**FIGURE 6.4** Documents by affiliation.

(31 articles), Nanyang Technological University (29 articles), Massachusetts Institute of Technology (28 articles), Georgia Institute of Technology (27 articles), Tsinghua University (27 articles), Prince Sattam Bin Abdulaziz University (27 articles), Zhejiang University (24 articles), and University of Chinese Academy of Sciences (23 articles) are the top ten affiliations that have contributed in this area (Figure 6.4).

China is the world's leading country which has supported researchers to publish in this area. China itself has published more than 613 publications across the globe. In addition, top countries have also published extensively in this field, such as the United States, which has 419 publications; India, 278 publications; the United Kingdom, 144 publications; Saudi Arabia, 115 publications; South Korea, 102 publications; Iran, 101 publications; Australia, 91 publications; Germany, 89 publications; and Canada, 87 publications (Figure 6.5).

The National Natural Science Foundation of China has provided the highest funding (more than 324) for this type of research. Other countries also offered funding for researchers to contribute to this field, such as the National Science Foundation (more than 92), the National Key Research and Development Program of China (more than 76), the Fundamental Research Funds for the Central Universities (more than 66), the National Research Foundation of Korea (more than 61), the Ministry of Science, ICT, and Future Planning (more than 41), the US Department of Energy (more than 32), the Engineering and Physical Sciences Research Council (more than 31), the European Commission (more than 30), and the Natural Sciences and Engineering Research Council of Canada (more than 29), as shown in Figure 6.6.

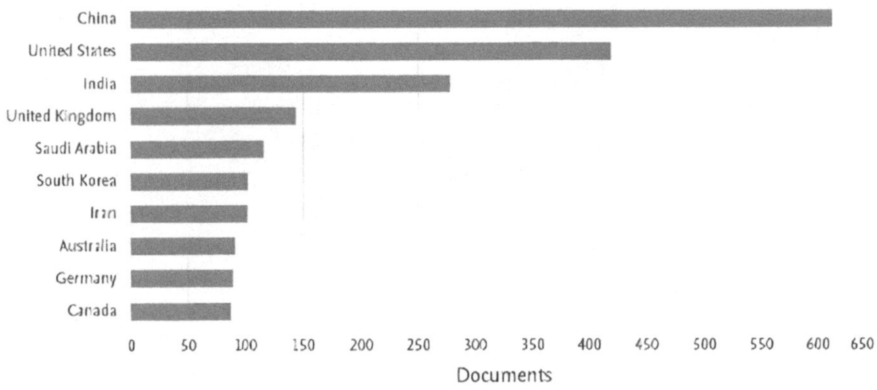


**FIGURE 6.5** Documents by country/territory.

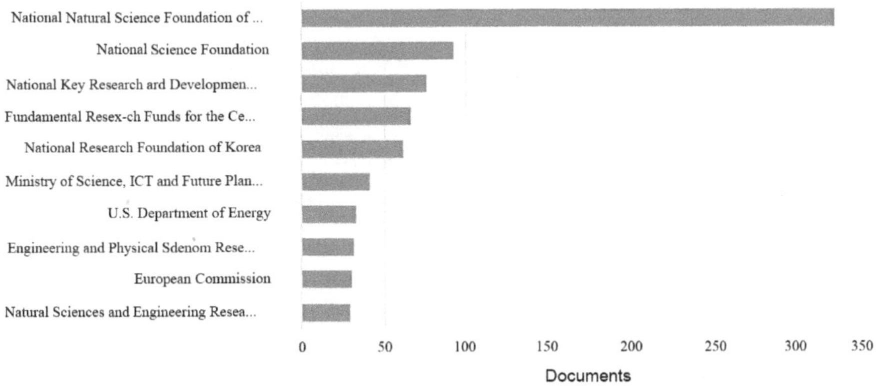


**FIGURE 6.6** Documents by funding sponsor.

Further research survey shows that from 1990 to the current year, February 2024, 59.6% of research articles (>1,215) were published in this domain. Review articles were also published in more than 554 (27.2%), in addition to conference papers (132, 6.5%), book chapters (80, 3.9%), and books (48, 2.4%), as shown in Figure 6.7. Editorials, notes, data papers, letters, and short surveys have also been involved in this field. Furthermore, a research survey shows that engineering fields had 1,201 publications (26.0%); materials science, 982 publications (21.2%); physics and astronomy, 446 publications (9.6%); computer science, 441 publications (9.5%); chemistry, 368 publications (8.0%); chemical engineering, 221 publications (4.8%); mathematics, 205 publications (4.4%); energy, 160 publications (3.5%); biochemistry, genetics, and molecular biology, 157 publications (3.4%); environmental science, 97 publications (2.1%); and other fields 346 publications (7.5%), published in this area (Figure 6.8).

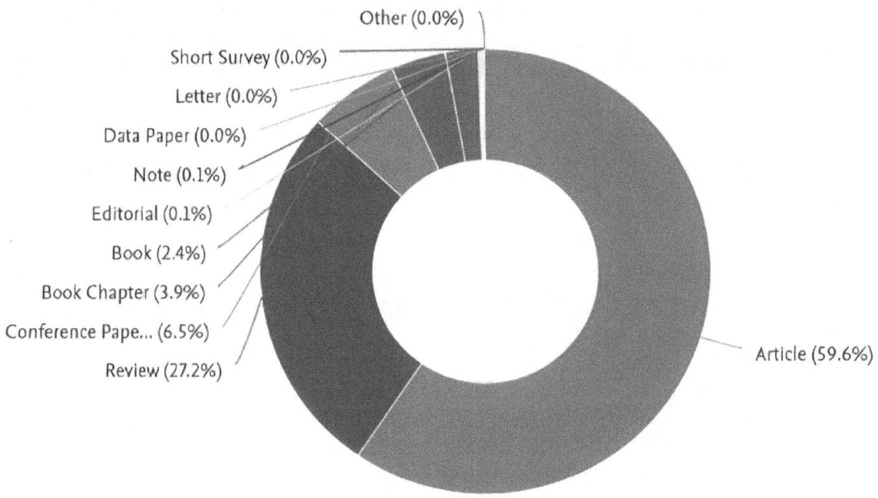


**FIGURE 6.7** Documents by type.

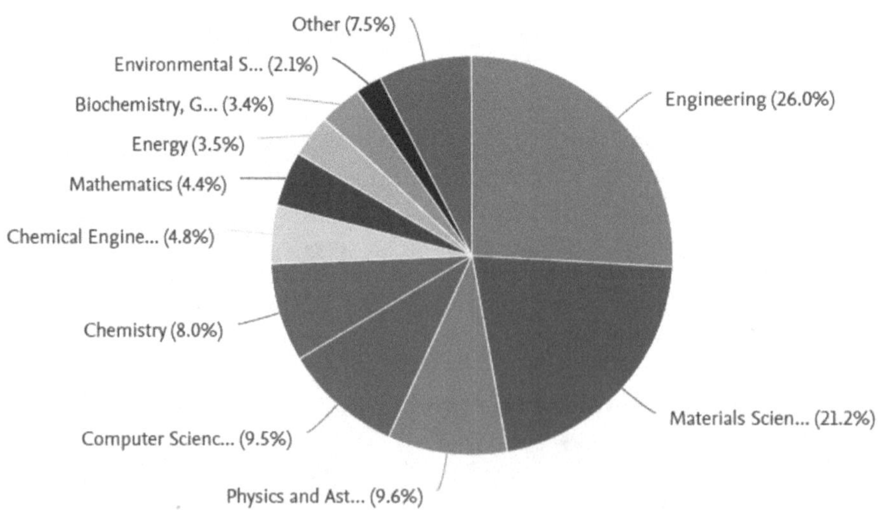


**FIGURE 6.8** Documents by subject area.

### 6.2.2 Discussion of Traditional Methods in Polymer Composite Research and Their Limitations

Understanding and producing polymer composite materials have been greatly aided by the use of traditional research methodologies. However, several limitations may prevent advancements in optimizing composite qualities and production procedures [42–46]. Here's a rundown of a few conventional/traditional techniques in polymer composite research and their limitations, as shown in Table 6.1.

Although polymer composite research has relied heavily on these conventional techniques, their shortcomings underscore the necessity for more sophisticated and data-driven methodologies. Literature surveys suggest ML and AI have emerged as promising methods in polymer composites to address these limits by offering more accurate, efficient, and adaptable solutions for material design, optimization, and manufacturing.

**TABLE 6.1**
**Conventional/Traditional Techniques in Polymer Composite Research and Their Limitations**

| Traditional Techniques | Limitations |
|---|---|
| Trial and error in experimentation | *Time-consuming.* It takes time to conduct many trials, which might not be feasible for developing sectors.<br>*Expensive.* Both labor and material costs can be high during the synthesis and testing of several formulations.<br>*Limited exploration.* This approach might not thoroughly explore the whole design space, which could result in the omission of ideal pairings. |
| Rule-of-thumb approaches | *Lack of precision.* Using rule-of-thumb methods to optimize composite qualities for specific applications may not be appropriate due to their lack of precision.<br>*Lack of data drivenness.* These approaches might not make the most of the data or adjust well to new knowledge. |
| Analytical modeling | Analytical models frequently use simple assumptions that might not adequately represent the complexities of real-world situations.<br>*Limited accuracy.* Predictions may be less-accurate in complicated composite systems with many variables. |
| Empirical correlations | *Specificity.* It's possible that correlations created for a particular set of materials or circumstances won't apply to other systems.<br>*Absence of generalization.* It's possible that these relationships don't translate well to a larger variety of composite formulations. |
| Manual process optimization | *Suboptimal solutions.* Manual optimization significantly depends on the knowledge of individual users; it may produce less-than-ideal results.<br>*Limited adaptability.* It can find it difficult to quickly adjust to shifts in the availability of materials or technological breakthroughs. |
| Limited use of data | *Information ignored.* When making decisions, pertinent information could be disregarded or not given enough thought.<br>*Lost opportunities.* Current datasets can yield additional insights using data-driven methods and machine learning. |
| Inadequate process monitoring | *Absence of real-time input.* The absence of real-time information from manual monitoring could result in potential flaws and differences in the finished output. |

## 6.3 APPLICATIONS OF ML AND AI IN POLYMER COMPOSITE DESIGN

### 6.3.1 Material Selection and Optimization Using AI and ML Algorithms

In materials science and engineering, material selection and optimization driven by the integration of AI and ML algorithms represent a paradigm change. Combining these cutting-edge technologies presents previously unheard-of chances to completely transform the conventional methods for identifying and customizing materials for uses. The capacity of ML algorithms to forecast and optimize material qualities is at the heart of this revolution. Researchers can make precise predictions thanks to regression models and neural networks, which have been trained on large datasets with various material compositions and properties. This feature simplifies the material selection by revealing how various aspects affect a material's ultimate qualities. Furthermore, generative models provide new opportunities for material design, such as generative adversarial networks (GANs). These models can produce new material structures with specified attributes by learning patterns from pre-existing information, encouraging creativity and innovation in materials science.

The materials informatics platforms further amplify the influence of AI and ML. These integrated systems simplify analyzing large, complicated datasets by fusing data from simulations, experiments, and extensive literature. This leads to a thorough understanding of material properties, which helps engineers and researchers choose and optimize materials with knowledge. One important component of AI and ML applications in materials research is overcoming the problem of small datasets. Transfer learning approaches alleviate data scarcity by enabling models pre-trained on large datasets to modify and improve their predictions for material systems. This method improves ML models' effectiveness and precision, mainly when gathering big datasets is difficult or time-consuming. Beyond predicting material qualities, AI and ML in material science also play an optimization role. Researchers can find the best mix of material attributes that satisfy criteria and limitations by using constraint-based optimization techniques, which are frequently combined with AI. This method makes it possible to explore the material design space more effectively and find compositions that satisfy several requirements simultaneously, including strength, weight, and cost.

AI is driving optimization in the field of manufacturing processes. Deep reinforcement learning algorithms can adaptively optimize manufacturing parameters, drawing inspiration from trial-and-error learning principles. It improves material performance overall and increases production process efficiency while also helping conserve resources. AI-driven digital twins in materials science bring another complexity to material optimization and selection. Researchers can simulate and forecast the behavior of materials under different conditions by building virtual reproductions of those materials. With this real-time insight into material performance, materials may be continuously optimized and adjusted to meet the exact requirements of the applications for which they are intended. The Materials Genome Initiative (MGI) exemplifies accelerating materials discovery through collaborative efforts, integrating AI and ML with materials research. To dramatically reduce the time needed to produce novel and sophisticated materials, MGI uses these technologies' analytical capabilities to decipher the intricate correlations between material composition and attributes.

Through data-driven decision support systems, AI and ML also aid decision-making. These systems let engineers and researchers make well-informed decisions by examining large datasets about material qualities, processing techniques, and performance in particular applications. It makes choosing materials more effective and efficient by lowering the dependence on gut feeling and subjective assessment. Furthermore, in the present day, the sustainability of materials has grown in importance. When evaluating the long-term environmental effects of materials, AI and ML are essential. These technologies allow for a comprehensive assessment of a material's ecological footprint by considering many aspects, such as extraction of raw materials, production procedures, and disposal at the end of its useful life. Consequently, this directs scholars and businesses toward more ecologically friendly and sustainable materials.

Materials science is entering a revolutionary era by incorporating AI and ML algorithms into the material selection and optimization procedures. These technologies provide many answers to the intricate problems of materials research, ranging from forecasting material properties to creating innovative designs, streamlining production procedures, and guaranteeing sustainability. The potential for innovation in material discovery and design seems limitless as the synergy between AI and ML and materials science continues to develop, pointing to a future where materials are not merely chosen but meticulously crafted to match the exact requirements of a wide range of applications.

### 6.3.2 Automated Process Optimization for Manufacturing Polymer Composites

Advanced technologies like AI and ML enable automated process optimization for polymer composite manufacturing, transforming composite production's productivity, sustainability, and quality. This innovative method improves several facets of the production process by utilizing computer models and data-driven insights. In automated process optimization, real-time control and monitoring are essential components. Data on temperature, pressure, curing times, and material composition are gathered via sensors and IoT devices integrated into the manufacturing environment. AI algorithms examine this streaming data to find trends and connections, allowing the system to adapt instantly. For example, if it detects departures from ideal conditions, the system can automatically adjust settings to guarantee reliable and superior composite manufacture.

ML algorithms are essential for anticipating and averting manufacturing flaws. These algorithms anticipate possible problems by evaluating past production outcome data and connecting it with process parameters. With proactive modifications made possible by this predictive capability, waste is reduced, rework is minimized, and the output of high-quality polymer composites is eventually improved. Moreover, optimization algorithms investigate the intricate design space of industrial processes and are frequently combined with AI. They iteratively test various parameter combinations to find the ideal set of parameters that satisfies requirements, like cutting production time, saving energy, or improving material qualities. This iterative methodology continuously enhances production for better efficiency and resource usage.

Process optimization is further enhanced using robotics and automation systems powered by AI. These machines are quick and accurate at handling materials, molding, and inspecting quality. Cobots, or collaborative robots, assist human operators in production, increasing productivity and security. Digital twins are also essential for automated process optimization. Before the start of the actual output, modeling and analysis are made possible by these virtual depictions of the manufacturing process. Manufacturers can reduce the need for expensive trial-and-error studies by simulating several scenarios, optimizing settings, and predicting outcomes by combining digital twins with AI. Beyond just increasing productivity right away, automated process improvement has many advantages. Modern production must consider sustainability, and AI can help reduce the adverse effects on the environment. Mechanical methods make possible more environmentally friendly and sustainable composite production, optimizing energy use, minimizing waste, and improving resource efficiency. However, an interdisciplinary and cooperative approach is necessary to apply automated process optimization successfully. Collaboration among materials scientists, engineers, data scientists, and automation specialists is essential for seamlessly integrating AI and ML technologies into current industrial processes. Furthermore, constant observation and algorithm modification are necessary to guarantee that the system adapts to shifting environmental conditions and industrial needs.

Automating process optimization for producing polymer composites by integrating AI and ML is revolutionary. This method advances sustainability objectives while simultaneously improving the productivity and caliber of composite manufacture. The manufacturing sector will profit from higher productivity, lower prices, and the capacity to satisfy the expanding need for high-performance polymer composites in various applications as these technologies develop.

### 6.3.3 Real-Time Quality Control and Defect Detection Using ML and AI

In polymer composite research, real-time quality control, and flaw identification with ML and AI herald a new era of unmatched precision and dependability in manufacturing. This revolutionary method's fundamental component incorporates AI algorithms into the industrial process's core structure. These algorithms enable manufacturing systems to continually monitor and evaluate the quality of polymer composites in real time. They have been trained on large datasets that cover a spectrum of material compositions, processing circumstances, and potential faults. Much data is captured by the sensors and IoT devices incorporated into the industrial environment. This data covers everything from material attributes and structural aspects to temperature and pressure. AI algorithms carefully examine this streaming data, which allows for the immediate identification of potential flaws and departures from ideal conditions.

Real-time defect detection has significant ramifications, especially in industries where accuracy and dependability are critical. Finding flaws early on could result in significant cost savings and improved product performance in sectors where polymer composites are essential, like electronics, automotive, and aerospace. Early identification allows for the quick implementation of corrective measures, which stop problems from getting worse, and the creation of subpar components results in a more productive and economical production line by reducing material waste and streamlining manufacturing. Furthermore, real-time quality control using AI's adaptive nature adds a dynamic component to the industrial ecosystem. Over time, these systems improve their flaw-detection capacities by learning and adapting continuously. The algorithms may adjust to variations in material qualities, manufacturing settings, and ambient conditions by utilizing ML models. Because of its flexibility, the defect detection system is guaranteed to remain reliable and efficient even when production demands change. Real-time quality control and defect detection are important because they improve the final product and radically alter the connection between production efficiency and quality assurance. AI-driven real-time monitoring is more proactive than traditional quality control techniques, which frequently rely on post-production inspection. Ensuring uniform quality across product lines not only improves the reputation of producers but also lowers the probability of defective products reaching the market.

Real-time quality control and defect identification by combining AI and ML signify a significant technological advancement in polymer composite research and manufacturing. Real-time data monitoring, analysis, and action enhancement improve manufacturing process accuracy and promote a continuous improvement culture. The combination of AI and polymer composite research promises to redefine quality, efficiency, and reliability standards in creating high-performance materials as companies increasingly adopt these advanced manufacturing processes.

## 6.4 CHALLENGES AND OPPORTUNITIES

### 6.4.1 Current Challenges in Implementing ML and AI in Polymer Composite Research

There are several challenges in using AI and ML in polymer composite research. Despite the promise of these technologies to transform production, characterization, and material design processes, researchers and engineers are having difficulty incorporating ML and AI into this specialized industry [47–50]. These are a handful of the present challenges, as shown in Table 6.2.

To overcome these obstacles, scientists, business partners, and legislators must work together to promote cooperation, harmonize procedures, and provide solutions that consider the particularities of polymer composite research.

### 6.4.2 Exploring Opportunities for Further Advancements and Overcoming Existing Limitations

Researchers and practitioners should consider the following tactics (as shown in Table 6.3) to investigate prospects for further breakthroughs and overcome the current constraints of AI and ML in polymer composite research.

**TABLE 6.2**
**Current Challenges in Implementing ML and AI in Polymer Composite Research**

| Current Challenges | Details |
|---|---|
| Quality and availability of data | *Restricted datasets.* Accurate and reliable model training may be hampered by the lack or incompleteness of high-quality datasets dedicated to polymer composites.<br>*Variety in data.* Comprehensive databases that capture different compositions, structures, and fabrication methods are necessary due to the diversity of polymer composite materials and production processes. |
| Model interpretability | *Absence of a black box.* Because many ML and AI models are black boxes with intrinsic complexity, it can be difficult to understand the logic behind predictions. Understanding model decisions is essential for validation and confidence in key applications. |
| Choosing and modifying algorithms | *Selection difficulty.* Given the various tasks involved in polymer composite research, including failure prediction, material design, and optimization, selecting the right ML methods can be challenging.<br>*Adaptation.* Current algorithms may need to be modified or adjusted to address the qualities and difficulties of polymer composites. |
| Absence of standardization | Standardized protocols lack uniform procedures for gathering data, preparing it for analysis, and assessing the model, making it challenging to work together and compare the outcomes of various research projects.<br>*Benchmarking.* To advance the field of polymer composite research, benchmark datasets and tasks must be established for assessing the effectiveness of ML and AI models. |
| Resources for computation | *High requirements for computation.* Large amounts of processing power are frequently needed to train complex ML and AI models. Smaller research groups or universities with restricted access to high-performance computing capabilities may find this to be an obstacle. |
| Domain expertise | *Interdisciplinary collaboration.* Materials scientists, chemists, engineers, and data scientists must work together to apply ML and AI in polymer composite research successfully. Although challenging, bridging the gap between these fields is necessary for significant advancements. |
| Dynamic nature of materials | *Changing properties.* Aging, usage, and environmental variables can all cause a polymer composite's properties to alter over time. ML models must consider such dynamic behavior to produce precise forecasts and insights. |
| Ethical considerations | *Fairness and bias.* It is imperative to ensure that AI and ML models do not reinforce biases in training data. It is necessary to address ethical issues surrounding the responsible application of these technologies in material research. |
| Noncompliance with regulations | Without standards to guarantee the dependability and safety of goods created utilizing ML and AI, regulatory requirements are required as these technologies are used increasingly in material design and manufacture. |
| Implementation expenses and challenges | *First outlay of funds.* The initial expenditures of setting up infrastructure, obtaining data, and training models can be high. Smaller research groups or institutions could struggle to set aside funds to apply AI and ML. |

Integrating AI and ML into polymer composite research can be refined to provide more robust models, more accurate forecasts, and faster breakthroughs in material design and manufacturing processes by proactively addressing these potentials and overcoming current obstacles.

**TABLE 6.3**
**Opportunities for Further Advancements and Overcoming Existing Limitations**

| Opportunities | Details |
|---|---|
| Improved information gathering and exchange | *Chance.* Work together to produce extensive and varied datasets encompassing several facets of polymer composites, such as structure, composition, and production factors.<br>*Getting past limitations.* Promote efforts to share data, and create venues where researchers can submit, retrieve, and certify datasets. Create uniform metadata and data formats to facilitate interoperability. |
| Combining models based on physics | *Possibility.* Integrate physics-based models with ML and AI to take advantage of the advantages of both techniques. Physics-based models provide essential insight, but ML and AI can identify intricate patterns and relationships in massive datasets.<br>*Overcoming limitations.* Create hybrid models that enhance interpretability and generalization by incorporating physics-based concepts into ML techniques. |
| Techniques for explainable AI (XAI) | *Chance.* Invest in explainable AI research to improve ML models' interpretability, making it more straightforward for engineers and researchers to comprehend and believe the predictions.<br>*Overcoming limitations.* To offer insights into the decision-making process and implement and promote XAI approaches, such as feature importance analysis and model-agnostic interpretability methods. |
| Tailored ML algorithms | *Prospect.* Create ML methods specifically adapted to the difficulties encountered in polymer composite research, considering the unique properties of the materials and production techniques utilized.<br>*Getting around restrictions.* Make research investments to modify current algorithms, or create new ones that consider the complexities of polymer composites, including managing a variety of material compositions and anticipating intricate material behaviors. |
| Interdisciplinary training and collaboration | *Chance.* Promote training initiatives and interdisciplinary collaborations that unite data scientists, engineers, chemists, and materials scientists.<br>*Overcoming obstacles.* Establish forums for information sharing, cooperative projects, and cooperative efforts that foster a typical comprehension of material and data science ideas. |
| Constant model modification and updating | *Chance.* Acknowledge the dynamic character of polymer composites, and put procedures in place for ongoing model modification and adaptation to changing material characteristics.<br>*Overcoming limitations.* To ensure the long-term relevance and accuracy of ML models, develop algorithms that can dynamically learn from and adapt to changes in material behavior over time. |
| Ethical and regulatory frameworks | *Chance.* Take the initiative to address ethical issues by incorporating the values of accountability, openness, and fairness into ML and AI applications for polymer composite research.<br>*Overcoming limitations.* Work with regulatory organizations to create moral norms and guidelines for the ethical application of ML and AI in material design and manufacture. Maintain constant communication to make sure that new ethical guidelines are followed. |
| Outreach and education | *Chance.* Encourage outreach and education initiatives to spread the word about the possibilities of ML and AI in polymer composite research, and to offer tools for developing one's skills.<br>*Getting past limitations.* Work with academic institutions, business associates, and trade associations to provide courses, workshops, and other materials that help close the knowledge gap between conventional materials science and the developing data-driven methodologies. |

## 6.5 FUTURE DIRECTIONS

### 6.5.1 Future Role of ML and AI in Polymer Composite Research

The potential applications of ML and AI in polymer composite research are exciting due to the increased interdisciplinary collaboration and continuous technological breakthroughs. Here are some speculative ideas about the possible future contributions of AI and ML to this field, as shown in Table 6.4.

Although these conjectures present a hopeful outlook for the future, it is crucial to acknowledge that obstacles like moral dilemmas, legal structures, and continuous technological progress will influence the course of ML and AI in polymer composite studies. To overcome obstacles and take advantage of new opportunities, researchers, industry stakeholders, and policymakers must work together to realize these ambitious goals. It is important to remember that real improvement in these areas would need continued study, technological advancements, and cooperative efforts between the scientific and industrial groups. Scholars and professionals in the domain will probably persist in investigating inventive uses of ML and AI to tackle the distinct obstacles linked to polymer composite investigations.

## 6.6 CONCLUSION

This study has examined how machine learning (ML) and artificial intelligence (AI) are integrated into polymer composites, illuminating these technologies' revolutionary effects on material design,

**TABLE 6.4**
**The Future Role of ML and AI in Polymer Composite Research**

| Future Idea | Details |
|---|---|
| Accurate material design | Evaluating enormous datasets to determine the best mixes of polymers, reinforcements, and additives for purposes, ML and AI systems may enable accurate material design. |
| Optimizing automated manufacturing | Theoretically, manufacturing processes might be automated and improved by AI-driven optimization algorithms that optimize variables like temperature, pressure, and curing time to increase the quality of the composite. |
| Modeling predictive material performance | ML models could be developed to forecast the long-term performance of materials while considering aging, usage, and environmental aspects. |
| Real-time quality control | In real-time, AI-based vision, systems and sensors may detect flaws or irregularities in polymer composite materials on the production line. |
| Accelerating materials discovery | By evaluating enormous databases of material properties and forecasting viable candidates for experimentation, ML algorithms may help discover innovative polymer composite materials. |
| Flexible and self-repairing substances | By continuously observing and reacting to changes in material conditions, ML and AI may aid in the creation of adaptable and self-healing polymer composites. |
| Cooperative human–machine research | More researchers are working with AI systems, with ML algorithms helping scientists analyze large, complicated datasets and suggest new lines of inquiry. |
| Tailoring for particular uses | By considering performance limits and requirements, ML algorithms may make it possible to customize polymer composites for certain purposes. |
| Responsible AI and ethical aspects | As ethical issues in AI applications become increasingly important, frameworks for the ethical and responsible deployment of ML algorithms in polymer composite research are being developed. |
| International cooperation and information exchange | AI platforms will enable researchers to collaborate more globally by facilitating sharing of data, models, and insights across geographic borders. |

production methods, and performance enhancement. The studies highlighted how these technologies may be used to forecast and improve the properties of polymer composites, which can lead to more specialized and efficient solutions in various industries. There is confidence in the precision and dependability of ML and AI applications in the field thanks to the benchmarking exercises and predictive model validation in this work. The best practices and guidelines significantly help practitioners, as they deploy these technologies for long-term, financially viable developments in polymer composite research. ML and AI have created new opportunities for polymer composite innovation, allowing scientists and engineers to go beyond conventional limits in materials science. Nonetheless, acknowledging the ever-evolving character of technology, the subsequent segment delineates prospective paths for forthcoming investigation and advancement in this fascinating convergence of artificial intelligence and polymer composites.

## ACKNOWLEDGMENTS

Rajesh Mahadeva thanks Khalifa University of Science and Technology, Abu Dhabi, UAE, for providing a research facility for this research.

## CONFLICT OF INTEREST

The authors have declared that no competing interests exist.

## DATA AVAILABILITY STATEMENT

The datasets generated during and/or analyzed during the current study are available from the corresponding author on reasonable request.

## SUBMISSION DECLARATION

The authors are declaring that the chapter is not under consideration for publication elsewhere. Its publication is approved by all authors and tacitly or explicitly by the responsible authorities where the work was carried out. If accepted, it will not be published elsewhere, including electronically in the same form, in English, or in any other language, without the written consent of the copyright holder.

### Abbreviations

| | |
|---|---|
| AFRP | aramid fiber–reinforced polymers |
| AI | artificial intelligence |
| CFRP | carbon fiber–reinforced polymers |
| FRP | fiber-reinforced polymers |
| GAN | generative adversarial networks |
| GFRP | glass fiber–reinforced polymers |
| MGI | materials genome initiative |
| ML | machine learning |
| X-AI | explainable AI |

## REFERENCES

[1] S. Cassola, M. Duhovic, T. Schmidt, D. May, Machine learning for polymer composites process simulation: A review, Compos B Eng 246 (2022) 110208. https://doi.org/10.1016/j.compositesb.2022.110208.

[2] F. Kibrete, T. Trzepieciński, H.S. Gebremedhen, D.E. Woldemichael, Artif icial intelligence in predicting mechanical properties of composite materials, J Compos Sci 7 (2023). https://doi.org/10.3390/jcs7090364.

[3] B. Liu, N. Vu-Bac, X. Zhuang, X. Fu, T. Rabczuk, Stochastic full-range multiscale modeling of thermal conductivity of polymeric carbon nanotubes composites: A machine learning approach, Compos Struct 289 (2022) 115393. https://doi.org/10.1016/j.compstruct.2022.115393.

[4] H.S. Gupta, S.K. Sethi, A. Verma, Applications of coating materials: A critical overview, Mater Horizon Nat Nanomater Part F 1086 (2023) 81–109. https://doi.org/10.1007/978-981-99-3549-9_4/COVER.

[5] A. Imam, S.K. Sethi, M.M.H. Tusher, M. Afroja Khatun, H.S. Gupta, A. Verma, Application of molecular dynamics simulations in coatings and composites, In *Dynamic Mechanical and Creep-Recovery Behavior of Polymer-Based Composites* (2024) 419–449. https://doi.org/10.1016/B978-0-443-19009-4.00022-9.

[6] H.S. Gupta, P. Kumar, A.K. Sethi, A. Verma, S.K. Sethi, Software-based simulations of the creep recovery model of polymer composites, In *Dynamic Mechanical and Creep-Recovery Behavior of Polymer-Based Composites* (2024) 271–289. https://doi.org/10.1016/B978-0-443-19009-4.00016-3.

[7] N.K. Guimard, K.K. Oehlenschlaeger, J. Zhou, S. Hilf, F.G. Schmidt, C. Barner-Kowollik, Current trends in the field of self-healing materials, Macromol Chem Phys 213 (2012) 131–143. https://doi.org/10.1002/MACP.201100442.

[8] M. Ammar, A. Haleem, M. Javaid, S. Bahl, A.S. Verma, Implementing Industry 4.0 technologies in self-healing materials and digitally managing the quality of manufacturing, Mater Today Proc 52 (2022) 2285–2294. https://doi.org/10.1016/J.MATPR.2021.09.248.

[9] S.K. Sethi, G. Manik, A combined theoretical and experimental investigation on the wettability of MWCNT filled PVAc-g-PDMS easy-clean coating, Prog Org Coat 151 (2021) 106092. https://doi.org/10.1016/j.porgcoat.2020.106092.

[10] S.K. Sethi, G. Manik, Recent progress in super hydrophobic/hydrophilic self-cleaning surfaces for various industrial applications: A review, Polym Plast Technol Eng 57 (2018) 1932–1952. https://doi.org/10.1080/03602559.2018.1447128.

[11] M. Singh, S.K. Sethi, G. Manik, Pressure-sensitive adhesives based on acrylated epoxidized linseed oil: A computational approach, Int J Adhes Adhes 112 (2022) 103031. https://doi.org/10.1016/J.IJADHADH.2021.103031.

[12] S.K. Sethi, R. Gogoi, G. Manik, Plastics in self-cleaning applications, In *Reference Module in Materials Science and Materials Engineering* (2021). https://doi.org/10.1016/B978-0-12-820352-1.00113-9.

[13] S.K. Sethi, S. Kadian, G. Manik, A review of recent progress in molecular dynamics and coarse-grain simulations assisted understanding of wettability, Arch Comput Methods Eng 1 (2022) 1–27. https://doi.org/10.1007/S11831-021-09689-1.

[14] R. Gogoi, S.K. Sethi, G. Manik, Surface functionalization and CNT coating induced improved interfacial interactions of carbon fiber with polypropylene matrix: A molecular dynamics study, Appl Surf Sci 539 (2021) 148162. https://doi.org/10.1016/j.apsusc.2020.148162.

[15] A. Kataria, A. Verma, S.K. Sethi, S. Ogata, Introduction to interatomic potentials/forcefields, Lect Notes Appl Comput Mech 99 (2022) 21–49. https://doi.org/10.1007/978-981-19-3092-8_2/COVER.

[16] S.K. Sethi, R. Gogoi, A. Verma, G. Manik, How can the geometry of a rough surface affect its wettability?: A coarse-grained simulation analysis, Prog Org Coat 172 (2022) 107062. https://doi.org/10.1016/J.PORGCOAT.2022.107062.

[17] A. Dhiman, A. Gupta, S.K. Sethi, G. Manik, G. Agrawal, Encapsulation of wax in complete silica microcapsules, J Mater Res 2022 (2022) 1–14. https://doi.org/10.1557/S43578-022-00865-Y.

[18] S. Chaturvedi, A. Verma, S.K. Sethi, S. Mavinkere Rangappa, S. Siengchin, Stalk fibers (rice, wheat, barley, etc.) composites and applications, In *Plant Fibers, Their Composites, and Applications* (2022) 347–362. https://doi.org/10.1016/B978-0-12-824528-6.00009-6.

[19] V. Khandelwal, S.K. Sahoo, A. Kumar, S.K. Sethi, G. Manik, Bio-sourced electrically conductive epoxidized linseed oil based composites filled with polyaniline and carbon nanotubes, Compos B Eng 172 (2019) 76–82. https://doi.org/10.1016/j.compositesb.2019.05.050.

[20] E. Champa-Bujaico, P. García-Díaz, A.M. Díez-Pascual, Machine learning for property prediction and optimization of polymeric nanocomposites: A state-of-the-art, Int J Mol Sci 23 (2022) 1–37. https://doi.org/10.3390/ijms231810712.

[21] A. Gaspar-Cunha, J.A. Covas, J. Sikora, Optimization of polymer processing: A review (part II-molding technologies), Materials 15 (2022) 1–20. https://doi.org/10.3390/ma15031138.

[22] M. Priyadharshini, D. Balaji, V. Bhuvaneswari, L. Rajeshkumar, M.R. Sanjay, S. Siengchin, Fiber reinforced composite manufacturing with the aid of artificial intelligence: A state-of-the-art review, Arch Comput Methods Eng 29 (2022) 5511–5524. https://doi.org/10.1007/s11831-022-09775-y.

[23] Z. Fu, W. Liu, C. Huang, T. Mei, A review of performance prediction based on machine learning in materials science, Nanomaterials 12 (2022) 1–19. https://doi.org/10.3390/nano12172957.

[24] R. Mahadeva, M. Kumar, S.P. Patole, G. Manik, Desalinatio n plant performance prediction model using grey wolf optimizer based ANN approach, IEEE Access 10 (2022) 34550–34561. https://doi.org/10.1109/ACCESS.2022.3162932.

[25] R. Mahadeva, M. Kumar, G. Manik, S.P. Patole, Modeling, simulation, and optimization of the membrane performance of seawater reverse osmosis desalination plant using neural network and fuzzy based soft computing techniques, Desalin Water Treat 229 (2021) 17–30. https://doi.org/10.5004/dwt.2021.27386.

[26] C. Schoenholz, N. Zobeiry, An accelerated process optimization method to minimize deformations in composites using theory-guided probabilistic machine learning, Compos Part A Appl Sci Manuf 176 (2024) 107842. https://doi.org/10.1016/j.compositesa.2023.107842.

[27] B. Bahtiri, B. Arash, S. Scheffler, M. Jux, R. Rolfes, A machine learning-based viscoelastic–viscoplastic model for epoxy nanocomposites with moisture content, Comput Methods Appl Mech Eng 415 (2023) 116293. https://doi.org/10.1016/j.cma.2023.116293.

[28] R. Mahadeva, M. Kumar, A. Goel, S.P. Patole, G. Manik, A novel $AGPSO_3$-based ANN prediction approach: Application to the RO desalination plant, Arab J Sci Eng (2023) 1–12. https://doi.org/10.1007/s13369-023-07631-0.

[29] R. Mahadeva, G. Manik, A. Goel, N. Dhakal, A review of the artificial neural network based modelling and simulation approaches applied to optimize reverse osmosis desalination techniques, Desalin Water Treat 156 (2019) 245–256.

[30] A. Ashok, A.S. Desai, R. Mahadeva, S.P. Patole, B. Pandey, N. Bhagat, Research network analysis and machine learning modeling on heusler alloys, Eng Sci 25 (2023). https://doi.org/10.30919/es954.

[31] A. Goel, R. Mahadeva, G. Manik, Analysis and optimization of parabolic trough solar collector to improve its optical performance, J Solar Energy Eng Trans ASME 145 (2023) 031009-1-031009–12. https://doi.org/10.1115/1.4055995.

[32] R. Mahadeva, G. Manik, O.P. Verma, S. Sinha, Modelling and simulation of desalination process using artificianeural network: A review, Desalin Water Treat 122 (2018) 351–364. https://doi.org/10.5004/dwt.2018.23106.

[33] R. Mahadeva, R. Mehta, G. Manik, A. Bhattacharya, An experimental and computational investigation of poly (Piperazineamide) thin-film composite membrane for salts separation from water using artificial neural network, Desalin Water Treat 224 (2021) 106–121. https://doi.org/10.5004/dwt.2021.27184.

[34] R.S. Chhabra, R. Mahadeva, G.D. Ransinchung, Unconfined compressive strength prediction of recycled cement-treated base mixes using soft computing techniques, Road Mater Pavement Des (2023) 1–15. https://doi.org/10.1080/14680629.2023.2199889.

[35] M. Hassani Niaki, M. Pashaian, M. Ghorbanzadeh Ahangari, Experimental assessment of the efficiency of deep learning method in predicting the mechanical properties of polymer concretes and composites, J Build Eng 76 (2023) 107107. https://doi.org/10.1016/j.jobe.2023.107107.

[36] C. Qiu, Y. Han, L. Shanmugam, Y. Zhao, S. Dong, S. Du, J. Yang, A deep learning-based composite design strategy for efficient selection of material and layup sequences from a given database, Compos Sci Technol 230 (2022) 109154. https://doi.org/10.1016/j.compscitech.2021.109154.

[37] R. Mahadeva, G. Manik, O.P. Verma, A. Goel, S. Kumar, Modelling and simulation of reverse osmosis system using PSO-ANN prediction technique, In *Advances in Intelligent Systems and Computing*, Springer, Singapore 1053 (2020) 1209–1219.

[38] R. Mahadeva, M. Kumar, V. Gupta, G. Manik, S.P. Patole, Modifie d whale optimization algorithm based ANN: A novel predictive model for RO desalination plant, Scient Rep 13 (2023). https://doi.org/10.1038/s41598-023-30099-9.

[39] A.K. Maurya, R. Mahadeva, G. Manik, S.P. Patole, An investigation into performance properties of sustainable polypropylene composites reinforced with basalt fiber and fly ash, Polym Compos 44 (2023) 5104–5120. https://doi.org/10.1002/pc.27476.

[40] E. Champa-Bujaico, A.M. Díez-Pascual, A. Lomas Redondo, P. Garcia-Diaz, Optimization of mechanical properties of multiscale hybrid polymer nanocomposites: A combination of experimental and machine learning techniques, Compos Part B Eng 269 (2024) 111099. https://doi.org/10.1016/j.compositesb.2023.111099.

[41] G. Yan, J. Li, A.H. Ali, T. Alkhalifah, F. Alturise, H.E. Ali, Innovative ANN hysteresis to predict hysteretic performance of composite reinforced concrete beam, Adv Eng Softw 176 (2023) 103373. https://doi.org/10.1016/j.advengsoft.2022.103373.

[42] A. Smeraldo, A.M. Ponsiglione, P.A. Netti, E. Torino, Artificial neural network modelling hydrodenticity for optimal design by microfluidics of polymer nanoparticles to apply in magnetic resonance imaging, Acta Biomater 171 (2023) 440–450. https://doi.org/10.1016/j.actbio.2023.09.029.

[43] Z. Liang, Z. Li, S. Zhou, Y. Sun, J. Yuan, C. Zhang, Machine-learning exploration of polymer compatibility, Cell Rep Phys Sci 3 (2022) 100931. https://doi.org/10.1016/j.xcrp.2022.100931.

[44] H. Li, X. Yin, L. Sha, D. Yang, T. Hu, Data-driven prediction model for high-strength bolts in composite beams, Buildings 13 (2023) 1–22. https://doi.org/10.3390/buildings13112769.

[45] H.E. Balcıoğlu, A. Ç. Seçkin, Comparison of machine learning methods and finite element analysis on the fracture behavior of polymer composites, Arch Appl Mech 91 (2021) 223–239. https://doi.org/10.1007/s00419-020-01765-5.

[46] C. Zhao, C.C. Chung, S. Jiang, M.M. Noack, J.H. Chen, K. Manandhar, J. Lynch, H. Zhong, W. Zhu, P. Maffettone, D. Olds, M. Fukuto, I. Takeuchi, S. Ghose, T. Caswell, K.G. Yager, Y. Chen, K. Chen-Wiegart, Machine-learning for designing nanoarchitectured materials by dealloying, Commun Mater 3 (2022) 1–12. https://doi.org/10.1038/s43246-022-00303-w.

[47] J. Preethikaharshini, K. Naresh, G. Rajeshkumar, V. Arumugaprabu, M.A. Khan, K.A. Khan, Review of advanced techniques for manufacturing biocomposites: Non-destructive evaluation and artificial intelligence-assisted modeling, J Mater Sci 57 (2022) 16091–16146. https://doi.org/10.1007/s10853-022-07558-1.

[48] A. Sharma, T. Mukhopadhyay, S.M. Rangappa, S. Siengchin, V. Kushvaha, *Advances in Computational Intelligence of Polymer Composite Materials: Machine Learning Assisted Modeling, Analysis and Design*, Springer Netherlands (2022). https://doi.org/10.1007/s11831-021-09700-9.

[49] W. Choi, R.C. Advincula, H.F. Wu, Y. Jiang, Artificial intelligence and machine learning in the design and additive manufacturing of responsive composites, MRS Commun 13 (2023) 714–724. https://doi.org/10.1557/s43579-023-00473-9.

[50] A.A. Stepashkin, S. Chavhan, S. V. Gromov, A. Khanna, V. V. Tcherdyntsev, D. Gupta, H. Mohammad, E. V. Medvedeva, N. Gupta, S.S. Alexandrova, ANN-based structure peciliaties evaluation of polymer composite reinforced with unidirectional carbon fiber, Alex Eng J 82 (2023) 218–239. https://doi.org/10.1016/j.aej.2023.09.062.

# 7 Exploring Polymer Composite Behavior through Advanced Computational Modeling

*Abhinav K. Singh, Manjinder Singh, Sachin Kumar, and Gaurav Manik*

## 7.1 INTRODUCTION

In the perplexing universe of materials science, the blueprint and optimization of polymer composites have become critical for various uses, starting from advanced airplane design to biomedical gadgets [1]. As engineers and scientists collectively look forward to pushing the limits of material development and processing, utilizing progress in computational material science is a potential methodology [2]. Molecular dynamics (MD) simulations stand apart as a strong method that offers into the behavior of materials at the atomic and mesoscale levels [3].

This chapter dives into the domain of MD simulations by polymer composites, unwinding the mind-boggling transaction of polymer chains and building up specialists at nanoscales [4]. With the capacity to reproduce the movement and communications of molecules after some time, MD simulation offers a virtual lab where scientists can notice, break down, and figure out the unique way to identify the properties and behavior of polymer composites with surprising accuracy [3, 5].

This chapter starts with an explanation of the basics of MD simulations, which includes the algorithm and its internal working mechanism at the atomic or molecular level, followed by some scientifically forced fundamental protocols for using MD simulations. This is something that should be understood by beginners before starting molecular modeling and simulations. After the basic explanation of MD simulations, the chapter moves toward the fundamentals of polymer composites, their potential application for the real-world examples of interesting investigations of polymer composites, and the basic rules that oversee their mechanical, thermal, and electrical properties. Subsequently, the role of molecular modeling and simulation in the determination of various parameters which affect the critical properties of polymer composites is described. This opens the door to a comprehensive discussion on unraveling the critical aspects of polymer composites and the effects of various parameters such as strain, temperature, and synthesis on their properties using MD simulations [6].

As we explore the complexities of MD simulations, we additionally investigate the difficulties and future possibilities of this state-of-the art procedure regarding the development of polymer composites via molecular dynamics. From addressing computational bottlenecks to upgrading the prescient abilities of reproductions, we look at the developing scene of MD simulations and their significance in supporting and developing cutting-edge materials and understanding their structure–property relationship [1].

## 7.2 MOLECULAR DYNAMICS FUNDAMENTALS

In the 1960s, there was a report on the first MD simulation. However, this kind of approach took over 50 years to become a reliable, widely used, and acknowledged instrument. Over the course of several decades, scientists have created and amassed a number of superior methods to enhance

 DOI: 10.1201/9781003476337-7

molecular simulations modeling. A few of these methods are now essential building blocks for the majority of widely used modern MD software programs. This segment provides further information on the fundamentals of MD simulation, emphasizing key methods, such as neighbor lists, PBCs, and time integration algorithms.

### 7.2.1 Basics of Molecular Dynamics Simulation

MD is a computational technique used in various scientific disciplines for several property estimations of materials. It simulates the behavior of a system of interacting atoms and molecules over time, involving particle modeling, force field calculations, integrations, time stepping, and ensembles. MD provides an insight into the dynamics and structural properties of systems and helps understand the structure–property relationships of developed or to-be-developed materials.

With the help of MD simulations, we can mimic the real behavior of the atom and molecules. For instance, if we have to study the properties of any polymer, we model its structure and simulate its interaction with similar kinds of chains surrounding it via force field and the integration of Newton's equation of motion at different properties, temperatures, and pressures, then predict.

### 7.2.2 Newton's Equations of Motion

After calculation of the potential energy function, with the forces between atoms used to calculate their force, and with help of the nature of force (attractive or repulsive), the next position of the atoms is predicted with the help of **Newton's equations of motion**.

Since the polymer chain contains a group of several atoms, we have to calculate the vector sum of the forces of all directions.

$$F_r = \left[(F_x)^2 + (F_y)^2 + (F_z)^2\right]^{1/2} \tag{7.1}$$

With the help of the net force calculation, we can easily predict the next position, and the same process will repeat several times.

**MD: Basic Algorithm**

This involves the following steps:

- Total time is divided into smaller time steps not greater than a few femtoseconds ($10^{-15}$ s).
- For different time steps:
  - The forces acting on every atom are calculated using a molecular mechanics force field.
  - The atoms are moved slightly according to the applied forces. The velocity and position of each atom are updated with the help of Newton's laws of motion.

### 7.2.3 Ensemble Methods: NVT and NPT

It is clear from an earlier discussion that MD simulation is a statistical technique. In classical statistical thermodynamics, average behaviors or qualities of the particles that make up a macroscopic system are the focus of investigation, rather than the individual particles themselves.

Making a lot of copies of the system, and finding the average values of these replicates, is the standard simulation procedure. Keep in mind that the average values derived from these duplicates lose significance if they are not all precisely the same at the atomic level. The similar values of several thermodynamic parameters, such as the total number of atoms, chemical potential, total energy, pressure, temperature, and volume, must be shared by all these duplicates. A statistical set is created by gathering these independent replicates to form a statistical ensemble.

In MD simulation, four significant statistical ensembles are commonly used. The first is the classical ensemble, where all system replicates have the same and constant number of atoms ($N$), system volume ($V$), and temperature ($T$). $N$, $V$, and the total energy $E$ are held constant in the second or microcanonical ensemble. The third ensemble, known as the isothermal-isobaric ensemble, maintains constants for $N$, $T$, and $P$. The grand canonical ensemble is the last one, when $V$, $T$, and the chemical potential are maintained at constant values.

## 7.3 SOLVING POLYMER COMPOSITE CHALLENGES USING MOLECULAR DYNAMICS

Since fiber-reinforced polymer (FRP) composites are now highly demanded in many engineering domains, it is crucial to explore and document their durability in terms of performance under service circumstances. Owing to its multifield coupling circumstances and heterogeneity, the durability of FRP composites has developed into a multifaceted, interdisciplinary, and intricate scientific issue. Researchers are employing top-down experiments and bottom-up multiscale simulations to study the degradation mechanisms and durability of FRP composites, a practice made possible by advances in nanotechnology and computer capacity. Here, we will discuss about the advancements made in the use of molecular and atomic simulations, as well as nano- and micromechanics related to the estimation and understanding of the mechanical and other important properties of FRP composites. We describe the use of multiscale techniques, in particular MD simulations, in the analysis of fiber-reinforced polymer composites and sketch its future to demonstrate how micro- and nanomechanics contribute to research on the properties of FRP composites.

### 7.3.1 Fiber-Reinforced Polymer Composite

In recent decades, engineers and scientists have turned to polymer composites as metal component substitutes because of the growing need for novel and innovative materials with attributes like high strength and low weight in the automotive and aerospace industries [7]. Due to its versatility as a reinforcing fiber, CF has made it possible to achieve these rigorous requirements. CF is widely used in many technical domains because of its desirable qualities, which include high strength, low density, high modulus, etc. [7]. In fiber-reinforced polymers (FRP), the reinforcing fiber functions as a discontinuous phase, and the base polymer as a continuous phase [8]. The composites can be either thermoset- or thermoplastic-based, depending on the kind of polymer. Furthermore, composites can be classified as continuous or short fiber–reinforced based on the length of the reinforcing fiber [4]. The production procedures for thermoset and thermoplastic-based composites differ significantly since the former needs to be cured, while the latter only needs to be melted and cooled [1, 2].

Scientists are trying to explain the associated macroscopic events from a microscopic perspective as experimental technology and modeling techniques below the microscale progress. The microstructure of FRP composites has been explored and described in investigations using transmission electron microscopy (TEM) and high-resolution TEM, as well as scanning electron microscopy (SEM) [9]. Microcracks, voids in a matrix, and fiber matrix interfaces have all been observed. The intricacies of their microstructure offer a natural point of reference for computer modeling and comprehension of the material degradation process. AFM has been used to examine the mechanical characteristics and microstructure of fiber-reinforced polymer (FRP) composites, including the sample's local modulus.

X-ray computed tomography (CT) may offer the interior pore distribution and non-invasively record the material's three-dimensional structure. Furthermore, it helps produce a three-dimensional model for mechanical performance study or simulation. Thermodynamic pressure and glass transition temperature ($T_g$) and differential scanning calorimetry (DSC) have been used to determine the characteristics of epoxy [10, 11]. The molecular structure and chemical constituents of

FRP composites can be characterized and analyzed using X-ray photoelectron spectroscopy (XPS), Fourier transform infrared spectroscopy (FTIR), and Raman spectroscopy under various environmental conditions [12]. These techniques include hydrogen bond and functional group analysis. These experimental techniques enable the comprehension of microscopic mechanisms from the atomic size to the macroscale by providing reference and comparison for modeling and simulation analysis, in addition to providing direct evidence for comprehending the mechanisms at the microscopic level.

### 7.3.2 Molecular Dynamics and Simulation Details

Atom-based molecular dynamics simulations are run to use commercial programs, and one such example involves using Materials Studio (MS) to understand the interactions between the interfaces in the CF (carbon fiber)–reinforced polypropylene (PP) composite [13]. For the calculation of potential energy of any simulated system, we often use equations of the COMPASS (condensed-phase optimized molecular potentials for atomistic simulation studies) force field of LJ (Lennard–Jones) potentials. Many researchers have, in the past, employed this force field for interfacial as well as bulk property estimations. Regarding the van der Waals (vdW) and electrostatic non-bonded interactions, a cut-off distance of around 12.5 Å is normally considered. Atom-based and Ewald summation techniques were used to calculate the vdW and electrostatic interactions, respectively. Using a time step of 1 fs (femto-second; $10^{-15}$ s), the equations of motion are integrated using the velocity–Verlet technique. The Nosé–Hoover thermostat may be used to ensure temperature control and uniformity. Periodic boundary conditions (PBC) are taken into consideration for every simulation cell. The three main steps in building such a composite structure are (1) initial structure preparation, (2) their equilibration, and (3) a post-equilibration (production) MD run for the estimate of desired attributes.

For purpose of exploration of MD application to composite simulations, CNT-coated carbon fiber and simulation of its properties are described in what follows. This involves the following six simple steps [14].

**Step 1:** Firstly, the initial chain configuration and the amorphous cell are constructed at an initial density of 0.4–0.5 g/cm$^3$. After that, an appropriate force field, such as COMPASS, is utilized and applied.

**Step 2:** In this step, geometry optimization of the initial created structure is done to remove the unnecessary and residual stresses in the system so as to find the minimum energy configuration of the chains and atoms.

**Step 3:** Subsequently, NVT dynamics is performed at a temperature ($T$) of 750 K for a time period ($t$) of 200 ps (picosecond) for sudden relaxation, and to equilibrate and randomize the molecules, constructed initially.

**Step 4:** Then, NVT dynamics is performed at a lower application temperature ($T$) of 298 K for $t$ = 300 ps, approximately, for equilibrating the amorphous cell at temperature ($T$) *of* 325 K.

**Step 5:** This is followed by NPT simulations performed at the temperature ($T$) of 298 K and pressure ($P$) at 1 atm for time ($t$) of 300 ps. Reduce the free volume by uniformly compressing amorphous cells, which was initially provided in the first step, so the system represents the real-time situation by attaining the equilibrated density.

The goal of such a simulation protocol adopted previously is to prevent the systems from overloading at the initial stage of the MD simulation. The systems were then put through a production run in order to obtain a sufficient number of frames, so that an accurate estimation of the desired attributes could be obtained.

**Step 6:** The developed equilibrated systems are then used to create a single-layered system for estimating the composite's mechanical characteristics and the fiber pullout strength (illustrated by Figure 7.1b).

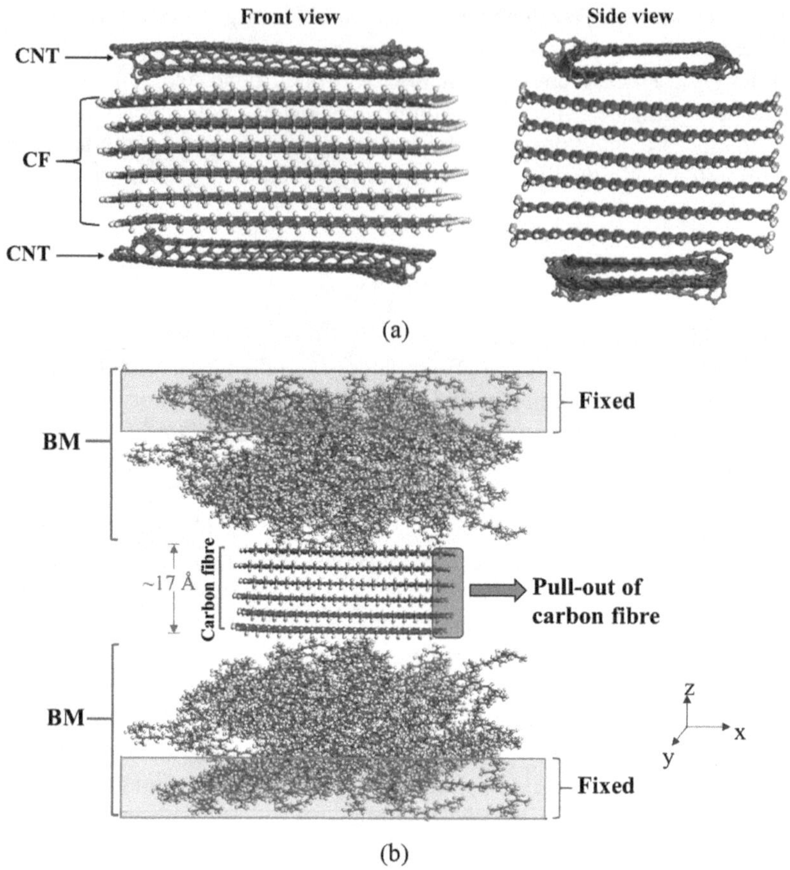


**FIGURE 7.1** Diagrams showing (a) the front and rear views of a carbon nanotube–coated carbon fiber that was first built and (b) a periodic amorphous cell that included BM/CF.

**Source: Reproduced with permission from Elsevier [14].**

### 7.3.3 Polymer Adhesive Composites

Due to their low surface free energy and the absence of polar functional groups, the majority of industrially applied polymer resins and composites have poor adhesion qualities by nature. The last ten years have seen a significant surge in research into polymer adhesion due to the increased demands of the aerospace and automotive sectors. For improved component and surface coating adherences, the current investigations into polymer adhesion have put particular emphasis on adhesion processes. An introduction to adhesion and explanations of adhesion phenomena are presented first. The adhesion processes of chemical bonding, thermodynamic adhesion, and mechanical coupling are then examined in light of recent studies. Adhesion promotion research is concentrated on chemical and plasma treatments, as well as direct adhesion measuring techniques.

Adhesion is the result of several processes. Since there are no local attractive forces, some of these adhesion effects are only secondary or perceived, yet the process nonetheless seems sticky globally. Thus, an adhesion formulation is not necessary for the local modeling of apparent adhesion processes. Global modeling is the only one that does. One can distinguish between the following mechanisms that contribute to adhesion [15].

**Chemical adhesion** results from weak hydrogen bonds or strong molecular connections, such as covalent and ionic bonds. Chemical adherence has a very narrow range—typically less than 1 nm.

**Van der Waals interaction** is the cause of dispersive adhesion. This is the result of molecules' (momentary) polarity acting in a quantum mechanical way. With a range of several nanometers, it is more extensive.

**Coulomb interaction** between oppositely charged bodies causes electrostatic attraction. Beyond the centimeter range, this contact can readily continue.

Make bond, such as a nail in a wall. Since there are just local compression and friction forces present and no local tensile interface forces, this is simply perceived adhesion.

**Diffusive adhesion** is brought on by molecular interlocking, such as that of polymer chains. Furthermore, this adherence is only visible.

In many applications, adhesion plays a significant, if not dominant, role—in debonding and delamination [16, 17], for instance, of thin films [18, 19]; peeling of adhesive strips, rubber adhesion, adhesion by capillary bridges, flow and aggregation of adhesive particles, and adhesion by rough surfaces and bonding; adhesive bonding technology; MEMS and NEMS (micro- and nano-electromechanical systems); and adhesion of insects and lizards, climbing robots, microfiber arrays, biomimetic and patterned adhesive surfaces, self-cleaning mechanisms, biofouling, membrane adhesion, adhesion of cells, and hemostatic platelet adhesion.

Pressure-sensitive adhesive (PSA) is another class of adhesives which adhere to surface with finger pressure without forming any covalent bond. MD simulations have played an important role in the past in optimizing PSA formulations, exploring new material and related applications. The development of bio-based PSAs has attracted immense interest from scientists globally owing to their renewability and sustainability [20, 21]. Singh et al. developed bio-derived components by modifying linseed oil and exploring their potential computationally to get green PSAs (Figure 7.2).

In this work, acrylated epoxidized linseed oil and acrylated epoxidized methyl ester have been obtained by modifying linseed oil, and these materials are further used as one of the components in developing bio-based PSAs. The potential of MD simulations has been explored in order to get various bulk, surface, and mechanical properties. For instance, density, solubility parameter, and

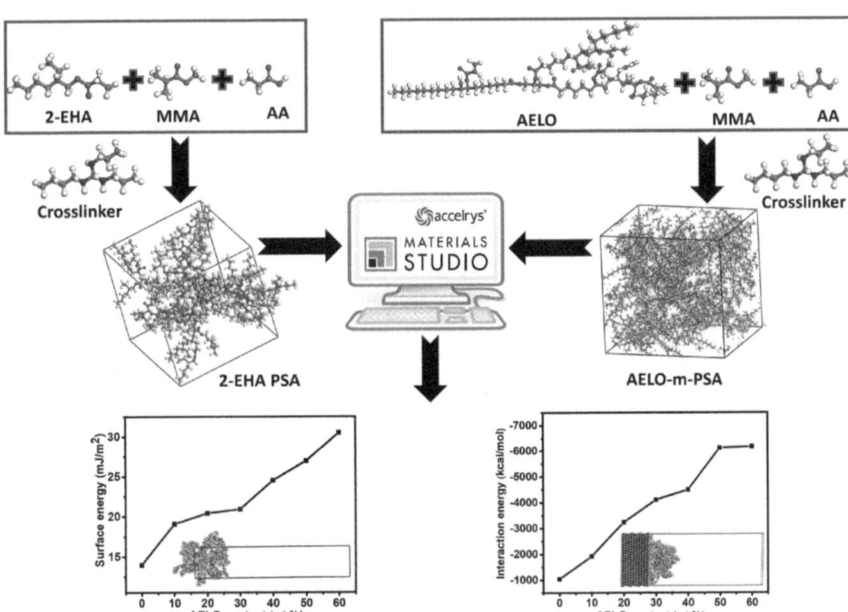


**FIGURE 7.2** Illustration of developing bio-based PSAs computationally and exploring their surface properties.

**Source: Reproduced with permission from Elsevier [22].**

glass transition temperature have been estimated and further compared with experimental values in order to validate the simulations protocol adopted for the present investigation. Further, surface properties, such as surface energy and interaction energy, have been calculated to understand the adhesion behavior of the developed PSAs, as shown in Figure 7.2. Also, shear and bulk modulus are important mechanical properties which can be calculated using MD simulations. Furthermore, MD has demonstrated great potential in optimizing the PSA formulations for desired applications [23–25]. Similarly, MD simulations have been employed likewise in fire-retardant studies to predict the sequencing of different layers by estimating interfacial adhesion between them [26].

### 7.3.4 Polymer Coating and Thin Films Composites

Composite coatings are being utilized more often to safeguard textured surfaces, apparel items, and other outside uses, like window glasses, automobiles, and so forth. New water-repelling polymer innovations are the fundamental building blocks of this advancement. Coatings had been more and more common for stylish clothes in between the 1950s and the 1960s, when polyurethanes replaced polyacrylates and polyvinyl acetate [27].

The most practical coating process should consider three key factors: (1) the type of substrate used for coating, (2) the formulation of the coating, and (3) the article's final use. This is because the interaction between the coating and the substrate determines its adherence to the substrate and, consequently, its endurance [28]. In addition, the coating's opposite face, which is exposed to the elements, promotes self-cleaning by super hydrophobic or super hydrophilic properties [29].

The hydrophobic treatment of metal or alloy surfaces has drawn a lot of interest recently from researchers who are looking to improve the materials' resistance to corrosive solutions. For example, Liu et al. [30] used a straightforward immersion approach to build a highly hydrophobic coating on a zinc substrate to form a corrosion barrier. $CF_3$ $(CF_2)_5$ $(CH_2)_2SiCl_3$, PFTS, hydrolyzed $^1H$,$^1H$,$^2H$, and $^2H$-perfluorooctyltrichlorosilane were dissolved in methanol and applied to the zinc substrate for five days. With such a modified surface, a WCA of 151° was produced. Using polypyrrole (PPy) and an ultrafast electrochemical technique, Hermelin et al. [31] covered a zinc electrode, producing a WCA of 125° in less than 3 sec by depositing 2 μm of PPy.

Similar to this, Wang et al. [32] used the silane sol-gel process to create a high-performance anticorrosion coating on a mild steel substrate. Using potentiostatic electrolysis, Wang et al. [33] created an extremely hydrophobic coating on metallic zinc. A zinc sample was utilized as the anode and a platinum wire was used as the cathode in a two-electrode cell. The treatment solution was an undiluted 0.1 M tetradecanoic acid/ethanol solution. At room temperature, a steady potential of 30 V was applied for 2 hr. The WCA was raised by this method from 62° to 152.5°. By exchanging laurate for ZnAl-LDH–$NO_3$ films via anion exchange, Zhang et al. [34] produced laureate-intercalated ZnAl layered double hydroxide (ZnAl-LDH–laurate) on a porous anodic alumina/aluminum (PAO/Al) substrate. With a WCA of 163°, the dual-scale roughness makes the material very hydrophobic, enhancing the resistance of substrate against corrosion [35].

Another recent interesting case in this direction by Sethi et al. [35] employed MD simulations to investigate poly(vinyl acetate)–poly(dimethyl siloxane)–based easy-clean coating to understand the surface behavior and substrate interaction (Figure 7.3).

### 7.3.5 Modeling and Simulation Details

Controlled use of Materials Studio 7.0 (MS) was made to run MD simulations. COMPASS was used as a force field to perform the simulations. For temperature control, Nosé–Hoover algorithm was used, whereas for the pressure control system, the Berendsen method was used. To evaluate the nonbonding interactions, nonbond cutoff distance of 15.5 Å was employed, while atom-based cutoff method was employed for van der Waals interactions. As seen in Figure 7.3, amorphous cells were created by creating chains with lengths greater than the critical chain length ($n > n_c = 20$ units), which was determined by their earlier investigation for all polymer cells. For all dynamics runs, the simulation time step was set to 1 fs, and the previously employed approach was utilized to equilibrate the resulting individual amorphous cells [36].

### 7.3.6 Polymer Membrane Composites

From molecular dynamics simulations, using water and oxygen as examples, these simple permeants (one hydrophilic and one hydrophobic) may be simulated by standard molecular dynamics throughout their entrance, translocation, and escape via membranes [3]. Permeability may be directly estimated using Fick's first law, transition rates, and a global Bayesian analysis of the inhomogeneous solubility–diffusion model [3]. Mesoscale simulations are necessary even for basic permeants, like oxygen and water. The results demonstrate the shortcomings of the homogeneous solubility diffusion model [27], validate the usefulness of compartmental and inhomogeneous solubility diffusion models, and emphasize the need for experiment comparison for both well-characterized membranes and simple solvent systems, like water/hexadecane. Included are subdiffusion, periodic boundary conditions, fractional viscosity dependence, and the most current discoveries in the area. Lastly, while enhanced diffusion treatments and sampling strategies considerably increase the possibilities of simulation-based systems, they do not immediately solve the critical requirement for multipole and polarizable force fields and constant pH techniques [37].

According to Figure 7.4, the experimental value transfer free energy ($\Delta Gw \rightarrow h$) for water to hexadecane is 5.9 kcal/mol at room temperature, which indicates hexadecane is not the suitable solvent model for the membrane's interior. Nonetheless, that deduction is predicated upon the assumption that the simulation's force field, or potential energy function, FF, produces the experimental value for the absorption of water into hexadecane. PMF of water from the water/hexadecane slab simulation must therefore be performed using the same software, strategies, and force field for the simulation of membrane to allow for an accurate correlation, as shown in a plot of Fw (z) at the bottom of Figure 7.4 and an image from one such simulation at the top. $\Delta Gw \rightarrow h = Fw(0) = 7.0 \pm 0.1$ kcal/mol is the free energy of transfer for water to hexadecane, which is 1 kcal/mol more from the experiment. Section 3 goes into additional detail about errors in the FF (as well as their fixes). The key finding here is that the simulated transfer of free energy of hexadecane (7.0 kcal/mol) and water to a POPC bilayer (6.8 kcal/mol) is almost comparable.

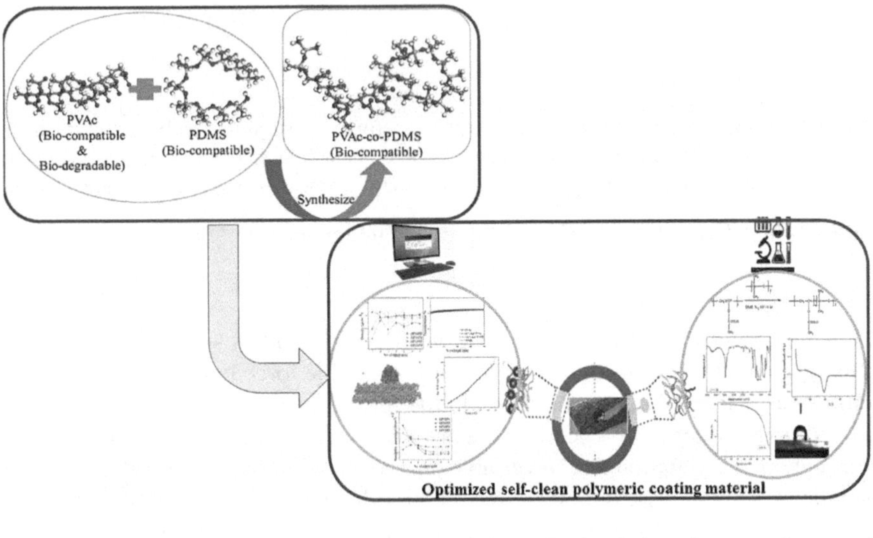


**FIGURE 7.3** Illustration of PVAc-co-PDMS developed using MD simulations for easy-clean coatings.

**Source: Reproduced with permission from Elsevier [13].**

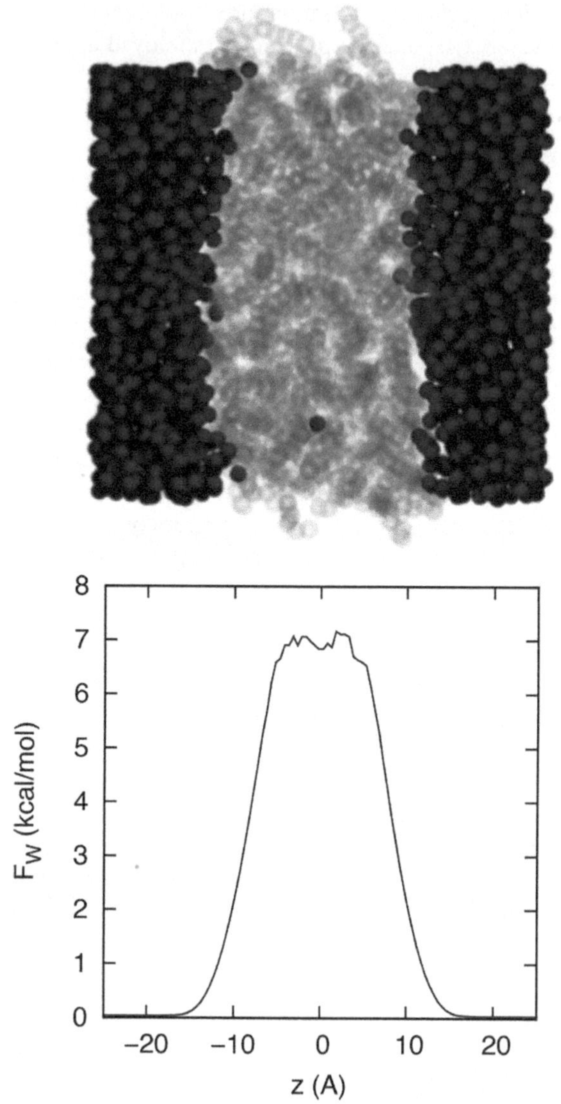


**FIGURE 7.4** Water and hexadecane (blue and gray, respectively) system for analyzing partition coefficients (top panel). One water molecule in the alkane phase is seen in this image. This particular combination is rare (roughly just 1 out of every 200 frames has water at the center of an 8 Å slab).

**Source: Reproduced with permission from ACS Publication [3].**

Since Newton's law is the basis for classical molecular dynamics simulations, the force field or potential energy function is necessary. Additive or polarizable force fields are possible. Each atom's partial atomic charge is fixed during the flight in the former case. As a result, a solute cannot adapt to various conditions (water to hydrocarbon, vacuum to water). In the latter, molecule dipoles might fluctuate throughout a simulation due to variable charge distributions. Although findings of simulations for smaller systems utilizing polarizable FF are also reported, the simulations of membrane analyzed here used additive FF. QM/MM refers to a class of hybrid approaches that use quantum mechanics to describe a portion of the system, such as an enzyme's active site or the movement of protons over a membrane [38].

Lately, researchers have got lots of help from superfast computers, better potential energy functions, and some other methodological advancements. Many tiny permeants no longer require the use of extra forces or different improved techniques to achieve membrane translocation, and their passive transport may already be directly detected in traditional MD simulations. Details from contemporary MD simulations help with experiment interpretation. For instance, Overton's rule shows remarkable accuracy across six orders of magnitude. However, because the acyl chain sections of the various bilayers are so similar, it is difficult to apply Overton's rule to explain tenfold variations in the permeability for water in those bilayers. Nagle and colleagues found a relationship between surface area per lipid and the correlation [39], and they suggested a model based on variations for the interface's structure. *Simulation* is a natural tool for refining such a model. Though the diffusion models, which begin with the Smoluchowski equation, are fundamental to permeation analysis, still it is crucial to understand that a diffusion equation can only roughly represent a solute's behavior in a homogenous molecular solvent. When considering the dynamics of a solute in a complex, diverse environment like a membrane, this approximation loses some of its validity. The results may be compared with that anticipated model to ascertain the dynamical model applications, including those that go beyond diffusion.

## 7.4 SUMMARY AND FUTURE SCOPE

In this book chapter, we have discussed the fundamentals of molecular dynamics and simulations. We have further elaborated how it works and how useful it is for polymer composite and allied fields since it mimics real-world problems and is easy to get results that are approximately 95% accurate, without preparing the real-time product or samples from synthetic processes. Also, it helps significantly save time and manpower. MD simulation is a very strong tool that can be employed in various polymer composite applications, like understanding the interaction of the polymer matrix with the fibers and the polymer composite, and the effect of reinforcement on the thermal and mechanical properties of the system.

In coatings and adhesive industries, it plays a crucial role in predicting their fundamental properties, like the interaction behavior of different polymers with different kinds of substrates, through the calculation of their interaction energy values. Similarly, we can also calculate the hydrophobicity extent of any coating by calculating the interaction energy of respective polymers with water and estimating the water content analysis. Nowadays, there are so many commercial software which can create the visual effect of these experiments close to real-world behavior. Similarly, we can use MD simulations in the study of cell membranes to understand the absorption or desorption of water molecules or ions and estimate their diffusion-related properties.

## ACKNOWLEDGMENTS

It is indeed a great pleasure and moment of immense satisfaction and opportunity for me to express my sense of deep gratitude toward my guide, Prof. Gaurav Manik, for his keen interest, valuable suggestions, and constant encouragement throughout the tenure of this work. I am indebted to him for his valuable guidance and untiring attention bestowed upon me right from the inception to the successful completion of this endeavor. I would like to express my sincere thanks to all faculty members, for guiding me in every inch of my research work, and the entire laboratory staff of the Polymer and Surface Engineering Department, for every possible help rendered by them.

## REFERENCES

[1] P. Van Der Ploeg and H. J. C. Berendsen, "Molecular dynamics of a bilayer membrane," *Mol. Phys.*, vol. 49, no. 1, pp. 233–248, 1983, https://doi.org/10.1080/00268978300101131.

[2] K. Lin and Z. Wang, "Multiscale mechanics and molecular dynamics simulations of the durability of fiber-reinforced polymer composites," *Commun. Mater.*, vol. 4, no. 1, pp. 1–16, 2023, https://doi.org/10.1038/s43246-023-00391-2.

[3] R. M. Venable, A. Krämer, and R. W. Pastor, "Molecular dynamics simulations of membrane permeability," *Chem. Rev.*, vol. 119, no. 9, pp. 5954–5997, 2019, https://doi.org/10.1021/acs.chemrev.8b00486.
[4] I. Volume, L. Sirovich, and J. E. Marsden, *Interdisciplinary Applied Mathematics Editors*, Vol. 8, 2009. [Online]. Available: http://www.springerlink.com/index/10.1007/978-0-387-75847-3.
[5] R. B. Martin, "Corrections 5709 3316 ± 450," p. 1964, 1964.
[6] D. Lau, O. Büyüköztürk, and M. J. Buehler, "Multiscale modeling of organic-inorganic interface: From molecular dynamics simulation to finite element modeling," *Mater. Res. Soc. Symp. Proc.*, vol. 1466, no. 1, pp. 44–49, Sep. 2012, https://doi.org/10.1557/OPL.2012.1427/METRICS.
[7] A. Hancock and L. Lin, "Challenges of UV curable ink-jet printing inks: A formulator's perspective," *Pigment. Resin. Technol.*, vol. 33, no. 5, pp. 280–286, 2004, https://doi.org/10.1108/03699420410560470.
[8] R. Gogoi, G. Manik, and B. Arun, "High specific strength hybrid polypropylene composites using carbon fibre and hollow glass microspheres: Development, characterization and comparison with empirical models," *Compos. Part B Eng.*, vol. 173, no. March, p. 106875, 2019, https://doi.org/10.1016/j.compositesb.2019.05.086.
[9] F. Guo, S. Al-Saadi, R. K. Singh Raman, and X. L. Zhao, "Durability of fiber reinforced polymer (FRP) in simulated seawater sea sand concrete (SWSSC) environment," *Corros. Sci.*, vol. 141, pp. 1–13, Aug. 2018, https://doi.org/10.1016/J.CORSCI.2018.06.022.
[10] B. Benmokrane, F. Elgabbas, E. A. Ahmed, and P. Cousin, "Characterization and comparative durability study of glass/vinylester, basalt/vinylester, and basalt/epoxy FRP bars," *J. Compos. Constr.*, vol. 19, no. 6, Dec. 2015, https://doi.org/10.1061/(ASCE)CC.1943–5614.0000564.
[11] ScienceDirect, "Purchase Laboratory assessment and durability performance of vinyl-ester, polyester, and epoxy glass-FRP bars for concrete structures." Accessed: Dec. 25, 2023. [Online]. Available: https://www.sciencedirect.com/getaccess/pii/S1359836816325045/purchase.
[12] A. H. I. Mourad, A. M. Beckry Mohamed, and T. El-Maaddawy, "Effect of seawater and warm environment on glass/epoxy and glass/polyurethane composites," *Appl. Compos. Mater.*, vol. 17, no. 5, pp. 557–573, Oct. 2010, https://doi.org/10.1007/S10443-010-9143-1.
[13] S. K. Sethi, U. Shankar, and G. Manik, "Fabrication and characterization of non-fluoro based transparent easy-clean coating formulations optimized from molecular dynamics simulation," *Prog. Org. Coat.*, vol. 136, no. June, p. 105306, 2019, https://doi.org/10.1016/j.porgcoat.2019.105306.
[14] R. Gogoi, S. K. Sethi, and G. Manik, "Surface functionalization and CNT coating induced improved interfacial interactions of carbon fiber with polypropylene matrix: A molecular dynamics study," *Appl. Surf. Sci.*, vol. 539, no. October 2020, p. 148162, 2021, https://doi.org/10.1016/j.apsusc.2020.148162.
[15] L. Afferrante, G. Carbone, G. Demelio, and N. Pugno, "Adhesion of elastic thin films: Double peeling of tapes versus axisymmetric peeling of membranes," *Tribol. Lett.*, vol. 52, no. 3, pp. 439–447, Dec. 2013, https://doi.org/10.1007/S11249-013-0227-6.
[16] O. Allix and P. Ladev~ze, "Interlaminar interface modelling for the prediction of delamination," *Compos. Struct.*, vol. 22, pp. 235–242, 1992.
[17] N. Point and E. Sacco, "A delamination model for laminated composites," *Int. J. Solids Struct.*, vol. 33, no. 4, pp. 483–509, 1996.
[18] J. M. Hendrickx, P. H. Geubelle, and N. R. Sottos, "A spectral scheme for the simulation of dynamic mode 3 delamination of thin films," *Eng. Fract. Mech.*, vol. 72, no. 12 spec. iss., pp. 1866–1891, 2005, https://doi.org/10.1016/J.ENGFRACMECH.2004.10.014.
[19] S. Roy, E. Darque-Ceretti, E. Felder, and H. Monchoix, "Cross-sectional nanoindentation for copper adhesion characterization in blanket and patterned interconnect structures: Experiments and three-dimensional FEM modeling," *Int. J. Fract.*, vol. 144, no. 1, pp. 21–33, Mar. 2007, https://doi.org/10.1007/S10704-007-9072-7.
[20] M. Singh, S. Kadian, and G. Manik, "Polymers in adhesive applications," *Encycl. Mater. Plast. Polym.*, vols. 1–4, pp. 370–381, Jan. 2022, https://doi.org/10.1016/B978-0-12-820352-1.00124-3.
[21] M. Singh and G. Manik, "Bio-based adhesives from plant oils," *Encycl. Green Mater.*, pp. 1–12, 2022, https://doi.org/10.1007/978-981-16-4921-9_128-1.
[22] M. Singh, S. K. Sethi, and G. Manik, "Pressure-sensitive adhesives based on acrylated epoxidized linseed oil: A computational approach," *Int. J. Adhes. Adhes.*, vol. 112, 2022, https://doi.org/10.1016/j.ijadhadh.2021.103031.
[23] M. Singh, S. K. Sethi, and G. Manik, "Pressure-sensitive adhesives based on acrylated epoxidized linseed oil: A computational approach," *Int. J. Adhes. Adhes.*, vol. 112, p. 103031, Jan. 2022, https://doi.org/10.1016/J.IJADHADH.2021.103031.

[24] M. Singh and G. Manik, "Computationally developed acrylated epoxidized methyl ester based pressure-sensitive adhesives," *Comput. Mater. Sci.*, vol. 228, p. 112329, Sep. 2023, https://doi.org/10.1016/J.COMMATSCI.2023.112329.

[25] M. Singh, S. K. Sahoo, and G. Manik, "Combined computational and experimental approach for bio-sourced monomers to design green pressure-sensitive adhesives," *Mol. Syst. Des. Eng.*, vol. 8, no. 12, pp. 1498–1511, Nov. 2023, https://doi.org/10.1039/D3ME00080J.

[26] A. Vishwakarma, M. Singh, B. Weclawski, V.J. Reddy, B.K. Kandola, G. Manik, A. Dasari, and S. Chattopadhyay, "Construction of hydrophobic fire retardant coating on cotton fabric using a layer-by-layer spray coating method," *Int. J. Biol. Macromol.*, vol. 223, pp. 1653–1666, 2022, https://doi.org/10.1016/j.ijbiomac.2022.10.231.

[27] A. K. Maurya, R. Gogoi, S. K. Sethi, and G. Manik, "A combined theoretical and experimental investigation of the valorization of mechanical and thermal properties of the fly ash-reinforced polypropylene hybrid composites," *J. Mater. Sci.*, vol. 56, no. 30, pp. 16976–16998, 2021, https://doi.org/10.1007/s10853-021-06383-2.

[28] S. K. Sethi and G. Manik, "Recent progress in super hydrophobic/hydrophilic self-cleaning surfaces for various industrial applications: A review," *Polym. Plast. Technol. Eng.*, vol. 57, no. 18, pp. 1932–1952, 2018, https://doi.org/10.1080/03602559.2018.1447128.

[29] P. Krajnc, D. Stefanec, J. F. Brown, and N. R. Cameron, "Aryl acrylate based high-internal-phase emulsions as precursors for reactive monolithic polymer supports," *J. Polym. Sci. Part A Polym. Chem.*, vol. 43, no. 2, pp. 296–303, 2005, https://doi.org/10.1002/pola.20501.

[30] K. Tanaka, K. Ohno, and T. Katayama, "Effects of PP modification and processing time on fiber/matrix interfacial strength for carbon fiber reinforced polypropylene," *WIT Trans. Built Environ.*, vol. 166, pp. 329–334, Nov. 2016, https://doi.org/10.2495/HPSM160301.

[31] Y. Arao, S. Yumitori, H. Suzuki, T. Tanaka, K. Tanaka, and T. Katayama, "Mechanical properties of injection-molded carbon fiber/polypropylene composites hybridized with nanofillers," 2013, https://doi.org/10.1016/j.compositesa.2013.08.002.

[32] H. Qian, A. Bismarck, E. S. Greenhalgh, and M. S. P. Shaffer, "Carbon nanotube grafted carbon fibres: A study of wetting and fibre fragmentation," https://doi.org/10.1016/j.compositesa.2010.04.004.

[33] P. Lv, Y.-Y. Feng, P. Zhang, H.-M. Chen, N. Zhao, and W. Feng, "Increasing the interfacial strength in carbon fiber/epoxy composites by controlling the orientation and length of carbon nanotubes grown on the fibers," 2011, https://doi.org/10.1016/j.carbon.2011.06.064.

[34] "Understanding molecular simulation," 2023, https://doi.org/10.1016/c2009-0-63921-0.

[35] S. K. Sethi, L. Soni, U. Shankar, R. P. Chauhan, and G. Manik, "A molecular dynamics simulation study to investigate poly(vinyl acetate)-poly(dimethyl siloxane) based easy-clean coating: An insight into the surface behavior and substrate interaction," *J. Mol. Struct.*, vol. 1202, p. 127342, 2020, https://doi.org/10.1016/j.molstruc.2019.127342.

[36] S. K. Sethi, L. Soni, and G. Manik, "Component compatibility study of poly(dimethyl siloxane) with poly(vinyl acetate) of varying hydrolysis content: An atomistic and mesoscale simulation approach," *J. Mol. Liq.*, vol. 272, pp. 73–83, 2018, https://doi.org/10.1016/j.molliq.2018.09.048.

[37] M. Singh, S. K. Sahoo, and G. Manik, "Combined computational and experimental approach for bio-sourced monomers to design green pressure-sensitive adhesives," *Mol. Syst. Des. Eng.*, 2023, https://doi.org/10.1039/d3me00080j.

[38] "Semiempirical quantum mechanical methods for noncovalent interactions for chemical and biochemical applications." Accessed: Dec. 25, 2023. [Online]. Available: https://pubs.acs.org/doi/epdf/10.1021/acs.chemrev.5b00584.

[39] J. C. Mathai, S. Tristram-Nagle, J. F. Nagle, and M. L. Zeidel, "Structural determinants of water permeability through the lipid membrane," *J. Gen. Physiol.*, vol. 131, no. 1, pp. 69–76, Jan. 2008, https://doi.org/10.1085/JGP.200709848.

# 8 Multiscale Modeling of Polymer Composites

*Sushanta K. Sethi*

## 8.1 INTRODUCTION

Polymer composites represent a class of materials that have garnered widespread attention and application across diverse industries due to their unique combination of polymers with reinforcing materials [1–7]. These composites offer a versatile platform, allowing engineers and scientists to tailor their properties for specific applications, thereby outperforming traditional materials. The remarkable success of polymer composites lies in their ability to synergistically combine the advantageous properties of different materials, resulting in enhanced strength, stiffness, and durability.

However, the behavior of polymer composites is inherently complex, characterized by a multitude of factors acting at various length scales. The performance of these materials is influenced not only by the macroscopic arrangement of components but also by interactions occurring at microscopic and nanoscopic levels. Understanding and predicting the intricate interplay of these factors are essential for optimizing composite performance, designing new materials, and ensuring reliability in real-world applications [8–11].

Utilizing multiscale modeling is a sophisticated computational approach that transcends the limitations of traditional single-scale models. Multiscale modeling recognizes that the behavior of polymer composites cannot be accurately captured by focusing solely on one length scale [12]. Instead, it integrates information from atomistic, microscopic, and macroscopic scales to provide a comprehensive understanding of the material's response under different conditions. This holistic modeling approach enables researchers to investigate into the nanoscale interactions between individual atoms, understand the mesoscale behavior of constituents, and predict the macroscale performance of the composite material.

Considering the intricate nature of polymer composites and the pivotal role of multiscale modeling, this chapter is driven by two overarching objectives. The primary aim is to articulate and elucidate the specific goals and scope associated with employing multiscale modeling techniques in the domain of polymer composites. By investigating into the intricacies of different length scales, the chapter seeks to establish a basic understanding of the complexities inherent in these materials. Through this exploration, readers will gain insights into the relevance requisite of the multiscale modeling tool for unraveling the mysteries of polymer composites. Secondly, this chapter aims to highlight the deep significance of multiscale modeling in advancing our comprehension of polymer composites. By seamlessly integrating information across scales, multiscale modeling not only enhances predictive accuracy but also opens avenues for innovative material design and optimization. Through a detailed discussion of the implications and potential applications, the chapter aims to promote an appreciation for the transformative impact that multiscale modeling can have on the field of polymer composites. Ultimately, it helps spark future research and inspires scientists to explore new areas in material science and engineering.

## 8.2 FUNDAMENTALS OF POLYMER COMPOSITES

Polymer composites, a key element in material science, bring together polymers and reinforcing materials to create versatile materials with enhanced properties [13, 14]. This section delves into


DOI: 10.1201/9781003476337-8

the fundamentals, starting with a comprehensive exploration of the definition and types of polymer composites.

### 8.2.1 Basic Overview and Type of Composite Materials

*Composite materials* are engineered combinations of two or more distinct materials with the aim of achieving unique and enhanced properties. In the context of polymer composites, polymers serve as the base matrix, while reinforcing materials, such as fibers or particles, are strategically incorporated to enhance strength. This combination imparts tailored characteristics that go beyond the individual properties of each component.

Polymer composites exhibit diverse characteristics based on the type of polymers and reinforcing materials used. They can be classified into various categories, including fiber-reinforced composites, particle-reinforced composites, and structural composites. Each classification offers distinct advantages, making polymer composites adaptable to a wide range of applications, from aerospace to automotive industries.

### 8.2.2 Properties of Polymer Composites

Understanding the properties of polymer composites is necessary in optimizing their performance for specific applications. Mechanical properties, such as strength and stiffness, are significantly improved through the incorporation of reinforcing materials, such as ZnO [15–17], CNT [18–21], graphene [22–24], carboxymethyl cellulose [6, 25] (CMC) [6, 25], etc. Thermal properties [26, 27], including conductivity and resistance to temperature variations, play a crucial role in applications exposed to varying environmental conditions. Additionally, electrical properties, such as conductivity or insulating capabilities, are tailored to meet the requirements of specific applications. The role of reinforcement in polymer composites is pivotal. Reinforcing materials, commonly in the form of fibers or particles, contribute to the enhancement of mechanical, thermal, and electrical properties. By strategically placing these reinforcements within the polymer matrix, the composite material gains increased strength, improved heat resistance, and better electrical conductivity. The interaction between the polymer matrix and reinforcing materials is finely tuned to achieve a synergistic effect, resulting in a composite material with superior performance compared to its individual components.

## 8.3 OVERVIEW OF MULTISCALE MODELING

Computational techniques play a crucial role in unravelling the fundamental behavior of polymer composites [1, 8, 28–30], contributing significantly to the optimization of their properties. These techniques help researchers simulate and analyze the intricate interactions between the polymer matrix and reinforcing materials, such as ZnO [31, 32], CNT [4, 28], graphene, and CMC. Through computational modeling, scientists can explore the impact of various factors on mechanical, thermal, and electrical properties, providing valuable insights into the performance of the composite materials. This in-depth understanding aids in the design of advanced composites with tailored properties, ensuring optimal performance in specific applications. Computational approaches also facilitate the prediction of material responses under different environmental conditions, guiding the development of robust and versatile polymer composites. In essence, the importance of computational techniques lies in their ability to accelerate the understanding and optimization of polymer composites, fostering innovation in materials science and engineering.

### 8.3.1 Definition and Scope

*Multiscale modeling*, in the context of polymer composites, refers to a sophisticated approach that considers various length and timescales inherent in these complex materials (illustrated in

Figure 8.1). Polymer composites exhibit hierarchical structures, where macroscopic properties are influenced by the arrangement and behavior of constituents at smaller scales. Understanding multiscale modeling involves capturing and predicting the interactions and phenomena occurring at different levels, from the molecular to the macroscopic scale. This approach is essential for gaining insights into the fundamental behavior of polymer composites, as it allows researchers to bridge the gap between the microscopic features of individual components and the overall performance of the composite material.

The importance of multiscale modeling becomes evident in its ability to unravel the intricacies of hierarchical structures within polymer composites. By simulating the behavior of individual polymer chains, fillers, and other components, researchers can predict how these interactions contribute to the mechanical, thermal, and electrical properties of the composite material. This comprehensive understanding is crucial for tailoring the properties of polymer composites to meet specific application requirements. Multiscale modeling provides a holistic view, considering both the micro and macro aspects, which is essential for optimizing the design and performance of advanced composite materials.

### 8.3.2 Approaches to Multiscale Modeling

Two primary approaches to multiscale modeling are the bottom-up and top-down methods. The bottom-up approach involves starting from the smallest scale, such as MD simulations, and gradually building up to larger scales, incorporating the interactions and behaviors observed at each level. This approach is well-suited for understanding the fundamental principles governing the material at the atomic and molecular levels. On the other hand, the top-down approach begins with macroscopic properties and breaks them down into smaller scales. Kumar et al. [33] investigated CNT-reinforced polymer nanocomposites as alternatives to conventional materials, emphasizing stringent

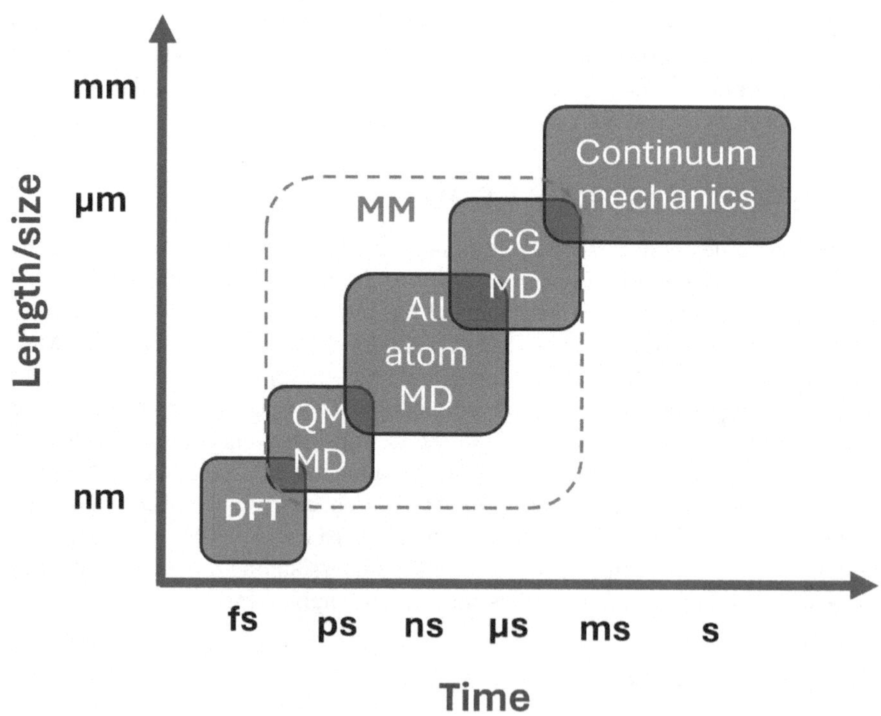


**FIGURE 8.1** Typical length and timescales utilized in simulations of polymer composites.

requirements. Their review explored the modeling techniques for understanding the mechanical behavior of CNTs, including continuum, quasicontinuum, and MD simulations. It advocates a multiscale modeling approach integrating various scales hierarchically or concurrently. The benefit lies in a more accurate and comprehensive understanding of mechanical properties, allowing a realistic representation of material behavior. This approach enhances insights into the average macroscopic response of polymer composites. This review article addressed current status, challenges, and future outlook, highlighting the potential of multiscale modeling in advancing the prediction of mechanical properties in advanced nanocomposites. This approach is effective in predicting the overall behavior of the material based on its macroscopic characteristics.

#### 8.3.2.1 Bottom-Up Approach

Vegt et al. [34] employed a bottom-up simulation approach utilizing a variant of the Markovian Mori–Zwanzig (MZ) dissipative particle dynamics (DPD) coarse-grain (CG) method. Their study aimed to enhance the accuracy of dynamic property predictions in molecular CG systems. Despite incomplete timescale separation, the MZ-DPD approach achieved quantitative accuracy in predicting diffusive dynamics in single-component liquids and polymer solutions. However, in simulating molecular penetrant diffusion in dense polymer networks, the agreement with the fine-grained model was not quantitative. The use of the MZ-DPD thermostat was noted by the authors to extend the applicability of molecular CG models, particularly in multicomponent systems where an accurate description of relative diffusion rates is crucial. The bottom-up simulation approach, guided by MZ-DPD, offered insights into improving the representation of dynamic properties in molecular CG systems, enhancing their applicability in diverse scenarios. In the same line, Canchaya et al. [35] presented a novel bottom-up method for optimizing dissipative CG models employing Bayesian optimization based on a CG reference trajectory from higher-resolution MD. Unlike force matching, this approach used averaged total forces, offering flexibility beyond pairwise interactions and estimating the friction parameter. Validated on a known system and tested on n-pentane, the method demonstrates excellent agreement in local molecular structure and equilibrium density. While dynamic properties were satisfactory, sensitivity to CG representation highlights its importance. The method emerges as a robust alternative to existing approaches, such as iterative Boltzmann inversion and Green–Kubo formulae, showcasing promise for optimizing dissipative CG models. Korvink et al. [36] introduced a CG approach for a 2D microscopic model of graphene, representing it in terms of "blobs" characterized by center-of-mass variables. The equations of motion for the CG variables adopt a DPD framework. Both the CG conservative forces and the friction in the DPD model are determined through a bottom-up procedure derived from MD simulations. The separation of timescales for blobs containing 24 and 96 carbon atoms allows the Markovian assumption in the DPD model to yield satisfactory results. Specifically, the MD velocity autocorrelation function of the blobs is well replicated by the DPD model when considering the effects of friction and noise. However, discrepancies in DPD cross-correlations between neighboring blobs were observed compared to MD results. Their study briefly outlines potential extensions to address the discrepancies, presenting insights into refining the CG model for improved accuracy in capturing dynamic correlations between adjacent blobs. Fang et al. [37] employed the generalized DPD equation, derived from the generalized Langevin equation, for simulating CG water cells. Mean force and friction coefficients for the DPD equation are directly obtained from MD simulations. However, Markovian approximations overestimate dissipative friction forces, impacting dynamic properties in CG water DPD simulations. To address this, they introduced a rescaling factor to the friction coefficients, accounting for non-Markovian dynamics and determined by matching water diffusivity. This semi-bottom-up mapping method ensures agreement between DPD-simulated and MD results for the radial distribution function, diffusion constant, and viscosity, bridging microscopic and mesoscopic levels with consistent length and timescales. Rousseau et al. [38] established literature procedures to construct CG conservative and friction forces for DPD simulations. Utilizing a bottom-up approach with large MD trajectories of n-pentane and n-decane, the CG model aims to semi-quantitatively reproduce structural and dynamical properties. They examined two CG levels (five and ten carbon atoms per DPD bead) and found an accurate reproduction of self-diffusion and viscosity coefficients for n-pentane, while the fully CG model for *n*-decane overestimates diffusion. However, when the n-pentane model was used for larger molecules (e.g., *n*-decane as a two-blob model), improved agreement with experimental data was achieved, indicating the transferability of the constructed force field to large macromolecular systems.

The bottom-up simulation approach has proven highly valuable in various studies, such as enhancing the accuracy of dynamic property predictions in molecular CG systems, optimizing dissipative CG models with Bayesian optimization, and constructing CG conservative and friction forces for DPD simulations. These approaches have demonstrated successful results in reproducing complex behaviors at different simulation length scales, showcasing the versatility and effectiveness of the bottom-up methodology in improving the representation of dynamic properties and optimizing models for diverse scenarios.

#### 8.3.2.2 Top-Down Approach

The top-down simulation approach involves starting with a detailed, high-resolution model and then simplifying or aggregating it to study larger systems or longer timescales. This method aims to capture macroscopic behaviors by deriving them from the detailed dynamics of the underlying microscale or mesoscale components. The advantages of the top-down simulation approach include efficiency in simulating larger systems, cost-effectiveness by avoiding excessive computational requirements, and the ability to study collective behaviors and emergent phenomena. Fundamentally, top-down simulations bridge the gap between the detailed understanding provided by microscopic models and the computational efficiency needed to simulate complex, real-world systems. Daoulas et al. [39] presented a review article focusing on a specialized class of hybrid models for polymeric materials. These particle-based models operate at the mesoscopic level, employing CG force fields derived from functionals of local-order parameters. These functionals, resembling those in classical DFT and phase-field simulations, enable the systematic and modular development of top-down, particle-based models with desired thermodynamic behavior. The unique structure of hybrid models facilitates efficient simulations using standard tools like Monte Carlo sampling and dynamical algorithms. Additionally, these models serve as a foundation for field-theoretical methods, including numerical self-consistent field theory, providing a comprehensive framework for studying polymeric materials.

Santiso et al. [40] utilized the top-down simulation approach and modeled polyester-based urethane thermosets for coating applications through MD simulations. Parameterizing force field interactions with a SAFT-$\gamma$ Mie approach, they developed effective models, implementing a theoretical model for experimental kinetic data. The simulations successfully captured cure kinetics, providing insights into molecular-level connections during cross-linking and yielding accurate predictions of glass transition temperatures ($T_g$) and thermomechanical properties that aligned well with experimental data. This study demonstrated the efficacy of the top-down simulation approach in accurately modeling and understanding the behavior of thermoset polymers. Li et al. [41] introduced an innovative artificial muscle made from polymeric fibers using a cost-effective "twist-insertion" procedure. A top-down multiscale modeling approach, from macroscale helical spring analysis to molecular-level chain interactions, showed excellent agreement with experimental results. Their investigation discussed into the influence of micro-, meso-, and macroscale parameters on actuation responses, providing insights for the optimal design of these advanced artificial muscles. Shahzadi et al. [42] employed a comprehensive stochastic multiscale modeling technique devised for predicting Young's modulus in nanoclay-reinforced polymers. By utilizing a top-down scanning approach, effective parameters at each scale were identified and categorized by the authors, covering nano-, micro-, meso-, and macroscales in a bottom-up modeling strategy. The modeling was conducted sequentially, with outputs from each scale serving as inputs for the next. Stochastic modeling captures process-induced uncertainties, considering random parameters, such as nanoclay morphologies, platelet numbers, inclusion size and location, spatial orientations, and non-uniform dispersion. The estimated results demonstrate strong agreement with experimental data from the literature.

Computational techniques and methods play a pivotal role in implementing multiscale modeling. MD simulations, finite element analysis (FEA), and continuum mechanics are commonly employed to capture the interactions at different scales. MD simulations provide insights into the atomic-level behavior of polymer chains and fillers, while FEA helps predict macroscopic properties by discretizing the material into finite elements (FE). Continuum mechanics bridges the gap between these scales, allowing for the integration of information obtained from various levels.

## 8.4 MICROSCALE MODELING

Microscale modeling is a crucial aspect of materials science and engineering that investigates into the intricate details of materials at the molecular or microstructural level. This level of analysis is particularly valuable for understanding complex systems like polymer composites, where the behavior of individual elements significantly influences the overall properties of the material.

### 8.4.1 Molecular Dynamics (MD) Simulation

MD simulation is a computational technique that allows researchers to study the movement and interactions of atoms and molecules over time. In the domain of polymer composites, MD simulation serves as a powerful tool to investigate the fundamental principles governing the behavior of polymers at the molecular level. Especially in the field of polymer composites, MD simulation is instrumental in unraveling the mysteries of composite materials. It facilitates a detailed examination of the interactions between the polymer matrix and reinforcing elements, such as nanoparticles or fibers. By simulating the motion and behavior of individual atoms and molecules, researchers can predict mechanical properties, thermal responses, and even chemical reactions within the composite material. Sethi et al. [10] utilized MD simulations to investigate the miscibility of polymer blends, specifically hydrophobic poly(dimethyl siloxane) (PDMS) with varying hydrolyzed grades of poly(vinyl acetate) (PVAc). Through MD simulations, their research estimated crucial parameters, like the Flory–Huggins interaction parameter ($\chi$) and $Tg$. The analysis revealed phase separation in certain blend systems and identifies factors affecting blend compatibility, such as hydrolysis degree, chain length, and PVAc content in PDMS. Notably, simulations indicate limited miscibility in short-chain systems and complete immiscibility in larger chain lengths. The research concludes by proposing optimal formulations, highlighting the pivotal role of MD simulations in understanding and optimizing polymer composite materials. Later, the same authors [43] investigated the challenge of poor substrate adhesion in hydrophobic coatings, proposing blends of PDMS with PVAc to enhance both hydrophobicity and adhesion. MD simulations models of such blends, revealing improved surface behavior, substrate adhesion, thermal stability, and transparency have been investigated. Their work suggested a 20 wt% PVAc (0% hydrolysis) in PDMS as a potential easy-clean coating material, showcasing the utility of simulations in predicting and optimizing coating properties. Gogoi et al. [29] investigated the interfacial interactions between carbon fiber (CF) and polypropylene (PP) matrix through MD simulations, focusing on pullout simulations. CF surface functionalization with –OH, –COOH, and $NH_2$ groups, along with the presence of CNTs, was explored. The composite with 6 wt% –$NH_2$-grafted CF exhibits a significant ~126% increase in interfacial shear strength (IFSS), while the one with CNT-coated CF and 6 wt% –$NH_2$-functionalized CF shows a ~112% increment. Both grafting and CNT coating contribute to improved Young's and shear modulus in the composites, highlighting potential enhancements in mechanical properties for CF-reinforced polypropylene composites. This level of granularity is particularly beneficial for optimizing composite properties, like strength, stiffness, and durability. Understanding the dispersion of reinforcing elements within the polymer matrix is crucial for tailoring composite materials for specific applications, such as aerospace, automotive, or biomedical engineering. MD simulations provide valuable insights into the structural aspects of polymers, aiding in the development of advanced materials with enhanced performance.

However, MD simulation of polymers is not without its challenges. One significant hurdle is the computational cost associated with modeling large polymer systems over extended periods. Despite improvements in algorithms and the advent of parallel computing, the demand for computational resources remains substantial. Accurate representation of intermolecular interactions and the incorporation of environmental factors such as temperature and pressure pose additional challenges. Developing reliable force fields that capture the complexity of polymer structures is an ongoing area of research. Despite these challenges, advancements in MD simulation techniques, including CG modeling and enhanced sampling methods, have broadened the scope of simulations, enabling researchers to explore complex phenomena with increased efficiency.

### 8.4.2 Finite Element Analysis (FEA) at the Microscale

FEA is a numerical method employed for predicting the behavior of structures by dividing them into smaller, manageable elements. At the microscale, FEA becomes a crucial tool for modeling the behavior of individual reinforcing elements within polymer composites, shedding light on their contribution to the overall mechanical properties.

In polymer composites, the role of reinforcing elements, such as fibers or nanoparticles, is pivotal in determining the material's mechanical properties. FEA at the microscale allows researchers to model the behavior of these individual elements under various loading conditions, including tension, compression, and shear. An accurate representation of the geometry, material properties, and boundary conditions of the reinforcing elements is essential for obtaining meaningful results. FEA provides valuable insights into stress distribution, deformation mechanisms, and failure modes at the microscale. This information is crucial for optimizing the design and manufacturing processes of composite materials to enhance their performance and reliability. In this line, Lee et al. [44] utilized FEA to investigate the impact of different interphase properties on the mechanical behavior of glass bead–filled polycarbonate (GB/PC) and interphase-coated glass bead–filled polycarbonate (GB/INT/PC) composites. Using a three-dimensional unit cell model, the study aimed to understand stress distribution, Young's modulus, and maximum stress concentration factor. FEA allowed the deduction of optimal interphase properties, providing valuable insights for reducing stress concentration and minimizing the drop in elastic modulus in the resulting composites. Similarly, Flöck et al. [45] employed FE contact technique to analyze steel ball indentation on a composite material with continuous carbon fibers in a poly(ether ether ketone) matrix. Two fiber orientations (normal and parallel) were investigated at a fiber volume fraction of 0.44. The FE analysis included both macro- and microcontact analyses, revealing subsurface stress-and-strain distribution. For normal fiber orientation, there is a high-shear-stress region, initiating interfacial failure below the surface. In parallel orientation, local plastic deformation occurs in the matrix with characteristic fiber bending. Their study provided an insight into the composite material's behavior during indentation.

One of the significant challenges in modeling polymer composites is bridging the gap between microscale and macroscale behavior [46]. While microscale FEA generates detailed information about the behavior of individual elements, integrating this data into macroscale models is essential for predicting the overall performance of the composite material. Lu et al. [46] highlighted the significance of polymer nanocomposites, emphasizing their enhanced properties through nanoparticle reinforcement in their review article. They discussed computational methods spanning molecular to macroscale, including MD, Brownian dynamics, and FEM. Applications in thermodynamics, kinetics, morphology, processing, and mechanical properties are detailed. The review discussed the crucial role of multiscale simulation strategies in advancing the understanding and predictive capabilities of polymer nanocomposites, aiming to stimulate further research in this field.

Techniques such as homogenization and multiscale modeling offer viable solutions for this integration. Homogenization methods aim to derive effective macroscopic material properties from microscale information, considering factors such as fiber orientation and interphase properties. Multiscale modeling approaches, on the other hand, provide a framework for simulating the material's behavior across different length scales. By combining microscale FEA with these integrated modeling techniques, researchers can gain a comprehensive understanding of polymer composites. This approach allows for accurate predictions of mechanical properties, durability, and failure behavior under real-world conditions, paving the way for the development of advanced composite materials tailored for specific applications in engineering and manufacturing.

## 8.5 MESOSCALE MODELING

Mesoscale modeling occupies a crucial position in the hierarchy of material modeling, offering insights into the behavior of materials at an intermediate scale between micro and macro. This section explores the techniques and methodologies employed in mesoscale modeling, focusing on representative volume element (RVE) modeling and homogenization techniques.

### 8.5.1 Representative Volume Element (RVE) Modeling

Representative volume element (RVE) modeling serves as a cornerstone in understanding mesoscale behavior [47]. The RVE represents a unit cell containing the essential microstructural features, allowing for the investigation of local interactions. This section delves into the challenges and considerations in defining an appropriate RVE for composite materials, particularly in polymer composites. Researchers must carefully select the size and shape of the RVE to capture the key features influencing material behavior. The simulation involves modeling the interactions between reinforcing elements, such as fibers or particles, and the surrounding matrix. This facilitates a detailed examination of stress distribution, deformation mechanisms, and failure initiation within the mesoscale structure. Feijóo et al. [47] presented a two-scale formulation for simulating failure in materials with a heterogeneous microstructure. Their model incorporated the nucleation of cohesive cracks at the microscale, influencing material instability at the macroscale. By utilizing the RVE concept, the model established a multiscale framework with kinematic information transfer and a variational principle. The novelty lies in addressing complex microstructure failure modes with multiple cohesive cracks and introducing tortuosity as a kinematic concept. Numerical examples demonstrate the model's capability in simulating realistic fracture problems, validated through comparisons with direct numerical solution (DNS) approaches. The study showcased the versatility and reliability of the multiscale technique in capturing intricate failure mechanisms in heterogeneous materials.

Beyond merely visualizing the mesoscale structure, extracting effective material properties is paramount. This involves determining how the local interactions within the RVE contribute to the overall mechanical behavior of the material. Simulation results provide valuable data for calculating effective properties, such as stiffness, strength, and thermal conductivity. The section explores the methodologies employed to extract these properties and their significance in informing macroscale models. Understanding the link between mesoscale interactions and macroscale behavior is crucial for accurate predictions and optimization of material performance. Similarly, Chen et al. [48] assessed the effective mechanical properties of CNT-based composites, utilizing a 3D nanoscale RVE and FEM. Formulas were derived based on elasticity theory to extract material constants under different loading conditions. The extended rule of mixtures was employed by the authors for comparison, focusing on estimating the effective Young's modulus in the axial direction of the RVE. Numerical examples using FEM demonstrate substantial load-carrying capacities of CNTs in a matrix, with a remarkable increase in composite stiffness (up to 0.7 and 9.7 times for 2% and 5% volume fractions, respectively). The findings were aligned with experimental results, highlighting the potential of CNTs for enhancing nanocomposite mechanical properties. This study provided valuable insights for material design and optimization.

### 8.5.2 Homogenization Techniques

Homogenization techniques play a pivotal role in bridging the gap between mesoscale and macroscale properties. This subsection provides an overview of various homogenization methods used for composite materials. These methods aim to capture the influence of mesoscale features on the overall response of the material. Common approaches include the Mori–Tanaka method, self-consistent schemes, and mean-field homogenization. Each method has its strengths and limitations, and the choice depends on the nature of the composite and the specific features under consideration.

The ultimate goal of homogenization is to establish a seamless connection between mesoscale and macroscale properties. This involves integrating the effective material properties obtained from mesoscale simulations into macroscale models. The section discusses the challenges and advancements in achieving this integration, emphasizing the importance of accurate homogenization for reliable predictions at larger scales. Successful linking of mesoscale and macroscale properties is critical for optimizing composite materials for specific applications, providing a comprehensive understanding of their performance under varying conditions.

Mergheim et al. [49] utilized computational homogenization for modeling nanosized reinforcements in composite materials. It introduces an extension based on interface energetics and a novel approach using continuously graded interphases. Through systematic evaluation, the latter proves effective in accurately modeling the mechanical behavior of nanocomposites with varying filler–matrix stiffness combinations, addressing limitations of standard methods. The technique contributes to advancing the understanding and modeling of complex nanocomposite materials. Kamiński et al. [50] investigated the influence of anisotropic distribution of reinforcing particles and stochastic interphases on the homogenized elastic characteristics of a carbon–polymer composite. Employing a probabilistic homogenization technique, including the stochastic perturbation method, Monte Carlo simulation, and a semianalytical approach, the researcher used FEA in ABAQUS for a composite model with two scales. Stochastic interface defects are modeled, and the three-component composite undergoes computational homogenization on the microscale. The study includes FEM error analysis, sensitivity assessment, and deterministic and probabilistic results, quantifying anisotropy using various indexes. A new tensor anisotropy index is proposed, effective for noncubic tensors, and comparisons with isotropic analyses demonstrate the anisotropy effect in composites with anisotropic distribution. The same author also investigated [51] the impact of interface defects on the stiffness tensor of a polymer-reinforced composite with carbon black particles using the stochastic FEM (SFEM). The defects are modeled as semi-ellipsoidal voids around the particles, and their effects are analyzed for different volume fractions. SFEM results are compared with classical Monte Carlo and semianalytical probabilistic FEM methods, revealing significant impacts of defects on composite stiffness even at low volume fractions.

## 8.6 CHALLENGES AND FUTURE DIRECTIONS IN MULTISCALE MODELING OF POLYMER COMPOSITES

The multiscale modeling of polymer composites has seen significant advancements, yet several challenges persist, and future directions are essential for continued progress in this field.

### 8.6.1 Current Challenges in Multiscale Modeling

#### 8.6.1.1 Computational Limitations

One of the primary challenges in multiscale modeling of polymer composites is computational limitations [52, 53]. As models aim to capture behaviors at various scales, computational demands increase exponentially. High-fidelity simulations, especially those involving intricate microstructures, require substantial computing resources, limiting the feasibility of real-time simulations and hindering the exploration of large design spaces. Addressing this challenge necessitates the development of efficient algorithms, parallel computing strategies, and optimization techniques to enhance the computational efficiency of multiscale models.

#### 8.6.1.2 Integration of Experimental Data into Models

Another critical challenge is the seamless integration of experimental data into multiscale models [54, 55]. Experimentally obtained data, such as material properties or failure mechanisms at different scales, are invaluable for model validation and calibration. However, effectively incorporating this data into computational models poses challenges related to data quality, consistency, and compatibility with simulation techniques. Bridging the gap between experimental findings and computational predictions requires robust methodologies for data assimilation, uncertainty quantification, and model validation to ensure the reliability and accuracy of multiscale simulations.

### 8.6.2 Future Directions and Emerging Trends

#### 8.6.2.1 Advances in Modeling Techniques

The future of multiscale modeling in polymer composites lies in the continual development of advanced modeling techniques. Integrating machine learning [55, 56], artificial intelligence, and data-driven approaches into multiscale models can significantly enhance their predictive capabilities. These techniques can adapt and learn from experimental data, improving model accuracy and reducing the need for extensive computational resources. Additionally, the incorporation of reduced-order models and surrogate modeling methods can expedite simulations without sacrificing accuracy, opening avenues for real-time multiscale simulations.

#### 8.6.2.2 Potential Breakthroughs in Multiscale Modeling of Polymer Composites

Anticipated breakthroughs in multiscale modeling of polymer composites include advancements in material informatics and digital twinning. Material informatics involves the systematic use of data-driven approaches to predict material properties and behaviors, offering a holistic understanding of complex composite systems [57]. Digital twinning, on the other hand, envisions creating virtual replicas of physical materials, enabling real-time monitoring and control of composite structures [58, 59]. These breakthroughs hold the promise of revolutionizing how we model, simulate, and understand polymer composites, unlocking new possibilities for material design and optimization.

## 8.7 SUMMARY AND CONCLUSION

In conclusion, this chapter has not only addressed the challenges and future directions of multiscale modeling in polymer composites but also provided a comprehensive exploration of the fundamental aspects of this field. Beginning with the basic fundamentals and properties of polymer composites, the discussion covered the essential components that form the backbone of understanding these materials. An overview of multiscale modeling was provided, highlighting the significance of bridging the micro- and macroscales for a holistic understanding. The various approaches to multiscale modeling, including the bottom-up and top-down approaches, were elucidated, emphasizing the versatility of techniques available. The discussion further delved into microscale modeling, employing MD simulation and FEA to understand intricate behaviors. Mesoscale modeling, using RVE modeling and homogenization techniques, was discussed in detail, providing insights into capturing the macroscopic properties of composites from microscale simulations. The inclusion of these topics enriches the conclusion, emphasizing the interconnectedness of fundamental knowledge with advanced modeling approaches, reinforcing the idea that a robust understanding of basic principles is crucial for the continued refinement and success of multiscale modeling in the domain of polymer composites.

## ACKNOWLEDGMENTS

The author thanks the Indian Institute of Technology–Dharwad for creating excellent research ethics and environment.

## REFERENCES

[1] H.S. Gupta, P. Kumar, A.K. Sethi, A. Verma, S.K. Sethi, Software-based simulations of the creep recovery model of polymer composites, Dynamic Mechanical and Creep-Recovery Behavior of Polymer-Based Composites (2024) 271–289. https://doi.org/10.1016/B978-0-443-19009-4.00016-3.

[2] H.S. Gupta, S.K. Sethi, A. Verma, Applications of coating materials: A critical overview, Materials Horizons: From Nature to Nanomaterials Part F 1086 (2023) 81–109. https://doi.org/10.1007/978-981-99-3549-9_4/COVER.

[3] A. Imam, S.K. Sethi, M.M.H. Tusher, M. Afroja Khatun, H.S. Gupta, A. Verma, Application of molecular dynamics simulations in coatings and composites, Dynamic Mechanical and Creep-Recovery Behavior of Polymer-Based Composites (2024) 419–449. https://doi.org/10.1016/B978-0-443-19009-4.00022-9.

[4] R. Gogoi, S.K. Sethi, G. Manik, Surface functionalization and CNT coating induced improved interfacial interactions of carbon fiber with polypropylene matrix: A molecular dynamics study, Applied Surface Science 539 (2021) 148162. https://doi.org/10.1016/j.apsusc.2020.148162.

[5] S. Chaturvedi, A. Verma, S.K. Sethi, S. Ogata, Defect energy calculations of Nickel, Copper and Aluminium (and their alloys): Molecular dynamics approach, Lecture Notes in Applied and Computational Mechanics 99 (2022) 157–186. https://doi.org/10.1007/978-981-19-3092-8_8/COVER.

[6] S.K. Sethi, S. Kadian, R. Gogoi, G. Manik, Layer-by-layer fabrication of self-cleaning superhydrophobic surface made from Carboxymethylcellulose and ZnO quantum dots: A combined experimental and computational study, Surfaces and Interfaces 37 (2023) 102752. https://doi.org/10.1016/J.SURFIN.2023.102752.

[7] A.K. Maurya, R. Gogoi, S.K. Sethi, G. Manik, A combined theoretical and experimental investigation of the valorization of mechanical and thermal properties of the fly ash-reinforced polypropylene hybrid composites, The Journal of Materials Science 56 (2021) 16976–16998. https://doi.org/10.1007/S10853-021-06383-2/FIGURES/10.

[8] A. Verma, N. Jain, S.K. Sethi, Modeling and simulation of graphene-based composites, Innovations in Graphene-Based Polymer Composites (2022) 167–198. https://doi.org/10.1016/B978-0-12-823789-2.00001-7.

[9] S.K. Sethi, M. Singh, G. Manik, A multi-scale modeling and simulation study to investigate the effect of roughness of a surface on its self-cleaning performance, Molecular Systems Design & Engineering 5 (2020) 1277–1289. https://doi.org/10.1039/d0me00068j.

[10] S.K. Sethi, L. Soni, G. Manik, Component compatibility study of poly(dimethyl siloxane) with poly(vinyl acetate) of varying hydrolysis content: An atomistic and mesoscale simulation approach, Journal of Molecular Liquids 272 (2018) 73–83. https://doi.org/10.1016/J.MOLLIQ.2018.09.048.

[11] A. Kataria, A. Verma, S.K. Sethi, S. Ogata, Introduction to interatomic potentials/force-fields, Lecture Notes in Applied and Computational Mechanics 99 (2022) 21–49. https://doi.org/10.1007/978-981-19-3092-8_2/COVER.

[12] V.A. Buryachenko, A. Roy, K. Lafdi, K.L. Anderson, S. Chellapilla, Multi-scale mechanics of nanocomposites including interface: Experimental and numerical investigation, Composites Science and Technology 65 (2005) 2435–2465. https://doi.org/10.1016/J.COMPSCITECH.2005.08.005.

[13] P. Peças, H. Carvalho, H. Salman, M. Leite, Natural fibre composites and their applications: A review, Journal of Composites Science 2 (2018) 66. https://doi.org/10.3390/jcs2040066.

[14] G.S. Mann, L.P. Singh, P. Kumar, S. Singh, Green composites: A review of processing technologies and recent applications, Journal of Thermoplastic Composite Materials (2018). https://doi.org/10.1177/0892705718816354.

[15] X. Jin, M. Deng, S. Kaps, X. Zhu, I. Hölken, K. Mess, R. Adelung, Y.K. Mishra, Study of tetrapodal ZnO-PDMS composites: A comparison of fillers shapes in stiffness and hydrophobicity improvements, PLoS One 9 (2014) e106991. https://doi.org/10.1371/journal.pone.0106991.

[16] Q. Chen, L.S. Schadler, R.W. Siegel, G.C. Irvin, Jr., ZnO/PMMA thin film nanocomposites for optical coatings, in: D.L. Andrews, Z. Gaburro, A.N. Cartwright, C.Y.C. Lee (Eds.), *International Society for Optics and Photonics*, 2003: p. 158. https://doi.org/10.1117/12.509647.

[17] D. Upadhaya, P. Kumar, D. Dhar Purkayastha, Superhydrophobic ZnO/$TiO_2$ heterostructure with significantly enhanced photocatalytic activity, Journal of Materials Science: Materials in Electronics 30 (2019) 10399–10407. https://doi.org/10.1007/S10854-019-01381-2/FIGURES/10.

[18] R. Gogoi, S.K. Sethi, G. Manik, Surface functionalization and CNT coating induced improved interfacial interactions of carbon fiber with polypropylene matrix: A molecular dynamics study, Applied Surface Science 539 (2021) 148162. https://doi.org/10.1016/J.APSUSC.2020.148162.

[19] S. Yang, J. Choi, M. Cho, Intrinsic defect-induced tailoring of interfacial shear strength in CNT/polymer nanocomposites, Composite Structures 127 (2015) 108–119. https://doi.org/10.1016/j.compstruct.2015.02.078.

[20] S. Yumitori, Y. Arao, T. Tanaka, K. Naito, K. Tanaka, T. Katayama, Increasing the interfacial strength in carbon fiber/polypropylene composites by growing CNTs on the fibers, WIT Transactions on Modelling and Simulation 55 (2013) 275–284. https://doi.org/10.2495/CMEM130221.

[21] N. Subramanian, A. Rai, A. Chattopadhyay, Atomistically derived cohesive behavior of interphases in carbon fiber reinforced CNT nanocomposites, Carbon 117 (2017) 55–64. https://doi.org/10.1016/j.carbon.2017.02.068.

[22] J.D. Afroze, L. Tong, M.J. Abden, Y. Chen, Multifunctional hierarchical graphene-carbon fiber hybrid aerogels for strain sensing and energy storage, Advanced Composites and Hybrid Materials 6 (2023) 1–13. https://doi.org/10.1007/S42114-022-00594-0/FIGURES/6.

[23] B. Kulyk, M.A. Freitas, N.F. Santos, F. Mohseni, A.F. Carvalho, K. Yasakau, A.J.S. Fernandes, A. Bernardes, B. Figueiredo, R. Silva, J. Tedim, F.M. Costa, A critical review on the production and application of graphene and graphene-based materials in anti-corrosion coatings, Critical Reviews in Solid State and Materials Sciences 47 (2021) 309–355. https://doi.org/10.1080/10408436.2021.1886046.

[24] M.J. Nine, M.A. Cole, L. Johnson, D.N.H. Tran, D. Losic, Robust superhydrophobic graphene-based composite coatings with self-cleaning and corrosion barrier properties, ACS Applied Materials & Interfaces 7 (2015) 28482–28493. https://doi.org/10.1021/acsami.5b09611.

[25] O.H. Lin, R.N. Kumar, H.D. Rozman, M. Azemi, M. Noor, Grafting of sodium carboxymethylcellulose (CMC) with glycidyl methacrylate and development of UV curable coatings from CMC-g-GMA induced by cationic photoinitiators, Carbohydrates Polymers 59 (2005) 57–69. https://doi.org/10.1016/j.carbpol.2004.08.027.

[26] S.K. Sahoo, V. Khandelwal, G. Manik, Development of completely bio-based epoxy networks derived from epoxidized linseed and castor oil cured with citric acid, Polymers for Advanced Technologies 29 (2018) 2080–2090. https://doi.org/10.1002/pat.4316.

[27] V. Khandelwal, S.K. Sahoo, A. Kumar, G. Manik, Study on the effect of carbon nanotube on the properties of electrically conductive epoxy/polyaniline adhesives, Journal of Materials Science: Materials in Electronics 28 (2017) 14240–14251. https://doi.org/10.1007/s10854-017-7282-y.

[28] S.K. Sethi, G. Manik, A combined theoretical and experimental investigation on the wettability of MWCNT filled PVAc-g-PDMS easy-clean coating, Progress in Organic Coatings 151 (2021) 106092. https://doi.org/10.1016/j.porgcoat.2020.106092.

[29] R. Gogoi, S.K. Sethi, G. Manik, Surface functionalization and CNT coating induced improved interfacial interactions of carbon fiber with polypropylene matrix: A molecular dynamics study, Applied Surface Science 539 (2021). https://doi.org/10.1016/j.apsusc.2020.148162.

[30] S.K. Sethi, S. Kadian, G. Manik, A review of recent progress in molecular dynamics and coarse-grain simulations assisted understanding of wettability, Archives of Computational Methods in Engineering 1 (2022) 1–27. https://doi.org/10.1007/S11831-021-09689-1.

[31] P. Marashizadeh, M. Abshirini, M. Saha, L. Huang, Y. Liu, Interfacial properties of ZnO nanowire-enhanced carbon fiber composites: A molecular dynamics simulation study, Langmuir 37 (2021) 7138–7146. https://doi.org/10.1021/ACS.LANGMUIR.1C00711/ASSET/IMAGES/LARGE/LA1C00711_0011.JPEG.

[32] S. Li, J. Hu, J. Li, J. Tian, Z. Han, X. Zhou, Y. Chen, Anisotropic wet etched silicon substrates for reoriented and selective growth of ZnO nanowires and enhanced hydrophobicity, Langmuir 27 (2011) 6549–6553. https://doi.org/10.1021/la201157m.

[33] G. Pal, S. Kumar, Modeling of carbon nanotubes and carbon nanotube–polymer composites, Progress in Aerospace Sciences 80 (2016) 33–58. https://doi.org/10.1016/J.PAEROSCI.2015.12.001.

[34] G. Deichmann, N.F.A. Van der Vegt, Bottom-up approach to represent dynamic properties in coarse-grained molecular simulations, Journal of Chemical Physics 149 (2018) 244114. https://doi.org/10.1063/1.5064369/197650.

[35] A. Dequidt, J.G. Solano Canchaya, Bayesian parametrization of coarse-grain dissipative dynamics models, Journal of Chemical Physics 143 (2015) 84122. https://doi.org/10.1063/1.4929557/73206.

[36] D. Kauzlariç, J.T. Meier, P. Español, S. Succi, A. Greiner, J.G. Korvink, Bottom-up coarse-graining of a simple graphene model: The blob picture, The Journal of Chemical Physics 134 (2011). https://doi.org/10.1063/1.3554395.

[37] L. Gao, W. Fang, Semi-bottom-up coarse graining of water based on microscopic simulations, The Journal of Chemical Physics 135 (2011). https://doi.org/10.1063/1.3658500.

[38] S. Trément, B. Schnell, L. Petitjean, M. Couty, B. Rousseau, Conservative and dissipative force field for simulation of coarse-grained alkane molecules: A bottom-up approach, The Journal of Chemical Physics 140 (2014). https://doi.org/10.1063/1.4870394.

[39] K.Ch. Daoulas, Top-down hybrid models of polymers, in: *Handbook of Materials Modeling*, 2020: pp. 1411–1430. https://doi.org/10.1007/978-3-319-44677-6_37.

[40] A.K. Pervaje, J.C. Tilly, A.T. Detwiler, R.J. Spontak, S.A. Khan, E.E. Santiso, Molecular simulations of thermoset polymers implementing theoretical kinetics with top-down coarse-grained models, Macromolecules 53 (2020) 2310–2322. https://doi.org/10.1021/ACS.MACROMOL.9B02255/ASSET/IMAGES/LARGE/MA9B02255_0012.JPEG.

[41] Q. Yang, G. Li, A top-down multi-scale modeling for actuation response of polymeric artificial muscles, Journal of the Mechanics and Physics of Solids 92 (2016) 237–259. https://doi.org/10.1016/J.JMPS.2016.04.007.

[42] R. Rafiee, R. Shahzadi, Predicting mechanical properties of nanoclay/polymer composites using stochastic approach, Composites Part B: Engineering 152 (2018) 31–42. https://doi.org/10.1016/J.COMPOSITESB.2018.06.033.

[43] S.K. Sethi, L. Soni, U. Shankar, R.P. Chauhan, G. Manik, A molecular dynamics simulation study to investigate poly(vinyl acetate)-poly(dimethyl siloxane) based easy-clean coating: An insight into the surface behavior and substrate interaction, The Journal of Molecular Structure 1202 (2020) 127342. https://doi.org/10.1016/j.molstruc.2019.127342.

[44] C.P. Tsui, C.Y. Tang, T.C. Lee, Finite element analysis of polymer composites filled by interphase coated particles, The Journal of Materials Processing Technology 117 (2001) 105–110. https://doi.org/10.1016/S0924–0136(01)01117–7.

[45] K. Váradi, Z. Néder, K. Friedrich, J. Flöck, Finite-element analysis of a polymer composite subjected to ball indentation, Composites Science and Technology 59 (1999) 271–281. https://doi.org/10.1016/S0266–3538(98)00066–9.

[46] Q.H. Zeng, A.B. Yu, G.Q. Lu, Multiscale modeling and simulation of polymer nanocomposites, Progress in Polymer Science 33 (2008) 191–269. https://doi.org/10.1016/J.PROGPOLYMSCI.2007.09.002.

[47] S. Toro, P.J. Sánchez, P.J. Blanco, E.A. De Souza Neto, A.E. Huespe, R.A. Feijóo, Multiscale formulation for material failure accounting for cohesive cracks at the macro and micro scales, The International Journal of Plasticity 76 (2016) 75–110. https://doi.org/10.1016/J.IJPLAS.2015.07.001.

[48] Y.J. Liu, X.L. Chen, Evaluations of the effective material properties of carbon nanotube-based composites using a nanoscale representative volume element, Mechanics of Materials 35 (2003) 69–81. https://doi.org/10.1016/S0167–6636(02)00200–4.

[49] P. Kumar, P. Steinmann, J. Mergheim, Enhanced computational homogenization techniques for modelling size effects in polymer composites, Computational Mechanics 68 (2021) 371–389. https://doi.org/10.1007/S00466-021-02037-X/FIGURES/14.

[50] D. Sokołowski, M. Kamiński, Homogenization of carbon/polymer composites with anisotropic distribution of particles and stochastic interface defects, Acta Mechanica 229 (2018) 3727–3765. https://doi.org/10.1007/S00707-018-2174-7/METRICS.

[51] D. Sokołowski, M. Kamiński, Computational homogenization of carbon/polymer composites with stochastic interface defects, Composite Structures 183 (2018) 434–449. https://doi.org/10.1016/J.COMPSTRUCT.2017.04.076.

[52] Q. Sun, G. Zhou, Z. Meng, M. Jain, X. Su, An integrated computational materials engineering framework to analyze the failure behaviors of carbon fiber reinforced polymer composites for lightweight vehicle applications, Composites Science and Technology 202 (2021) 108560. https://doi.org/10.1016/J.COMPSCITECH.2020.108560.

[53] P. Sheng, J. Zhang, Z. Ji, An advanced 3D modeling method for concrete-like particle-reinforced composites with high volume fraction of randomly distributed particles, Composites Science and Technology 134 (2016) 26–35. https://doi.org/10.1016/J.COMPSCITECH.2016.08.009.

[54] E. Van Der Giessen, P.A. Schultz, N. Bertin, V.V. Bulatov, W. Cai, G. Csányi, S.M. Foiles, M.G.D. Geers, C. González, M. Hütter, W.K. Kim, D.M. Kochmann, J. Llorca, A.E. Mattsson, J. Rottler, A. Shluger, R.B. Sills, I. Steinbach, A. Strachan, E.B. Tadmor, Roadmap on multiscale materials modeling, Modelling and Simulation in Materials Science and Engineering 28 (2020) 043001. https://doi.org/10.1088/1361–651X/AB7150.

[55] M. Alber, A. Buganza Tepole, W.R. Cannon, S. De, S. Dura-Bernal, K. Garikipati, G. Karniadakis, W.W. Lytton, P. Perdikaris, L. Petzold, E. Kuhl, Integrating machine learning and multiscale modeling: Perspectives, challenges, and opportunities in the biological, biomedical, and behavioral sciences, NPJ Digital Medicine 2 (2019) 1–11. https://doi.org/10.1038/s41746-019-0193-y.

[56] K.K. Bejagam, Y. An, S. Singh, S.A. Deshmukh, Machine-learning enabled new insights into the coil-to-globule transition of thermosensitive polymers using a coarse-grained model, Journal of Physical Chemistry Letters 9 (2018) 6480–6488. https://doi.org/10.1021/acs.jpclett.8b02956.

[57] S. Ramakrishna, T.Y. Zhang, W.C. Lu, Q. Qian, J.S.C. Low, J.H.R. Yune, D.Z.L. Tan, S. Bressan, S. Sanvito, S.R. Kalidindi, Materials informatics, The Journal of Intelligent Manufacturing 30 (2019) 2307–2326. https://doi.org/10.1007/S10845-018-1392-0/FIGURES/3.

[58] E.H. Baalbergen, J.A. de Marchi, R. Klerx, Unleashing the potentials of digital twinning in the production of composite aircraft components, Journal of Physics: Conference Series 2526 (2023) 012046. https://doi.org/10.1088/1742–6596/2526/1/012046.

[59] S. Mihai, M. Yaqoob, D.V. Hung, W. Davis, P. Towakel, M. Raza, M. Karamanoglu, B. Barn, D. Shetve, R.V. Prasad, H. Venkataraman, R. Trestian, H.X. Nguyen, Digital twins: A survey on enabling technologies, challenges, trends and future prospects, IEEE Communications Surveys and Tutorials 24 (2022) 2255–2291. https://doi.org/10.1109/COMST.2022.3208773.

# 9 Advanced Characterization Techniques for Polymer Composites

*Nisha Tamta*

## 9.1 INTRODUCTION

In today's marketplace, a wide range of goods are offered in accordance with our everyday requirements. These goods are constructed from a variety of materials, including metals, ceramics, polymers, and their composites. Polymer composites are materials comprised of a polymer matrix reinforced with fibers, particles, or other additives to enhance their thermal, mechanical, and electrical properties, for example. The exceptional ability of polymer composites to combine the qualities of the matrix and filler materials, leading to improved characteristics that surpass those of individual components, has drawn considerable scientific attention. Modern structures benefit greatly from the use of refined composite materials, as is well-known for their high specific stiffness and strength, low weight, and resistance to corrosion. The ideal composites resist heat, corrosion, and cyclic loading very well, making them incredibly durable materials. The performance of these composites depends heavily on the distribution, orientation, and interfacial adhesion of the reinforcing elements within the polymer matrix. Polymer composites are commonly used in aerospace applications, such as in rocket motor cases, ultra-high-performance pressure vessels, external fuel tanks for combat aircraft, and helicopter rotor blades, to name but a few. A few examples of their uses in many commercial applications include tables, chairs, tennis rackets, bicycle frames, high-performance tubular items used in the offshore oil and gas industry, and in vehicle bumpers, leaf springs, connecting rods, etc. [1–3].

The production of polymer composite materials involves a number of intricate steps, and at every turn, flaws or imperfections are introduced, such as improper material handling tactics and environmental contaminants. A fundamental prerequisite for the development of different materials depends on a solid understanding of the connections between structure and composition, and properties and behavior. Further, incorporating this knowledge to create a material with the required properties and knowing the failure mechanisms of the new material are also another essential requirement. The term "characterization" encompasses all essential measures taken to analyze different factors that influence the characteristics of polymer composites. In order to ensure the qualitative properties of novel composite material, various advanced techniques are devised. Advanced characterization techniques are significantly known for analyzing and optimizing these intricate structures, including morphology, composition, and properties of polymer composites, which facilitate their design, fabrication, and enhancement of performance.

The general functionality and structural integrity of fiber-reinforced polymer composites (FRPC) are impacted by a variety of defects ranging in size from small-fiber faults to large-impact damage, resulting to serious safety ramifications. Throughout a composite's life cycle, dependable and efficient nondestructive testing (NDT) techniques are of great importance to reduce maintenance expenses and safety hazards. In order to characterize polymer composites without causing the material any harm, nondestructive testing (NDT) techniques are considered as vital tools [4–6]. The structural and chemical properties of polymer composites can be fully investigated

 DOI: 10.1201/9781003476337-9

with the use of advanced microscopy and spectroscopy techniques [7, 8]. These techniques allow researchers to examine polymer composites at various scales in terms of their composition, morphology, and performance. Additionally, in situ and operando characterization techniques can be potential approaches to elucidate the real-time monitoring of the behavior of polymer composites under various circumstances. These techniques provide insights into dynamic processes, allowing researchers to examine changes in structure, morphology, and properties as they occur. All these methods offer important new perspectives on the performance, quality, and structural integrity of polymer composites. This study aims to elaborate various advanced characterization techniques with respect to polymer composites.

## 9.2 NONDESTRUCTIVE TESTING (NDT) TECHNIQUES FOR POLYMER COMPOSITES

The term *nondestructive testing* (NDT) refers to a broad variety of techniques employed for testing, characterizing, or analyzing materials, components, or systems properties for specific variations, defects, and discontinuities without causing harm [9]. In polymer composite materials, throughout their entire life cycle, including the stages of material processing, component manufacturing, and in-service, noticeable defects and damage can be seen at any duration of time. A systematic classification of defects into matrix, fiber, and matrix/fiber interface defects has been done by certain studies based on the composition of fiber-reinforced polymer composites (FRPC) [10, 11]. Defects in the matrix generally arise from voids and porosities caused by inadequate polymer infusion, polymer-rich areas because of random fiber distribution, and impact-induced matrix cracking [11, 12]. Fiber defects might include uneven fiber distribution across the surface and misalignment of fiber orientation due to inappropriate material layup methods, fiber waviness [13], fiber wrinkling [14], and fiber breaking from intense impact. The primary causes of defects at the matrix–fiber interface include regions where the fiber and the matrix are not effectively bonded, along with occurrences of delamination between layers and separation in laminates bonded with adhesives [15]. NDT techniques are applied during the fabrication or postfabrication process and for fiber-reinforced polymer composites (FRPC) products already in use. Utilizing NDT techniques during the fabrication process determines the product's quality and the period of time it will last before failing. In the operational phases, NDT techniques are employed to examine emerging defects or damage and evaluate the attributes of structure to ensure safety during service. Integral to the structural health monitoring (SHM) system, NDT techniques are helpful in monitoring damage progression and estimating the remaining service life also [4]. Classifying NDT techniques poses a considerable challenge due to numerous criteria and considerations. Various methods are commonly employed to categorize these techniques, taking into account factors such as the testing procedure, the sensor's position with respect to the composite material's surface, safety considerations at the inspection site, and nature of the output signal [5]. NDT techniques leverage different frequency spectrum segments in the output signal for the identification and characterization of defects [6].

The classification of NDT techniques relies on the operational frequency and data processing approaches used in examining the output signal, encompassing three primary categories: acoustic wave–based, imaging-based, and electromagnetic-based methods. Figure 9.1 illustrates a broad range of NDT methods that cover eight NDT techniques, namely, acoustic emission (AE), ultrasonic testing (UT), terahertz (THz), infrared thermography (IRT), eddy current testing (ECT), digital image correlation (DIC), shearography, and X-ray computed tomography (CT) testing; these all are included in the mentioned three groups.

In the following sections, all these eight NDT techniques used in finding the defects and damages in the polymer composites will be discussed.

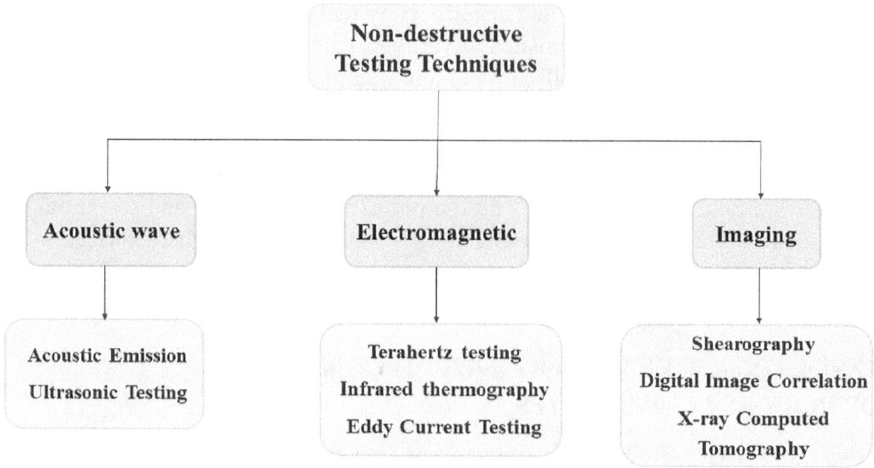


**FIGURE 9.1** Nondestructive testing techniques for polymer composites.

### 9.2.1 Acoustic Wave–Based NDT Technique

The techniques based on acoustic wave use sound waves to inspect and evaluate the integrity of materials without causing any damage to the material being tested. Sound waves consist of mechanical vibration that traverses through a medium, whether it be solid, water, or air. These methods are widely employed in various industries for quality control, safety assessments, and maintenance inspections. NDT techniques based on acoustic waves involve the application of various modes of energy emission on a sample of polymer composite. The subsequent analysis focuses on examining the acoustic waves received due to phenomena like reflection, refraction, transmission, and scattering, with the goal of identifying and detecting defects or damage. This section covers acoustic emission testing (AET) and ultrasonic testing (UT), which are prevalent NDT techniques based on acoustic waves.

#### 9.2.1.1 Acoustic Emission Testing (AET)

Throughout history, acoustic emission (AE) has been recognized as one of the most efficient techniques for assessing mechanical damage in composite materials. AET monitors the release of transitory acoustic waves produced by the rapid energy release within a material. When damage occurs in a polymer composite material, it creates temporary and localized changes in stored elastic energy. This energy release generates stress waves, leading to outcomes such as breaking fibers, cracking in the matrix, separation (debonding), delamination, etc. NDT techniques based on acoustic emission identify and monitor these abrupt releases of stress waves using arrays of extremely sensitive sensors or transducers [16].

In Figure 9.2, the basic functioning of an AET system is illustrated. In this representation, a sample containing cracks is exposed to specific loading conditions, causing the cracks to produce stress waves. These stress waves are subsequently recorded by the AE sensors installed on the surface of the sample. The recorded signals are then subjected to analysis, extracting information that reveals the extent of damage within the structure. The AET technique enables users to not only discern the type of failure but also pinpoint the origin of the generated crack. However, this process necessitates the use of multiple sensors, and composites are primarily anisotropic; this introduces additional complexity. Therefore, while a single AE sensor is adequate for detecting damage, it falls short in precisely identifying the location of the damage [17].

The AET possesses distinct characteristics, such as: the signals, specifically stress waves, originate from the tested sample itself, differing from other NDT methods, where signals come externally;

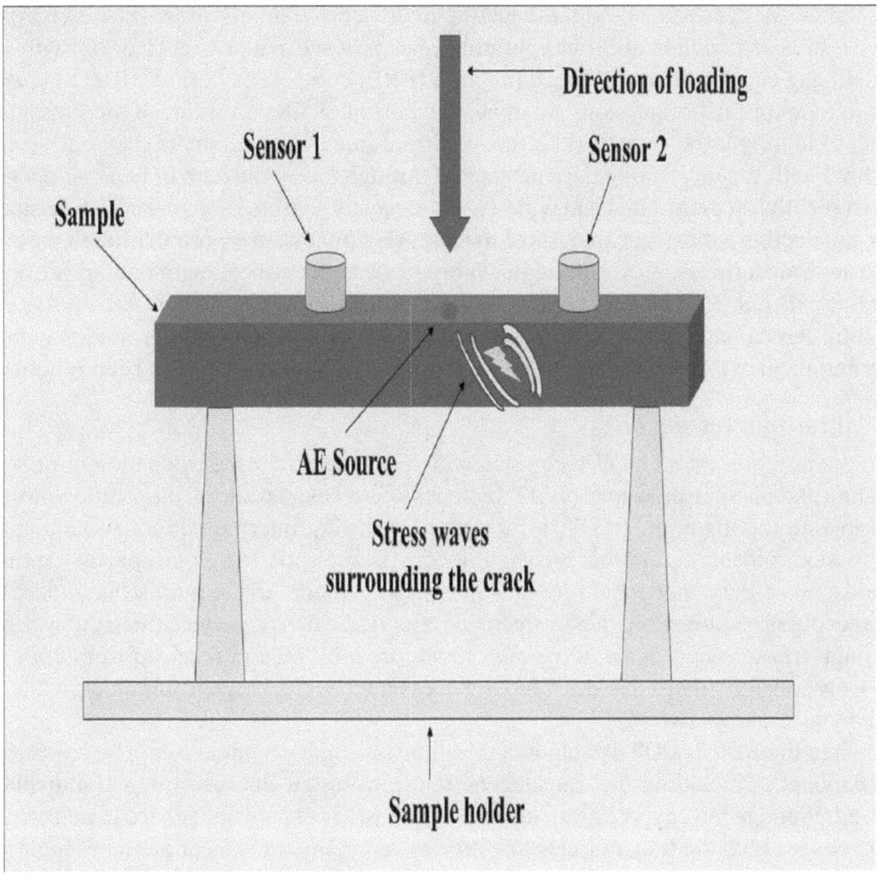


**FIGURE 9.2** Acoustic emission testing system.

**Source: Based on [5].**

the recorded data typically involve strain or displacement rather than focusing on geometric defects; and it observes dynamic processes within a material, allowing the tracking of specific defect development, thereby providing significant advantages in fatigue tests. The AET-based NDT method can be identified in a wide range of frequencies, from several Hz, which includes infrasonic waves, to several MHz, which encompasses ultrasonic waves. However, the predominant share of the released energy falls within the range of 1 kHz to 1 MHz. Output signals from acoustic emission can be broadly classified into three main types: bursts, continuous, and mixed [18]. Transient signals generated by the formation of damage (fiber breaking and delamination) are referred to as *bursts*. On the other hand, the continuous signal is a result of overlapping bursts, making it challenging to identify each burst individually. Factors like rubbing and electrical noise can produce continuous AE. The mixed-signal type incorporates both burst and continuous signals, representing the common signal encountered during service. The utilization of the AET method dates back to the early 1950s, marked by Kaiser's pioneering use of electronic instrumentation to identify audible sounds resulting from the tensile deformation of metallic specimens [19]. Kaiser's observation regarding the impact of sample stress history on AE production was coined as the "Kaiser effect" [20]. In the 1970s, AET found its initial application in the examination of composite materials and has since gained widespread use in diverse facets of composite research [21].

Reports indicated that AET can identify fatigue cracks, tiny fractures within the matrix, fiber fractures, and matrix–fiber interface delamination and debonding in the composite materials [22].

Presently, the most effective method for detecting microscopic damage mechanisms entails employing acoustic emission signals and incorporating unsupervised pattern recognition with finite element modeling (FEM) in fiber-reinforced polymer (FRP) composites [23]. AET technique was also employed to monitor the progression of damage in short glass fiber–reinforced thermoset composite panels, and to highlight the versatility of the employed integrity assessment methodology, AE data were captured with varying timing parameters and through the application of two distinct transducer types—specifically, medium-band and wide-band frequency sensors [24]. In the FRP composites, the benefits of conducting a thorough analysis of the raw AE signal were reported through the application of fast Fourier transform (FFT), resulting in a more intricate identification of damage throughout the entire loading process [25]. The application of AET in natural fiber/polymer (flax/epoxy) composite for evaluating the damage based on frequency over time, associating failure modes with standard damage criteria and real-time damage levels during flexural loading, has also been reported [26].

#### 9.2.1.2 Ultrasonic Testing (UT)

UT stands out as a prominent NDT technique widely employed for the evaluation of polymer composites. The fundamental principle of UT includes the transmission of ultrasonic waves through a transducer into the material. In UT, information regarding internal defects is acquired through the transmission, reflection, and backscattering of pulsed elastic waves, primarily captured using two methods: pulse-echo mode and through-transmission mode. In the pulse-echo mode, transmission and reception of ultrasonic waves occur on a single side for material detection. Conversely, in the through-transmission mode, ultrasonic waves are emitted and received from both sides of a material. Consequently, the pulse-echo mode necessitates just a single transducer, while through-transmission necessitates two [4].

As shown in Figure 9.3, a UT system in a through-transmission mode comprises of various functioning components, including the transducers, the transmitter, the receiver, and a display device. The transmitter and receiver serve as an electronic apparatus capable of generating electrical pulses with high voltage. With the transmitter as the driving force, the transducer produces high-frequency ultrasonic energy into the material, leading to its propagation in waveform.

If there is an interruption in the wave path, like a crack, some of the energy reflects back from the surface of the flaw. The transducer transforms the reflected wave signal into an electrical signal, subsequently displayed on a viewing device. This method relies on the use of high-frequency sound waves beyond the audible range to inspect and characterize materials without causing any damage. Operating across a broad frequency spectrum, spanning from 20 kHz to over 1 GHz, UT is employed in various industrial applications. However, within the domain of fiber-reinforced polymer composites (FRPC), the predominant frequency range for most industrial UT applications typically falls between 0.5 and 10 MHz [27].

In most cases of NDT techniques, the three distinct types of elastic waves are usually used: volume waves, which are further categorized into compression (longitudinal) and shear (transversal) waves; surface waves, also known as Rayleigh waves; and guided waves, like Lamb waves in plates. Moreover, a variety of operational modes are employed in the measurements, encompassing A-scan (representation of amplitude over time), B-scan (amplitude image resulting from unidirectional scanning), C-scan (amplitude image from bidirectional scanning), D-scan (time-of-flight image from bidirectional scanning), etc. [5]. Over the course of numerous years, UT has showcased its ability to qualitatively and quantitatively assess and analyze damage and defects in composite materials. The C-scan approach in ultrasonic testing stands out as a highly significant and comprehensive NDT technique.

In the literature, it has been reported that the C-scan method in ultrasonic testing (UT) has been employed for the assessment of artificial defects in laminated composite material [28] and cracks induced by thermal stress in carbon fiber–epoxy composites [29]. For assessing the extent of damage (delamination) in carbon fiber laminate composite resulting from impact-induced damage, both double-through transmission and pulse-echo methods immersion ultrasonic C-scans were

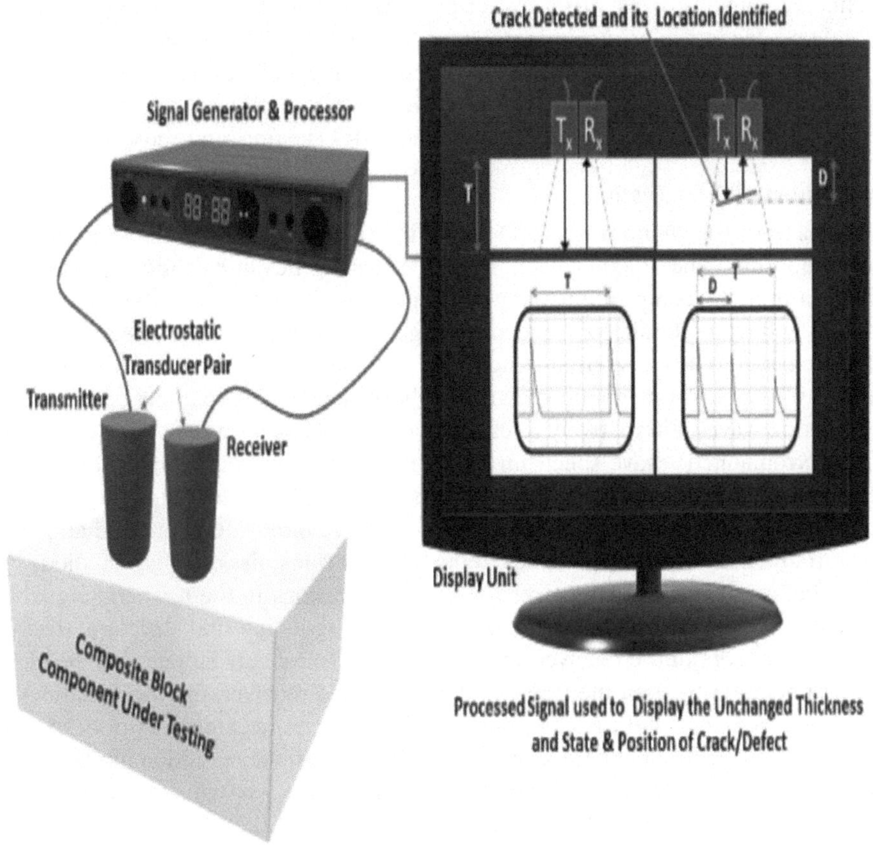


**FIGURE 9.3** Ultrasonic testing (UT) system.

**Source: Based on [6].**

employed [30]. Ultrasonic C-scan and recurrence analysis were employed for the identification of actual defects within the glass fiber–reinforced polymer (GFRP) and carbon fiber–reinforced polymer (CFRP) composites [31]. For investigating the varying levels of porosity in composite panel, A-scan and C-scan techniques in ultrasonic testing (UT) were successfully employed [32]. Their successful analysis involved obtaining data on porosity content and depth through A-scan, histogram, and C-scan techniques, all of which were aligned with the obtained results from destructive testing (DT). CFRP laminates exhibiting surface impact damage and internal delamination were examined using visual and optical observation, along with an inspection conducted through ultrasonic C-scan [33]. UT for polymer composites is particularly advantageous due to its capability to identify various types of defects, including delamination, voids, and fiber misalignment, with high precision. Additionally, UT can assess the material's thickness, ensuring uniformity and structural integrity. UT's ability to provide real-time and detailed information makes it a crucial tool in industries such as aerospace, automotive, and construction, where polymer composites play a pivotal role.

### 9.2.2 Electromagnetic-Based NDT Technique

The NDT technique utilizing electromagnetism for polymer composites involves the use of electromagnetic fields to evaluate the material's integrity and detect potential defects without causing damage. This technique utilizes various electromagnetic principles to examine the composition, structure, and possible irregularity within polymer composites, offering a non-invasive and efficient means of quality control and assessment in the manufacturing and maintenance of composite materials. This section explores some of the valuable electromagnetic-based techniques used in NDT, such as terahertz (THz) testing, infrared thermography (IRT) testing, and eddy current (EC) testing.

#### 9.2.2.1 Terahertz (THz) Testing

Terahertz (THz) testing is an innovative NDT technique that holds great promise for the assessment of polymer composites. The unique advantage of THz testing lies in its ability to offer comprehensive details regarding the composition and potential defects in materials without causing any damage. Operating within the terahertz frequency range, this method utilizes electromagnetic waves to inspect and characterize the internal structure of polymer composite materials. Terahertz (THz) waves are part of the electromagnetic spectrum, whose frequencies span between microwave and infrared. Terahertz waves, alternatively referred to as submillimeter waves, T-rays, or T-waves, are part of the electromagnetic wave spectrum, with frequencies ranging from 300 GHz to 10 THz. The wavelengths of radiation in the terahertz band span from 30 μm to 1 mm. Hence, they are also recognized as submillimeter waves [34]. Within the NDT technique, terahertz radiation finds application in two traditional testing approaches: reflection and transmission. Both methods necessitate a terahertz radiation source and a receiver to capture alterations in the terahertz signal caused by absorption and reflection within the tested sample, including its internal structural discontinuities. In the reflection mode, both the receiver and the radiation source are situated on the identical side of the tested sample, whereas in the transmission mode, the receiver and the radiation source are positioned on opposite sides of the tested sample, and it yields more extensive information regarding delaminations and cracks in fiber composites as compared with the reflection mode [35].

Figure 9.4 depicts the fundamental elements of a terahertz (THz) system in the reflection and transmission mode. This shows that the tested sample is positioned in the holder, ensuring that the front surface is perpendicular to the terahertz source. The configuration also includes a receiver that measures the incoming radiation. Notably, the arrangement of the radiation source and the receiver is strategically designed to facilitate movement along the examined sample by utilizing the holding device.

This system generates brief THz waves that interact with various phases within a material (such as defect or damage and inclusions). Then, analyzing the reflected or transmitted THz waves reveals the material's internal characteristics. THz-testing NDT technology offers the ability for contactless defect detection with high precision and effective penetration for nonconductive materials, including ceramics, plastics, GFRP, and certain composite materials [36, 37]. Conversely, THz waves are unable to permeate through metal and other conducting materials.

THz waves enable the detection of various defects, including delamination, voids, insufficient glue, resin deficiencies, and uneven reinforcement in the composite [38]. Solid pyrotechnics, such as rocket fuels, belong to the category of materials that can transmit terahertz radiation. Hence, initial findings on the potential use of terahertz radiation for detecting defects in composite pyrotechnic materials have been introduced in the reported study [37]. Terahertz radiation was employed to identify the defects in specific components composed of aramid fiber–reinforced composites utilized in military applications for safeguarding against ballistic threats [35]. Employing a top-tier THz scanner enabled a comprehensive examination of the surveyed regions, confirming the presence of imperfections, like cracks, material absence, delamination, and internal shrinkage. This emphasizes the strong effectiveness of THz radiation in overseeing composite materials designed for military purposes. Furthermore, THz technology emerges as a hopeful instrument for accurately electromagnetically detecting and evaluating delamination or internal voids concealed due to the existence of water in GFRP used in marine environments [39]. THz reflective imaging has also been greatly employed to characterize defects and damage, both on the surface and the subsurface of woven laminates made from carbon fiber–reinforced composites [40].

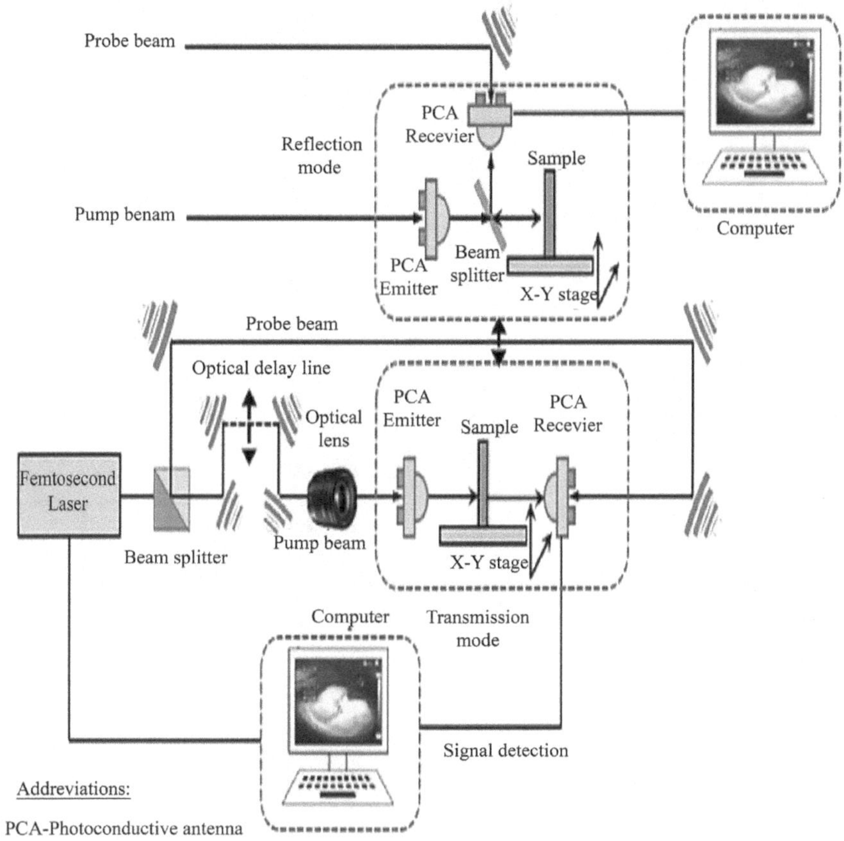


**FIGURE 9.4** Terahertz (THz) testing system.

**Source: Based on [5].**

### 9.2.2.2 Infrared Thermography (IRT) Testing

Infrared thermography (IRT) testing technique for polymer composites involves the utilization of infrared radiation to evaluate and analyze the material's thermal behavior. IRT is a noncontact testing method, alternatively referred to as thermal imaging or thermography testing, primarily employed for detecting and characterizing surface defects or flaws in the composites. This technique relies on the fact that different materials emit, absorb, and conduct heat in distinct ways. By applying heat to the surface or capturing the natural thermal patterns, infrared thermography enables the detection of anomalies, such as defects, delamination, or variations in composition within polymer composites. This is achieved by imaging and analyzing the thermal patterns exhibited by the composite structure, as depicted in Figure 9.5. In the context of IRT testing, a fundamental requirement is that the intact material possess consistently uniform and well-documented thermal properties, including thermal conductivity, specific heat, and thermal diffusivity, included in other factors. IRT testing is commonly categorized into two types: passive IRT and active IRT [41, 42]. In passive thermography, an infrared imaging device is used to measure the material's thermal variations without the involvement of outer thermal sources. This approach is often utilized for materials not in thermal balance, where the noteworthy features exhibit a temperature that deviates from the surrounding atmosphere, either being higher or lower than the background temperature [5]. Passive IRT technique is capable of detecting defects that are either superficial or located a few millimeters deep [41].

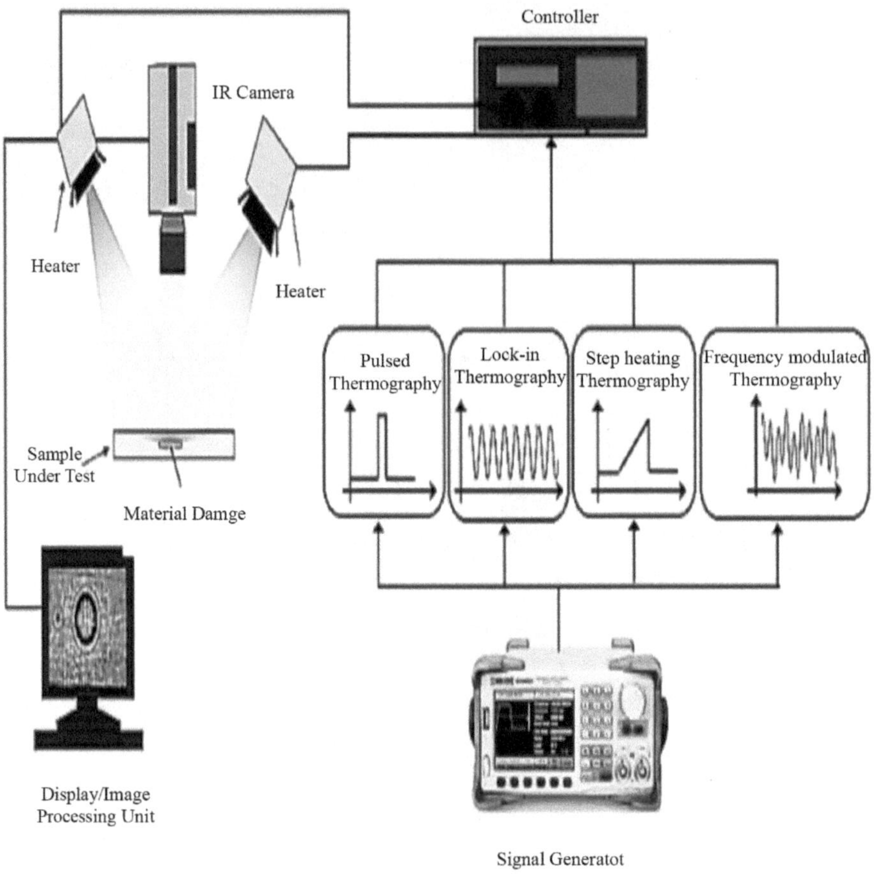


**FIGURE 9.5** Infrared thermography (IRT) testing system.

**Source: Based on [5].**

In contrast to passive IRT, active IRT employs an external heat source to induce thermal variations in the material under examination, generating a contrast in temperature between the background and the noteworthy features. The quantification of defect depth predominantly relies on the test structure's ability to conduct heat from its outer surface to its inside, along with its subsequent reflective reversal. The active approach is commonly employed, especially when the examined components are generally in thermal balance with the environment around them [5, 43]. Based on the excitation source, the AIRT (active infrared thermography) can be broadly categorized into various types, including optical thermography (involving noncontact pulsed, lock-in, and step thermography), ultrasonic mechanical thermography, laser infrared thermography, eddy current thermography, cycled loading thermography, etc. [41, 44, 45].

Over the last few decades, there has been an extensive exploration of IRT applications for detecting and evaluating defects in composite materials. It has been reported that subsurface defects in a reference Plexiglas sample were quantitatively assessed using measurement techniques such as pulse-phase and lock-in IRT [44]. IRT has been evolved into a widely embraced nondestructive testing (NDT) method for inspecting heritage structure, allowing real-time temperature measurement in a noncontact manner. It has been reported that IRT images have been utilized to capture and visualize moisture damage, cracks, defects in the medium, and other issues in heritage structure [46]. Pulse IRT and ultrasonic IRT have been used for detecting damage in low-velocity-impact-induced

CFRP [47]. From the obtained results, pulse IRT is found better suited for detecting delamination damage, particularly in cases involving extensive areas and non-joint delamination. On the other hand, ultrasonic infrared thermography was found better in identifying tiny damages, like joint delamination, matrix cracking, and fiber breakage. The effectiveness of passive IRT has been assessed in relation to its potential for in situ impact event monitoring in thin-walled FRP composites [48]. The examination of damage evolution in composites during open-hole fatigue (OHF) tests has been reported using infrared thermography [49]. In the realm of composite materials and structures, IRT has effectively demonstrated its ability to detect debonding, and delamination, damage caused by impacts, voids, cracks in the matrix, and fiber breakage.

#### 9.2.2.3 Eddy Current (EC) (IRT) Testing

Eddy current (EC) testing involves inducing electrical currents in a conductive material, generating magnetic fields that can reveal defects or variations in the material. As shown in Figure 9.6, an alternating current is run by means of a probe or a coil during an EC test [50]. This induces a primary magnetic field surrounding the coil. Upon bringing the coil near a polymer composite material, eddy currents are induced within the conductive components of the material. In polymer

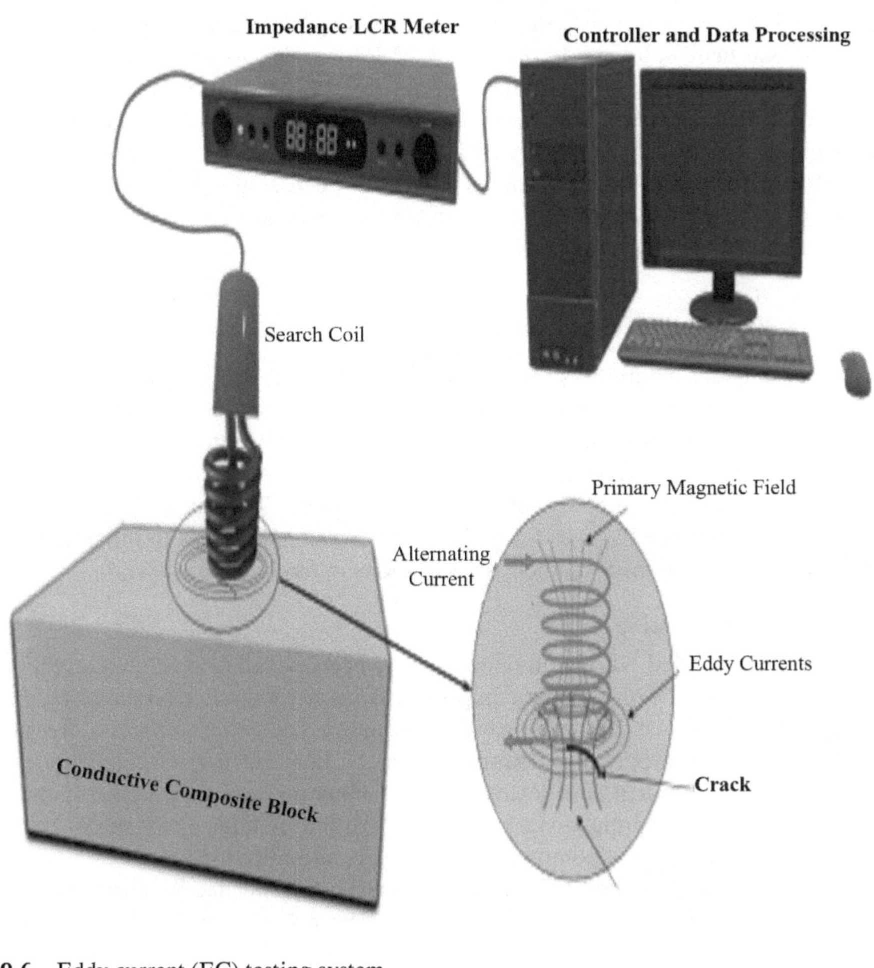


**FIGURE 9.6** Eddy current (EC) testing system.

**Source: Based on [6].**

composites, this is often due to the presence of conductive additives or fibers. Then, a secondary magnetic field is created by these eddy currents that are passing through the conducting material and opposing the primary magnetic field. Any variations in the material's magnetic permeability and electrical conductivity, caused by defects, delamination, or changes in composition, influence the eddy current response. By analyzing the alterations in the eddy current signals, this testing method enables the identification of subsurface flaws, delamination, voids, or other irregularities within polymer composites. The ability to perform these assessments without causing damage to material makes eddy current testing an invaluable tool for quality control, structural integrity assessment, and maintenance of polymer composite structures in various industries.

Various inspection techniques are accessible within such a kind of examination, including array scanning, near-field testing, pulsed eddy current, multifrequency remote field techniques, as well as single-probe testing across a broad frequency spectrum [51]. The choice of technique depends on the specific excitation signal and detection principles applied. Also, the output from an EC testing probe can be affected by numerous additional factors, underscoring the necessity for inspectors to consistently mitigate or eliminate undesirable influences to ensure the precise characterization of flaws in composites. The key factors influencing eddy current inspection encompass the testing piece's magnetic permeability and electrical conductivity, the skin effect that affects the testing piece's current distribution, the edge effect, the coil fill factor for encircling probes, the liftoff between the coil sensor and the tested piece, as well as considerations related to signal-to-noise ratio and phase lag [50]. The high-frequency eddy current testing method is recognized as a promising NDT tool for carbon fiber–reinforced polymers (CFRP). For the inspection of CFRP using eddy current testing, a high-frequency transmitter–receiver (T-R) probe has been chosen [52]. An NDT technique based on eddy currents has been introduced, allowing the visualization of 1D and 2D flaws in carbon fiber–reinforced polymer (CFRP), where the corresponding electronic circuit block relied on a Wheatstone bridge for signal acquisition and processing [53]. Research also expanded the application of EC testing to insulators and non-uniformly conductive materials and established the feasibility of detecting defects in CFRP [54].

### 9.2.3 Imaging-Based NDT Technique

The evaluation of polymer composites through imaging-based NDT techniques has become a crucial aspect of ensuring their structural integrity and performance. Various imaging methods are employed for nondestructive assessments, offering valuable insights into the internal characteristics of these advanced materials. Imaging-based NDT technique involves the use of digital image correlation (DIC), shearography, and X-ray computed tomography (CT), which will be discussed in the following section. These methods provide detailed cross-sectional images of the internal structure, helping identify defects, fiber misalignment, or variations in material density.

#### 9.2.3.1 Shearography Testing

*Shearography* is a specialized NDT technique that has been established to be incredibly successful for evaluating the integrity of polymer composites. This method relies on the principles of laser interferometry to detect surface deformation caused by internal defects or structural irregularities in the composite material. Shearography, also recognized as shear electronic speckle interferometry, holds a distinct position in optical flaw detection. This technique involves the examination of material surfaces, followed by the application of stress to detect internal defects by measuring the strain field, enabling the identification of common issues like delaminations [55].

In a shearograph testing, as shown in Figure 9.7, the sample surface is illuminated by a broad laser beam, and the interference pattern is captured using a CCD camera. When coherent light reflects from a rough surface, a grainy speckle structure appears in the camera image. The optical arrangement of the shearograph produces two transversely shifted images that interfere with each other. A CCD camera then records and captures the interferometric pattern, producing a fringe

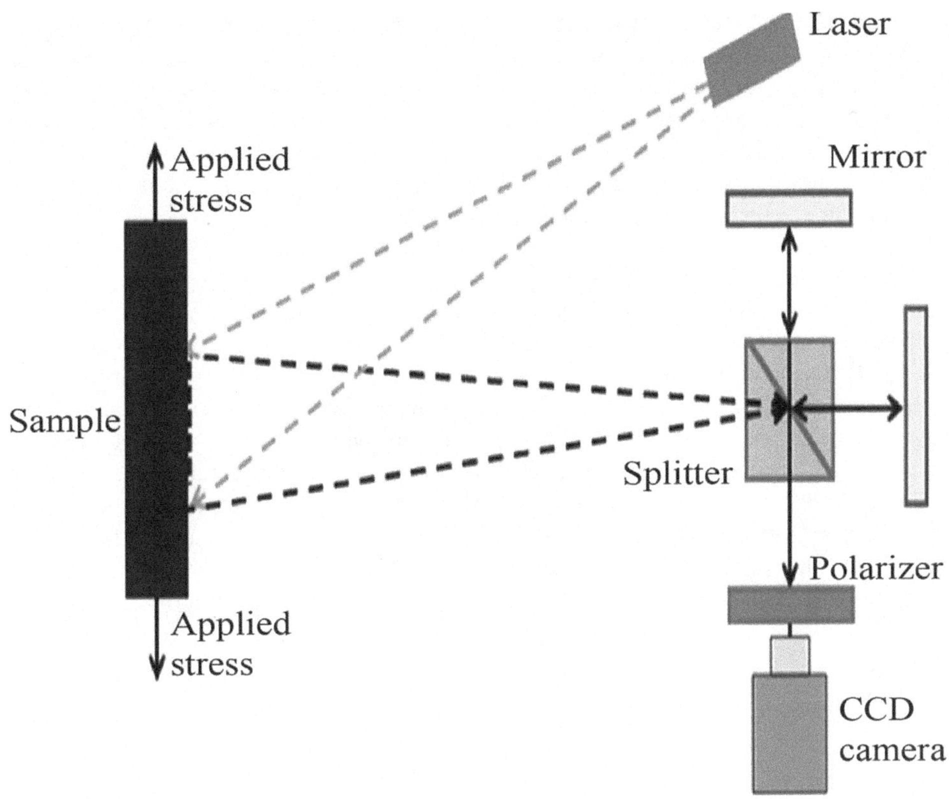


**FIGURE 9.7** Shearography testing system.

**Source: Based on [4].**

pattern that contains details regarding the structure. To derive information about the strain field, two measurements of an object are necessary—one in the unloaded state, and another one in the loaded state. To induce a load on the object, devices are employed to introduce disturbance through mechanical bending, reduced pressure (vacuum), microwaves, thermal, or sound effects, etc. [4, 55]. The advantages of shearography in NDT for polymer composites include its ability to inspect large areas quickly and its sensitivity to a wide range of structural anomalies. Additionally, shearography is a noncontact NDT technique, ensuring that the inspected materials remain undamaged during the testing process.

Digital shearography is recognized for its effectiveness in inspecting and evaluating materials, particularly in identifying disbands and delaminations in honeycomb structures and composite materials. A digital shearography methodology was introduced recently for measuring the depth and size of the tiniest noticeable imperfection during varying loading intensities [56]. The methodology begun with the formulation of a thin plate theory–based mechanical model to compute the anticipated bending of defects in close proximity to the surface. This model established relationships among defect-induced deformation, defect size, depth, load, and material properties. Subsequently, in this study, a connection was identified between the relative deformation assessed through shearography and the deformation caused by a defect, considering optimized shearing amount and digital shearography sensitivity. The utilization of both finite element methods (FEM) and experimental tests has been documented to evaluate shearography's defect detection capabilities when inspecting thick laminates of GFRP and found that using shearography is an effective NDT tool for thick composite materials [57]. In this study, a trusted thermal-mechanical model was developed by calculating equivalent mechanical and thermal properties, and experimental shearography testing was used to evaluate the model's accuracy. Another reported study provided valuable insights into the efficacy of the digital shearography method application for identifying subsurface defects in high-pressure composite tubes [58]. Shearography is particularly well-suited for detecting defects, such as delaminations, disbonds, and other internal inconsistencies in polymer composites. It provides rapid and real-time results, making it a valuable tool for quality control and integrity assessments in manufacturing and maintenance processes.

#### 9.2.3.2 Digital Image Correlation (DIC) Testing

DIC is an adaptable and powerful NDT technique that is progressively employed for assessing polymer composite materials. DIC is an optical, noncontact method that involves capturing two-dimensional (2D) or three-dimensional (3D) digital images of an object's surface. Subsequently, it performs the analysis of an image to extract comprehensive deformation data and measurements across the entire field [59, 60]. As shown in Figure 9.8, this method utilizes high-resolution digital images to analyze the surface deformation and displacement of composite materials, offering valuable insights into their mechanical behavior and structural integrity. DIC possesses the capability to capture images of deformed surfaces, facilitating precise analysis of materials subjected to stress under load. The process begins with capturing a reference image prior to loading, and then a sequence of images is taken throughout the deformation. Distinct dot patterns can be observed

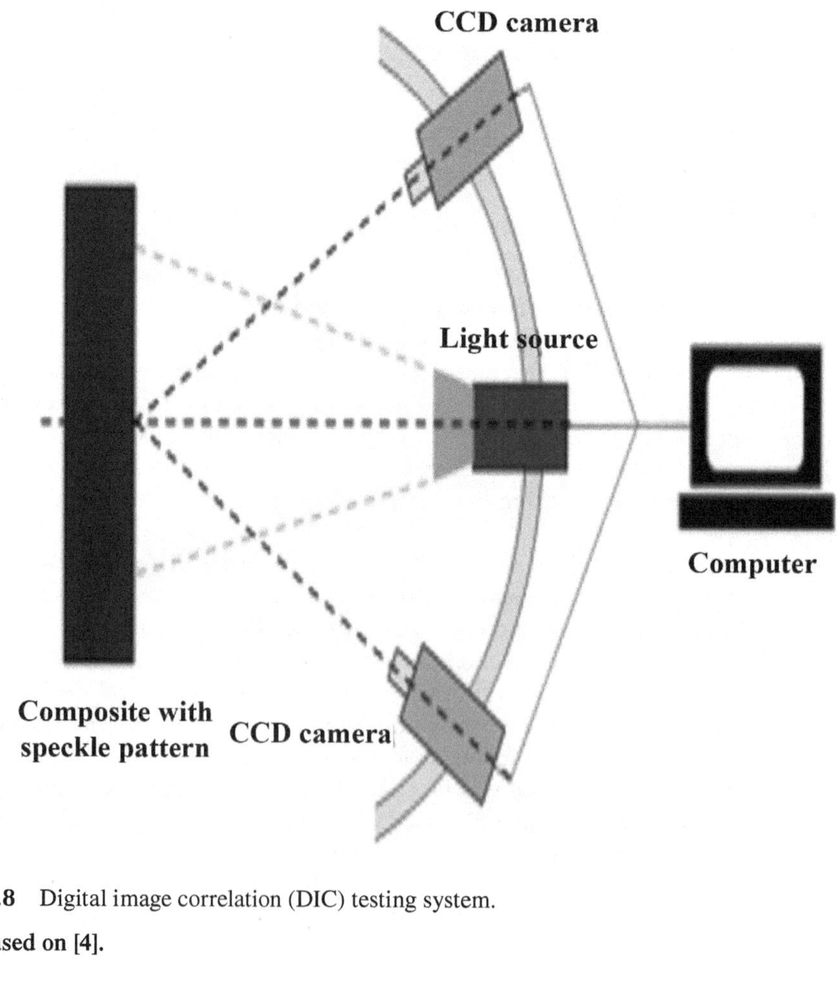


**FIGURE 9.8** Digital image correlation (DIC) testing system.

**Source: Based on [4].**

in deformed images when compared to the original, undeformed reference image. Calculating the dissimilarity in these patterns involves performing pixel correlation between any deformed image and the reference image, allowing for the computation of full-field displacement measurements [59, 61].

Deriving the strain distribution involves applying derivatives to the displacement field. Various methods, such as lines, grids, dots, etc., can be employed on the specimen surface to achieve this. The effectiveness of the results relies primarily on two factors: (1) the quality of the speckle pattern on the specimen and (2) the resolution of the camera [59, 60].

DIC testing allows for the characterization of material properties, including modulus of elasticity, tensile stress, and other mechanical parameters. This information is crucial for optimizing the design and performance of polymer composite structures. For investigating tensile properties, three methods—experimental methods, strain gauge measurements, and DIC methods—have been employed to find out the stress and strain values [62]. It was observed that the DIC method provided slightly higher values than the experimental method did, making it a recommended approach for approximating tensile properties. DIC analyzed surface images captured during mechanical testing of aramid and carbon fiber–reinforced polymer specimens to obtain strain data [63]. The obtained results in the reported study show that the deformation localization in aramid fiber composite occurred in bundle areas due to the pliability of fibers and weaving, while in carbon fiber composite, strain localizations occurred at low loads due to misalignments and lack of preloading. The use of digital image correlation (DIC) has also been explored to detect artificial damage in the structures of FRP composite crucial for personal safety [64]. The results indicated that the DIC method effectively identified approximate delamination locations by analyzing strain maps, where zones with distinct strain values indicated the presence of damage. The fracture toughness of short carbon fiber (SCF)–reinforced thermoplastic composites has been evaluated using DIC [65]. In this study, DIC was employed to analyze the full-field strain around the crack tip, revealing a notable reduction in strain levels with the addition of SCFs.

#### 9.2.3.3 X-Ray Computed Tomography (CT) Testing

The technique based on X-ray computed tomography is an exceptional method which uses radiographic penetrating imaging. This technique is gaining significance, especially in applications where the three-dimensional (3D) aspects of phenomena are crucial or when monitoring the development of critical features is essential, whether during the manufacturing process or while in service. This exceptional imaging technique offers the capability to examine structures with remarkable precision in terms of size and volume and is particularly advantageous for inspecting the structural integrity of intricate geometries [66, 67]. The resolution of this technique is inversely proportional to the measured volume. As a result, standard CT, micro-CT, and nano-CT techniques have been devised to enhance the resolution of feature sizes, though at the expense of the 3D volume that can be captured in the imaging process. Using X-rays, the X-ray CT method creates precise cross-sectional pictures of the internal structure of the composite material. The operational configuration of X-ray CT is depicted in Figure 9.9.

There are two main steps involved in the X-ray CT imaging process. First, a number of 2D radiographs are captured from various lighting angles. Following this, a computed reconstruction algorithm is used to create a stack of cross-sectional slices based on these 2D projections (radiographs) of the specimen, and the digital 3D grayscale representation provides information about the internal structure [68].

Computed tomography (CT) has been employed to assess the structural homogeneity of hemp fiber (HMPF)–reinforced poly(3-hydroxybutyric-co-3-hydroxyvaleric acid) (PHBV) (HMPF/PHBV) biocomposite and to evaluate the arrangement and orientation of fibers on the sample cross section under different injection rates [69]. The obtained results showed that, in addition to the amount of porosity observed on the material's cross section, the computed tomography images revealed how hemp fibers were dispersed and aggregated within the PHBV matrix. Micro–computed tomography

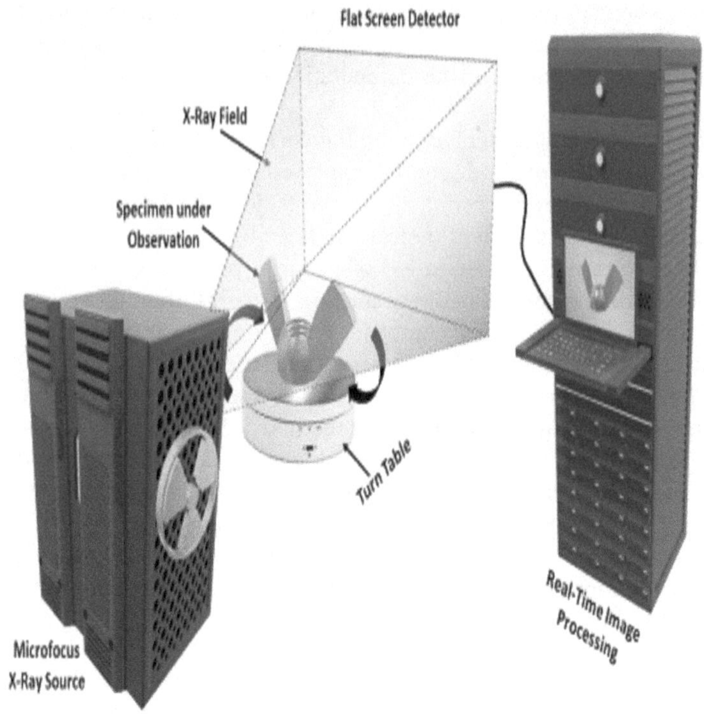


**FIGURE 9.9** X-ray computed tomography (CT) testing system.

Source: Based on [6].

(μCT) characterization and micromechanical analysis both have been employed to examine the evolution of cryogenic damage in the laminates of CFRP [70]. Analysis of defects, including porosity, inadequate fiber disorientation, and fiber impregnation with the polymer, in the onyx/glass- and the onyx/carbon fiber–reinforced composites has been reported through the implementation of an X-ray CT testing [71]. The obtained results revealed that as the number of continuous fiber reinforcements increases, so does the overall pore content's tendency to increase. Low-velocity impact (LVI) tests at three energy levels, followed by compression-after-impact (CAI) tests, were conducted for the carbon/epoxy composite laminate, and for studying the damage, X-ray computed tomography was employed [72]. It has been observed in the obtained result that even with barely visible impacts, the compressive strength has been reduced significantly. Failure mode was observed as buckling in the damaged zone, particularly at the back face of the composite, where damage was more. The size of the damaged zone was increased with higher impact energy, diminishing the CAI strength of the composite laminate.

## 9.3 ADVANCED MICROSCOPY AND SPECTROSCOPIC TECHNIQUES FOR POLYMER COMPOSITES

Polymer composites, renowned for their diverse applications and unique material properties, undergo intricate analyses through advanced microscopy and spectroscopic characterization techniques. This exploration encompasses cutting-edge methods that examine the microstructural and molecular intricacies of polymer composites. Through advanced microscopy, such as electron and atomic force microscopy, and spectroscopic techniques like infrared, Raman spectroscopy, etc., researchers gain unprecedented insights into the composition, morphology, and performance of polymer composites. This comprehensive examination elucidates the significant contributions of advanced techniques to the evolving landscape of polymer composite research and development. This section explores the application of these advanced characterization techniques in the analysis of polymer composites.

## 9.3.1 Microscopy Techniques

Polymer composites benefit from various microscopy techniques that offer detailed insights into their microstructure and properties. The following section will discuss the key microscopy characterization techniques employed for polymer composites, including transmission electron microscopy (TEM) and atomic forced microscopy (AFM).

### 9.3.1.1 Transmission Electron Microscopy (TEM)

TEM typically encompasses electron diffraction and imaging methods facilitated by the use of a transmission electron microscope [73]. The transmission electron microscope operates similarly to a light microscope, but it uses an electron beam for illumination instead of light [74]. Transmission electron microscopes (TEMs) are widely regarded as highly effective and versatile instruments for analyzing materials across a range of spatial dimensions. They excel in characterizing materials, spanning from the atomic scale to the expanding realm of the "nano" regime (from less than 1 nm to 100 nm) and extending to the micrometer level and beyond [75]. A transmission electron microscope (TEM) comprises essential components that collaborate to facilitate its imaging capabilities. Fundamentally, a TEM employs a filament of tungsten as the source of electron, which generates concentrated beam of electrons. This beam undergoes manipulation and control through electron lenses, encompassing condenser lenses, for focusing onto the specimen, and objective lenses, for magnification. The specimen stage serves as a platform for positioning the thin specimen, crucial for enabling electron passage. Electromagnetic lenses further refine the focus of the electron beam, while a vacuum system establishes an air-molecule-free environment to prevent electron scattering. The condenser aperture regulates both the angle and strength of the electron beam prior to reaching the specimen, and the specimen holder ensures proper positioning. An objective aperture restricts electron angles, enhancing image contrast. Projector lenses magnify the transmitted electron image onto a detector, often a digital camera, which, along with an image recording system, captures and stores high-resolution images. Collectively, these components empower TEM to scrutinize the intricate details of materials at the atomic and nanoscale levels.

For many years, the transmission electron microscope (TEM) has been widely employed in examining the morphology of synthetic polymers. Its enduring importance stems from the unique capacity of these materials to self-organize hierarchically across various length scales, generating morphologies that contribute to advancements in functionality and application. TEM remains a crucial tool for studying these intricate structures. Notably, advancements in instrumentation over the last decade have expanded the accessibility of both established and emerging transmission electron microscopy (TEM) imaging techniques to a growing array of applications-focused research teams globally [76].

A rising number of researchers are now leveraging these methods to quantitatively analyze polymer morphology [76]. In a more recent study, transmission electron microscope (TEM) has been used for characterization of a pliable electrochemical sensor constituted of polymer poly(butylene adipate-co-terephthalate) (PBAT) and carbon nitrite dots (CNDs) in order to detect emerging pollutants (i.e., cadmium and copper) in water [77]. In addition, a significant amount of work has reported on the usefulness of TEM in studying the pore size distribution maps of SBA-15 (mesoporous material) in polymer composites [78] and in the creation and analysis of a fire-resistant geopolymer reinforced with fibers based on metakaolin [79]. In general, transmission electron microscopy (TEM) stands as a pivotal tool in diverse scientific disciplines, offering unparalleled insights into the nanoscale and atomic structures of materials. The versatility of TEM, coupled with its high spatial resolution and analytical capabilities, positions it as an indispensable tool for researchers across a multitude of disciplines, driving innovation and expanding the boundaries of our understanding of the microscopic world.

#### 9.3.1.2 Atomic Force Microscopy (AFM)

Atomic force microscopy (AFM) is a characterization and imaging approach which provides a detailed understanding of surface structure and mechanical specifications of materials at atomic level. AFM works on the concept that utilizes a sharp tip. The working principle of AFM involves scanning the surface of a sample in a grid manner by employing a sharp tip affixed to a spring cantilever. Figure 9.10 shows that between the tip and the sample, an interatomic space is formed, and when the tip connected to end of the spring cantilever is placed in this space, an interatomic potential is generated between atoms both present on the tip and the surface of the sample as well [80]. When the interaction between the tip and the surface takes place, oscillation in the cantilever is induced due to these interatomic potentials responding to variations in the surface contours. As a result, topographical characteristics of the surface can be systematically mapped by means of measuring the deflection of the cantilever. A laser beam reflecting off the rear of the cantilever in a set of photodetectors is employed to measure the deflection.

AFM can offer comprehensive information regarding roughness, morphology, and mechanical characteristics of materials at the nanoscale. It is a valuable tool for understanding the surface characteristics and behavior of polymer composite materials, making it useful for both

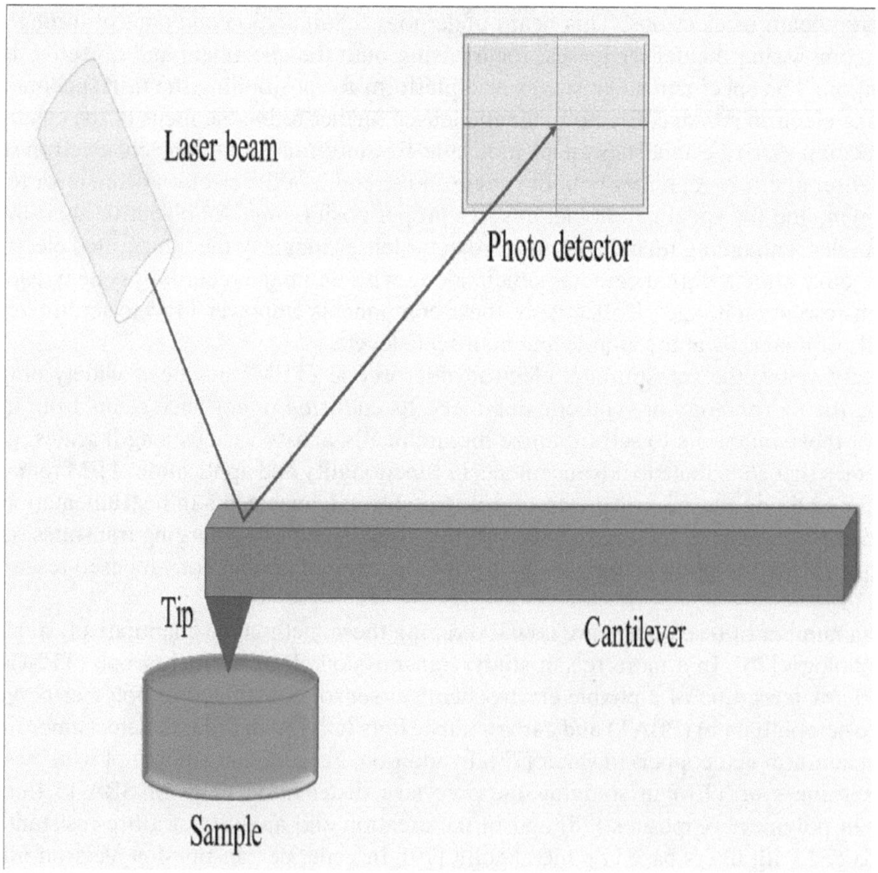


**FIGURE 9.10** Atomic force microscopy (AFM) setup.

research and quality control purposes. In the reported literature, middle lamellae from various plant fibers commonly employed for composite reinforcement were scrutinized using atomic force microscopy (AFM) [81]. This study utilized the peak-force quantitative nanomechanical property mapping (PF-QNM) approach to assess the indentation modulus of these layers. The results underscored the efficacy of AFM PF-QNM in mechanically characterizing and comparing the rigidity of minute regions at micro- as well as nanoscale, specifically within cell walls and middle lamellae of plants. In another investigation, structures resembling scales were observed from topographical images of AFM, which suggest that the primary cell wall of kenaf fiber treated with alkali (AT-KF) contains cellulose microfibrils [82]. In comparison to untreated fiber, AT-KF showed smooth fiber surface on account of its lower average surface roughness. In order to enhance their adhesive bonding characteristics, the surfaces of wood plastic composites (WPC) and continuous glass fiber–reinforced plastic (FRP) were treated with forced atmospheric plasma treatment (FAPT). AFM was also employed to assess the roughness of surface of wood plastic composite before and after undergoing forced atmospheric plasma treatment (FAPT) (FAPT) [83]. The results from this study indicate that there were no significant differences in roughness after the treatment, a noticeable observation, considering the absence of high temperatures generated during FAPT. Furthermore, AFM assessments, employing a silicon-tipped probe, revealed a significant up to 56% increase in adhesive force interactions on the surfaces of WPC after the application of FAPT.

### 9.3.2 Spectroscopic Techniques

Spectroscopy is a predominantly utilized technique to examine molecular behavior both pre- and post-fabrication of polymer composites. These techniques utilize the interaction of materials with electromagnetic radiation across different frequency ranges, including UV, infrared, and X-ray regions. Various spectroscopic techniques employed for the analysis of polymer composites will be discussed in this section, such as Fourier transmission infrared (FTIR) spectroscopy, X-ray diffraction (XRD), and Raman spectroscopy.

#### 9.3.2.1 Fourier Transmission Infrared (FTIR) Spectroscopy

For polymer composite materials, FTIR is used to recognize the chemical functional groups present in them by means of exposing the material to a beam of IR radiation and measuring the absorption of infrared radiation by each bond in the molecule. As a result, a spectrum is obtained, which is represented as (%) transmittance vs. wave number ($cm^{-1}$) [84]. The fundamental working principle of FTIR is based on subjecting the sample to a broad spectrum of infrared light, inducing unique vibrations and rotations within the molecular bonds of the material. As the sample interacts with the infrared radiation, specific frequencies are absorbed that generate a distinct molecular fingerprint. The subsequent passage of transmitted or reflected light through an interferometer, followed by Fourier transformation, unveils an infrared spectrum. The obtained peaks in this spectrum correspond to the characteristic vibrational modes of the sample's molecular bonds, enabling the identification of functional groups and chemical bonds.

This technique collectively contributes to a comprehensive knowledge of the structural, functional, and chemical facets of polymer composites. The FTIR analysis of glass/epoxy composite laminates revealed that new functional groups were not added to the surface of the glass fibers by plasma treatment [85]. Consequently, the strength of the chemical bond between the epoxy resin and the glass fibers did not increase. In composites reinforced with natural fibers, FTIR is employed to analyze groups, including carboxyl (-C = O), hydroxyl (-OH), and other groups also. As reported in the study, FTIR has been used for composites developed by chemically functionalized styrene acrylonitrile (CF-SAN) polymer with kenaf fiber (KNF) and established that esterification and hydrogen bonding impart adhesion and bonding in the CF-SAN/KNF composites [86]. In another study, FTIR has been employed for the *Kigelia pinnata* fruit (KPF) fiber–reinforced vinyl ester composites and confirmed that steam explosion combined with silane grafting enhances the interfacial adhesion between the matrix and KPF fibers by promoting the synthesis of Si-O-Si bonds in the interfacial bonding structure [87].

#### 9.3.2.2 Raman Spectroscopy

Raman spectroscopy (Figure 9.11) is a nondestructive analytical technique that gives a comprehensive account on molecular interactions, chemical composition, phase, and polymorphic characteristics of diverse materials [88, 89]. It is based on the principle of Raman scattering, which is characterized by the scattering of photons by molecules in an excited state with higher energy levels. Sir C. V. Raman was awarded the Nobel Prize in physics in 1930 for his pioneering work on light scattering and the discovery of the Raman effect [89].

Raman spectroscopy is a potent analytical tool. Furthermore, it becomes highly impactful when used in collaboration with other methods, including microscopy, reaction kinetics, and in-depth analysis of polymers. It allows for both qualitative and quantitative assessments, as well as detailed examinations of the structure and morphology of materials [90]. In a recent study, Raman spectroscopy has been applied in the formulation and optimization of a unique material for bioelectronic applications composed of poly(3,4-ethylenedioxythiophene) doped with dodecyl benzenesulfonic acid (PEDOT:DBSA) [91]. In addition, Raman spectroscopy has also been used to examine the desired functional group and molecular status in synthesized compounds [92] and the orientation of crystals in different fibers [93].

#### 9.3.2.3 X-Ray Diffraction (XRD)

For the analysis of materials exhibiting a crystalline structure, the XRD method is employed widely. It furnishes details about structures, phases, crystal orientations, and various structural parameters, including average grain size, strain, crystallinity, and crystal defects. XRD operates based on the concept of constructive interference. When monochromatic X-rays strike a crystalline material, they undergo diffraction due to the periodic arrangement of atoms in the crystal lattice. The basic principle of XRD involves exposing a crystalline sample to X-rays, typically generated by a cathode

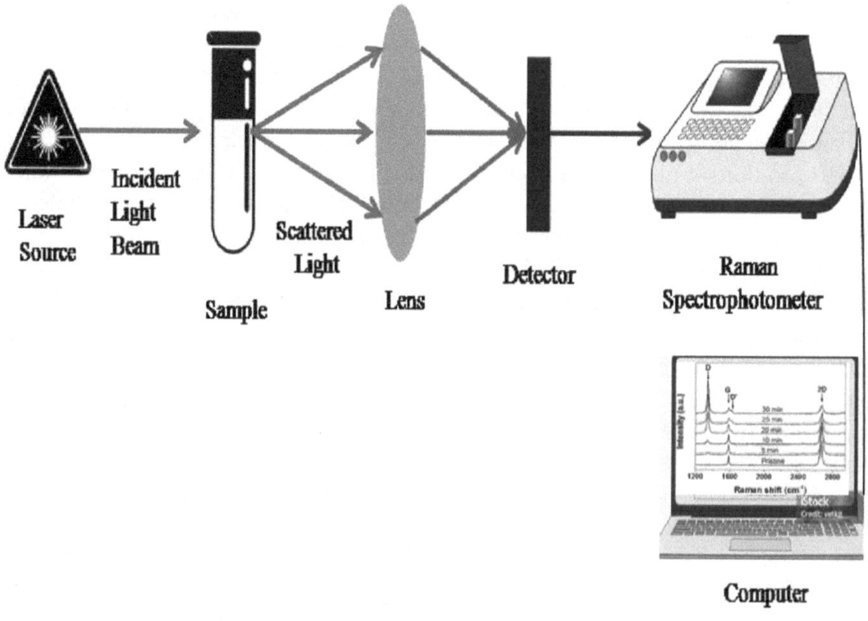


FIGURE 9.11 Raman spectroscopy setup.

ray tube [94]. These X-rays undergo constructive interference when interacting with the crystalline structure, resulting in a diffraction pattern. A monochromator is used to filter the X-rays, ensuring a specific wavelength, and a collimator focuses and narrows the beam. The sample, securely positioned in a holder, interacts with the X-rays, leading to diffracted rays. Finally, a detector captures the diffracted X-rays, converting the signal into measurable data.

XRD techniques prove valuable in the examination of semicrystalline polymers and the reorganization of crystalline phases (polymorphism) within polymers. It stands as a key method for determining the degree of crystallinity in polymers, and its composites, because polymers do not exhibit 100% crystalline nature. In XRD, the diffracted X-rays create a distinctive pattern that corresponds to the crystal structure, providing valuable insights into atomic arrangements and spacing. This distinctive technology is instrumental in identifying both the crystal structure type and the separation distance between the polymeric chains [95]. In polymer composites, the interphase structure (crystal form and orientation) of polyamide 6 (PA6) composites reinforced with high-modulus carbon fiber (HMCF) and standard-modulus carbon fiber (SMCF) has been revealed by using microbeam wide-angle X-ray diffraction (μWAXD) [96]. Based on the μWAXD findings, it was found that in PA6/SMCF and PA6/HMCF composites, the α phase is the primary crystalline fraction. Along with this, in the PA6/HMCF composite, there is an observed slight increase in the α crystalline percentage at the transcrystalline layer.

The XRD technique is also used to identify the crystallinity of natural fibers in natural FRP composites. In the reported literature, XRD was employed to study the chemical properties of hemp fiber (HF)–reinforced polypropylene (PP) composites compatibilized with maleic anhydride (MAPP) [97]. The incorporation of MAPP and HF led to the elimination of peaks associated with the beta phase in pure PP, as demonstrated by the analysis of XRD. HF outperformed other natural fibers in terms of crystallinity, measuring 82.10%, and they also had a significant molecular orientation angle (MFA) of 6.06°, which made them an excellent choice for engineering applications. Furthermore, the measured crystallite size of 32.49 nm was comparatively large. To determine the suitability of *Glycyrrhiza glabra* fibers (GGFs) for natural FRP composites, their characterization using XRD and other methods has also been reported [98]. With a 35.86% crystal index, the XRD characterization analyses of these GGF fibers confirmed the presence of cellulose.

## 9.4 IN SITU AND OPERANDO CHARACTERIZATION TECHNIQUES FOR POLYMER COMPOSITES

Polymer composites represent a diverse array of materials used in different industries, on account of their tailored properties and adaptability. Therefore, it becomes crucial to understand and analyze their behavior in natural operating conditions in order to enhance their potential and reliability in real-world applications. In this regard, in situ and operando characterization techniques are very helpful to understand the dynamic behavior of polymer composites during interactions with altering environmental conditions and throughout mechanical or chemical processes.

*In situ techniques* refer to the analytical study of materials under realistic or operating conditions, such as temperature, pressure, or in the presence of gases. Since the materials are subjected to these conditions, this technique helps to observe the changes in the properties and behavior of materials. Similarly, operando methods are a subdivision of in situ techniques used for characterizing materials under conditions relevant to their actual use in a device or process. These focus on studying the material while it is actively participating in a chemical reaction or an electrochemical process. Operando techniques offer significant understanding of catalysts, battery materials, and other materials used in energy conversion and storage devices. A number of in situ and operando characterization techniques, including microscopy (TEM and AFM) and spectroscopy (Raman, FTIR, and XRD), have been widely employed to study the microstructural changes [99], compression damage behavior [100, 101], stress transfer [102], and chemical structure variations [103] in polymer composites during processing or under specific environmental conditions.

## 9.5 CONCLUSIONS

The prominent progress in the field of advanced characterization techniques for polymer composites has offered considerable understanding on structures and behavior of these materials. Various nondestructive testing techniques as discussed in this study, including AE, UT, THz, IRT, ECT, shearography, DIC, and X-ray CT testing, have provided precise information about subsurface defects, deformation, and internal structures. Furthermore, these techniques have significantly enhanced knowledge on the different aspects of polymer composites by means of analyzing their mechanical properties, strain distribution, and damage progression.

In addition, microscopic and spectroscopic techniques, such as TEM, AFM, FTIR, Raman, and XRD, have played decisive roles in elucidating the characteristics (morphological, chemical, and mechanical properties) of polymer composites. The advancements in these techniques have made them persuasive tools for characterizing crystalline structures and phases of polymer composites, thus imparting indispensable information about their crystallographic features. Along with this, in situ and operando techniques have contributed to a deeper understanding of their dynamic responses by means of assessing the real-time monitoring of materials under natural and practical operating conditions.

## ACKNOWLEDGMENTS

The author acknowledges the Indian Institute of Technology–Roorkee for providing eminent facilities and environment for conducting research work. Special thanks to the Ministry of Human Resource Development for providing financial assistance in the form of a fellowship.

## REFERENCES

[1] Dinița, A., Ripeanu, R.G., Ilinca, C.N., Cursaru, D., Matei, D., Naim, R.I., Tanase, M., & Portoacă, A.I. (2023). Advancements in fiber-reinforced polymer composites: A comprehensive analysis. *Polymers*, *16*, 2.

[2] Jain, P., & Jariwala, H. (2022). A comprehensive review on natural fiber reinforced polymer composites and its applications. *Trends in Applications of Polymers and Polymer Composites*, *4–1*.

[3] Sajan, S., & Philip Selvaraj, D. (2021). A review on polymer matrix composite materials and their applications. *Materials Today: Proceedings*, *47*, 5493–5498.

[4] Chen, J., Yu, Z., & Jin, H. (2022). Nondestructive testing and evaluation techniques of defects in fiber-reinforced polymer composites: A review. *Frontiers in Materials*, *9*, 986645.

[5] Nsengiyumva, W., Zhong, S., Lin, J., Zhang, Q., Zhong, J., & Huang, Y. (2021). Advances, limitations and prospects of nondestructive testing and evaluation of thick composites and sandwich structures: A state-of-the-art review. *Composite Structures*, *256*, 112951.

[6] Gupta, R., Mitchell, D., Blanche, J., Harper, S., Tang, W., Pancholi, K., Baines, L., Bucknall, D.G., & Flynn, D. (2021). A review of sensing technologies for non-destructive evaluation of structural composite materials. *Journal of Composites Science*, *5*(12), 319.

[7] Naik, N., Shivamurthy, B., Thimmappa, B.H.S., Gupta, A., Guo, J.Z., & Seok, I. (2022). A review on processing methods and characterization techniques of green composites. *Engineered Science*, *20*, 80–99.

[8] Bokobza, L. (2017). Spectroscopic techniques for the characterization of polymer nanocomposites: A review. *Polymers*, *10*(1), 7.

[9] Cartz, L. (1995). *Nondestructive testing*. OSTI ID:260617. U.S. Department of Energy Office of Scientific and Technical Information, United States. OSTI.GOV.

[10] Zhou, X.Y., Qian, S.Y., Wang, N.W., Xiong, W., & Wu, W.Q. (2022). A review on stochastic multiscale analysis for FRP composite structures. *Composite Structures*, *284*, 115132.

[11] Irving, P.E., & Soutis, C. (2019). *Polymer composites in the aerospace industry*. 2nd edition. Britain: Woodhead Publishing.

[12] Heslehurst, R.B. (2014). *Defects and damage in composite materials and structures*. Florida: CRC Press.

[13] Pain, D., & Drinkwater, B.W. (2013). Detection of fibre waviness using ultrasonic array scattering data. *Journal of Nondestructive Evaluation*, *32*, 215–227.

[14] Çınar, K., & Ersoy, N. (2015). Effect of fibre wrinkling to the spring-in behaviour of L-shaped composite materials. *Composites Part A: Applied Science and Manufacturing*, *69*, 105–114.

[15] Senthil, K., Arockiarajan, A., Palaninathan, R., Santhosh, B., & Usha, K.M. (2013). Defects in composite structures: Its effects and prediction methods—A comprehensive review. *Composite Structures*, *106*, 139–149.

[16] Scruby, C.B. (1987). An introduction to acoustic emission. *Journal of Physics E*, *20*, 946–953.

[17] Ghadarah, N., & Ayre, D. (2023). A review on acoustic emission testing for structural health monitoring of polymer-based composites. *Sensors*, *23*(15), 6945.

[18] Unnþorsson, R. (2013). Hit detection and determination in AE bursts. In *Acoustic emission: Research and applications*. London, UK: InTech.

[19] Kaiser J. (1950). *An investigation into the occurrence of noises in tensile tests or a study of acoustic phenomena in tensile tests*. PhD Dissertation, Technische Hochschule Munchen, Munich, 5.

[20] Tensi, H.M. (2004). The Kaiser-effect and its scientific background. *Journal of Acoustic Emissions*, *22*, S1–S16.

[21] Dahmene, F., Yaacoubi, S., & Mountassir, M.E.L. (2015). Acoustic emission of composites structures: Story, success, and challenges. *Phys Procedia*, *70*, 599–603.

[22] Gholizadeh, S. (2016). A review of non-destructive testing methods of composite materials. *Procedia Structural Integrity*, *1*, 50–57.

[23] Brunner, A.J. (2018). Identification of damage mechanisms in fiber-reinforced polymer-matrix composites with Acoustic Emission and the challenge of assessing structural integrity and service-life. *Construction and Building Materials*, *173*, 629–637.

[24] Nazaripoor, H., Ashrafizadeh, H., Schultz, R., Runka, J., & Mertiny, P. (2022). Acoustic emission damage detection during three-point bend testing of short glass fiber reinforced composite panels: Integrity assessment. *Journal of Composites Science*, *6*(2), 48.

[25] Samborski, S., & Korzec, I. (2023). Application of the acoustic emission technique for damage identification in the fiber reinforced polymer composites. *Advances in Science and Technology Research Journal*, *17*(1), 210–221.

[26] Sivagangai, T., Rao, K.V.S.N., Bharath, M., Vamsi, V., & Babu, V.D. (2022). Failure characterization of flax/epoxy composite materials with filler under acoustic wave. *International Journal of Vehicle Structures & Systems (IJVSS)*, *14*(6).

[27] Duchene, P., Chaki, S., Ayadi, A., & Krawczak, P. (2018). A review of non-destructive techniques used for mechanical damage assessment in polymer composites. *Journal of Materials Science*, *53*(11), 7915–7938.

[28] Hasiotis, T., Badogiannis, E., & Tsouvalis, N. (2011). Application of ultrasonic C-scan techniques for tracing defects in laminated composite materials. *Strojniski Vestnik*, *57*, 192–203. https://doi.org/10.5545/sv-jme.2010.170.

[29] Shiino, M.Y., Faria, M.C.M., Botelho, E.C., & Oliveira, P.C.D. (2012). Assessment of cumulative damage by using ultrasonic C-Scan on carbon fiber/epoxy composites under thermal cycling. *Materials Research*, *15*, 495–499. https://doi.org/10.1590/S1516–14392012005000062.

[30] Fromme, P., Endrizzi, M., & Olivo, A. (2018). Defect imaging in composite structures. In *AIP conference proceedings*. AIP Publishing LLC.

[31] Ciecieląg, K., Kęcik, K., Skoczylas, A., Matuszak, J., Korzec, I., & Zaleski, R. (2022). Non-destructive detection of real defects in polymer composites by ultrasonic testing and recurrence analysis. *Materials*, *15*(20), 7335.

[32] Alarifi, I.M., Movva, V., Rahimi-Gorji, M., & Asmatulu, R. (2019). Performance analysis of impact-damaged laminate composite structures for quality assurance. *Journal of the Brazilian Society of Mechanical Sciences and Engineering*, *41*, 1–16.

[33] Jia, Z., Garnier, C., Pastor, M.L., Rousseau, J., & Gong, X. (2023). Experimental investigation of impacted multidirectional laminates under compressive static and cyclic loading. *Composite Structures*, *322*, 117335.

[34] Besson, A., & Minasyan, A. (2017). Terahertz imaging for composite non-destructive testing. In *15th Asia pacific conference for non-destructive testing*, Singapore, 28–30.

[35] Strąg, M., & Swiderski, W. (2023). Defect detection in aramid fiber-reinforced composites via terahertz radiation. *Journal of Nondestructive Evaluation*, *42*(1), 9.

[36] Zhong, S. (2019). Progress in terahertz nondestructive testing: A review. *Frontiers of Mechanical Engineering, 14*, 273–281.

[37] Hlosta, P., Nita, M., Powala, D., & Swiderski, W. (2022). Terahertz radiation in non-destructive testing of composite pyrotechnic materials. *Composite Structures, 279*, 114770.

[38] Chady, T., & Schabowicz, K. (2016). Non-destructive testing of fiber-cement boards using time domain terahertz spectroscopy. *Nondestructive Testing and Diagnostics, 1–2*, 62–66.

[39] Ibrahim, M.E., Headland, D., Withayachumnankul, W., & Wang, C.H. (2021). Nondestructive testing of defects in polymer–matrix composite materials for marine applications using terahertz waves. *Journal of Nondestructive Evaluation, 40*(2), 37.

[40] Dong, J., Pomarède, P., Chehami, L., Locquet, A., Meraghni, F., Declercq, N.F., et al. (2018). Visualization of subsurface damage in woven carbon fiber-reinforced composites using polarization-sensitive terahertz imaging. *NDT & E International, 99*, 72–79.

[41] Mouahid, A. (2018). Infrared thermography used for composite materials. In *MATEC web of conferences* (Vol. 191, p. 00011). EDP Sciences.

[42] Theodorakeas, P., Cheilakou, E., Ftikou, E., & Koui, M. (2015, October). Passive and active infrared thermography: An overview of applications for the inspection of mosaic structures. In *Journal of physics: Conference series* (Vol. 655, No. 1, p. 012061). IOP Publishing.

[43] Fernandes, H., Zhang, H., & Maldague, X. (2015). An active infrared thermography method for fiber orientation assessment of fiber-reinforced composite materials. *Infrared Physics & Technology, 72*, 286–292.

[44] Montanini, R. (2010). Quantitative determination of subsurface defects in a reference specimen made of Plexiglas by means of lock-in and pulse phase infrared thermography. *Infrared Physics & Technology, 53*(5), 363–371.

[45] Favro, L., Thomas, R., Han, X., Ouyang, Z., Newaz, G., & Gentile, D. (2001). Sonic infrared imaging of fatigue cracks. *Internatioanl Journal of Fatigue, 23*, 471–476.

[46] Ghosh, D., Gupta, H., Mittal, A.K., & Shekhar, R. (2017). Inspection of heritage structure using infrared thermography. In *Conference and exhibition of the Indian Society for NDT (December)*, Chennai, Tamil Nadu, India.

[47] Li, Y., Zhang, W., Yang, Z.-W., Zhang, J.-Y., & Tao, S.-J. (2016). Low-velocity impact damage characterization of carbon fiber reinforced polymer (CFRP) using infrared thermography. *Infrared Physics & Technology, 76*, 91–102.

[48] Popow, V., Vogtmann, J., & Gurka, M. (2022). In-situ characterization of impact damage in carbon fibre reinforced polymers using infrared thermography. *Infrared Physics & Technology, 122*, 104074.

[49] Pothnis, J.R., Hajagolkar, A.K., Anilchandra, A.R., Das, R., & Gururaja, S. (2023). Open-hole fatigue testing of UD-GFRP composite laminates containing aligned CNTs using infrared thermography. *Composite Structures, 324*, 117557.

[50] Garcia-Martin, J., Gomez-Gil, J., & Vazquez-Sanchez, E. (2011). Non-destructive techniques based on eddy current testing. *Sensors, 11*(3), 2525–2565.

[51] Ali, K.B., Abdalla, A.N., Rifai, D., & Faraj, M.A. (2017). Review on system development in eddy current testing and technique for defect classification and characterization. *IET Circuits, Devices & Systems, 11*(4), 338–351.

[52] Wu, D., Cheng, F., Yang, F., & Huang, C. (2019). Non-destructive testing for carbon-fiber-reinforced plastic (CFRP) using a novel eddy current probe. *Composites Part B: Engineering, 177*, 107460.

[53] Kacprzak, D., & Tuleasca, I. (2017). Wheatstone bridge approach to the inspection of composite material. In *2017 eleventh international conference on sensing technology (ICST)*, IEEE, 1–6.

[54] Hamill, L., Emerson, J., McGushion, K., & Nutt, S. (2018). Low frequency eddy current testing of insulators and composites. *Journal of Nondestructive Evaluation, 37*, 58.

[55] Ivanov, A.D., Minaev, V.L., & Vishnyakov, G.N. (2019). A shearograph for nondestructive testing of products obtained by additive technologies. *Instruments and Experimental Techniques, 62*, 871–875.

[56] Guo, B., Zheng, X., Gerini-Romagnoli, M., & Yang, L. (2023). Digital shearography for NDT: Determination and demonstration of the size and the depth of the smallest detectable defect. *NDT & E International, 139*, 102927.

[57] Tao, N., Anisimov, A.G., & Groves, R.M. (2022). Shearography non-destructive testing of thick GFRP laminates: Numerical and experimental study on defect detection with thermal loading. *Composite Structures, 282*, 115008.

[58] Xie, X., & Zhou, Z. (2018). Shearographic nondestructive testing for high-pressure composite tubes. SAE Technical Paper 2018-01-1219. https://doi.org/10.4271/2018-01-1219.

[59] McCormick, N., & Lord, J. (2010). Digital image correlation. *Materials Today, 13*(12), 52–54.

[60] Pan, B., Qian, K., Xie, H., & Asundi, A. (2009). Two-dimensional digital image correlation for in-plane displacement and strain measurement: A review. *Measurement Science and Technology, 20*(6), 062001.

[61] Ab Ghani, A.F., Ali, M.B., Dhar Malingam, S., & Mahmud, J. (2016). Digital image correlation (DIC) technique in measuring strain using opensource platform Ncorr. *Journal of Advanced Research in Applied Mechanics, 26*, 10–21.

[62] Quanjin, M., Rejab, M.R.M., Halim, Q., Merzuki, M.N.M., & Darus, M.A.H. (2020). Experimental investigation of the tensile test using digital image correlation (DIC) method. *Materials Today: Proceedings, 27*, 757–763.

[63] Eremin, A., Burkov, M., Luybutin, P., & Kononova, A. (2023). Mechanical behavior of carbon and aramid fiber reinforced composites with various layups under tension. *Procedia Structural Integrity, 50*, 65–72.

[64] Szebenyi, G., & Hliva, V. (2019). Detection of delamination in polymer composites by digital image correlation: Experimental test. *Polymers, 11*(3), 523.

[65] Shi, J., Zong, X., Jiang, W., Yao, R., & Zheng, J. (2023). Evaluation of the fracture toughness of short carbon fiber reinforced thermoplastic composites. *The Journal of Engineering Materials and Technology, 146*(2), 021002.

[66] Garcea, S.C., Wang, Y., & Withers, P.J. (2018). X-ray computed tomography of polymer composites. *Composites Science and Technology, 156*, 305–319.

[67] Kastner, J. (2016). Special issue on the 6th conference on industrial computed tomography 2016 (iCT2016). *Case Studies in Nondestructive Testing and Evaluation, 6*, 2–3.

[68] Withers, P.J., Bouman, C., Carmignato, S., Cnudde, V., Grimaldi, D., Hagen, C.K., Manley, M., Pliess, A.D., & Stock, S.R. (2021). X-ray computed tomography. *Nature Reviews Methods Primers, 1*(1), 18–21.

[69] Frącz, W., Janowski, G., Pruchniak, M., & Walek, L. (2021). The use of computed tomography in the study of microstructure of molded pieces made of poly (3-hydroxybutyric-co-3-hydroxyvaleric acid) (PHBV) biocomposites with natural fiber. *Polymers, 13*(17), 2942.

[70] Li, Y., Meng, J., Niu, G., Yang, H., Wang, P., Lei, H., & Fang, D. (2023). Cryogenic damage mechanisms of CFRP laminates based on in-situ X-ray computed tomography characterization. *Composites Science and Technology*, 110413.

[71] Pace, F., Stamopoulos, A.G., Eckl, M., Senck, S., & Glinz, J. (2023). Analysis of the manufacturing porosity in fused filament fabricated onyx/long fiber reinforced composites using X-ray computed tomography. *Journal of Non destructive Evaluation, 42*(4), 86.

[72] Anuse, V.S., Shankar, K., Velmurugan, R., & Ha, S.K. (2022). Compression-after-impact analysis of carbon fiber reinforced composite laminate with different ply orientation sequences. *International Journal of Impact Engineering, 167*, 104277.

[73] Zuo, J.M., & Spence, J.C.H. (2017). *Advanced transmission electron microscopy* (p. 741). New York: Springer Science + Business Media. https://doi.org/10.1007/978-1-4939-6607-3.

[74] Noorunnisa Khanam, P., & Al Ali AlMaadeed, M. (2015). Processing and characterization of polyethylene-based composites. *Advanced Manufacturing: Polymer & Composites Science, 1*(2), 63–79. https://doi.org/10.1179/2055035915Y.0000000002.

[75] Williams, D.B., & Carter, C.B. (2009). *Transmission electron microscopy: A textbook for materials science* (p. 779). https://doi.org/10.1007/978-0-387-76501-3.

[76] Libera, M.R., & Egerton, R.F. (2010). Advances in the transmission electron microscopy of polymers. *Polymer Reviews, 50*(3), 321–339. https://doi.org/10.1080/15583724.2010.493256.

[77] Maciel, C.C., de Barros, A., Mazali, I.O., & Ferreira, M. (2023). Flexible biodegradable electrochemical sensor of PBAT and CNDs composite for the detection of emerging pollutants. *Journal of Electroanalytical Chemistry, 940*, 117491. https://doi.org/10.1016/j.jelechem.2023.117491.

[78] Mao, T., Shi, X., Liu, M., Su, Q., Cheng, Y., Fang, C., & Luo, X. (2023). Study on pore size distribution method of SBA-15 in polymer composites. *Materials Today Communications, 35*, 105592. https://doi.org/10.1016/j.mtcomm.2023.105592.

[79] Akarken, G., & Cengiz, U. (2023). Fabrication and characterization of metakaolin-based fiber reinforced fire resistant geopolymer. *Applied Clay Science, 232*, 106786. https://doi.org/10.1016/j.clay.2022.106786.

[80] Jagtap, R.N., & Ambre, A.H. (2006). Overview literature on atomic force microscopy (AFM): Basics and its important applications for polymer characterization. *Indian Journal of Engineering & Materials Sciences, 13*, 368–384.

[81] Melelli, A., Arnould, O., Beaugrand, J., & Bourmaud, A. (2020). The middle lamella of plant fibers used as composite reinforcement: Investigation by atomic force microscopy. *Molecules, 25*(3), 632.

[82] Razaina, M.T., Ariawan, D., & Mohd Ishak, Z.A. (2016). Surface characterization of alkali treated kenaf fibers by XPS and AFM. *Key Engineering Materials*, *694*, 29–33.
[83] Oporto, G.S., Douglas, J.G., George, B., & Neivandt, D.J. (2009). Forced air plasma treatment (FAPT) of hybrid wood plastic composite (WPC)-fiber reinforced plastic (FRP) surfaces. *Composite Interfaces*, *16*(7–9), 847–867.
[84] Khan, S.A., Khan, S.B., Khan, L.U., Farooq, A., Akhtar, K., & Asiri, A.M. (2018). Fourier transform infrared spectroscopy: Fundamentals and application in functional groups and nanomaterials characterization. In S. Sharma (Ed.), *Handbook of materials characterization*. Cham: Springer. https://doi.org/10.1007/978-3-319-92955-2_9
[85] Ahangar, M., & Saeedifar, M. (2023). The effect of plasma treatment on the mode II interlaminar fracture toughness of Glass/Epoxy laminates. *Theoretical and Applied Fracture Mechanics*, *127*, 104093.
[86] Tamta, N., & Palsule, S. (2024). Kenaf fiber reinforced chemically functionalized styrene-acrylonitrile composites. *Journal of Thermoplastic Composite Materials*, *37*(1), 5–27.
[87] Vishal, K., Rajkumar, K., Sabarinathan, P., & Arun, A. (2023). Mechanical and thermal characteristics of steam-exploded silane grafted Kigelia Pinnata fruit (KPF) fiber reinforced vinyl ester polymer composites. *Polymer Composites*, *1*(1), 1–17.
[88] Ahlawat, N. (2014). Raman spectroscopy: A review. *International Journal of Computer Science and Mobile Computing*, *3*(11), 680–685.
[89] Vaskova, H. (2011). A powerful tool for material identification: Raman spectroscopy. *International Journal of Mathematical Models and Methods in Applied Sciences*, *5*(7), 1205–1212.
[90] Gerrard, D.L., & Birnie, J. (1992). Raman spectroscopy. *Analytical Chemistry*, *64*(12), 502–513.
[91] Tumova, S., Maleckova, R., Kubac, L., Akrman, J., Enev, V., Kalina, L., & Weiter, M. (2023). Novel highly stable conductive polymer composite PEDOT: DBSA for bioelectronic applications. *Polymer Journal*, 1–13.
[92] Nayak, D., & Choudhary, R.B. (2024). ZnS-integrated polypyrrole and emeraldine polyaniline composites for enhance photophysical and electrochemical performance for an active layer in an optoelectronic device. *Journal of Molecular Structure*, *1299*, 137135.
[93] Uematsu, H., Ishikawa, M., Yamaguchi, A., Sugihara, S., Nishitsuji, S., Nishimura, F., Yamane, M., Kawabe, K., Ozaki, Y., & Tanoue, S. (2024). Enhancement of flexural properties of carbon fiber-reinforced polyamide 6 via oriented crystallization of polyamide 6 among carbon fibers. *Composites Part A: Applied Science and Manufacturing*, *176*, 107837.
[94] Bunaciu, A.A., Udriştioiu, E.G., & Aboul-Enein, H.Y. (2015). X-ray diffraction: Instrumentation and applications. *Critical Reviews in Analytical Chemistry*, *45*(4), 289–299.
[95] Chowdhury, G., Kruczek, B., & Matsuura, T. (Eds.). (2001). *Polyphenylene oxide and modified polyphenylene oxide membranes: Gas, vapor, and liquid separation*. Springer Science & Business Media.
[96] Kobayashi, D., Hsieh, Y.T., & Takahara, A. (2016). Interphase structure of carbon fiber reinforced polyamide 6 revealed by microbeam X-ray diffraction with synchrotron radiation. *Polymer*, *89*, 154–158.
[97] da Silveira, P.H.P.M., Santos, M.C.C.D., Chaves, Y.S., Ribeiro, M.P., Marchi, B.Z., Monteiro, S.N., Gomes, A.V., Tapanes, N.D.L.C.O., Pereira, P.S.D.C., & Bastos, D.C. (2023). Characterization of thermo-mechanical and chemical properties of polypropylene/hemp fiber biocomposites: Impact of maleic anhydride compatibilizer and fiber content. *Polymers*, *15*(15), 3271.
[98] Ovalı, S. (2023). Characterization of lignocellulosic glycyrrhiza glabra fibers as a potential reinforcement for polymer composites. *Journal of Thermoplastic Composite Materials*, 08927057231151928.
[99] Ishikawa, T., Tezura, M., & Kizuka, T. (2020). Direct observation of microstructural fracture dynamics in carbon fiber reinforced plastics via in situ transmission electron microscopy. *Composites Science and Technology*, *198*, 108264.
[100] Hapke, J., Gehrig, F., Huber, N., Schulte, K., & Lilleodden, E.T. (2011). Compressive failure of UD-CFRP containing void defects: In situ SEM microanalysis. *Composites Science and Technology*, *71*(9), 1242–1249.
[101] Hamdi, K., Moreau, G., & Aboura, Z. (2021). Digital image correlation, acoustic emission and in-situ microscopy in order to understand composite compression damage behavior. *Composite Structures*, *258*, 113424.
[102] Mu, M., Osswald, S., Gogotsi, Y., & Winey, K.I. (2009). An in situ Raman spectroscopy study of stress transfer between carbon nanotubes and polymer. *Nanotechnology*, *20*(33), 335703.
[103] Watanabe, R., Sugahara, A., Hagihara, H., Mizukado, J., & Shinzawa, H. (2020). In situ Fourier transform infrared spectroscopic imaging for elucidating variations in chemical structures of polymer composites at the matrix–filler interface during reactive processing. *Macromolecules*, *53*(24), 10711–10717.

# 10 Polymer Nanocomposites
## *Synthesis, Characterization, and Applications*

*Kannan PK, G Bhavani, Praveen Eliyas, and Mariappan Anandkumar*

## 10.1 INTRODUCTION

Polymer nanocomposites have emerged as one of the most promising materials in the field of nanotechnology. These nanocomposites possess unique physical and chemical properties that can be exploited in a wide range of applications. For the past five decades, research was focused on macrocomposites, such as filler-reinforced polymer–matrix composites with filler or fiber diameter length scales in micrometers. Polymer composites are commercially produced for a wide range of uses, including in vehicles, sporting goods, and aircraft components. We have discovered polymer nanocomposites over the last 20 years, where the length scale of the reinforcement (nanomaterials) is on the nanometer scale [1]. Polymer nanocomposites have the potential to have a substantial influence on growth at all levels of the global economy in the 21st century.

Polymer nanocomposites (PNs) are made up of a polymeric material (e.g., thermoplastics, thermosets, or elastomers) and a nanoscale reinforcing substance (nanoparticle). Due to the nanosize of the filler, it offers a very large interfacial area per unit volume. Also, the electron's wavelike properties dominate once the material's dimension is shrunk to nanosize. Also, such structures offer an interesting possibility to control fundamental properties, such as melting point, charge storage capacity, magnetic properties, etc., without altering the material's compositions [1]. The nanofillers (nanomaterials) in the polymer nanocomposites have ultralarge interfacial area per unit volume, which at low concentrations have a significant impact on the physical properties of the polymer nanocomposites.

This is not possessed in the conventional macrocomposites. Furthermore, we can fabricate and process many polymeric nanocomposites via a similar way of conventional polymer-based macrocomposites, indicating their particular attractiveness from a manufacturing point of view. Due to the nanodimension, polymer nanocomposites offer various advantages, like enhanced mechanical properties, controllable optical and electrical properties, enhanced thermal conductivity, ablation and chemical resistance, etc.; it can also have a significant impact on the processing parameters due to increased viscosity, non-uniform dispersion, diminished optical and dimensional stability, etc.

### 10.1.1 Advantages of Polymer Nanocomposites over Macrocomposites

An ideal polymer nanocomposite should have the following attributes:

1. *Uniform distribution of the nanofillers into the polymer matrix.* This leads to enhanced interfacial contact between the polymer matrix and the nanofiller, which enhances the physical properties of the polymer. Achieving this is highly difficult due to agglomerates that can form easily due to the highly reactive nanofillers.
2. *Enhanced evaluation of the nanofiller dispersion in the polymer matrix.* Mechanical and thermal properties are strongly related to how the nanofiller is dispersed with the polymer matrix.

DOI: 10.1201/9781003476337-10

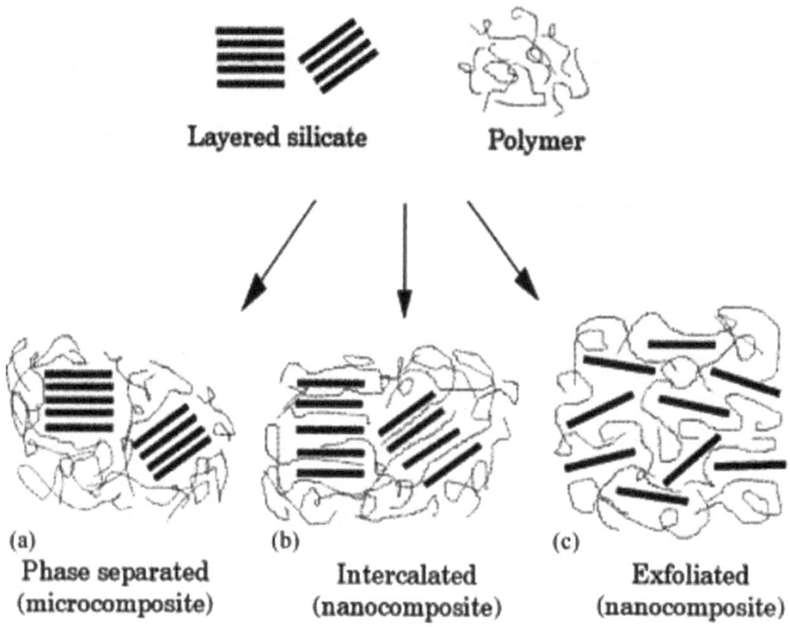


**FIGURE 10.1** Classification of polymer nanocomposites based on the dispersion of the fillers in the polymer matrix.

**Source: Based on [2].**

For example (Figure 10.1), in case of layered silicates, when dispersed in a polymer matrix, depending upon how the nanosilicates disperse within the polymer matrix, it can exist in:

1. Phase separated, from where the silicates do not interact with the polymer matrix.
2. Intercalated and exfoliated nanocomposites, where the polymer diffuses into the nanosilicates or individual layers of the nanosilicate disperse into the polymer matrix.

The properties of phase-separated composites are similar to that of the polymer phase, but in case of intercalated and exfoliated composites, the properties are enhanced significantly, owing to the increased interfacial contact between the polymer matrix and the nanofillers [2].

The properties of polymer nanocomposite can be tuned by varying the synthesis technique and surface treatments and altering the crystallinity, molecular weight, and chemistry of the polymerization.

### 10.1.2 Filler and Reinforcement

Fillers and reinforcements are used to change and improve the physical and mechanical properties of plastics. In general, reinforcing fillers increases the mechanical properties of polymer composites, while particular fillers of various types enhance a particular property.

In particular, if a filler is added only to enhance the mechanical property alone, then such a filler can be called as reinforcement. They are generally added in large concentrations to enhance the mechanical properties of the polymer. Examples: carbon fiber, clay, etc.

Besides that, a filler that does not only enhance the mechanical strength of the polymer but also helps enhance other properties, like thermal stability, impact resistance. Examples: minerals, antioxidants, metal fillers, etc.

## 10.2 COMMON NANOCOMPOSITE TECHNOLOGIES

Polymer nanocomposites can be created by combining various commercially available nanoparticle kinds into the polymer matrix. The selection of the nanoparticle required to provide the intended result must be determined by the researcher based on the application. Even though the nanofillers can significantly enhance mechanical, thermal, optical, and electrical properties, the addition of a filler at higher concentrations can also have non-uniform dispersion due to the viscosity increment [3]. So a proper control on the concentration, type, and dispersion should be achieved in order to utilize the composite for the desired application. The following are the most commonly used/utilized nanofillers.

### 10.2.1 Carbon-Based Nanofillers

CNTs are ultrathin carbon fibers with a diameter in nanometer and length in micrometer. CNTs were first discovered in 1991 and have since been used in a variety of applications. The structure of carbon nanotubes is composed of enrolled graphite sheets, which are sheets of carbon atoms arranged in a planar hexagonal lattice. The nanotube can be divided into multiwall (MWC) and single-wall (SWC) nanotubes, depending on the preparation method. SWC nanotubes are composed of a single layer of graphene rolled up into a seamless cylinder. MWC nanotubes consist of two concentric cylinders of graphene sheets coiled around a central hollow core. In addition to their exceptional electrical and conducting properties, carbon nanotubes also have excellent mechanical properties. The elastic modulus of a carbon nanotube is in the order of 1 TPa, and the tensile strength of a single-carbon nanotube can reach up to 300 GPa [4].

Due to their outstanding properties, there has been a lot of interest in polymer nanocomposites with CNT. Carbon nanotube incorporation into polymer matrices has been proposed as a strategy to produce composite materials with electrical characteristics and superior mechanical and thermal characteristics. These fillers are materials that are difficult to disperse in a polymeric matrix [5]. Problems of agglomeration are commonly observed during processing because of the low polymer-to-CNT interaction (see Figure 10.2).

Another major advantage of carbon nanotubes is that it can be functionalized very easily, and such functionalization can significantly alter the dispersion of CNT into the filler matrix. By treating carbon nanotubes with a mixture of nitric and sulfuric acid, carboxylic acid groups can be obtained on the surface, thus resulting in the uniform dispersion of CNT and enhanced electrical properties [7]. In addition to obtaining nanocomposites using a single polymer matrix, the mixture of two different polymer matrices into polymer blends enhances the distribution of CNT into the matrix.

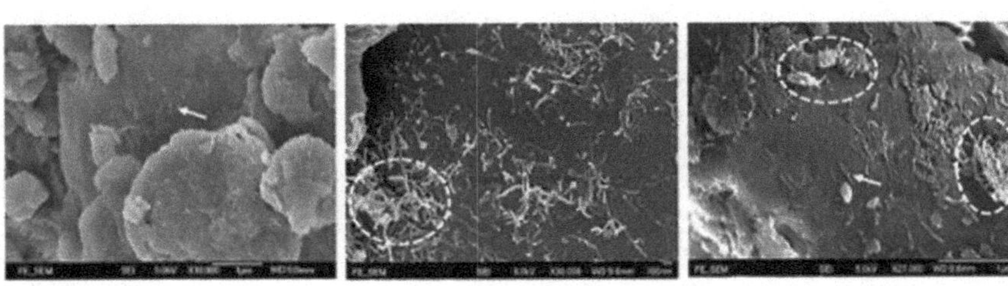

(a) (b) (c)

**FIGURE 10.2** (a) CNTs embedded on AZ61 surface after 30 min. (b) Agglomerates of CNT in AZ61/CNT. (c) Agglomerates remain after 60 min milling.

**Source: Based on [6].**

### 10.2.2 Nanosilica

A patent on the high-temperature hydrolysis process of metallic oxides to produce extremely fine particle oxides by Degassa was further modified into large-scale production of silicon dioxide, aluminum oxide, and titanium dioxide. Silicon dioxide manufactured using this route led to the product AEROSIL [8].

Amorphous, highly scattered, and ultrapure silica known as AEROSIL is created when silicon tetrachloride is hydrolyzed at a high temperature in an oxyhydrogen gas flame. It is a fluffy white powder made up of primary particles with a spherical shape. The average primary particle size of AEROSIL OX 50 is 40 nm, whereas the size of both AEROSIL 300 and 380 is 7 nm. The main particles have no pores and are spherical in shape. Agglomerates are formed when the main particles in the flame combine to create aggregates, which then reversibly unite to form agglomerates. The SEM micrograph of AEROSIL 300 is displayed in Figure 10.3, where the main particles, aggregates, and agglomerates are all easily visible. Based on the AEROSIL grade, the principal particle average sizes range from 7 to 40 nm.

The extremely tiny particles result in a huge specific surface area, ranging from around 50 $m^2/g$ (AEROSIL OX 50 has an average primary particle size of 40 nm) to 380 $m^2/g$ (AEROSIL 380 has an average primary particle size of 7 nm). The magnitude of the surface area is easily shown by comparing 20 g of AEROSIL 200 (average primary particle size is 12 nm) to the surface area of an American football field. AEROSIL is entirely made of amorphous silicon dioxide, and it begins to sinter and transform into glass at 1,200°C.

AEROSIL is utilized effectively in a variety of applications, including reinforcing fillers, thickening and thixotropic agents, antisettling agents, and free-flow aids.

AEROSIL is a submicrometer amorphous silica that has been used effectively in liquid systems for decades as a thickening agent and thixotrope. It is also employed in the modification of the rheological characteristics of epoxy resins. It also acts as a rheological assist in providing epoxy systems with effective, stable thickening and thixotropy. For rheology control, AEROSIL-fumed silica is commonly used in silicone rubber, coatings, plastics, printing inks, adhesives, lubricants, lotions, ointments, and toothpaste. Nelson and colleagues utilized this nanoparticle with polystyrene in a recent study, revealing a considerable improvement in polystyrene flammability.

### 10.2.3 Layered Nanoclays

*Nanoclays* are a type of material composed mostly of layered silicates or clay minerals with residues of metal oxides and organic substances. Clay minerals are hydrous aluminum phyllosilicates that can include varying quantities of iron, magnesium, alkali metals, alkaline earths, and other

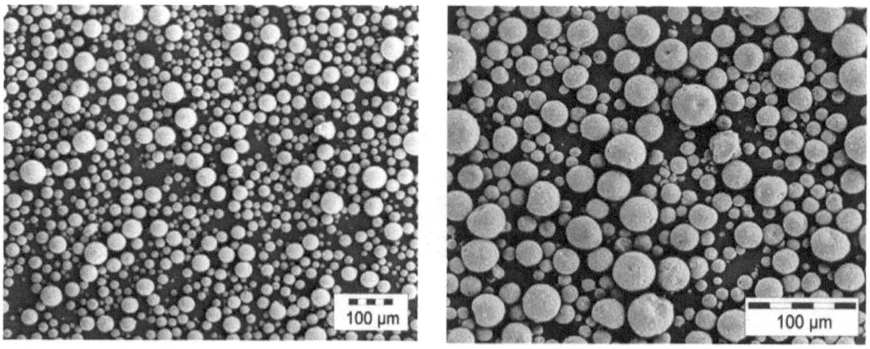


**FIGURE 10.3** FESEM images of AEROSIL 300.

**Source: Based on [9].**

cations [10]. It is reported that exfoliated clay in polymer results in a considerable improvement in stiffness, fire retardancy, and barrier characteristics, beginning with a very low nanoparticle volume percentage [11].

Because of their lamellar structure and large specific surface area (750 $m^2/g$), clays have been proven to be efficient reinforcing fillers for polymers [12]. Because of their inexpensive cost, swelling qualities, and high cation exchange capacities, smectite clays are a needed choice for the creation of polymer nanocomposites. These clays include montmorillonite, saponite, laponite, hectorite, sepiolite, and vermiculite. Because of its extensive availability, well-known intercalation/exfoliation chemistry, high surface area, and reactivity, montmorillonite is the most commonly utilized clay in polymer nanocomposites. Montmorillonite (MMT) is made up of two tetrahedral silica sheets with an alumina octahedral sheet in the center (2:1 layered structure), and hydrated exchangeable cations occupy the gaps between the lattices, as seen in Figure 10.4.

Since most polymers are hydrophobic and natural montmorillonite is hydrophilic, the clay surface needs to be altered in these situations to produce organic clay. After alteration, the clay that was formerly hydrophilic became organophilic [12]. To extend the clay galleries of smectite clays, organic ions are substituted for inorganic exchange cations on the gallery surfaces. This makes it easier for either of the polymer chains to intercalate, or penetrate, into the gallery area. The method of silanization is another form of modification that has been used for nanoclays. Through reactive extrusion, the addition of organosilanes to the clay enhanced covalent connections between the polymer and the clay, resulting in robust interactions between the clay and the matrix.

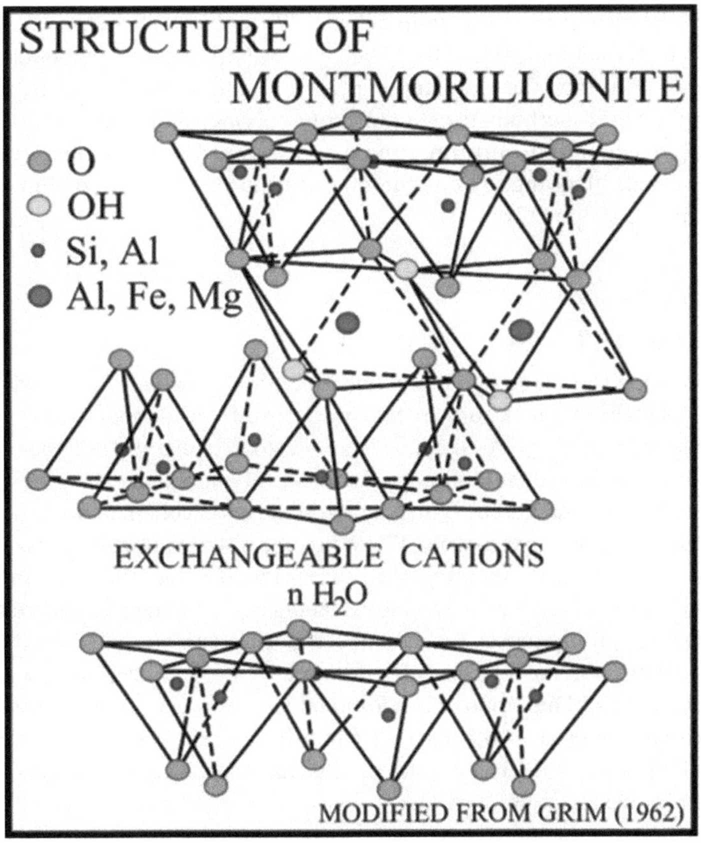


**FIGURE 10.4** Crystal structure of montmorillonite.

**Source: Based on [13].**

### 10.2.4 Metal Hydrides

The use of metal hydride (MH) alloys in hydrogen technology has piqued the curiosity of researchers [14]. These materials react reversibly with hydrogen, allowing them to be effectively used in solid-state gas storage. It is projected to improve, even if an increase in undesirable degradative events (such as oxidation) may reduce the material's total storage capacity. Furthermore, from a technical standpoint, the presence of unconfined nanoparticles within the device might obstruct gas passage through the material. In any case, storing hydrogen as a chemical molecule in MH beds looks to be a viable, cost-effective, and safe approach of storing hydrogen in the near future [15].

Besides the aforementioned categories, several other nanofillers, such as nanocellulose, titanium dioxide, zeolites, etc., are also tested as fillers, and polymer nanocomposites are successfully fabricated.

## 10.3 SYNTHESIS OF POLYMER NANOCOMPOSITES

*Polymer nanocomposites* are materials composed of polymers and nanoscale filler materials. Nanofillers are added to the polymer matrix during the production of polymer nanocomposites to improve the characteristics of the final product. Crucial factors should be considered while fabricating polymer nanocomposites to comprehend the characteristics and traits of the newly manufactured materials, such as (1) the chemical structure of the polymer, (2) mass of the polymer, (3) chemical solubility of the polymer, (4) polymer thermal stability, (5) semicrystallinity of polymer, (6) chemical structure of the nanoparticle, (7) active surface area of the nanoparticle, and (8) dispersion of the nanoparticle [16]. The synthesis of polymer nanocomposites involves incorporating nanofillers into the polymer matrix to enhance the properties of the resulting material. There are several methods for synthesizing polymer nanocomposites, and these methods can be broadly categorized into three main types: (1) in situ polymerization, (2) solution dispersion, and (3) melt blending. The technique used to prepare the polymer nanocomposite also affects its ultimate shape [17].

### 10.3.1 In Situ Polymerization

#### 10.3.1.1 Sol-Gel Method

This method involves the formation of a colloidal suspension of nanoparticles, which is then mixed with monomers. Polymerization occurs in the presence of the nanoparticles, forming polymer chains with embedded nanofillers. As the solvents used to dissolve the polymers are very hazardous, in situ polymerization is a suitable strategy for those polymers that cannot be synthesized safely and affordably using solution processes. This method achieves excellent nanoparticle dispersion and distribution throughout the polymer matrix [17]. When the lower-molecular-mass monomer trickles between the interlayers and causes swelling, the process involves the filler swelling in the monomer solution or liquid monomer (Figure 10.5). The advantages of this technique are that through this method, thermoplastic- and thermoset-based nanocomposites can be synthesized.

Additionally, it allows polymers to be grafted onto the filler surface, improving the final nanocomposite's characteristics. This approach additionally allows for the development of similarly partially exfoliated structures due to improved nanofiller dispersion and intercalation within the polymeric matrix [18]. This is likely because smaller nanoparticles have higher surface energies, promoting further particle aggregation [19].

#### 10.3.1.2 Solution Dispersion

The solution dispersion approach works effectively when the nanofillers swell and the polymer is solvent-soluble. The nanofillers are dissolved in a specific solvent, such as water, toluene, or chloroform, during this procedure. The polymeric chains intercalate, and the solvents dissolve into the interlayer of nanofillers, when the nanofiller and polymer are combined in the solution. Polymer nanocomposites are formed when the solvent is withdrawn while the exfoliated/intercalated structure remains. This procedure includes vigorous agitation, regulated solvent evaporation, and composite film casting to disperse the nanofiller in the polymer solution (Figure 10.6). Shear mixing, magnetic stirring, reflux, and sonication produce agitation [20]. This method makes it simple and

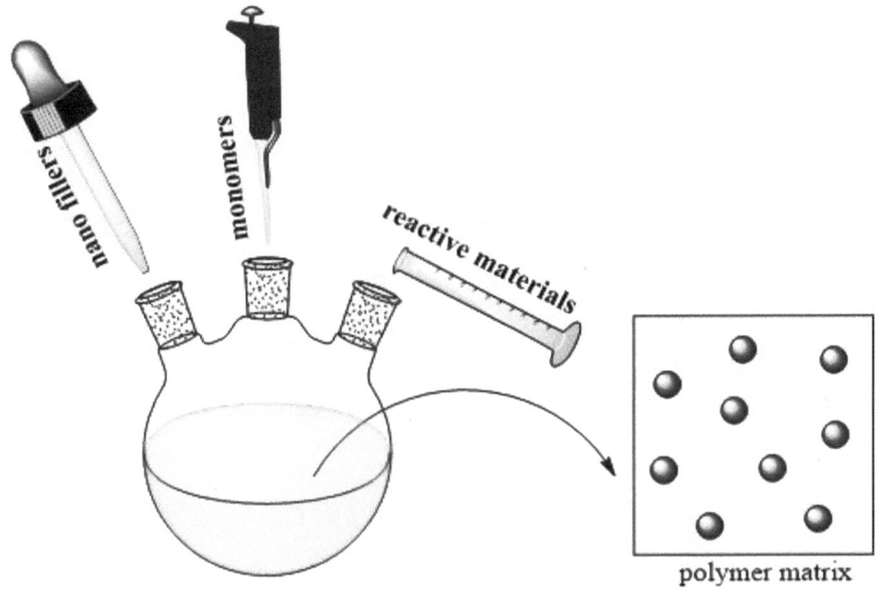


**FIGURE 10.5** Sol-gel technique: in situ polymerization method.

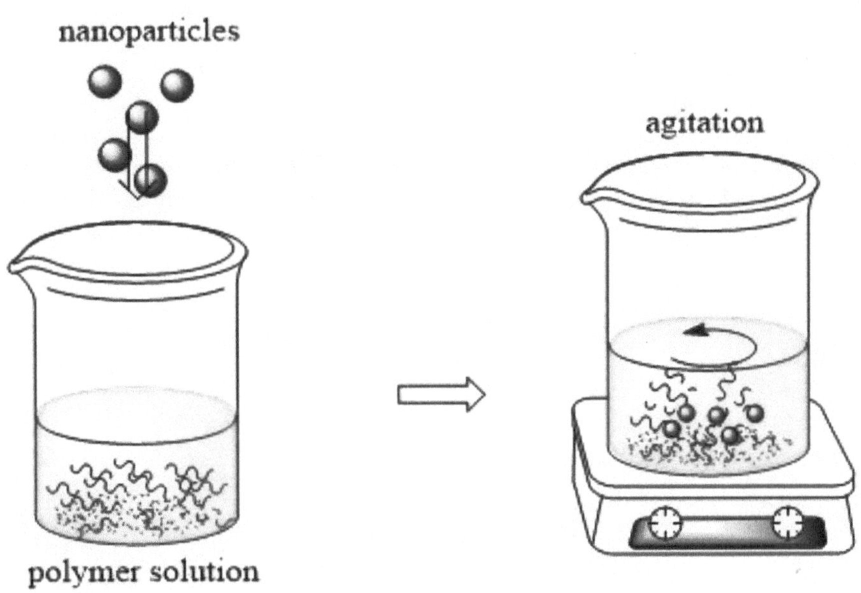


**FIGURE 10.6** Scheme for fabricating polymer nanocomposites through solution dispersion.

fast to produce superior nanocomposites. When a less-hazardous solvent, such as water, acetone, alcohol, or chloroform, is employed, this approach works better [21]. Because of the nanoparticles' improved interaction with the solvent and polymer during this procedure, varying numbers of nanoparticles may be spread throughout the polymer [17]. In this procedure, the mixture becomes polymerized in an atmosphere of nitrogen and dried in a vacuum, and then an initiator and a chain transfer agent are added [19].

### 10.3.2 Melt Blending

A standard method to manufacture polymer nanocomposites is melt blending, which is particularly effective for getting nanofillers to disperse within a polymer matrix. The polymer and nanofillers are molten during this process so that they may be thoroughly combined. The main objective is to guarantee that the nanofillers are well-distributed throughout the polymer matrix, since this might improve the final nanocomposite material's mechanical, thermal, and barrier qualities. There are various types of blending of polymer nanocomposites, such as solution intercalation/exfoliation, melting-mixing, twin-screw extrusion, etc.

*Solution intercalation/exfoliation.* Nanofillers and the polymer are dispersed in a solvent. The solvent is then removed, and the remaining mixture is melt-blended to achieve a uniform dispersion of nanoparticles within the polymer matrix. Exfoliation occurs when individual nanoparticles are dispersed between polymer chains.

*Melting-mixing.* The nanofillers and the polymer are mixed in the molten state in this method. The mixture is then cooled to solidify the nanocomposite material.

*Twin-screw extrusion.* This is a continuous processing method, where polymer pellets and nanofillers are fed into a twin-screw extruder. The mechanical action of the screws facilitates the dispersion of nanoparticles in the polymer melt.

Melt-blending is the most widely used technique to produce a polymer nanocomposite because it is a flexible and responsive process that can be utilized at the industrial level without requiring solvents [22]. Due to the absence of solvents and monomers, this technique for producing polymer nanocomposite is economical and environmentally friendly. In addition, this technique is simple to apply and efficient [23]. As shown in the scheme in Figure 10.7, a die with the appropriate cross section may modify the material's initial physical characteristics while maintaining regulated pressure and temperature [24]. The many characteristics involved in blending processes, such as viscosity and how it fluctuates with temperature and shear rate, elasticity, extensional flow, and material slippage over hot metal surfaces, must be closely investigated and controlled [25]. The essential advantage of this procedure over previous methods is that no solvent is needed. Since this method facilitates nanoparticle agglomeration more quickly than other approaches, spreading the amounts of nanoparticles requires constant supervision. Because the melting and degradation points of certain biopolymers and natural polymers are so similar, care should be taken while using this technology.

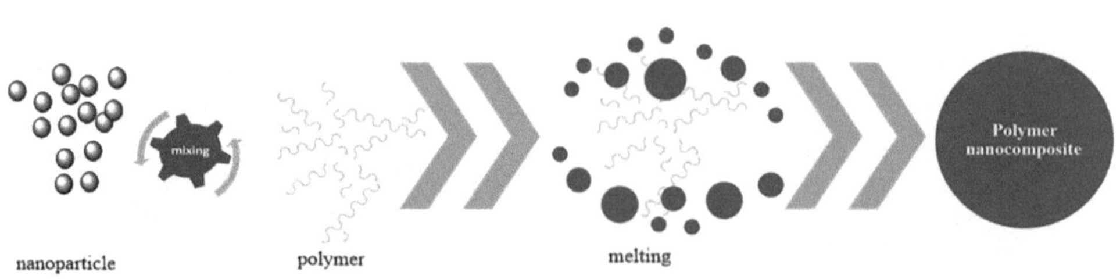


**FIGURE 10.7** Scheme of obtaining polymer nanocomposites by melt-blending.

### 10.3.3 Other Methods

#### 10.3.3.1 Inclusion Complex Formation

Nanofillers can form inclusion complexes with polymers, enhancing compatibility and dispersion in the polymer matrix.

#### 10.3.3.2 Layer-by-Layer Assembly

This technique involves the alternate deposition of the polymer and the nanofiller layers to build up a multilayered structure.

The choice of synthesis method depends on factors such as the type of polymer, the nanofiller's nature, and the nanocomposite material's desired properties. Each method has advantages and limitations, and researchers continue exploring and developing new approaches for synthesizing polymer nanocomposites.

## 10.4 CHARACTERIZATION OF NANOCOMPOSITES

Various features of polymer nanocomposites are understood by characterization. Features like quality of dispersion, along with orientation of the filler in the polymer matrix, how parameters of methodology of synthesis affect the morphology, the impact of filler surface, and analysis of their potential applications, can be done through characterization. Normally, more than one technique of characterization is needed to characterize the nanocomposites, such as thermogravimetric analysis, differential scanning calorimetry (thermal analysis), transmission electron microscopy, scanning electron microscopy, X-ray diffraction (structural and morphological analysis), nuclear magnetic resonance, IR spectroscopy, Raman spectroscopy, X-ray photoelectron spectroscopy, dielectric relaxation spectroscopy, atomic force microscopy, etc. In addition to the aforesaid, mechanical performance, fire behavior, barrier performance, biodegradability, rheological properties, PVT characterization, tribological behavior, etc. are also done. However, lack of sensitivity of the analysis and limits of the equipment can lead to wrong conclusions about the nanocomposite structure. Hence, careful analysis should be done before interpretation of results.

### 10.4.1 X-Ray Diffraction (XRD)

X-ray diffraction (XRD) is a widely used method for investigating the spacing between clay layers due to its simplicity and availability. The analysis of XRD data reveals characteristic peaks for clays and organoclays, indicating their regular layered structures. The peak position, expressed as $2\theta$ angles, corresponds to platelet separation or d-spacing in the clay structure. An increase in d-spacing is observed when polymers intercalate into the interlayer spacing between clay platelets.

The shift of XRD peaks to lower diffraction angles ($2\theta$) indicates an increase in d-spacing. Intercalation suggests that the polymer has entered the space between clay platelets, though stacking is still present. The disappearance of intensity peaks suggests exfoliation. In the case of exfoliation, the polymer has entered and separated the individual clay platelets, leading to the absence of characteristic peaks. Sharma [26] demonstrated an increase in d-spacing values, indicating successful intercalation. Chow's [27] observations point to the complete disappearance of the characterization peak of OMMT in epoxy/glass fiber/OMMT, indicating exfoliation. TEM is also utilized for direct visualization of clay exfoliation at the nanometer scale. It provides a more direct way to examine whether the clay is intercalated or exfoliated. The dispersion quality of nanoparticles in a matrix can be observed directly, providing insights into the states of clay exfoliation.

The combination of XRD and TEM techniques allows researchers to characterize and visualize the structural changes in clay materials.

### 10.4.2 Wide-Angle X-Ray Diffraction (WAXRD)

The presence or absence of peaks in the diffractogram does not solely indicate the state of nanocomposites. Other factors, such as the concentration and orientation of aluminosilicates, also play a role.

The observation of a peak does not exclude the possibility of exfoliated parts, and its absence does not necessarily mean complete exfoliation. WAXRD analysis is employed to determine the increase in interlayer spacing of fillers after compounding with polymers. This increase is associated with the shifting of diffraction peaks to lower angles. However, WAXRD does not provide quantitative information about the extent of intercalation and exfoliation. Wide-angle X-ray diffraction is used to characterize the alignment of cellulose nanocrystals (CNC) within composites. It is employed to measure differences in diffracted X-ray intensities for a given diffraction plane within the cellulose crystal structure. This measurement is done as a function of orientation within the composite sample geometry.

Small-angle X-ray analysis is mentioned as an alternative for analyzing materials in a low-diffraction-angle range, which is not achievable with wide-angle X-ray techniques.

### 10.4.3 Infrared Spectroscopy

IR spectrum is used to assess the composition, structural changes, and interactions within polymer nanocomposites. By comparing experimental spectra with reference spectra of individual components, they can gain insights into the nature of the composite and the success of the nanomaterial dispersion within the polymer matrix.

It is important to note that interpreting IR spectra requires expertise, and researchers often use complementary techniques, such as X-ray diffraction or microscopy, to corroborate their findings. Additionally, advancements in techniques like Fourier transform infrared (FTIR) spectroscopy have improved the sensitivity and resolution of IR analyses for nanocomposite studies.

FT-IR Raman and FT-IR spectroscopies play crucial roles in vibrational characterization, providing valuable insights into the composition, structure, and functionality of nanoparticles. Mostly, the combination of XPS and FT-IR was employed to gain a comprehensive understanding of the functionality of small nanoparticles and their interactions with substrates.

### 10.4.4 Microscopy

Scanning electron microscopy (SEM) is used for imaging the surface features of polymer nanocomposite materials.

The behavior of larger, more complex structural formations influences mechanical properties. These formations, referred to as supermolecular structures, include interface boundaries that are expected to be the starting points of chemical reactions within the polymer as well as potential hubs for fracture formation and early degradation. Large-scale organized structure occurrences have been discovered to be characteristic of both crystalline and amorphous polymers.

It is important to remember that all these structures are composed of individual polymeric molecules, despite the intricate morphology of structural creation in polymers.

Electron microscopy revealed that "neck" formation in polymers involves a jump-wise transition from one supermolecular structure to another. The transition is characterized by a sharp interfacial boundary observable on a microscopic scale. Poly(methyl methacrylate) is an example where a sharp boundary is observed between isotropic and oriented parts of the specimen.

The formation of a "neck" during deformation of a large spherulite of isotactic poly(methyl methacrylate) is described, with a discernible boundary between unchanged and oriented portions. In addition, secondary formations can be observed due to recrystallization of the oriented parts separated by sharp boundary lines.

Transmission electron microscopy (TEM), atomic force microscopy (AFM), focused ion beam microscopy, and confocal laser scanning microscopy are used in studying carbon nanotubes in polymer matrices. These techniques indeed play crucial roles in understanding the microstructural features, dispersion state, and interfacial adhesion of nanomaterials in different environments.

Transmission electron microscopy (TEM) provides detailed information about microstructural features and dispersion state. TEM faces challenges which include the need for improved image contrast, especially for carbon nanotubes in polymer matrices, and determination of size distribution of suspended carbon nanotube objects is critical.

Atomic force microscopy (AFM) is important for characterizing the size distribution of carbon nanotubes. It is capable of imaging and characterizing structures in various environments, including liquid, ambient, and vacuum. Focused ion beam microscopy is introduced for detailed imaging. Confocal laser scanning microscopy is successfully applied to evaluate the dispersion state of carbon nanotubes in polymer matrices. Scanning probe microscopy (SPM), including AFM and SEM, excels at providing surface information with high resolution. While they offer excellent surface imaging capabilities, they may not inherently provide detailed information about the dispersion of nanomaterials within a sample.

Due to the low-atomic-number elements in cellulose nanocomposites, electron scattering in TEM is weak, leading to poor contrast. Hence, in those cases, the mass-thickness contrast mechanism is utilized by staining the thin specimen with heavy metals like uranyl acetate to improve contrast in TEM images.

The properties of nanoparticle-reinforced polymer composites are directly affected by the distribution of nanoparticles within the matrix. The ability to characterize nanoparticle distribution, whether along a surface or when embedded within a polymer matrix, is crucial for understanding composite properties and the mechanisms of property enhancement.

TEM is highlighted as the most suitable microscopy method for analyzing the thickness of nanocomposite materials, provided that the distribution and dispersion of cellulose nanotubes (CNs) are of interest. Bondeson and Oksman [28] studied the impact of adding a surfactant on the dispersion and distribution of cellulose nanocrystals (CNC) in a polylactic acid (PLA) matrix.

TEM is well-suited for high-spatial-resolution imaging, making it effective for studying features at the submicron scale. However, challenges include the need for thin sample sections, potential artifacts during sample preparation, and the necessity for attention to imaging conditions for proper characterization. When dealing with nanocomposites or materials at the nanoscale, researchers often employ a combination of techniques to overcome the limitations of individual methods.

Optical microscopy (OM) has limitations when it comes to characterizing nanoscale materials within nanocomposites. The diffraction limit of light restricts the resolution of conventional optical microscopes, making it challenging to resolve features at the nanoscale. However, OM can still be valuable for certain aspects of characterization, especially when dealing with macroscopic composite structures or specific optical properties.

Polarized optical microscopy, a specialized form of optical microscopy, can provide information about the crystallization behavior of materials within a composite. This technique takes advantage of the birefringent properties of crystalline structures. It can be particularly useful for studying the orientation and alignment of crystalline regions, including those involving carbon nanotubes (CN) within the composite.

Atomic force microscopy serves as a versatile tool for investigating the alignment of cellulose nanocrystals within nanocomposites. The combination of qualitative description and quantitative measurements, facilitated by image analysis and post-processing techniques, offers a comprehensive understanding of CNC distribution and alignment. This information is crucial for correlating the nanocomposite's structural features with its mechanical and functional properties, providing insights into how the arrangement of nanocrystals influences material behavior.

#### 10.4.4.1 EDX

EDX, when coupled with FE-SEM or TEM, is a powerful technique for characterizing the elemental composition of nanoparticles. By focusing the electron beam on individual nanoparticles, researchers can obtain detailed information about the elemental constituents, and the energy of emitted X-rays is used to confirm and quantify the presence of specific elements. This analytical approach is widely utilized in scientific research to enhance the understanding of the composition of nanomaterials.

#### 10.4.4.2 XPS

X-ray photoelectron spectroscopy (XPS), also known as electron spectroscopy for chemical analysis (ESCA), is indeed a highly sensitive technique commonly used for characterizing the surface composition of materials, including nanocomposites. It is widely used to determine the bonding nature and the percentage of elements present in nanoparticles within nanocomposites.

XPS is particularly useful for surface analysis, providing information about the composition of the top few nanometers of a material. XPS data are typically represented as a plot of X-pivot electrons' binding energy (eV) versus the Y-axis plot electrons.

Each chemical element exhibits a unique binding energy value in XPS spectra.

It is used to analyze nanocomposites, providing information about the chemical composition, bonding, and electronic states of the elements involved.

The method of quantitative visualization of Young's modulus of soft materials by atomic force microscopy (AFM) was presented by Kim et al. [29]. The authors measured the Young's modulus values of various polymeric materials at each pixel within the entire scanning area and implemented the quantitative visualization of the mechanical properties.

### 10.4.5 Mechanical Characterization

Specimens were clamped between an actuating force gauge and a stationary ground clamp in mechanical characterization. The samples were pulled apart, and force vs. displacement data were collected (Figure 10.8). The force and displacement data were used to calculate stress-and-strain values.

Stress–strain curves provided valuable information, including Young's modulus and ultimate tensile strength. As shown in Figure 10.9, the tensile strengths of specific curves of pure polyimide: 26.64 MPa; 10-wt% nanocomposite: 24.5 MPa; 15-wt% nanocomposite: 9.12 MPa were measured. Pure polyimide exhibited the highest average ultimate tensile strength (29.77 MPa): 10% nanoparticle samples showed a reduced tensile strength of 24.42 MPa, and 15% nanoparticle samples had a further reduced strength of 13.31 MPa. This reduction in strength at higher concentrations is attributed to potential stress concentration around the particles. The Young's modulus of the films was calculated, and it was found that 0% nanoparticle samples had an average Young's modulus of 629.4 MPa, and 10% concentration samples exhibited an increased modulus of 841.35 MPa. Also, 15% concentration samples experienced a reduction in modulus to 587.22 MPa.

The trend observed aligns with the general behavior of nanomaterial-filled materials: an increase in Young's modulus at low concentrations and a decrease at higher concentrations.

This change in tensile strength and Young's modulus can be explained by the influence of nanoparticle concentration on stress distribution and overall material behavior (Figure 10.10). In the example of the mechanical property characterization for polypropylene nanocomposites generated with an imidazolium-modified filler, the modification of polypropylene with the imidazolium-modified filler affects its mechanical properties. While there is an increase in tensile modulus due to load transfer, the incomplete exfoliation of the filler and the presence of thicker tactoids contribute to a decrease in yield stress. This underscores the importance of both macroscopic mechanical testing and microscopic characterization to fully understand the behavior of nanocomposites.

FIGURE 10.8 Polyimide–$SiO_2$ nanocomposite undergoing tensile testing—mechanical characterization.

Source: Based on [30].

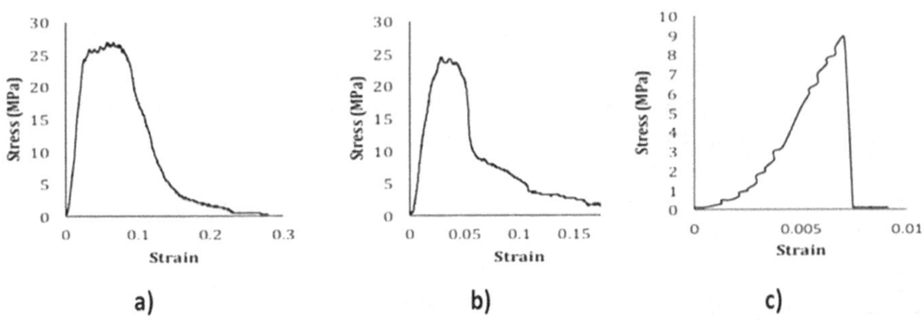


FIGURE 10.9 Examples of stress–strain curves of three samples: (a) 0% nanoparticle concentration, (b) 10% nanoparticle concentration, and (c) 15% nanoparticle concentration.

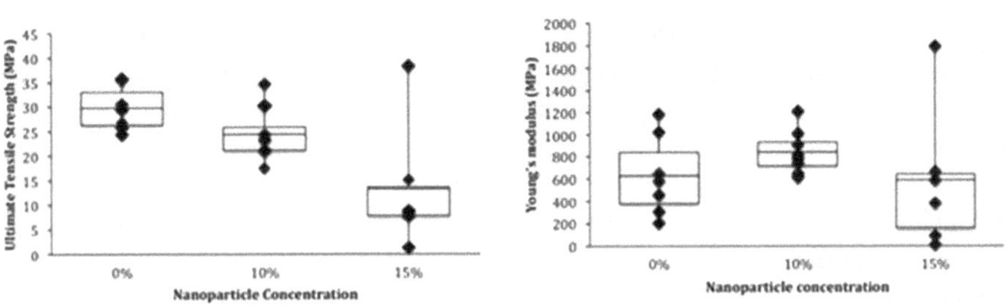


FIGURE 10.10 Box plot comparing the ultimate tensile test and the Young's modulus of three samples having 0%, 10%, and 15% nanoparticle concentrations.

Source: Based on [30].

A generalized approach for mechanical properties evaluation of polymer nanocomposites with spherical fillers based on a three-phase series-parallel model was presented by Martinez-Garcia et al. [31]. In the proposed model, authors considered percolation and the glassy interphase between the filler and the matrix. The proposed model has been validated with experimental data obtained for various polymer nanocomposites.

Visco et al. [32] studied polyurethane foams modified with carbon nanofibers (CN) to enhance the selectivity and mechanical durability of PU foams in relation to oils. They revealed that the incorporation of CN leads to an increase in the hydrophobic behavior and a good oleophilic character of the composite sponges. The best properties were found for composites containing 1 wt% of nanofiller.

### 10.4.6 Electrical Characterization

To determine the dielectric strength of the material, a Vitrek 955i Hi-Pot testing system can be used. The system has a built-in safety feature to determine the exact voltage at which the material fails. Two wires from the system are attached to the electrodes that are contained within a 3D-printed safety enclosure. The testing sample is then sandwiched between the electrodes, and the voltage is increased from 0 V at a rate of 500 V/s, until a current spike is detected. Once this occurs, material breakdown is achieved, the voltage is recorded, and the system ceases to apply electricity. Testing was conducted on three different sample concentrations, 0%, 10%, and 15% in nanoparticle weight percentages [30]. The base polyimide had the highest average dielectric strength of 86.8 kV/mm. The 10% nanoparticle concentration samples and the 15% concentration had average dielectric breakdown values of 34.0 kV/mm and 40.6 kV/mm, respectively. The decrease in the breakdown voltage as nanoparticle concentration increases is backed up by the fact that polyimide has a higher dielectric strength than $SiO_2$.

### 10.4.7 Thermal Analysis

Thermal analysis techniques, such as thermogravimetric analysis (TGA), differential thermal analysis (DTA), and differential scanning calorimetry (DSC), play a crucial role in understanding the thermal behavior of polymer nanocomposites. Thermogravimetric analysis (TGA) measures the change in weight of a sample as a function of temperature or time. It is often used to study decomposition processes, determine composition, and analyze thermal stability.

Differential thermal analysis (DTA) measures the temperature difference, $\Delta T$, between a sample and an inert reference material as a function of temperature. It detects changes in heat content.

It is useful for identifying phase transitions, reactions, and studying thermal events in materials.

Differential scanning calorimetry (DSC) is a modified form of DTA designed to quantitatively measure enthalpy changes ($\Delta H$) in a sample as a function of temperature or time.

DSC is widely used in polymer science to study phenomena like glass transition temperature ($T_g$), melting points, crystallization, and reaction kinetics.

DSC in polymer nanocomposites helps in determining the $T_g$ of polymers and their composites. $T_g$ is a critical parameter representing the transition from a rigid to a more flexible state.

An increase in $T_g$ values in DSC curves indicates the presence of inorganic materials in the polymer matrix. This is valuable information for understanding the effects of nanocomposite fillers on the polymer's thermal properties. Understanding $T_g$ is crucial for predicting a material's performance at different temperatures. The presence of inorganic materials can enhance thermal stability and provide insights into the composite's behavior. DSC allows researchers to quantify heat changes associated with transitions, aiding in the optimization of polymer processing conditions. DSC is a valuable tool for studying the thermal properties of polymers and nanocomposites. The technique allows researchers to investigate the influence of nanofillers on crystallization behavior and provides information on the interactions between the polymer matrix and fillers. Additionally, when fillers are organically modified, DSC aids in understanding the thermal characteristics associated with the organic layers.

This thermal analysis technique, when coupled with other characterization methods, contributes to a comprehensive understanding of the structure and properties of polymer nanocomposites, enabling researchers to tailor materials for specific applications.

## 10.5 APPLICATION OF NANOCOMPOSITES

It has become apparent that nanocomposites are one of the most promising materials in the field of engineering because of their unique properties. Typically, these materials are composed of nano-sized particles or fibers that are embedded in a matrix, which leads to enhanced mechanical, electrical, and thermal performance as well. The use of nanocomposites has found numerous applications in a variety of domains over the past few years, including aeronautics, automotive, electronics, biomedical, food packaging, and solar energy. There are many industries where nanocomposites have the potential to revolutionize the way they are done, making them more cost-efficient, eco-friendly, and efficient than ever before. There is also a possibility that they will create new jobs and stimulate the growth of the economy.

### 10.5.1 Automobiles

It is common for parts of an automobile to wear out and be worn down as a result of their heavy use. As a result of their high fracture resistance, nanocomposites are ideal candidates for use as automotive parts. As a result of the use of nanocomposites, the parts where high efficiency is required can be made of stronger materials. Automobile manufacturers are working to develop technologies that can reduce pollution cost-effectively because pollution is increasing rapidly. Therefore, polymeric nanocomposites have gained a lot of popularity in recent years [33].

Nanocomposites have been utilized in most of the components, including bumpers and fenders, headlight lenses, wheel covers, door panels, etc. S. A. Pradeep et al. [34] have clearly summarized the various polymer composites utilized in the various components of an automotive vehicles, and they are enclosed in what follows.

**TABLE 10.1**
**Polymer Nanocomposites Utilized in Automobile Industries**

| Component | Material | Automaker |
|---|---|---|
| Wheel covers | ASA (acrylonitrile styrene acrylate) | Mercedes-Benz |
| | 30% mineral-filled PA-6 | SEAT |
| | ABS (acrylonitrile butadiene styrene) | Ford Mondeo, Fiesta; Nissan; Toyota; Renault |
| Lighting house lenses | Lens: polycarbonate (PC) | Ford |
| | Reflector: unsaturated polyester (UP) | |
| | Lens: PC | Volkswagen |
| | Reflector: electroplated ABS/PC blend, PPS (poly[p-phenyl sulfide]) | |
| | Reflectors: PEI (polyetherimide), PPS | BMW |
| Body | CFRP (carbon fiber–reinforced plastics) | BMW |
| Door panels | Glass fiber–PUR (polyurethane) | Volkswagen |
| | Glass fiber–PP (polypropylene); Kenal–PP | Ford |
| | Thermoplastic olefin (TPO) | Dodge |
| | Jute–epoxy | Mercedes-Benz |
| | Flax/sisal–PUR | Audi |

### 10.5.2 Aircraft Components

In the field of aircraft components, nanocomposites play a crucial role when it comes to improving the performance and efficiency of those components. Because nanomaterials have a high strength-to-weight ratio, they can be used to fabricate lightweight structural components, thereby reducing carbon dioxide emissions and improving fuel efficiency in vehicles. Moreover, nanocomposites are more efficient when it comes to distributing and controlling electrical energy, thanks to their enhanced electrical conductivity, which permits their application in aircraft.

It is also highly recommended that nanocomposites be used for aerospace applications due to their enhanced corrosion resistance. The presence of nanoparticles within the matrix of a material protects it against corrosion caused by the environment, including humidity, oxygen, and acids, which can cause corrosion under adverse conditions. As a result of this extended lifespan, aircraft components will be more reliable and durable, thus reducing maintenance costs. Some of the polymer composites that are currently employed in aircrafts are listed in what follows [35].

### 10.5.3 Automotive

The automotive industry has incorporated nanocomposites into its design and manufacturing processes in order to enhance vehicle performance and durability. By using lightweight nanocomposites, vehicles can be made lighter and more fuel-efficient, resulting in a reduction in carbon dioxide emissions and an improved fuel efficiency. It is also possible to produce lighter and smaller components with nanocomposites due to their high strength-to-weight ratio.

Further, the improved thermal management of nanocomposites makes them ideal for use in automotive batteries and thermal management systems due to their improved thermal management. It is a matrix that is embedded with nanoparticles which provide excellent insulation properties, as they do not allow any heat to transfer between the exterior and the interior of the vehicle. In addition to reducing energy consumption, this enhanced thermal management increases the level of battery reliability, thereby extending the battery's overall life cycle.

### 10.5.4 Biomedical

A significant amount of attention has been given to nanocomposites in biomedical applications due to the unique properties of these materials and their potential to simplify a number of health-care challenges. As a result of the combination of the nanoscale properties of nanoparticles with the macroscale properties of conventional materials, these materials provide enhanced performance and versatility over conventional materials alone.

A number of nanocomposites have been shown to be promising for imaging and diagnosis in medicine. Due to the unique properties of nanoparticles, for example, their magnetic properties and their optical properties, innovative imaging techniques can be developed using these nanoparticles.

**TABLE 10.2**
**Polymer Composites Utilized in Aircrafts**

| Material | Model | Advantages |
|---|---|---|
| Hybrid aluminum/glass-reinforced plastic system (GLARE) | A380 | Reduced weight, increased damage tolerance, and improved fatigue life |
| CFRP (carbon fiber–reinforced plastic) for wings | A350XB | 50% fewer structure maintenance tasks |
| Victrex–PEEK (polyetheretherketone) | Airbus A350–900 | Maximum geometrical moment of inertia, effective structural load transfer |

In addition to being capable of magnetic imaging, nanocomposites can also be used to track the motion of nanoparticles in the body using magnetic imaging. It is also possible to use them as a means of optical imaging, due to their fluorescence properties, which allows them to be used as a real-time monitoring device for biological processes in a variety of situations.

Furthermore, nanocomposites have the potential to be used in the field of tissue engineering in addition to their therapeutic and diagnostic applications. There is a growing body of research on the use of scaffolds for the development of new tissues and organs, which can provide a biocompatible and supportive environment for cell growth and differentiation. Depending on the composition of the nanocomposites, they can mimic the architecture and composition of natural tissues, which enhances their mechanical properties as well as promotes their integration with the surrounding tissue. It is likely that nanocomposites will be used in a wide range of biomedical applications over the next decade, and we can expect even greater advancements as research and development in this field continue.

### 10.5.5 Electronics

Since the introduction of nanocomposites into electronic devices, they have gained significant attention due to their promising applications. There are several unique properties of nanoparticles that make them ideal components of electronic devices because of their high surface area, enhanced electrical conductivity, and optical properties.

The development of lightweight and flexible electronics is one of the areas where nanocomposites have been extensively explored. As a result of unique combinations of matrix and nanoparticles, nanocomposites that are suitable for flexible applications can improve the mechanical properties of electronic devices.

Additionally, nanocomposite materials are also being explored for their potential in energy storage devices. In order to enhance the performance of energy storage devices, nanoparticles with a high specific surface area and a high storage capacity can be incorporated into the matrix, such as carbon nanotubes or graphene. As a result of their ability to store and release energy, these nanoparticles are highly suitable for application in energy storage systems, such as batteries and supercapacitors, that are designed for energy storage. The future of electronic devices is likely to be shaped by nanocomposites in an increasingly prominent role, with research and development in this field continuing rapidly.

### 10.5.6 Food Packaging

Nanocomposites are a good choice for food packaging because of their barrier properties, which are one of their most appealing features. It has been shown that the addition of nanoparticles to the polymer matrix can improve the resistance of the material to moisture, oxygen, and gases. As a result, food products can remain fresh and fresh for longer periods of time by preventing spoilage and oxidation of the product.

The use of nanocomposites can also enhance the quality of food by improving food safety. It is possible to incorporate antimicrobial agents into polymer matrixes, such as silver ions, through nanoparticles. In many cases, these agents can kill bacteria, fungi, and viruses, thus reducing the risk of foodborne illnesses in humans and animals. In addition, nanocomposites can be used to increase the packaging's ability to prevent the migration of harmful substances from the packaging to the food by significantly improving its barrier properties.

Nanocomposites are a promising material for food packaging, but they have some limitations, regardless of their advantages. A challenge associated with nanoparticle dispersion in the polymer matrix is that an uneven distribution of the particles can lead to a reduction in the performance of the system. As well as this, there are some toxicological issues that need to be addressed when it comes to nanoparticles in food packaging. Nanocomposites have great potential for food packaging, but further research is needed in order to overcome some of the limitations and ensure their safety.

### 10.5.7 Solar Energy

There has been a great deal of attention paid to nanocomposites in the solar energy sector because of their enhanced optical properties. As a result of the nanoparticles that are present within the matrix, sunlight can be efficiently collected and converted to electricity in the subsequent stages. A nanoparticle has the ability to modify the refractive index of a material as well as its absorption coefficient, resulting in an increase in light capture and a reduction in reflection. With the improved photoconversion efficiency of solar cells and solar panels, the performance of solar cells and solar panels will be improved, making them more cost-effective and energy-efficient.

In addition, nanocomposites have the advantage of being highly durable, which makes them suitable for long-term applications in solar energy systems. In the presence of nanoparticles, the material is protected from environmental degradation, which reduces the need for regular maintenance and replacement costs, as a result of the nanoparticles in the matrix. As a result of this extended lifespan, solar energy systems will be more reliable and sustainable, resulting in them being more economical to operate.

## 10.6 CONCLUSION

Due to their advanced design options, polymer nanocomposite materials act as alternatives to traditional materials, attributed to careful matrix reinforcement selection, easy constituent access, lightweight nature, and outstanding physical, thermal, electrical, and mechanical properties. This comprehensive review focuses on composite materials, elaborating the various synthesis and characterization routes that are utilized for evaluating the physical, mechanical, electrical, and thermal characteristics. The review delves into detailed discussions on commonly used synthetic reinforcing materials, like carbon, silica, nanoclays such as MMT, and metal hydrides., emphasizing their impact on the modulus, strength, and fatigue resistance of the polymers. Additionally, an in-depth review on the applications of polymer composites on various streams of engineering, like automobile, aircraft, biomedical, food packaging, and energy applications, has been provided.

## ACKNOWLEDGMENTS

The authors thank PSG Institute of Technology and Applied Research and South Ural State University for their support.

## REFERENCES

[1] Bitinis N, Hernandez M, Verdejo R, Kenny JM, Lopez-Manchado MA. Recent advancesin clay/polymer nanocomposites. Advanced Materials. 2011;23:5229–5236.
[2] Alexandre M, Dubois P. Polymer-layered silicate nanocomposites: Preparation, properties and uses of a new class of materials. Materials Science & Engineering R: Reports. 2000;28:1–63.
[3] Müller K, Bugnicourt E, Latorre M, Jorda M, Echegoyen Sanz Y, Lagaron J, et al. Reviewon the processing and properties of polymer nanocomposites and nanocoatings andtheir applications in the packaging, automotive and solar energy fields. Nanomaterials. 2017;7:74–121.
[4] Monemian S, Jafari SH, Khonakdar HA, Goodarzi V, Reuter U, Pötschke P. MWNT-filledPC/ABS blends: Correlation of morphology with rheological and electrical response. Journal of Applied Polymer Science. 2013;130:739–748.
[5] Liu L, Grunlan JC. Clay assisted dispersion of carbon nanotubes in conductive epoxynanocomposites. Advanced Functional Materials. 2007;17:2343–2348.
[6] Rubel RI, Ali MH, Jafor MA, Alam MM. Carbon nanotubes agglomeration in reinforced composites: A review. AIMS Materials Science. 2019;6–5:756.
[7] Goyanes S, Rubiolo GR, Salazar A, Jimeno A, Corcuera MA, Mondragon I. Carboxylationtreatment of multiwalled carbon nanotubes monitored by infrared and ultraviolet spectroscopies and scanning probe microscopy. Diamond and Related Materials. 2007;16:412–417.

[8] Technical Bulletin AEROSIL® No. 27, Degussa AG, D-63403 Hanau-Wolfgang, Germany, 10/2001.
[9] AEROPERL®Granulated fumed oxides: Technical Information 1341.
[10] Wei J, Aboud MFA, Shakir I, Tong Z. Graphene oxide-supported organo-montmorillonite composites for the removal of Pb(II), Cd(II), and As(V) contaminants from water. ACS Applied Nano Materials 2020;3–1:806–813.
[11] Zaïri F, Gloaguen JM, Naït-Abdelaziz M, Mesbah A, Lefebvre JM. Study of the effect ofsize and clay structural parameters on the yield and post-yield response of polymer/clay nanocomposites via a multi-scale micromechanical modelling. Acta Materialia. 2011;59:3851–3863.
[12] Ray SS. Recent trends and future outlooks in the field of clay-containing polymer nanocomposites. Macromolecular Chemistry and Physics. 2014;215:1162–1179.
[13] https://serc.carleton.edu/details/images/13660.html.
[14] Goodell PD. Stability of rechargeable hydriding alloys during extended cycling. Journal of the Less Common Metals. 1984;99:1–14.
[15] Kikkinides ES, Georgiadis MC, Stubos AK. On the optimization of hydrogen storage inmetal hydride beds. International Journal of Hydrogen Energy. 2006;31:737–751.
[16] Khan I, Saeed K, Ali N, Zada N, Khan A, Ali F, et al. *Polymer nanocomposites: An overview*, INC, 2022. https://doi.org/10.1016/B978-0-323-91611-0.00017-7.
[17] Okamoto M. Polymer nanocomposites. Intechopen. 2023;4:457–479. https://doi.org/10.3390/eng4010028.
[18] Fawaz J, Mittal V. Synthesis of polymer nanocomposites: Nanocomposite method. Synth. Tech. Polym. Nanocomposite. (2015) 1–30. https://application.wiley-vch.de/books/sample/3527334556_c01.pdf.
[19] Fu S, Sun Z, Huang P, Li Y, Hu N. Some basic aspects of polymer nanocomposites: A critical review. Nano Materials Science. 2019;1:2–30. https://doi.org/10.1016/j.nanoms.2019.02.006.
[20] Rane AV, Kanny K, Abitha VK, Thomas S. *Methods for synthesis of nanoparticles and fabrication of nanocomposites*, Elsevier Ltd., 2018. https://doi.org/10.1016/B978-0-08-101975-7.00005-1.
[21] Lovell PA, Schork FJ, Fundamentals of emulsion polymerization. Biomacromolecules. 2020;21:4396–4441. https://doi.org/10.1021/acs.biomac.0c00769.
[22] Ávila-Orta CA, González-Morones P, Agüero-Valdez D, González-Sánchez A, Martínez-Colunga JG, Mata-Padilla JM, Cruz-Delgado VJ. Ultrasound-assisted melt extrusion of polymer nanocomposites. Nanocomposites: Recent Evolution. 2019. https://doi.org/10.5772/intechopen.80216.
[23] Moja TN, Bunekar N, Mishra SB, Tsai TY, Hwang SS, Mishra AK. Melt processing of polypropylene-grafted-maleic anhydride/Chitosan polymer blend functionalized with montmorillonite for the removal of lead ions from aqueous solutions. Science Reports. 2020;10:1–14. https://doi.org/10.1038/s41598-019-57079-2.
[24] Ghaleb ZAA, Jaafar M, Rashid AA. *Fabrication methods of carbon-based rubber nanocomposites and their applications*, Elsevier Inc., 2019. https://doi.org/10.1016/B978-0-12-817342-8.00003-2.
[25] Li L, Hui Zhou C, Ping Xu Z. Self-nanoemulsifying drug-delivery system and solidified self-nano-emulsifying drug-delivery system. Nanocarriers Drug Delivery Nanoscience. 2018:421–449. https://doi.org/10.1016/B978-0-12-814033-8.00014-X.
[26] Sharma B, Mahajan S, Chhibber R, Mehta R. Glass fiber reinforced polymer-clay nanocomposites: processing, structure and hygrothermal effects on mechanical properties. Procedia Chemistry. 2012;4:39–46.
[27] Chow W. Water absorption of epoxy/glass fiber/organo-montmorillonite nanocomposites. eXPRESS Polymer Letters. 2007;1(2):104–108.
[28] Bondeson D, Oksman K. Dispersion and characteristics of surfactant modified cellulose whiskers nanocomposites. Composite Interfaces. 2007;14:617–630.
[29] Kim S, Lee Y, Lee M, An S, Cho SJ. Quantitative visualization of the nanomechanical young's modulus of soft materials by atomic force microscopy. Nanomaterials. 2021;11:1593. https://doi.org/10.3390/nano11061593.
[30] Nalbach JR, Rowan University, Theses and Dissertations. *Design and characterization of polymer nanocomposites for engineering applications*, 2018.
[31] Martinez-Garcia JC, Serraïma-Ferrer A, Lopeandía-Fernández A, Lattuada M, Sapkota J, Rodríguez-Viejo J. A generalized approach for evaluating the mechanical properties of polymer nanocomposites reinforced with spherical fillers. Nanomaterials. 2021;11:830. https://doi.org/10.3390/nano11040830.
[32] Visco A, Quattrocchi A, Nocita D, Montanini R, Pistone A. Polyurethane foams loaded with carbon nanofibers for oil spill recovery: Mechanical properties under fatigue conditions and selective absorption in oil/water mixtures. Nanomaterials. 2021;11:735. https://doi.org/10.3390/nano11030735.

[33] Maurya AK, Manik G. Advances towards development of industrially relevant short natural fiber reinforced and hybridized polypropylene composites for various industrial applications: A review. Journal of Polymer Research 2023;30:47. https://doi.org/10.1007/s10965-022-03413-8.
[34] Pradeep SA, Iyer RK, Kazan H, Pilla S. Automotive applications of plastics: Past, present, and future. In *Applied plastics engineering handbook* (Second Edition), William Andrew Publishing, 2017, 651–673, ISBN 9780323390408. https://doi.org/10.1016/B978-0-323-39040-8.00031-6.
[35] Soutis C. Aerospace engineering requirements in building with composites. In Philip Irving, Constantinos Soutis, eds., *Polymer composites in the aerospace industry* (Second Edition), Woodhead Publishing, 2020, 3–22, ISBN 9780081026793. https://doi.org/10.1016/B978-0-08-102679-3.00001-0.

# 11 Nanocomposites
## *Synthesis, Characterization, and Applications*

*Pankaj Kumar and Ganesh Sahadeo Meshram*

### 11.1 INTRODUCTION

Nanotechnology is the subject of research concerned with nanomaterials or an applied science that deals with matter's size control at the molecular and atomic scale [1]. It is possible to categorize it according to its dimensions, such as quantum dots, particles, and spheres with three dimensions less than 100 nm; fibers, tubes, and plates with two dimensions less than 100 nm; and nanowires and nanorods with one dimension less than 100 nm [2, 3]. Nanomaterials can be divided into three categories based on their phase composition: matrix composites, multiple composites, and single-phase solids.

*Nanocomposites* are materials that incorporate nanoscale reinforcements into a matrix material, resulting in a combination of properties that are distinct from those of the individual components shown in **Figure 11.1**. In nanocomposites, the nanoscale reinforcements typically have at least one dimension less than 100 nm [4–12]. Nanocomposites show rare properties that ascend from their large surface area, small size, and the relations of phases at their interfaces. It has been stated that when the particle size is less than a specific threshold, referred to as "the critical size," changes in the particle properties can be seen [13].

There are various types of nanocomposites, but one common example involves incorporating nanoparticles into a polymer matrix [14, 15]. The most commonly used nanoparticles in these composites include carbon nanotubes, graphene, clay nanoparticles, and metal oxides. Even at minimal nanofiller loading, polymer nanocomposites exhibit improved characteristics, making them an intriguing and quickly expanding class of new materials. Because of these qualities, polymer nanocomposites are seen as promising materials for a wide range of applications, such as in food packaging [16], medicine [17], aerospace, electronics [18], catalysis [19], automotive, and environmental protection [20].

It can be further divided into inorganic–organic hybrid polymer nanocomposites, ceramic–polymer nanocomposites, polymer matrix nanocomposites, and polymer–layered silicate nanocomposites. Among the most studied kinds of nanocomposites are polymer ones. They involve dispersing nanoscale additives, such as clay nanoparticles, within a polymer matrix. This can improve the mechanical and thermal properties of the polymer. The aerospace, automotive, marine, infrastructure, military, and other industries have seen a surge in the use of composites due to the commercial significance of polymers [21]. Any composite material's ability to perform throughout use is crucial, since it determines how well items work in outside settings. Because of their useful lifetime, upkeep, and replacement requirements, polymeric materials' longevity is a natural concern for many applications [22, 23].

Carbon nanotubes (CNTs) are widely used in nanocomposites due to their exceptional thermal, mechanical, and electrical properties [24, 25]. Incorporating CNTs into polymers, metals, or ceramics can result in composites with superior performance. Graphene is another popular nanomaterial for composites [26]. Graphene composites can exhibit excellent conductivity, mechanical strength, and flexibility. Similar to microcomposites, materials for nanocomposite structures can be categorized into three groups based on the matrix materials they contain: ceramic, metal, and polymer matrix nanocomposites.

DOI: 10.1201/9781003476337-11

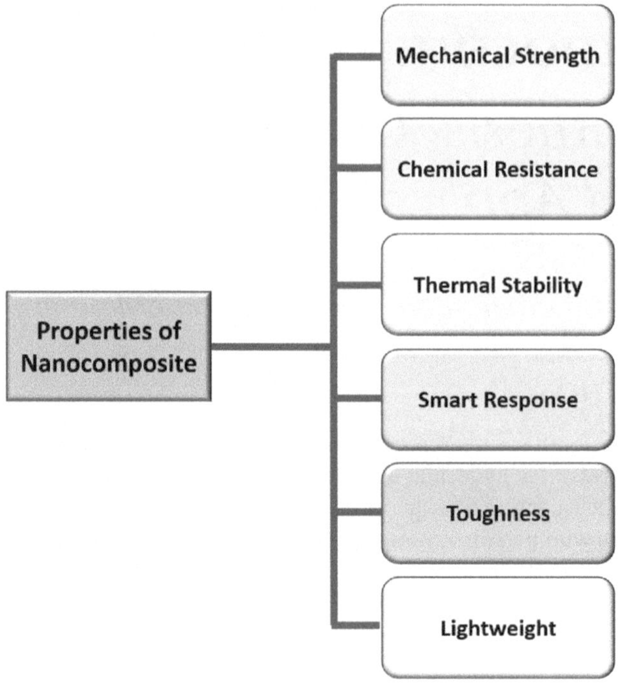


**FIGURE 11.1** Properties of nanocomposites.

### 11.1.1 Ceramic Matrix Nanocomposites

Ceramic matrix nanocomposites (CMNCs) are advanced materials where nanoscale reinforcements are incorporated into a ceramic matrix. These nanocomposites are designed to harness the unique properties of ceramics while benefiting from the improved mechanical, thermal, and electrical properties offered by the nanoscale additives. A variety of ceramic matrices, such as nitrides (like silicon nitride), oxides (like alumina and zirconia), carbides (like silicon carbide), and others, can serve as the foundation for CMNCs. The application and desired features depend on the matrix. Nanoscale reinforcements commonly used in CMNCs include nanoparticles, nanotubes, and nanofibers [27–29]. Examples include alumina nanoparticles, silicon carbide nanotubes, and carbon nanofibers. The incorporation of nanoscale reinforcements enhances the mechanical properties of ceramics, including increased hardness, strength, and fracture toughness. This improvement is attributed to the reinforcing effect of nanoparticles, which can hinder crack propagation and improve resistance to deformation [30].

Ceramic matrix nanocomposites often exhibit enhanced thermal conductivity due to the high thermal conductivity of some nanomaterials. Additionally, certain nanofillers can introduce electrical conductivity to ceramic matrices. CMNCs find applications in high-performance engineering components, cutting tools, biomedical implants, thermal barrier coatings, and aerospace components. The combination of high strength, hardness, and resistance to high temperatures makes them suitable for demanding environments. Nanoscale reinforcements contribute to toughening mechanisms in ceramics, such as crack deflection, crack bridging, and grain boundary pinning. These mechanisms improve the material's ability to withstand stress and prevent catastrophic failure.

It is difficult to disperse nanoscale reinforcements evenly throughout the ceramic matrix. The homogeneity and characteristics of the resultant nanocomposite can be impacted by nanoparticle agglomeration. The production of CMNCs may involve high processing costs, and researchers are exploring cost-effective methods to make these materials more viable for industrial applications [31, 32]. At the forefront of materials science research, ceramic matrix nanocomposites are being developed to optimize their properties for a variety of applications, particularly in industries where high-performance materials are essential [33].

### 11.1.2 Metal Matrix Nanocomposites

Advanced materials known as metal matrix nanocomposites (MMNCs) are made of metallic matrices with nanoscale reinforcements integrated into them. The purpose of these nanocomposites is to fuse the improved mechanical, thermal, and occasionally electrical capabilities offered by the nanoscale additions with the desired qualities of metals [34–36]. Common metal matrices for nanocomposites include aluminum, titanium, magnesium, copper, and their alloys. The intended uses and characteristics of the final nanocomposite determine the best metal matrix. Various nanoscale reinforcements can be incorporated into metal matrices [37]. Common types include nanoparticles (e.g., carbon nanotubes, alumina), nanofibers (e.g., carbon nanofibers), and nanolaminates. Several methods are employed to synthesize metal matrix nanocomposites, including powder metallurgy, in situ fabrication, and mechanical alloying.

Certain nanocomposites are intended to show better resistance to corrosion than conventional metal alloys, which qualifies them for use in harsh settings. It is difficult to disperse nanoscale reinforcements evenly throughout the metal matrix. Processing parameters, agglomeration, and interfacial bonding can all have an impact on a nanocomposite's overall performance. Researchers are looking on ways to make these materials more economically viable for widespread industrial uses, as the cost of making MMNCs may be a factor.

### 11.1.3 Polymer Matrix Nanocomposites

*Polymer nanocomposites* (PMNCs) are materials in which nanoscale fillers, such as nanoparticles, graphene, metal–metal oxide, and nanotubes, are dispersed within a polymer matrix. The incorporation of these nanofillers imparts unique properties to the polymer, enhancing mechanical, electrical, thermal, and barrier properties [38–40]. Montmorillonite clay is commonly used due to its high aspect ratio and surface area. The exfoliation of clay layers within the polymer matrix enhances mechanical properties [41]. Carbon nanotubes with one or more walls can be very strong and electrically conductive. Graphene oxide or single-layer graphene sheets have superior electrical, mechanical, and thermal properties. Zinc oxide, titanium dioxide, and silica are examples of

**TABLE 11.1**
**Different Types of Nanocomposites and Their Examples**

| Nanocomposite | Examples | References |
|---|---|---|
| Ceramic matrix nanocomposite | $Si_3N_4$/SiC | [44] |
| | $Al_2O_3$/SiC | [45] |
| | MgO/SiC | [46] |
| | $Al_2O_3$/$ZrO_2$ | [47] |
| Metal matrix nanocomposite | Ag/Au | [48] |
| | Cu/Nb | [49] |
| | Al/AlN | [50] |
| | Al/SiC | [51] |
| Polymer matrix nanocomposite | PMMA/$SiO_2$ | [52] |
| | Polyimide/$SiO_2$ | [53] |
| | Poly(amide-imide)/$TiO_2$ | [54] |
| | Polycarbonate/$SiO_2$ | [55] |

metallic and oxide nanoparticles that can improve characteristics, including strength, UV resistance, and thermal stability [42].

Polymer nanocomposites can be synthesized using the various techniques mentioned earlier, including in situ polymerization, solution casting, melt-mixing, and others [43]. The choice depends on the polymer and nanofiller types, as well as the desired properties of the resulting nanocomposite. Different types of nanocomposites and their examples are shown in Table 11.1. Polymer nanocomposites continue to be an active area of research, driven by the desire to create materials with enhanced performance for various industrial applications. The optimization of synthesis methods and understanding of the fundamental interactions between the polymer matrix and nanofillers are critical for advancing the field.

## 11.2 SYNTHESIS

The synthesis of nanocomposites involves combining nanoparticles with a matrix material to create a material that exhibits enhanced properties compared to the individual components. There are several techniques for synthesizing nanocomposites, and the choice of method depends on the type of nanoparticles and matrix material. Here are some common synthesis techniques:

### 11.2.1 Ceramic Matrix Nanocomposites

*Powder metallurgy.* Nanoparticles are mixed with ceramic powders, and the mixture is sintered to achieve a dense nanocomposite. This method is effective for oxides, nitrides, and carbides [31, 56, 57].

*Chemical vapor deposition (CVD).* Gas-phase precursors react to form a ceramic matrix with embedded nanoparticles. This method allows precise control over the composition and structure of the nanocomposite [58].

*In situ fabrication.* Nanoparticles are formed within the ceramic matrix during processing, often involving chemical reactions. This method offers control over nanoparticle size and distribution [59, 60].

### 11.2.2 Metal Matrix Nanocomposites

*Powder metallurgy.* Nanoparticles are mixed with metal powders and then consolidated and sintered to form a dense nanocomposite. This method is versatile for various metal matrices [56, 61, 62].

*In situ fabrication.* Nanoparticles are formed within the metal matrix during manufacturing, often through chemical reactions. This approach offers control over the interface between the metal and the nanofillers [34, 63, 64].

*Mechanical alloying.* High-energy ball milling is used to achieve a uniform dispersion of nanoparticles within the metal matrix. This method is effective for achieving fine particle sizes [65].

### 11.2.3 Polymer Matrix Nanocomposites

*In situ polymerization.* A widely employed method involves the simultaneous polymerization of monomers and the formation of nanoparticles within the polymer matrix. This approach ensures a strong interfacial interaction between the polymer and nanofillers [66, 67].

*Solution casting.* Nanoparticles are dispersed in a solvent along with the polymer, followed by casting and solvent evaporation. This method is effective for both thermoplastic and thermosetting polymers [68, 69].

*Melt-mixing.* Commonly used for thermoplastic polymers, melt-mixing involves dispersing nanofillers into the polymer matrix at elevated temperatures above the melting point. Intensive mixing ensures a homogeneous distribution [70, 71].

## 11.3 CHARACTERIZATION

Characterizing nanocomposites involves using various techniques to understand and analyze their properties at the nanoscale, shown in Figure 11.2. Here are some common characterization techniques for nanocomposites:

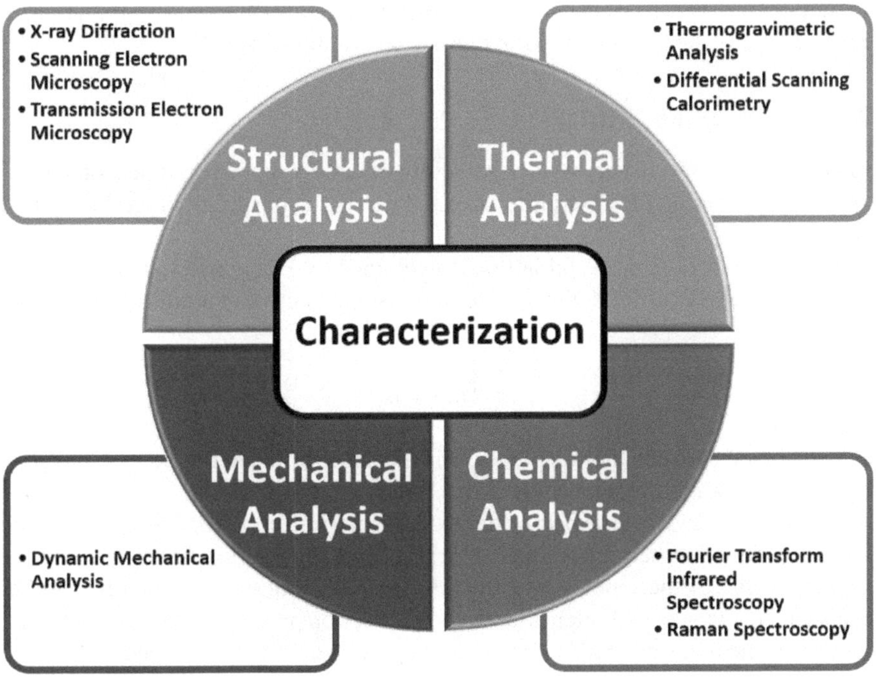


**FIGURE 11.2** Characterization techniques of nanocomposites.

### 11.3.1 Scanning Electron Microscopy (SEM)

High-resolution pictures of the surface morphology of nanocomposites can be obtained using SEM. It allows researchers to visualize the distribution of nanoparticles within the matrix and assess the overall structure of the material. Jamalpour et al. investigated the fabrication of supramolecular polymer nanocomposite foams. The study examined the effect of supramolecular interactions on the thermomechanical properties and morphology of the foams. The result has been investigated using SEM and TEM. The results showed that increasing nanoparticle loading improves the mechanical properties and foamability of the system up to an optimum concentration, beyond which strong interactions between fillers result in decreased performance [72].

### 11.3.2 Transmission Electron Microscopy (TEM)

TEM provides even higher resolution than SEM and is particularly useful for studying the internal structure of nanocomposites at the nanoscale. Yang et al. revealed information about particle dispersion, size, and interfacial interactions. TEM pictures of Ni/NiO polymer nanocomposites were examined after they were synthesized at various reaction temperatures. The study reveals that the formation of NiO nanoparticles at 400°C and 500°C is slow due to coagulation. Aggregation between nanoparticles occurs at higher temperatures, facilitating particle diffusion and reducing polymer–metal interaction energy. Agglomerates are observed at all temperatures, indicating slow coalescence [73].

### 11.3.3 Atomic Force Microscopy (AFM)

AFM is used to study the surface topography and mechanical properties of nanocomposites at the nanoscale. It can provide information on particle distribution, surface roughness, and other surface characteristics. Gopal et al. investigated the AFM images that reveal the micromorphology and roughness of nanocomposites. The surface showed a network of multiwalled carbon nanotubes (MWCNT), uniform carbon nanotubes (CNT) dispersion, and no apparent aggregation. The nanocomposite had a smooth surface, with a roughness value of 32.4 nm [74].

### 11.3.4 X-Ray Diffraction (XRD)

XRD is employed to analyze the crystalline structure of nanocomposites. It helps identify the crystal phases present, determine the crystallite size, and assess the degree of crystallinity in nanocomposite materials [75].

The scattering of X-rays in a crystal lattice is explained by Bragg's law, which relates the angle of incidence, the wavelength of the X-rays, and the distance between atomic planes. The equation is given by

$$2dSin\theta = n\lambda$$

where $d$ is the distance between atomic planes, $\theta$ is the angle of incidence, $n$ is an integer (the order of the diffraction), and $\lambda$ is the wavelength of the X-rays.

When X-rays strike a crystal at a specific angle, they interfere constructively, leading to a diffraction pattern. The diffraction pattern consists of bright spots (diffraction peaks) and dark areas. The positions and intensities of these peaks provide information about the crystal lattice and its orientation. By measuring the angles and intensities of the diffraction peaks, researchers can determine the crystal structure of the material. This involves solving mathematical equations and using computer software to interpret the diffraction data.

### 11.3.5 Fourier Transform Infrared Spectroscopy (FTIR)

FTIR is used to study the chemical composition and bonding characteristics of nanocomposites. It can provide information on functional groups, polymer matrix interactions, and any chemical changes occurring in the material. Each type of chemical bond (e.g., C–H, O–H, C=O) has characteristic vibrational frequencies, making FTIR a valuable tool for identifying functional groups in a sample. The interferogram is a time-domain signal that contains information about the sample's absorption at different wavelengths. Fourier transform is used to convert this signal into a frequency-domain spectrum. The resulting spectrum represents the intensity of absorbed infrared light at different wavelengths, providing detailed information about the sample's molecular composition. Q. Chen et al. used FTIR spectroscopy to determine hydrogen bonding between graphene oxide and a poly(lactic acid) polymer matrix [76].

### 11.3.6 Raman Spectroscopy

Information regarding molecular vibrations in nanocomposites is obtained using Raman spectroscopy. It can be applied to investigate the bonding patterns of nanoparticles and evaluate the way in which they interact with the matrix. In Raman spectroscopy, a monochromatic light source, typically a laser, is directed onto a sample. Elastic scattering, also known as Rayleigh scattering, occurs for the majority of incident photons and has the same energy as the incident photons. However, a small fraction of the scattered light undergoes inelastic scattering, resulting in photons with different energies due to interactions with molecular vibrations, rotations, or other modes. Stokes scattering occurs when a photon gains energy during the scattering process and the scattered light has a lower frequency (longer wavelength) than the incident light. Antistokes scattering occurs when a photon loses energy during the scattering process and the scattered light has a higher frequency (shorter wavelength) than the incident light. The Raman spectrum is a plot of intensity (y-axis) versus Raman shift (x-axis), where the Raman shift represents the energy difference between the incident and scattered photons. The peaks in the spectrum correspond to specific vibrational or rotational modes of the molecules in the sample [77].

These characterization techniques, when used collectively, offer a comprehensive understanding of the structural, morphological, thermal, mechanical, and chemical properties of nanocomposites, aiding researchers in optimizing their design and performance for specific applications.

## 11.4 APPLICATIONS

Nanocomposites find applications in various industries, including automotive, aerospace, medicine, packaging, and electronics. For example, they can be used to manufacture lightweight yet strong materials for structural components in vehicles or aircraft. Figure 10.3 depicts the schematics of the uses of nanocomposites.

### 11.4.1 Automotive

Nanocomposites have found various applications in the automotive industry due to their unique properties, which can enhance the performance, durability, and efficiency of automotive components. Nanocomposites, particularly those incorporating carbon nanotubes or graphene, can be used to develop lightweight and strong materials. These components are necessary to minimize the overall weight of cars, which increases their fuel economy and lowers their emissions [78–80].

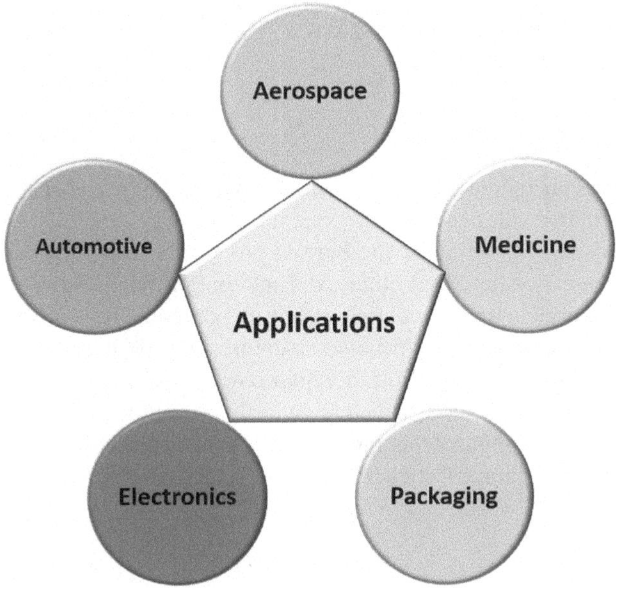


**FIGURE 11.3** Applications of nanocomposites.

Glass fiber–reinforced polyamide 66 (PA66), which possesses excellent properties but is limited in application because of its high cost, was employed in a study. The goal of this investigation was to find out whether microcellular injection molding could lower component costs without sacrificing material properties. The study examined the impact of processing variables on the morphology and performance of glass fiber–foamed PA66 +30% components. Key factors affecting the shape of molded components were found using an analysis of variance (ANOVA). For homogeneous foamed components with superior mechanical properties, ANOVA indicates that high gas injection pressure, large part thickness, and injection temperature of 300°C are necessary [81].

### 11.4.2 Aerospace

Nanocomposites have gained significant attention in the aerospace industry for their unique properties that can improve the performance, durability, and safety of aircraft components. Nanocomposites, especially those incorporating carbon nanotubes, graphene, or nanocellulose, can be used to manufacture lightweight and strong structural materials for aircraft components. This lowers overall weight, increases fuel economy, and lowers operating expenses. Molybdenum disilicide nanoparticles dispersed within an aluminum matrix were found to have good wear resistance, which is crucial for guaranteeing that the components of an aeronautical system do not begin to deteriorate under prolonged use [82]. The aerospace industry continues to invest in research and development to explore the full potential of nanocomposites, aiming to improve the efficiency, safety, and environmental impact of aircraft and spacecraft.

### 11.4.3 Medicine

Nanocomposites have shown great promise in various medical applications, offering unique properties that can enhance drug delivery, imaging, diagnostics, and the development of biomedical devices. Nanocomposites can be designed to encapsulate drugs and deliver them to specific target sites in a controlled and sustained manner. This improves the therapeutic efficacy of medications while minimizing side effects. Various nanomaterials, such as liposomes, polymer-based nanoparticles, and dendrimers, are used in drug delivery systems [83–87]. Almond gum–poly(acrylamide)/Ag hydrogel nanocomposites, which have antibacterial properties, were created by Syed et al. [88]. Magnetic nanoparticles, for example, can be guided to tumor sites using external magnetic fields for enhanced drug delivery and imaging [89].

### 11.4.4 Packaging

Nanocomposites have been increasingly employed in the packaging industry due to their unique properties that can enhance the performance, barrier properties, and sustainability of packaging materials. Nanocomposites can enhance the barrier properties of packaging materials, providing better resistance to gases, moisture, and other external contaminants. This is crucial for extending the shelf life of food products and protecting sensitive contents from environmental factors [90, 91]. According to Gorrasi et al., fully exfoliated structures exhibit nearly two orders of magnitude improvement in water vapor barrier qualities compared to tactoid and intercalated montmorillonite/polycaprolactone composite structures, which show only marginal improvement [92]. Jacquelot et al. reported a PE-based nanocomposite using oxidized paraffins as compatibilizers showed notable improvement in gas barrier characteristics (almost 100% for $CO_2$ and 60% for $O_2$) [93].

### 11.4.5 Electronics

Nanocomposites play a significant role in the electronics industry, contributing to the development of advanced materials and devices with improved performance, miniaturization, and functionality

[94]. According to Yoo et al., a unique technique for producing multifunctional hybrid nanocomposites involves dispersing the high-density polyethylene (HDPE) polymer matrix after impregnating CB surfaces with nanosized MgO getter particles. It was attempted to satisfy as many requirements as possible for the battery packaging of electric vehicles without imposing trade-offs. The mechanical strength, water vapor transmission rate, and thermal conductivity of carbon-supported MgO polymer composites were also examined in this work [95]. The integration of nanocomposites in electronics enables the creation of more efficient, compact, and versatile electronic devices. Ongoing research in this field continues to explore novel nanomaterials and their applications, driving advancements in electronic technology.

## 11.5 CHALLENGES AND FUTURE DIRECTIONS

### 11.5.1 Challenges

While nanocomposites exhibit remarkable properties, challenges remain:

*Dispersion and agglomeration.* Achieving a homogeneous dispersion of nanofillers within the matrix is critical, as agglomeration scan negatively impacts performance.

*Processing costs.* Some synthesis methods may involve high processing costs, and efforts are ongoing to develop cost-effective manufacturing processes.

*Environmental and toxicological considerations.* The potential environmental impact and toxicity of nanocomposites are areas of active research aiming to ensure their safe and sustainable use.

There is intense competition in the automotive industry when it comes to using environmentally friendly components, according to reports. Due to ongoing environmental demands, efforts have been undertaken recently to produce hybrid composites that combine plant fiber and aluminum to make lighter, more fuel-efficient automobiles. However, a number of published papers have questioned the skyrocketing cost of raw materials used in the manufacturing process. The advantages and disadvantages of the nanocomposites are listed in Table 11.2.

In nanocomposites, it is challenging to accomplish directed nanoparticle assembly, uniform dispersion, and interfacial adhesion between components [96]. The viscoelasticity caused by flow is highly dependent on the processing conditions [97]. Agglomeration of nanoparticles is caused by particle–particle interactions [98]. These are all reasonable methods. It is time-consuming and costly to maintain phase co-continuity without sacrificing the directed nature of the particle–matrix assembly by employing a solvent–built nanoparticle diffusion technique that leads to solvent evaporation. These techniques are currently used by aerospace businesses [99].

### 11.5.2 Future Directions

*Optimizing synthesis techniques.* Continuous refinement of synthesis methods to enhance control over nanocomposite properties.

**TABLE 11.2**
**Advantages and Disadvantages of the Nanocomposites**

| Advantages | Disadvantages |
|---|---|
| Superior structural and thermal stability | Non-uniform distribution |
| Improved mechanical properties | High agglomeration |
| Heat resistance | High processing cost |

*Multifunctional nanocomposites.* Designing nanocomposites with multiple functionalities to meet diverse application requirements.

*Scale-up and industrial adoption.* Addressing challenges related to scalability and encouraging the industrial adoption of nanocomposites.

## 11.6 CONCLUSION

In conclusion, the field of nanocomposites represents a dynamic and promising avenue in materials science, offering a broad range of possibilities for tailoring and enhancing material properties. The synthesis of nanocomposites involves a variety of techniques, including in situ polymerization, solution casting, and powder metallurgy, each with its unique advantages and challenges. Achieving a homogeneous dispersion of nanoscale reinforcements within the matrix and optimizing interfacial interactions remain critical aspects of successful synthesis.

Characterization techniques play a pivotal role in unravelling the complex structures of nanocomposites. Advanced analytical tools, such as TEM, SEM, XRD, and spectroscopy, provide valuable insights into the morphology, structure, and properties of these materials. Emphasizing the importance of comprehensive characterization, including the interfaces and interphases, is vital for tailoring nanocomposites to specific applications.

The applications of nanocomposites are diverse and impactful, spanning across industries. From aerospace and automotive engineering to electronics and biomedical devices, these materials have demonstrated significant improvements in mechanical strength, thermal conductivity, electrical properties, and corrosion resistance. In summary, this chapter highlights the current state of knowledge in nanocomposites, offering a comprehensive understanding of synthesis methods, characterization techniques, and diverse applications. The interdisciplinary nature of this field underscores its importance, making it an exciting area of research and development, with significant potential for future advancements in materials science and technology.

## ACKNOWLEDGMENTS

I would like to thank Dr. Sushanta K. Sethi for providing me an opportunity to contribute to this book chapter. I would like to express my heartfelt gratitude to my co-authors for their invaluable guidance and mentorship throughout the writing process of this chapter. Their expertise and support were instrumental in shaping the content and direction of the work.

## REFERENCES

[1] M. Al-Haik, C.C. Luhrs, M.M. Reda Taha, A.K. Roy, L. Dai, J. Phillips, S. Doorn, Hybrid carbon fibers/carbon nanotubes structures for next generation polymeric composites, J. Nanotechnol. 2010 (2010) 860178. https://doi.org/10.1155/2010/860178.

[2] D.M. Marquis, E. Guillaume, C. Chivas-Joly, Properties of nanofillers in polymer, in: J. Cuppoletti (Ed.), IntechOpen, Rijeka, 2011, 11. https://doi.org/10.5772/21694.

[3] L. Sang, Y. Zhao, C. Burda, $TiO_2$ nanoparticles as functional building blocks, Chem. Rev. 114 (2014) 9283–9318. https://doi.org/10.1021/cr400629p.

[4] A. Dalton, S. Collins, E. Muñoz, et al., Super-tough carbon-nanotube fibres, Nature 423, 703 (2003). https://doi.org/10.1038/423703a.

[5] R. Roy, R.A. Roy, D.M. Roy, Alternative perspectives on "quasi-crystallinity": Non-uniformity and nanocomposites, Mater. Lett. 4 (1986) 323–328. https://doi.org/10.1016/0167-577X(86)90063-7.

[6] S.K. Sethi, L. Soni, G. Manik, Component compatibility study of poly(dimethyl siloxane) with poly(vinyl acetate) of varying hydrolysis content: An atomistic and mesoscale simulation approach, J. Mol. Liq. 272 (2018) 73–83. https://doi.org/10.1016/j.molliq.2018.09.048.

[7] S.K. Sethi, L. Soni, U. Shankar, R.P. Chauhan, G. Manik, A molecular dynamics simulation study to investigate poly(vinyl acetate)-poly(dimethyl siloxane) based easy-clean coating: An insight into the surface behavior and substrate interaction, J. Mol. Struct. 1202 (2020) 127342. https://doi.org/10.1016/j.molstruc.2019.127342.

[8] S.K. Sethi, U. Shankar, G. Manik, Fabrication and characterization of non-fluoro based transparent easy-clean coating formulations optimized from molecular dynamics simulation, Prog. Org. Coatings 136 (2019) 105306. https://doi.org/10.1016/j.porgcoat.2019.105306.
[9] S.K. Sethi, G. Manik, A combined theoretical and experimental investigation on the wettability of MWCNT filled PVAc-g-PDMS easy-clean coating, Prog. Org. Coatings 151 (2021) 106092. https://doi.org/10.1016/j.porgcoat.2020.106092.
[10] S.K. Sethi, M. Singh, G. Manik, A multi-scale modeling and simulation study to investigate the effect of roughness of a surface on its self-cleaning performance, Mol. Syst. Des. Eng. 5 (2020) 1277–1289. https://doi.org/10.1039/d0me00068j.
[11] S. Chaturvedi, A. Verma, S.K. Sethi, S. Mavinkere Rangappa, S. Siengchin, Chapter 15: Stalk fibers (rice, wheat, barley, etc.) composites and applications, in: S. Mavinkere Rangappa, J. Parameswaranpillai, S. Siengchin, T. Ozbakkaloglu, H.B.T.-P.F. Wang their Composites, and Applications (Eds.), Text. Inst. B. Ser., Woodhead Publishing, 2022, 347–362. https://doi.org/10.1016/B978-0-12-824528-6.00009-6.
[12] R. Gogoi, S.K. Sethi, G. Manik, Surface functionalization and CNT coating induced improved interfacial interactions of carbon fiber with polypropylene matrix: A molecular dynamics study, Appl. Surf. Sci. 539 (2021) 148162. https://doi.org/10.1016/j.apsusc.2020.148162.
[13] Y.H. Choa, J.K. Yang, B.H. Kim, Y.K. Jeong, J.S. Lee, T. Nakayama, T. Sekino, K. Niihara, Preparation and characterization of metal/ceramic nanoporous nanocomposite powders, J. Magn. Magn. Mater. 266 (2003) 12–19. https://doi.org/10.1016/S0304–8853(03)00450–5.
[14] U. Shankar, S.K. Sethi, B.P. Singh, A. Kumar, G. Manik, A. Bandyopadhyay, Optically transparent and lightweight nanocomposite substrate of poly(methyl methacrylate-co-acrylonitrile)/MWCNT for optoelectronic applications: An experimental and theoretical insight, J. Mater. Sci. 56 (2021) 17040–17061. https://doi.org/10.1007/s10853-021-06390-3.
[15] V. Khandelwal, S.K. Sahoo, A. Kumar, S.K. Sethi, G. Manik, Bio-sourced electrically conductive epoxidized linseed oil based composites filled with polyaniline and carbon nanotubes, Compos. Part B Eng. 172 (2019) 76–82. https://doi.org/10.1016/j.compositesb.2019.05.050.
[16] J.R.A. Pires, C. Rodrigues, I. Coelhoso, A.L. Fernando, V.G.L. Souza, Current applications of bionanocomposites in food processing and packaging, Polymers (Basel). 15 (2023). https://doi.org/10.3390/polym15102336.
[17] A. Bharadwaz, A.C. Jayasuriya, Recent trends in the application of widely used natural and synthetic polymer nanocomposites in bone tissue regeneration, Mater. Sci. Eng. C 110 (2020) 110698. https://doi.org/10.1016/j.msec.2020.110698.
[18] Y. Abdullahi Hassan, H. Hu, Current status of polymer nanocomposite dielectrics for high-temperature applications, Compos. Part A Appl. Sci. Manuf. 138 (2020) 106064. https://doi.org/10.1016/j.compositesa.2020.106064.
[19] N. Ahmad, J. Anae, M.Z. Khan, S. Sabir, X.J. Yang, V.K. Thakur, P. Campo, F. Coulon, Visible light-conducting polymer nanocomposites as efficient photocatalysts for the treatment of organic pollutants in wastewater, J. Environ. Manage. 295 (2021) 113362. https://doi.org/10.1016/j.jenvman.2021.113362.
[20] S.R. Yashas, B. Shahmoradi, K. Wantala, H.P. Shivaraju, Potentiality of polymer nanocomposites for sustainable environmental applications: A review of recent advances, Polymer (Guildf). 233 (2021) 124184. https://doi.org/10.1016/j.polymer.2021.124184.
[21] A.K. Mohanty, M. Misra, L.T. Drzal, Natural, fibers, biopolymers and biocomposites, 2009. https://api.semanticscholar.org/CorpusID:137373984.
[22] K.R. Reddy, Polypropylene clay nanocomposites BT: Handbook of polymernanocomposites, in: J.K. Pandey, K.R. Reddy, A.K. Mohanty, M. Misra (Eds.), Processing, Performance and Application: Volume A: Layered Silicates, Springer, Berlin, Heidelberg, 2014, 153–175. https://doi.org/10.1007/978-3-642-38649-7_2.
[23] M. Alexandre, P. Dubois, Polymer-layered silicate nanocomposites: Preparation, properties and uses of a new class of materials, Mater. Sci. Eng. R Rep. 28 (2000) 1–63. https://doi.org/10.1016/S0927–796X(00)00012–7.
[24] A.K. Jagadeesan, K. Thangavelu, V. Dhananjeyan, Chapter 2: Carbon nanotubes: Synthesis, properties and applications, in: P. Pham, P. Goel, S. Kumar, K. Yadav (Eds.), IntechOpen, Rijeka, 2020. https://doi.org/10.5772/intechopen.92995.
[25] R. Dubey, D. Dutta, A. Sarkar, P. Chattopadhyay, Functionalized carbon nanotubes: Synthesis, properties and applications in water purification, drug delivery, and material and biomedical sciences, Nanoscale Adv. 3 (2021) 5722–5744. https://doi.org/10.1039/d1na00293g.
[26] S. Rathinavel, K. Priyadharshini, D. Panda, A review on carbon nanotube: An overview of synthesis, properties, functionalization, characterization, and the application, Mater. Sci. Eng. B 268 (2021) 115095. https://doi.org/10.1016/j.mseb.2021.115095.

[27] A.V. Belyakov, Carbon nanotubes for the synthesis of ceramic matrix composites (cleaning, dispersion, surface modification) (review), Refract. Ind. Ceram. 60 (2019) 92–100. https://doi.org/10.1007/s11148-019-00315-z.
[28] A. Tiwari, M. Michalska-Domanska, S.J. Dhoble, Chapter 10: Nanocomposite-based functional materials: Synthesis, properties, and applications, in: S.J. Dhoble, A. Nande, N.T. Kalyani, A. Tiwari, A.K.B.T.-F.M. from C. Arof Inorganic, and Organic Sources (Eds.), Woodhead Publ. Ser. Electron. Opt. Mater., Woodhead Publishing, 2023, 317–365. https://doi.org/10.1016/B978-0-323-85788-8.00016-1.
[29] S. Maitra, J. Roy, Chapter 3: Nanoceramic matrix composites: Types, processing, and applications, in: I.M.B.T.-A. in C.M.C. Second E. Low (Ed.), Woodhead Publ. Ser. Compos. Sci. Eng., Woodhead Publishing, 2018, 27–48. https://doi.org/10.1016/B978-0-08-102166-8.00003-7.
[30] K. Ravichandran, P.K. Praseetha, T. Arun, S. Gobalakrishnan, Chapter 6: Synthesis of nanocomposites, in: S. Mohan Bhagyaraj, O.S. Oluwafemi, N. Kalarikkal, S.B.T.-S. of I.N. Thomas (Eds.), Micro Nano Technology, Woodhead Publishing, 2018, 141–168. https://doi.org/10.1016/B978-0-08-101975-7.00006-3.
[31] P. Palmero, Structural ceramic nanocomposites: A review of properties and powders' synthesis methods, Nanomater. (Basel, Switzerland) 5 (2015) 656–696. https://doi.org/10.3390/nano5020656.
[32] N. Sharma, T. Saxena, S.N. Alam, B.C. Ray, K. Biswas, S.K. Jha, Ceramic-based nanocomposites: A perspective from carbonaceous nanofillers, Mater. Today Commun. 31 (2022) 103764. https://doi.org/10.1016/j.mtcomm.2022.103764.
[33] P.H.C. Camargo, K.G. Satyanarayana, F. Wypych, Nanocomposites: Synthesis, structure, properties and new application opportunities, Mater. Res. 12 (2009) 1–39. https://doi.org/10.1590/S1516–14392009000100002.
[34] T.K. Kotteda, M. Kumar, P. Kumar, R.B.R. Chekuri, metal matrix nanocomposites: Future scope in the fabrication and machining techniques, Int. J. Adv. Manuf. Technol. (2022). https://doi.org/10.1007/s00170-022-09847-0.
[35] W. Chen, T. Yang, L. Dong, A. Elmasry, J. Song, N. Deng, A. Elmarakbi, T. Liu, H.B. Lv, Y.Q. Fu, Advances in graphene reinforced metal matrix nanocomposites: Mechanisms, processing, modelling, properties and applications, Nanotechnol. Precis. Eng. 3 (2020) 189–210. https://doi.org/10.1016/j.npe.2020.12.003.
[36] S. Pan, T. Wang, K. Jin, X. Cai, Understanding and designing metal matrix nanocomposites with high electrical conductivity: A review, J. Mater. Sci. 57 (2022) 6487–6523. https://doi.org/10.1007/s10853-022-07010-4.
[37] P.K. Rohatgi, B.F. Schultz, J.B. Ferguson, Synthesis and properties of metal matrix nanocomposites (MMNCS), syntactic foams, self lubricating and self-healing metals BT, in: F. Marquis (Ed.), Proceedings of the 8th Pacific Rim International Congress on Advanced Materials and Processing, Springer International Publishing, Cham, 2016, 1515–1524.
[38] K.S. Usha Devi, D. Ponnamma, V. Causin, H.J. Maria, S. Thomas, Enhanced morphology and mechanical characteristics of clay/styrene butadiene rubber nanocomposites, Appl. Clay Sci. 114 (2015) 568–576. https://doi.org/10.1016/j.clay.2015.07.009.
[39] T.D. Fornes, D.R. Paul, Modeling properties of nylon 6/clay nanocomposites using composite theories, Polymer (Guildf). 44 (2003) 4993–5013. https://doi.org/10.1016/S0032–3861(03)00471–3.
[40] D.L. VanderHart, A. Asano, J.W. Gilman, Solid-state NMR investigation of paramagnetic nylon-6 clay nanocomposites. 2. Measurement of clay dispersion, crystal stratification, and stability of organic modifiers, Chem. Mater. 13 (2001) 3796–3809. https://doi.org/10.1021/cm011078x.
[41] R. Das, S. Borthakur, A. Charles, M. Hazarika, P. Borgohain, T.R. Borah, L. Saikia, P. Dutta, J. Kalita, A novel montmorillonite clay based bio-nanocomposite as an emerging biocontrol agent against stored grain pulse beetle, Hybrid Adv. 5 (2024) 100138. https://doi.org/10.1016/j.hybadv.2023.100138.
[42] A. Chaudhuri, A. Joydhar, Graphene nanocomposites and applications in electrochemical energy storage materials, Mater. Today Proc. 64 (2022) 1569–1581. https://doi.org/10.1016/j.matpr.2022.02.226.
[43] E.I. Akpan, X. Shen, B. Wetzel, K. Friedrich, Chapter 2: Design and synthesis of polymer nanocomposites, in: K. Pielichowski, T.M.B.T.-P.C. with F.N. Majka (Eds.), Micro Nano Technology, Elsevier, 2019, 47–83. https://doi.org/10.1016/B978-0-12-814064-2.00002-0.
[44] A. Sawaguchi, K. Toda, K. Niihara, Mechanical and electrical properties of silicon nitride–silicon carbide nanocomposite material, J. Am. Ceram. Soc. 74 (1991) 1142–1144. https://doi.org/10.1111/j.1151–2916.1991.tb04357.x.
[45] Y. Sun, L. Zheng, S. Zhang, X. Luo, F. Liu, Y. Li, In situ synthesis of $Al_2O_3$–SiC powders via molten-salt-assisted aluminum/carbothermal reduction method, Ceram. Int. 48 (2022) 24471–24475. https://doi.org/10.1016/j.ceramint.2022.05.061.

[46] S.R. Hariharan, S. Sakthi, S. Mahendran, S.T. Jayasuthakar, Influence of MgO & SiC on mechanical and wear characteristics of AA6061 hybrid matrix composites synthesized by stir casting, J. Appl. Res. Technol. 21 (2023) 162–168. https://doi.org/10.22201/icat.24486736E.2023.21.2.1618.
[47] I. Koltsov, G. Kimmel, S. Stelmakh, K. Sobczak, W. Lojkowski, The new nano-enabled phase map of $ZrO_2$-$Al_2O_3$, Sci. Rep. 9 (2019) 1–9. https://doi.org/10.1038/s41598-019-42058-4.
[48] W. Chen, J. Zhang, W. Cai, Sonochemical preparation of Au, Ag, Pd/$SiO_2$ mesoporous nanocomposites, Scr. Mater. 48 (2003) 1061–1066. https://doi.org/10.1016/S1359–6462(02)00635–8.
[49] F. Heringhaus, H.-J. Schneider-Muntau, G. Gottstein, Analytical modeling of the electrical conductivity of metal matrix composites: Application to Ag–Cu and Cu–Nb, Mater. Sci. Eng. A 347 (2003) 9–20. https://doi.org/10.1016/S0921–5093(02)00590–7.
[50] M. Balog, P. Krizik, P. Svec, L. Orovcik, Industrially fabricated in-situ Al-AlN metal matrix composites (part A): Processing, thermal stability, and microstructure, J. Alloys Compd. 883 (2021) 160858. https://doi.org/10.1016/j.jallcom.2021.160858.
[51] M.T. Sijo, K.R. Jayadevan, Analysis of stir cast aluminium silicon carbide metal matrix composite: A comprehensive review, Procedia Technol. 24 (2016) 379–385. https://doi.org/10.1016/j.protcy.2016.05.052.
[52] D. Boulerba, A. Zoukel, Poly(methyl methacrylate)/$SiO_2$ nanocomposites: Effects of the molecular interaction strength on thermal properties, Polym. Polym. Compos. 29 (2021) S49–S56. https://doi.org/10.1177/0967391120985710.
[53] H. Shao, Z. Fei, X. Li, Z. Zhang, S. Zhao, K. Li, Z. Yang, Polyimide/$SiO_2$ composite aerogels with excellent thermal and sound insulation properties prepared by confined filling method, Mater. Lett. 354 (2024) 135402. https://doi.org/10.1016/j.matlet.2023.135402.
[54] Z. Rafiee, E. Zare, Optically active poly(amide–imide)/$TiO_2$ nanocomposites containing amino acid moieties: Synthesis and properties, Amino Acids 47 (2015) 2293–2302. https://doi.org/10.1007/s00726-015-2008-z.
[55] S. Mallakpour, M. Naghdi, Polymer/$SiO_2$ nanocomposites: Production and applications, Prog. Mater. Sci. 97 (2018) 409–447. https://doi.org/10.1016/j.pmatsci.2018.04.002.
[56] V. Shrivastava, G. Kumar Gupta, H. Jain, D.V. Mangal, P. Singh, I.B. Singh, Novel synthesis approach of making efficient nanocomposite via powder metallurgy route: Study of microstructure and mechanical properties, Manuf. Lett. 31 (2022) 36–39. https://doi.org/10.1016/j.mfglet.2021.06.001.
[57] X. Zhang, Y. Xu, M. Wang, E. Liu, N. Zhao, C. Shi, D. Lin, F. Zhu, C. He, A powder-metallurgy-based strategy toward three-dimensional graphene-like network for reinforcing copper matrix composites, Nat. Commun. 11 (2020) 1–13. https://doi.org/10.1038/s41467-020-16490-4.
[58] G.L. Vignoles, Chapter 8: Chemical vapor deposition/infiltration processes for ceramic composites, in: P.B.T.-A. in C.M. and P.D. Boisse (Ed.), Woodhead Publishing, 2015, 147–176. https://doi.org/10.1016/B978-1-78242-307-2.00008-7.
[59] M. Balestrat, A. Lale, A. Bezerra, V. Proust, E. Wasan Awin, R. Machado, P. Carles, R. Kumar, C. Gervais, S. Bernard, In-situ synthesis and characterization of nanocomposites in the Si-Ti-N and Si-Ti-C systems, Molecules 25 (2020) 5236. https://doi.org/10.3390/molecules25225236.
[60] S. Mondal, P. Mondal, D.P. Mishra, Research progress on ceramic nanomaterials reinforced aluminum matrix nanocomposites, Mater. Sci. Technol. (United Kingdom) 39 (2023) 1841–1857. https://doi.org/10.1080/02670836.2023.2187153.
[61] S.M. Almotairy, N.H. Alharthi, H.F. Alharbi, H.S. Abdo, Superior mechanical performance of inductively sintered Al/SiC nanocomposites processed by novel milling route, Sci. Rep. 10 (2020) 10368. https://doi.org/10.1038/s41598-020-67198-w.
[62] A. Saxena, K.K. Saxena, V.K. Jain, S.K. Rajput, B.N. Pathak, A review of reinforcements and process parameters for powder metallurgy-processed metal matrix composites, Mater. Today Proc. (2023). https://doi.org/10.1016/j.matpr.2023.02.227.
[63] M. Malaki, W. Xu, A.K. Kasar, P.L. Menezes, H. Dieringa, R.S. Varma, M. Gupta, Advanced metal matrix nanocomposites, Metals (Basel). 9 (2019). https://doi.org/10.3390/met9030330.
[64] V.K. Thakur, Materials Horizons: From Nature to Nanomaterials Series Editor, Springer, 2021. http://www.springer.com/series/16122.
[65] C. Suryanarayana, A.A. Al-Joubori, Z. Wang, Nanostructured materials and nanocomposites by mechanical alloying: An overview, Met. Mater. Int. 28 (2022) 41–53. https://doi.org/10.1007/s12540-021-00998-5.
[66] L. Mrah, R. Meghabar, In situ polymerization of styrene–clay nanocomposites and their properties, Polym. Bull. 78 (2021) 3509–3526. https://doi.org/10.1007/s00289-020-03274-5.

[67] H. Mao, X. Wang, Use of in-situ polymerization in the preparation of graphene/polymer nanocomposites, New Carbon Mater. 35 (2020) 336–343. https://doi.org/10.1016/S1872–5805(20)60493–0.
[68] M.K. Vyas, A. Chandra, Role of organic/inorganic salts and nanofillers in polymer nanocomposites: Enhanced conduction, rheological, and thermal properties, J. Mater. Sci. 53 (2018) 4987–5003. https://doi.org/10.1007/s10853-017-1912-x.
[69] C.M. Hussain, S. Thomas, Handbook of Polymer and Ceramic Nanotechnology: Volume 1, 2, 2021. https://doi.org/10.1007/978-3-030-40513-7.
[70] G. Zhang, T. Wu, W. Lin, Y. Tan, R. Chen, Z. Huang, X. Yin, J. Qu, Preparation of polymer/clay nanocomposites via melt intercalation under continuous elongation flow, Compos. Sci. Technol. 145 (2017) 157–164. https://doi.org/10.1016/j.compscitech.2017.04.005.
[71] C.A. Ávila-Orta, P. González-Morones, D.A. Valdez, A. González-Sánchez, J.G. Martínez-Colunga, J.M. Mata-Padilla, V.J. Cruz-Delgado, Chapter 9: Ultrasound-assisted melt extrusion of polymer nanocomposites, in: S. Sivasankaran (Ed.), IntechOpen, Rijeka, 2018. https://doi.org/10.5772/intechopen.80216.
[72] S. Jamalpour, S.R. Ghaffarian, A. Jangizehi, Effect of matrix-nanoparticle supramolecular interactions on the morphology and mechanical properties of polymer foams, Polym. Int. 67 (2018) 859–867. https://doi.org/10.1002/pi.5536.
[73] Y. Yang, M. Romano, G. Feng, X. Wang, T. Wu, S. Holdren, M.R. Zachariah, Growth of sub-5 nm metal nanoclusters in polymer melt aerosol droplets, Langmuir 34 (2018) 585–594. https://doi.org/10.1021/acs.langmuir.7b02900.
[74] N. Gopal, S. Kumar, R. Sahney, Towards the development of flexible carbon nanotube-parafilm nanocomposites and their application as bioelectrodes, RSC Adv. 11 (2021) 34193–34205. https://doi.org/10.1039/d1ra01840j.
[75] V. Abhilash, N. Rajender, K. Suresh, Chapter 14: X-ray diffraction spectroscopy of polymer nanocomposites, in: S. Thomas, D. Rouxel, D.B.T.-S. of P.N. Ponnamma (Eds.), William Andrew Publishing, 2016, 410–451. https://doi.org/10.1016/B978-0-323-40183-8.00014-8.
[76] Q. Chen, J.D. Mangadlao, J. Wallat, A. De Leon, J.K. Pokorski, R.C. Advincula, 3D printing biocompatible polyurethane/poly(lactic acid)/graphene oxide nanocomposites: Anisotropic properties, ACS Appl. Mater. Interfaces 9 (2017) 4015–4023. https://doi.org/10.1021/acsami.6b11793.
[77] C. Velasco-Santos, A.L. Martinez-Hernandez, W. Brostow, V.M. Castaño, Influence of silanization treatment on thermomechanical properties of multiwalled carbon nanotubes: Poly(methylmethacrylate) nanocomposites, J. Nanomater. 2011 (2011). https://doi.org/10.1155/2011/928659.
[78] A.G. Arsha, V. Manoj, L.A. Raag, M.G. Akhil, T.P.D. Rajan, Chapter 9: Metallic nanocomposites for automotive applications, in: H. Song, T.A. Nguyen, G. Yasin, N.B. Singh, R.K.B.T.-N. in the A.I. Gupta (Eds.), Micro Nano Technology, Elsevier, 2022, 163–198. https://doi.org/10.1016/B978-0-323-90524-4.00009-8.
[79] A. Elmarakbi, W. Azoti, Chapter 1: State of the art on graphene lightweighting nanocomposites for automotive applications, in: F. Marotti de Sciarra, P.B.T.-E.C. Russo Predictive Mechanical and Thermal Modeling of Nanostructures and Their Polymer Composites (Eds.), Micro Nano Technology, Elsevier, 2018, 1–23. https://doi.org/10.1016/B978-0-323-48061-1.00001-4.
[80] A.S.H. Makhlouf, H. Soliman, Chapter 18: Effect of nano-additives ($Al_2O_3$ and NaF) on the performance of ceramic coatings formed by microarc oxidation on magnesium alloys, in: A.S.H. Makhlouf, D.B.T.-H. of N. and N.C. and M. Scharnweber (Eds.), Butterworth-Heinemann, 2015, 389–401. https://doi.org/10.1016/B978-0-12-799947-0.00018-3.
[81] V. Volpe, S. Lanzillo, G. Affinita, B. Villacci, I. Macchiarolo, R. Pantani, Lightweight high-performance polymer composite for automotive applications, Polymers (Basel) 11 (2019). https://doi.org/10.3390/polym11020326.
[82] M. Sameezadeh, M. Emamy, H. Farhangi, Effects of particulate reinforcement and heat treatment on the hardness and wear properties of AA 2024-$MoSi_2$ nanocomposites, Mater. Des. 32 (2011) 2157–2164. https://doi.org/10.1016/j.matdes.2010.11.037.
[83] M. Attia, R.D. Glickman, G. Romero, B. Chen, A.J. Brenner, J.Y. Ye, Optimized metal-organic-framework based magnetic nanocomposites for efficient drug delivery and controlled release, J. Drug Deliv. Sci. Technol. 76 (2022) 103770. https://doi.org/10.1016/j.jddst.2022.103770.
[84] Talinungsang, D. Upadhaya, P. Kumar, D.D. Purkayastha, Superhydrophilicity of photocatalytic ZnO/$SnO_2$ heterostructure for self-cleaning applications, J. Sol-Gel Sci. Technol. 92 (2019) 575–584. https://doi.org/10.1007/s10971-019-05127-8.

[85] D. Upadhaya, Talinungsang, P. Kumar, D.D. Purkayastha, Tuning the wettability and photocatalytic efficiency of heterostructure ZnO-$SnO_2$ composite films with annealing temperature, Mater. Sci. Semicond. Process. 95 (2019) 28–34. https://doi.org/10.1016/j.mssp.2019.02.009.

[86] W.M. Kedir, G.F. Abdi, M.M. Goro, L.D. Tolesa, Pharmaceutical and drug delivery applications of chitosan biopolymer and its modified nanocomposite: A review, Heliyon 8 (2022) e10196. https://doi.org/10.1016/j.heliyon.2022.e10196.

[87] D. Upadhaya, P. Kumar, D. Dhar Purkayastha, Superhydrophobic ZnO/$TiO_2$ heterostructure with significantly enhanced photocatalytic activity, J. Mater. Sci. Mater. Electron. 30 (2019) 10399–10407. https://doi.org/10.1007/s10854-019-01381-2.

[88] I. Syed Ahamed Hussain, V. Jaisankar, An eco-friendly synthesis, characterization and antibacterial applications of novel almond gum: Poly(acrylamide) based hydrogel silver nanocomposite, Polym. Test. 62 (2017) 154–161. https://doi.org/10.1016/j.polymertesting.2017.06.021.

[89] C.A.S. Ballesteros, D.S. Correa, V. Zucolotto, Polycaprolactone nanofiber mats decorated with photoresponsive nanogels and silver nanoparticles: Slow release for antibacterial control, Mater. Sci. Eng. C 107 (2020). https://doi.org/10.1016/j.msec.2019.110334.

[90] P. Kumar, P.S. Prasad, J. David, M. Guerrero, Different Curing Methods Used for Polymer Composites, Springer Nature, Singapore, 2024. https://doi.org/10.1007/978-981-97-0888-8.

[91] P.S. Prasad, J. David, M. Guerrero, P. Kumar, Natural and Synthetic Fiber-Filled Polymer Composites Used as Anticorrosive Materials, Springer Nature, Singapore, 2024. https://doi.org/10.1007/978-981-97-0888-8.

[92] G. Gorrasi, M. Tortora, V. Vittoria, E. Pollet, B. Lepoittevin, M. Alexandre, P. Dubois, Vapor barrier properties of polycaprolactone montmorillonite nanocomposites: Effect of clay dispersion, Polymer (Guildf). 44 (2003) 2271–2279. https://doi.org/10.1016/S0032–3861(03)00108–3.

[93] E. Jacquelot, E. Espuche, J.F. Gérard, J. Duchet, P. Mazabraud, Morphology and gas barrier properties of polyethylene-based nanocomposites, J. Polym. Sci. Part B Polym. Phys. 44 (2006) 431–440. https://doi.org/10.1002/polb.20707.

[94] P. Kumar, Recent Progress in Computational, Springer Nature, Singapore, 2023. https://doi.org/10.1007/978-981-99-3549-9.

[95] J.E. Yoo, V. Roev, J. Bae, D.S. Yoon, S.D. Kim, E.S. Lee, Multifunctional hybrid polymer nanocomposites for automotive-battery packaging, J. Appl. Polym. Sci. 137 (2020) 1–9. https://doi.org/10.1002/app.49059.

[96] T. Ramanathan, A.A. Abdala, S. Stankovich, D.A. Dikin, M. Herrera-Alonso, R.D. Piner, D.H. Adamson, H.C. Schniepp, X. Chen, R.S. Ruoff, S.T. Nguyen, I.A. Aksay, R.K. Prud'Homme, L.C. Brinson, Functionalized graphene sheets for polymer nanocomposites, Nat. Nanotechnol. 3 (2008) 327–331. https://doi.org/10.1038/nnano.2008.96.

[97] C. Wan, B. Chen, Reinforcement and interphase of polymer/graphene oxide nanocomposites, J. Mater. Chem. 22 (2012) 3637–3646. https://doi.org/10.1039/c2jm15062j.

[98] D.W. Schaefer, R.S. Justice, How nano are nanocomposites?, Macromolecules 40 (2007) 8501–8517. https://doi.org/10.1021/ma070356w.

[99] M. Yoonessi, Y. Shi, D.A. Scheiman, M. Lebron-Colon, D.M. Tigelaar, R.A. Weiss, M.A. Meador, Graphene polyimide nanocomposites: Thermal, mechanical, and high-temperature shape memory effects, ACS Nano 6 (2012) 7644–7655. https://doi.org/10.1021/nn302871y.

# 12 Coating Techniques for Interfacial Adhesion Improvement in Polymer Composites

*Alisan Imam, Md Mahamud Hasan Tusher, Shourav Biswas, Mosa. Afroja Khatun, and Akib Jabed*

## 12.1 INTRODUCTION

Optimizing the interfacial adhesion between the polymer matrix and reinforcing fillers is critical to developing polymer composites as high-performance materials. The overall mechanical, thermal, and durability properties of the composite material are significantly influenced by the interphase area where these two components interact. Many coating approaches have become useful options for augmentation in resolving the problems caused by insufficient interfacial adhesion.

The coating methods designed to increase interfacial adhesion in polymer composites are thoroughly examined in this chapter. This conversation attempts to provide a comprehensive grasp of the various techniques used in this subject, ranging from the basic ideas guiding interfacial interactions to state-of-the-art coating techniques. We explore their consequences for the mechanical strength, thermal stability, and overall performance of polymer composites by looking at both traditional and novel coating techniques. The objective of this chapter is to provide significant insights for academics and practitioners working toward the development of advanced composite materials by illuminating the real-world application of various coating methods through case studies and realistic examples.

## 12.2 PLASMA SURFACE MODIFICATION

### 12.2.1 Understanding the Impact of Plasma Treatment on Polymer Composites

Plasma treatment is a surface modification method. It has a substantial effect on the characteristics of polymer composites. Different properties, such as adhesion, wettability, and surface chemistry, can change due to the plasma treatment of polymer composites. Increasing adhesion with different materials results from introducing various functional groups to the polymer surface using plasma treatment [1]. The wettability of polymer composites is enhanced by plasma treatment, which is good for better coating adherence [2]. The adherence of polymer matrix and reinforcing components (such as fibers or fillers) in composite materials is enhanced by using plasma treatment [3, 4]. The polymer surface can be cleaned, and further impurities can be diminished using plasma treatment, improving the interface's overall quality [5]. Plasma treatment can improve the mechanical properties of polymer composites. Tensile strength and modulus of shear strength are among them [6, 7]. The compatibility of different polymer components in a composite material can be enhanced by plasma treatment [8, 9].

 DOI: 10.1201/9781003476337-12

### 12.2.2 Effects on Interfacial Adhesion

Adhesion qualities may be influenced by dehydrogenation, and surface etching occurs by hydrogen plasma treatment [10, 11]. Polymer surfaces are always cleaned and activated the surface by argon plasma treatment, which increases adhesion [12, 13]. By adding different groups like amino groups, nitrogen plasma treatment can exchange the surface chemistry of polymers and enhance adhesion [14, 15]. To increase adhesion, oxygen-containing functional groups are added to the polymer surface by using oxygen plasma treatment [16]. Atmospheric pressure plasma techniques have several advantages, including cost-effectiveness and ease of equipment use. Additionally, they can improve adhesive qualities. These are the effects of using different types of plasma treatments on interfacial adhesion.

### 12.2.3 Practical Consideration for Implementing Plasma Surface Modification

There are a large number of methods available for changing the surface characteristics of polymer composites. The reason for these treatments is to enhance the interfacial adhesion of polymers with fillers or fibers. Vacuum environments are used for low-pressure plasma treatments. They can enhance stickiness and wettability by adding functional groups to the polymer surface [17]. *Pressure in the atmosphere.* Plasma therapies give advantages in terms of cost and scalability. They are carried out at or close to atmospheric pressure. It also enhances surface qualities for stronger and tougher adherence [18]. A dielectric barrier is used in DBD plasma treatments. It produces plasma. Surface energy is improved, and polar functional groups are introduced effectively [19]. In a low-pressure gas environment, a voltage is applied. It is applied across two electrodes. It will create plasma for use in glow–discharge plasma therapies. Surface activation and adhesion will increase [20]. The plasma produced away from the substrate is called remote plasma treatment. The polymer surface can be modified and enhanced by this process [21].

## 12.3 ELECTROLESS DEPOSITION

### 12.3.1 Principles of Electroless Deposition in Polymer Composites

The process of depositing a metal or other material onto a substrate without the need for an external electrical power source is called electroless deposition. It is also known as chemical or autocatalytic deposition. This method is used in the field to enhance the properties or some new functions of polymer composites.

**Principle:** An autocatalytic chemical reaction is applied in electroless deposition to decrease metal ions on the substrate surface. Without an external electrical power source, metal can be deposited on the substrate surface. It is a reducing agent in the solution. It catalyzes the reduction of metal ions. There are two primary phases of the electroless deposition process. These are nucleation and growth. Nucleation is the first thing to develop, and small metal clusters form; then growth is the enlargement of these clusters. It will happen into a continuous metal film. The polymer composite surface must be activated. Then, before deposition, it will act as a catalyst for the electroless reaction. There are common surface activation methods, such as sensitization, activation, and acceleration steps. It may involve chemical treatments [22]. Metals like nickel, copper, gold, and silver are deposited. Because each has unique, different qualities like conductivity, resistance to corrosion, or catalytic activity, the electroless deposition process is affected by process variables. These variables are bath composition, temperature, pH, and deposition time. These parameters must be optimized. It is for the coatings on the polymer composite to be uniform, unique, and adherent [23]. The overall performance of the composite material depends on several things, like good adhesion between the deposited layer and the polymer matrix. Bonding is facilitated by surface activation, appropriate cleaning, and optimal deposition conditions. It can be applied to the polymer matrix to enhance conductivity, corrosion resistance, or other desired characteristics [24].

### 12.3.2 Metal Selection and Its Influence on Adhesion Improvement

For polymer composites, choosing the right metal for electroless deposition is necessary. It is essential for increasing adhesion. The adhesion strength can be influenced by the interaction between different metals and the polymer matrix.

Adhesion is affected by the surface energy and affinity of the deposited metal for the polymer matrix. Bonds are generally formed by metals, because it has higher surface energy and a good affinity for the surface of polymers [22]. The metal should be compatible with the particular polymer matrix because of good interfacial bonding. The presence of functional groups on the polymer surface, thermal expansion coefficients, and chemical composition all influence compatibility. The adhesion strength may be influenced by the mechanical properties of the deposited metal layer, such as its elasticity and hardness. More adhesion may result from a metal layer [24]. The adhesion interface gives long-term durability, and it is dependent on the deposited metal's resistance to corrosion. The metal layer keeps its integrity over time, and finally, corrosion resistance stops the adhesion from deteriorating. To enhance adhesion, metals with particular electrical or catalytic properties can be selected for the required functionalities. Silver and gold, for instance, are known to have good conductivity. Adhesion may be influenced by the metal layer's thickness. Good interfacial bonding may be affected by thin layers, but decreasing adhesion and stress concentrations may result from overly thick layers [25]. Adhesion is strengthened by metals. It will encourage interdiffusion and intermixing with the polymer chains. A stronger link between the metal layer and the polymer composite is influenced by this molecular-level interfacial interaction [25].

### 12.3.3 Challenges and Solutions

Polymer composites have several uses since their qualities allow for flexibility, strength, and a light weight. The characteristics of polymer composites can be greatly enhanced when metal coatings are added to surfaces utilizing the electroless deposition process, which does not require an external power source. Nevertheless, these composite materials can occasionally provide many problems for the deposition process. These include surface contamination, uneven coating, and poor adherence. Researchers and engineers may be able to improve the electroless deposition method for polymer composites and enable the development of new functional materials with enhanced performance attributes by comprehending these issues and putting potential answers into practice [26].

#### 12.3.4.1 Challenges

Functional groups are frequently absent from polymer composites. Its low surface energies and the necessity for good metal adhesion stem from this. A breakdown in adhesion at the polymer–matrix interface may result in delamination and decrease coating longevity. Surface activation techniques like chemical etching, plasma treatment, or surface functionalization of the composite can enhance adhesion. Surface flaws in polymer composites can lead to uneven coating thickness and inadequate coverage. Sometimes, the substrate surface is not properly wettable by the plating solution, which might exacerbate problems with coating uniformity. Deposition parameters must be carefully chosen to get a uniform coating. The temperature, agitation, and composition of the bath are these parameters [27]. Impurities on the composite surface, such as oils, dust, or release agents, might degrade the coating quality and metal adherence. Effective cleaning methods, including ultrasonic, solvent, and detergent washing, are required to get rid of surface contaminants. Maintaining a clean deposition environment and using caution when handling substrates can reduce processing contamination. Certain polymer matrices may not respond well to all plating solutions. These can result in adhesion issues or deterioration of the composite material. Plating solutions must be precisely selected for the polymer composite, taking into consideration factors like pH, temperature, and chemical interactions to achieve better compatibility [28].

#### 12.3.4.1 Solutions

By increasing the surface energy and roughness of polymer composites via plasma treatment, chemical etching, or surface functionalization, metal adhesion can be improved. Pretreatment techniques should be adjusted to provide the desired surface characteristics without compromising the integrity of the composite material. Systematic testing is necessary to identify the optimal deposition parameters, which include bath composition, temperature, and agitation, to gain consistent coating thickness and coverage. Throughout the process, reproducibility and consistency can be preserved by regularly checking and modifying the deposition settings. Specific cleaning methods, such as solvent, detergent, and ultrasonic cleaning, must be enhanced for polymer composites to eliminate surface contaminants. Processing contamination can be reduced by following cleaning protocols and best practices in the deposition setting. The right plating and deposition solutions that are especially made to work with polymer composites need to consider things like pH, temperature, and chemical interactions. The needs of applications using polymer composites can be met by custom plating solutions made by materials scientists and electroplating experts. Examples of post-treatment techniques that can improve the adhesion and performance of coatings applied to polymer composites include annealing and surface modification [29]. Estimating the post-treatment effects on composite qualities, including mechanical, electrical, and corrosion resistance, is necessary to improve the coated composite material's overall performance [30].

The challenges associated with electroless deposition on polymer composites impede the full potential of this method to improve composite materials. It is possible to resolve these problems and apply superior metal coatings on polymer composites. These issues can be resolved with the right surface preparation, optimization of the deposition parameter, and development of tailored solutions. The continued research and collaboration of materials scientists, engineers, and businesses enable the creation of innovative solutions for a variety of applications [31].

## 12.4 SILANE COUPLING AGENTS

Silane coupling agents play an important role in the preparation of polymer composites, acting as a molecular bridge between organic materials, such as polymers, and inorganic materials, such as fillers [32]. In general, silane coupling agents are organosilicon compounds containing two functional groups of differing reactivity. The general structure of silane coupling agents is $(X)_3$-Si-R-Y, where $X$ represents a functional group that can be hydrolyzed by water or moisture to produce silanol, which may then connect with inorganic materials, and $Y$ represents a functional group that can either react with synthetic resin or other polymers or make hydrogen bonds to become soluble in them [33, 34]. This significantly enhances the performance of polymer composite materials, including their physical, electrical, thermal, and optical characteristics [35, 36].

In general, the interaction between silane coupling agents and polymer composite entails a number of crucial stages:

First Stage: *Application of silane coupling agent.* The polymer composite is coated with the silane coupling agent using techniques such as spraying, brushing, or dip-coating.

Second Stage: *Adsorption onto the surface.* A homogenous and thin coating of silane molecules is adsorbed on the surface of the composite. The adsorption process is affected by the characteristics of both the polymer matrix and the reinforcing elements.

Third Stage: *Hydrolysis.* The silane coupling agent generally begins with hydrolysis, when the silane molecule undergoes a reaction with water to generate silanol groups (Si-OH). Typically, an acid or a base serves as a catalyst for this reaction.

Fourth Stage: *Condensation.* Silanols undergo condensation (aging) during the hydrolysis process. At this point, it is important to limit condensation in order to keep the silanols available for adsorption onto the hydroxyl groups in the fibers. The regulation of the silanol condensation rate can be achieved through the manipulation of the hydrolysis system's pH.

Fifth Stage: *Chemical interactions with the polymer matrix.* Covalent bonds are formed when the activated siloxane forms bridges with organofunctional groups on the polymer matrix.

Final Stage: *Adherence to reinforcement or filler materials.* Other hydrolyzable functional groups, such as an alkoxy group, react with reinforcement or filler materials and are converted into covalent bonds and liberating water [33, 37–39].

## 12.5 POLYMER–MATRIX COMPOSITE

Fibers serve as the reinforcing medium in polymer–matrix composites (PMCs), which are made of a polymer resin matrix. Because of their room temperature characteristics, ease of production, and affordability, these materials are employed in a greatest variety of composite applications and in the largest quantities. This section discusses the several classifications of PMCs based on the types of reinforcement (glass, carbon, and aramid), their uses, and the different polymer resins that are used [40].

### 12.5.1 Glass Fiber–Reinforced Polymer (GFRP) Composites

Fiberglass, the most widely produced type of composite, is essentially composed of continuous or discontinuous glass fibers enclosed in a polymer matrix. For various reasons, glass is a popular material for fiber-reinforcing cement.

1. It may be effortlessly incorporated into molten fibers with excellent strength.
2. It is widely accessible and may be effectively manufactured into a glass-reinforced plastic through a range of composite production methods.
3. Despite its relatively high specific strength, the fiber creates a very high composite when it is implanted in a plastic matrix.
4. The composite is suitable in a range of harsh situations due to its chemical inertness when combined with the different plastics [41].

### 12.5.2 Carbon Fiber–Reinforced Polymer (CFRP) Composites

The most widely utilized reinforcement in advanced polymer–matrix composites (i.e., non-fiberglass) is carbon, a high-performance fiber material. Here are the reasons behind this:

1. Among all materials for reinforcing fibers, carbon fibers have the highest specific modulus and strength.
2. They are strong and maintain their high tensile modulus at high temperatures; nevertheless, there may be an issue with high-temperature oxidation.
3. Carbon fibers are unaffected by a wide range of solvents, acids, bases, or moisture when they are at room temperature.
4. Because of the variety of mechanical and physical properties that fibers display, composites containing them can be manufactured to have particular qualities.

5. Relatively cheap and efficient methods of manufacturing fiber and composites have been established [42].

### 12.5.3 Aramid Fiber–Reinforced Polymer Composites

In composites with polymer matrices, aramid fibers are most frequently utilized; typical matrix materials include polyesters and epoxies. The fibers may be treated using the majority of standard textile procedures because they are moderately ductile and flexible. Typical uses for these aramid composites include pressure tanks, athletic goods, tires, ropes, ballistic products (bulletproof jackets and armor), and automotive brake and clutch linings and gaskets where asbestos was previously used. Thus, it is possible to compare these three materials' mechanical properties in both longitudinal and transverse orientations [43].

## 12.6 CHEMICAL VAPOR DEPOSITION (CVD)

A unique amorphous, hydrogenated carbon (a-C: H) coating vessel from Plasma Electronic, GmbH, will be used to investigate surface modification and coating in this technology [44]. To enhance the adhesion of an a-C: H coating, the vessel specifically used an undercoating based on silicon. Chemical vapor deposition using plasma activation is the deposition technique used (PACVD). The vessel creates a voltage bias of about 500 V in a regulated low atmosphere of reaction gasses. An even, steady, high-energy plasma of reaction elements forms on the whole cathode surface when the steel to be coated comes into contact with the cathode target. Chemical vapor deposition (CVD) can take place at much lower temperatures due to the kinetics of the plasma. The properties of the coating vary based on the reaction gas used and other processing parameters [45, 46]. Since the process operates at 200°C, there shouldn't be any thermal effects on the material. The standard processing conditions were established to use a tetramethyl silane reaction gas to deposit an adhesive layer of silicon, silicon carbide, and carbon at a thickness of 0.5 μm [47].

After that, the procedure switches the adhesive layer gas's partial pressure to the top layer gas's constant pressure. Methane gas was used to deposit the top layer, which has a thickness of 2.0 μm. Some samples with only the silicon-rich layer were produced using a modified procedure, and they were compared to samples that were normally processed, fully coated, and uncoated. The adhesive layer was examined independently of the entire carbon coating since prior studies [45, 48] have demonstrated that silicon's reactivity is advantageous for forming reactive groups that would stick to both the iron and the polymer matrix material [49].

### 12.6.1 Effects on Interfacial Adhesion

The analysis in this chapter focused on polymer functionalization. The most widely used polymer is PP, which can have its polar functional groups added and steel adhesion improved by treating it with maleic anhydride. Along with being affordable, PP is easy to work with and has strong mechanical qualities. Its high polar group concentration makes it an excellent option for forming adhesion with metal surfaces, even though its poor mechanical properties are mitigated by the polymer's increased wettability. The major failure of CNT epoxy's attempt to improve adhesion was caused by agglomeration and the formation of surface voids during curing.

Surface treatment with silicon provides improved adhesion between metal and polymer surfaces. To be more exact, this strategy appears to be a workable and effective means of promoting enhanced adhesion between steel and PPMA.

On the other hand, investigations for this kind of surface treatment showed no relationship between surface energy and binding strength. It is possible that diffusivity is the mechanism of adhesion between coated steel and PPMA in the case of coated steel, as wetting tension on the steel surface is not indicative of potentially improved adhesion. It was discovered that the kind of polymer and bond strength had a directly proportionate relationship. This suggests that the magnitude of the contact angle predicts the adhesion with steel, with polar groups on the polymer surface dominating adhesion. Additionally, for the polymers in the study, wettability has a greater influence on adhesion to steel.

## 12.7 SURFACE MODIFICATION WITH BIOMOLECULES

### 12.7.1 Introduction to Using Biomolecules for Surface Modification

Surface modification through the use of biomolecules is an exciting and quickly developing topic with applications in materials science, biotechnology, medicine, and other fields [50]. Surfaces can be tailored for improved functionality, biocompatibility, and targeted interactions using biomolecules, such as proteins, peptides, nucleic acids, and carbohydrates. Here are some common strategies and examples of using biomolecules for surface modification:

*Protein coating.* This method encourages particular cell attachment and improves biocompatibility. Example: Immobilizing fibronectin or collagen on surfaces to support cell adhesion and tissue growth [51].

*Peptide functionalization.* This technique creates selective binding sites for specific molecules or cells. Example: Immobilizing a cell-binding peptide on a surface for targeted cell adhesion [51].

*DNA and RNA modification.* Modification of DNA and RNA produces biofunctional surfaces for gene delivery or biosensors. Example: Immobilizing DNA probes on a surface for selective DNA or RNA detection [52].

*Carbohydrate coating.* This technique enhances biocompatibility and prevents nonspecific protein adsorption. Example: Coating surfaces with polysaccharides to resist fouling and improve material performance [53].

*Enzyme immobilization.* Enzyme immobilization allows for the regulated facilitation of enzymatic processes. Example: Immobilizing enzymes on surfaces for biosensors or biocatalysis [54].

*Lipid bilayer formation.* This produces lipid-layered biomimetic surfaces suitable for sensing or medication administration. Example: Self-assembling lipid bilayers on surfaces for controlled release of bioactive compounds [55].

*Antibody immobilization.* Antibody immobilization is a technique used to create highly selective biosensors or diagnostic platforms. Example: immobilizing antibodies on a surface for the detection of specific antigens [56].

*Bioconjugation techniques.* This is applied for covalently linking biomolecules to surfaces for stability and functionality. Example: using techniques like click chemistry to attach biomolecules to surfaces with high precision.

*Bacterial or cellular surface modification.* This is used in creating biofunctional interfaces for various applications. Example: modifying the surface of bacteria or cells to enhance their binding to specific substrates.

*Nanoparticle functionalization.* This technique creates hybrid biomolecule–nanoparticle systems for targeted drug delivery. Example: coating nanoparticles with peptides or antibodies for targeted cellular interactions [57].

### 12.7.2 Specific Interaction between Biomolecules and Polymer Matrices

Particular interactions between biomolecules and polymer matrices are important in several domains, including tissue engineering, medication delivery, and the creation of biosensors. The resultant biomaterial's performance, functionality, and stability may be affected by these interactions [58].

The following particular interactions between biomolecules and polymer matrices are frequently observed:

*Hydrogen bonding.* Hydrogen bonding is a prevalent interaction between polar groups in biomolecules (e.g., amino and hydroxyl groups) and polymer matrices [59]. Example: polymers with hydrogen-bonding moieties (e.g., polyvinyl alcohol) can interact with biomolecules like proteins and nucleic acids, contributing to the overall stability of the biomaterial.

*Electrostatic interactions.* Electrostatic interactions involve the attraction or repulsion of charged groups in biomolecules and polymer matrices [60]. Example: positively charged polymers (e.g., polyethyleneimine) can form complexes with negatively charged biomolecules (e.g., DNA or RNA), facilitating gene delivery in biomedical applications.

*Van der Waals forces.* Weak, noncovalent forces that arise due to transient fluctuations in electron distribution [60]. Example: Van der Waals forces contribute to the interaction between hydrophobic regions in polymers and lipids, proteins, or other biomolecules in drug delivery systems.

*Covalent bonding.* Formation of strong chemical bonds between functional groups in biomolecules and polymer matrices. Example: bioconjugation reactions involve covalent bonding, allowing biomolecules (e.g., antibodies or enzymes) to be covalently attached to polymer surfaces, enhancing stability and functionality.

*Ionic interactions.* Interaction between charged groups leading to the formation of ion pairs [61]. Example: polyelectrolytes (positively or negatively charged polymers) can interact with oppositely charged biomolecules, influencing the structure and properties of the resulting complex.

*Host–guest interactions.* Specific molecular recognition events between host and guest molecules. Example: cyclodextrin-containing polymers can form inclusion complexes with guest biomolecules, providing a controlled release mechanism in drug delivery.

*Coordination bonds.* Formation of bonds through coordination between metal ions and specific functional groups [62]. Example: metal-containing polymers can coordinate with biomolecules, offering potential applications in catalysis or imaging.

*Conformational matching.* Alignment of molecular structures that allows complementary fitting. Example: polymer scaffolds with specific shapes and conformations can match the structures of biomolecules, influencing cellular interactions in tissue engineering.

*Biomineralization.* Formation of mineral structures on or within polymer matrices in the presence of biomolecules [63]. Example: polymer matrices can interact with biomolecules to induce the formation of mineralized structures, mimicking natural processes in bone regeneration.

Comprehending and managing these interplays are essential for creating biomaterials with customized characteristics for particular uses. These interactions are often used by researchers to develop biomimetic materials with improved biocompatibility, regulated release, and specific capabilities.

### 12.7.3 Biologically Inspired Coatings and Their Potential for Adhesion Improvement

Biologically inspired coatings are designed to increase the adherence of surfaces by utilizing natural processes and structures seen in live organisms. These coatings frequently use natural phenomena as inspiration to accomplish certain goals. Among them are:

*Taking after gecko's model of adhesion.* The sticky qualities of gecko feet, which make use of van der Waals forces and hierarchical structures, served as inspiration [17]. Producing artificial adhesives with micro- and nanostructures to improve adherence on a range of substrates is the coating technique in this case. Improved adhesion on challenging surfaces, such as smooth or rough substrates, is the possible outcome in this scenario.

*The lotus effect for antiadhesion.* The superhydrophobicity of lotus leaves, which reduces the adherence of water droplets and dirt particles, served as inspiration [64]. Creating hydrophobic coatings with self-cleaning capabilities is the coating approach. Liquid, pollutant, or fouling agent adherence is prevented, making cleaning and maintenance simpler as the potential effect.

*Protein-based layers for adhesion of cells.* Extracellular matrix proteins that facilitate cell attachment in live tissues serve as an inspiration. The coating approach involves immobilizing biomimetic proteins, such as collagen or fibronectin, on surfaces. Improved cell adherence on implants or biomaterials for tissue engineering applications might be the potential impact.

*Mimicry of marine adhesive.* Source of inspiration for its marine creatures' sticky secretions, such as those of mussels. The coating approach involves creating artificial adhesives using proteins inspired by mussels and catechol chemistry [65]. Improved adhesion on damp or submerged surfaces is a potential impact, having applications in maritime and health-care environments.

*Coatings inspired by spider silk.* The suppleness and power of the spider silk serve as inspiration [66], coating methodologies using biomimetic spider silk proteins to create artificial coatings. The potential effect is improved mechanical strength and adhesion, beneficial for applications such as medical sutures or textiles.

*Biopolymer antifouling coatings.* It is inspired by the antifouling characteristics of several sea species. Creating coatings that are resistant to biofouling using biopolymers is the coating approach [67]. Reduced microbial adherence might have the effect of avoiding biofouling on surfaces such as medical equipment or ship hulls.

*Adhesives inspired by plants.* The adhesive qualities of materials derived from plants serve as inspiration. Plant-based adhesion processes provide inspiration for the coating approach to synthetic adhesive design. Improved adhesion on porous surfaces is a possible impact that might have uses in woodworking or construction.

*Ant adhesion strategies.* It is inspired by the sticky qualities of ant feet, which include sticky pads and claws [68]. Creating coatings with microstructures akin to ant adhesion processes is known as the coating approach. Better stickiness on various surfaces, especially in microscale applications, is the potential impact.

Coatings with biological inspiration have the power to transform adhesion technologies by offering creative answers to particular problems in a range of sectors. These coatings provide a viable path forward for adhesion-enhancement tactics, be it improving cell adhesion in biomedical applications or reducing fouling in marine settings.

## 12.8 GRAFTING TECHNIQUES

### 12.8.1 Graft Polymerization Methods for Surface Modification

With the surface modification process known as "graft polymerization," polymer chains are covalently bonded (or "grafted") onto a substrate to change its surface characteristics. This technique is often used to improve a surface's adhesion, biocompatibility, or functioning in a variety of applications. Although this is a rather new technology, polymer surface grafting offers a versatile means for providing existing polymers with new functionalities, such as hydrophilicity, adhesion, biocompatibility, conductivity, antifogging, antifouling, and lubrication [69]. Here are several graft polymerization methods for surface modification:

#### 12.8.1.1 Initiator-Driven Graft Polymerization

The basic idea behind initiator-driven graft polymerization is that initiators—which can be chemically or physically bonded to the substrate—produce ions or radicals that start the polymerization process [70].

Essentially, there are two methods:

*Surface-initiated atom transfer radical polymerization (SI-ATRP).* Surface-initiated ATRP allows for controlled radical polymerization.
Other controlled radical polymerization methods, such as nitroxide-mediated polymerization (SI-NMP) and reversible addition-fragmentation chain transfer (SI-RAFT), are included in surface-initiated controlled radical polymerization (SI-CRP) [70].

Applications include controlled medication release, responsive surfaces, and biomaterials.

#### 12.8.1.2 Plasma-Induced Graft Polymerization

The underlying idea here is that polymerization is started on the substrate by radicals or ions produced by plasma treatment. The methods utilized are:

*Plasma polymerization.* In a plasma discharge, monomers are polymerized and then deposited onto a substrate.
*Plasma grafting.* After exposing the substrate to plasma, monomers are added to facilitate grafting [71].

Applications include adhesion enhancement, surface hydrophilization, and biomaterial modification.

#### 12.8.1.3 Photoinitiated Graft Polymerization

The principle of this method is that photoinitiators absorb light and generate radicals or ions, initiating polymerization [72]. The techniques are:

*UV-initiated graft polymerization.* Uses ultraviolet light.
*Visible light–initiated graft polymerization.* Uses visible light and specific photoinitiators.

Applications include surface patterning, responsive surfaces, and photoactive materials.

#### 12.8.1.4 Enzyme-Catalyzed Graft Polymerization

The idea is, monomer polymerization onto a substrate is catalyzed by enzymes [72]. Methodologies employed in this:

*HPP (horseradish peroxidase)–mediated HRP.* Catalyzes the creation of radicals necessary for polymerization.
*Laccase-initiated polymerization.* To start polymerization, laccase oxidizes monomers.

Applications include the immobilization of enzymes, biocompatible surfaces, and bioactive coatings.

#### 12.8.1.5 Metal-Catalyzed Graft Polymerization

Fundamentals are polymerization processes that are catalyzed by metal complexes. The techniques are:

*Atom transfer radical polymerization (ATRP).* In this process, radical polymerization is regulated by metal catalysts.
*Ring-opening metathesis polymerization (ROMP).* Cyclic monomer opening is triggered by metal catalysts.

Applications include bioconjugation, responsive materials, and controlled polymerization.

Graft polymerization techniques provide diversity in surface customization for a variety of applications, such as bioactive coatings, responsive materials, and biomedical implants. The intended use, the substrate material, and the desired surface qualities all influence the technique choice.

### 12.8.2 Impact of Grafted Polymer Chains on Adhesion

Grafted polymer chains affect adhesion qualities in a big way, affecting how materials cling to one another and interact with surfaces. Grafted polymer chains' impact on adhesion may be divided into many important categories:

*Improved adhesion strength.* Grafted polymer chains can enhance adhesion strength by providing additional interaction points between the adhering surfaces. The increased contact area and intermolecular forces contribute to stronger adhesion.

*Increased surface energy.* Grafted polymers can alter the surface energy of a material [73]. If the grafted chains are polar or hydrophilic, they can increase the surface energy, promoting better wetting and adhesion of liquids or other materials.

*Reduced interfacial energy.* The presence of grafted polymer chains can reduce the interfacial energy between two materials. This is particularly beneficial in preventing the adhesion of undesired substances, such as fouling or contaminants, on surfaces.

*Tailored interactions.* Grafted polymers can be designed with specific functional groups or chemical moieties that facilitate tailored interactions with adhering materials. This specificity allows for controlled and selective adhesion.

*Prevention of interfacial fracture.* Grafted polymer chains can act as energy dissipation mechanisms, preventing interfacial fracture by absorbing mechanical energy during adhesion processes [74]. This can improve the overall durability and resilience of adhesive interfaces.

*Enhanced flexibility and conformability.* The flexibility of grafted polymer chains allows them to conform to surface irregularities, improving contact between adhering materials [73]. This property is particularly beneficial in achieving adhesion on rough or uneven surfaces.

*Antifouling properties.* Grafted polymers with specific functionalities can create antifouling surfaces, preventing the adhesion of proteins, cells, or other biological substances [75]. This is particularly important in biomedical applications and marine environments.

*Controlled release.* Adhesives or active substances can be released under regulated conditions thanks to grafted polymer coatings. This regulated release mechanism might be very important for applications like transient adhesion or medication administration.

*Temperature- and pH-responsiveness.* Responsive grafted polymers that undergo conformational changes in response to temperature or pH variations can alter their adhesion properties accordingly. This responsiveness allows for dynamic control over adhesion strength.

*Biocompatibility.* Grafted polymer chains can be engineered to be biocompatible, making them suitable for applications in biological environments [76]. This is essential in biomedical applications, where adhesion to biological tissues or implants is desired.

In summary, grafted polymer chains play a versatile role in modulating adhesion properties, offering opportunities for tailoring surfaces based on specific requirements in diverse applications. Designing and manipulating the properties of grafted polymers offer a potent tool for attaining the required adhesion results in a variety of fields, such as manufacturing, materials science, and medicine.

### 12.8.3 Tailoring Interfacial Properties through Grafting for Improved Performance

Grafting may effectively tailor interfacial characteristics, improving material performance in a range of applications [77]. In this method, surfaces or interfaces are modified by grafting polymer chains on them, changed from their original configuration. Key elements for customizing interfacial characteristics for enhanced performance by grafting include the following:

*Improved bonding and adhesion.* Polymers can be grafted onto surfaces using surface-initiated polymerization processes [78]. The result is better material adhesion and bonding, which improves mechanical characteristics and structural integrity.

*Regulated wettability.* The grafting approach modifies surface energy by introducing hydrophilic or hydrophobic polymer chains. Controlled wettability has an impact on contact angles, wetting behavior, and interactions with liquids. This is important for applications such as biomedical devices, coatings, and surfaces that clean themselves.

*Decreased wear resistance and friction.* The introduction of lubricating polymer chains or additives to lower friction is known as the grafting approach [78]. Its effects include increased durability, decreased friction, and wear resistance in tribological applications like gears and bearings.

*Biocompatibility and bioadhesive properties.* Biocompatible polymers are attached to surfaces via the grafting approach. It has the effect of improving medical implant biocompatibility and lowering immunogenicity. Grafted polymers have the potential to enhance bioadhesion in tissue engineering and medication delivery.

*Bioresistance and antifouling.* In order to stop pollutants or biological materials from adhering, the grafting approach incorporates antifouling polymer chains. Reduced fouling on surfaces is one of its effects; this is crucial for food processing, biomedical equipment, and marine coatings.

*Interfaces that are responsive to external stimuli.* The grafting approach introduces stimuli-responsive polymers that alter their structure. Its impact is dynamic control over interfacial characteristics, opening up possibilities for sensors, smart surfaces, and medication delivery, among other uses.

*Tailored electrical and optical properties.* The grafting approach modifies electrical or optical properties by attaching conducting or optically active polymers [79]. Its effects include improved optical qualities, increased conductivity, and the creation of cutting-edge materials for photonics, electronics, and sensors.

*Controlled release and drug delivery.* Grafting is the process of incorporating drug-loaded polymer chains onto surfaces for controlled release and medication delivery. One of its effects is the controlled release of therapeutic substances from surfaces, which makes it possible for medical coatings or implants to administer drugs continuously.

*Resistance to corrosion.* The introduction of inhibitors or corrosion-resistant polymers is known as the grafting approach. Its effect is enhanced corrosion resistance, especially in metal-based materials utilized in industrial or infrastructural applications.

*Enhanced thermal stability.* The attachment of thermally stable polymer chains is the grafting approach [79]. Better heat resistance renders materials appropriate for high-temperature uses, such as those in automobile or aircraft components.

Through grafting, interfacial characteristics may be modified to match individual application needs and offer a flexible way to enhance material performance across a wide variety of sectors. It is possible to create new materials with better functionality and performance by designing and controlling the interfacial properties.

## 12.9 NANOCOMPOSITE COATINGS

### 12.9.1 Incorporating Nanoparticles for Enhanced Adhesion

The performance of adhesives and coatings may be increased by using nanoparticles, a technology that is widely employed in many different sectors. Nanoparticles can effectively change the adhesive characteristics because of their tiny size and special qualities. Some methods for adding nanoparticles to improve adhesion are as follows:

*Surface modification.* Certain chemical groups can be added to nanoparticles to functionalize them and increase their adherence to various surfaces [80]. For instance, silane-functionalized nanoparticles are frequently employed to improve surface adherence.

*Reinforcement of the matrix.* Introducing nanoparticles into the adhesive matrix can reinforce the material, improving its mechanical properties and adhesion strength [80]. Graphene, carbon nanotubes, and silica are examples of common reinforcing nanoparticles.

*Increased surface roughness.* Nanoparticles can be used to increase the surface roughness of a material [81], enhancing the mechanical interaction between the adhesive and the substrate and adding more contact sites.

*Improved wetting properties.* Nanoparticles can modify the wetting properties of the adhesive, ensuring better contact with the substrate. This is particularly important when dealing with low-surface-energy materials.

*Electrostatic interactions.* Nanoparticles with specific charge properties can be employed to exploit electrostatic interactions between the adhesive and the substrate [82], enhancing adhesion.

*Barrier properties.* Nanoparticles can act as barriers against environmental factors [83], inhibiting the adhesive interface's deterioration. This is essential for uses where the glue is subjected to adverse environments.

*Functionalized nanoparticles for specific applications.* In certain applications, tailoring nanoparticles with certain functions might improve adherence. Magnetic nanoparticles, for example, can be applied to magnetic fields to promote adherence [84].

*Improved flexibility and toughness.* Rubbery or elastomeric nanoparticles can be added to adhesives to increase their toughness and flexibility, increasing their resistance to environmental strains and stresses.

*UV-curable nanoparticles.* The curing process and adherence of UV-curable adhesives can be improved by adding nanoparticles that can absorb UV radiation and start polymerization [85].

*Biocompatible nanoparticles.* In biomedical applications, biocompatible nanoparticles can be added to adhesives to improve tissue compatibility and adherence [86].

It is crucial to take the application's particular needs, dispersion uniformity, and compatibility with the adhesive matrix into account while adding nanoparticles. In order to attain the intended enhancements in adhesion qualities, appropriate testing and optimization are frequently required.

### 12.9.2 Dispersion Techniques and Their Impact on Composite Properties

The performance of composites, especially those containing nanoparticles or reinforcing elements, is greatly influenced by dispersion methods. Improved mechanical, thermal, and other qualities of the composite result from correct dispersion [87], which guarantees that the reinforcing components are spread uniformly throughout the matrix. The following are several dispersion methods and how they affect the characteristics of composites:

*Mechanical mixing.* In mechanical mixing, the matrix and reinforcing elements are blended using tools like ball mills, three-roll mills, or high-speed mixers. The mechanical characteristics of the composite are improved by homogeneous dispersion, which is attained by adequate mechanical mixing. Nevertheless, damage or nanoparticle aggregation may result from extreme shear pressures [88].

*Ultrasonication.* Using high-frequency sound waves, ultrasonication breaks apart agglomerates and distributes nanoparticles throughout the matrix. The process of ultrasonication is useful for dispersing tiny particles, which improves uniformity and mechanical strength. To prevent harming the nanoparticles or changing the matrix, it should be closely regulated.

*Stirring and stir bar.* A uniform dispersion can be achieved by using a stir bar to agitate the composite mixture. Even if stirring is an easy method, it could not work as well as other methods to distribute materials evenly, particularly when it comes to nanoparticles. Time and agitation speed are critical to this method's effectiveness.

*Solvent mixing.* The process of solvent mixing entails dissolving nanoparticles in a solvent prior to combining them with the matrix. Although careful solvent cleanup is necessary, this approach can provide good dispersion. Compatibility with the matrix material is important since residual solvent may impact the composite qualities.

*In situ polymerization.* During polymerization, the reinforcing material is created inside the polymer matrix. A solid link between the matrix and the reinforcing material is ensured by in situ polymerization, improving the composite's qualities. Nanoparticles are frequently incorporated into polymer matrices using this technique.

*Electrospinning.* In this process, a polymer melts or solution is spun into nanofibers using an electric field. Nanofibers may be uniformly dispersed via electrospinning, which improves the composite's mechanical qualities, surface area, and functions [89].

*Melt mixing.* It is the process of combining the molten polymer matrix with the reinforcing elements. This technique works well with thermoplastic matrices. Due to some nanoparticles' propensity to clump together while melted, it may be less successful for some nanoparticles than others in terms of dispersion.

*Chemical surface modification.* Applying certain functional groups to the surface of nanoparticles to improve their compatibility with the matrix. Better dispersion and improved composite qualities result from surface modification [89], which also enhances the interaction between the matrix and reinforcing material.

Depending on the kind of composite, the type of reinforcing components, and the required end-product qualities, a particular dispersion process may be used. Achieving the necessary performance gains in composite materials requires careful management and optimization of the dispersion processes.

### 12.9.3 Performance Improvements and Limitations of Nanocomposite Coatings

Nanocomposite coatings, which involve the incorporation of nanoparticles into coating materials, have gained significant attention for various applications due to their unique properties. Here are some aspects related to the performance improvements and limitations of nanocomposite coatings:

#### 12.9.3.1 Performance Improvements

*Mechanical strength.* Nanocomposite coatings frequently show increased wear resistance, hardness, and mechanical strength [90]. Nanoparticles have the capacity to strengthen the coating matrix, which produces coatings with increased durability.

*Corrosion protection.* When compared to conventional coatings, nanocomposite coatings may offer superior corrosion protection. By acting as barriers, the nanoparticles can stop corrosive substances from penetrating and stop the substrate underneath from corroding.

*Thermal stability.* The use of nanoparticles has the potential to improve coatings' thermal stability, increasing their resistance to elevated temperatures [91]. This is especially crucial in situations where coatings are subjected to high temperatures.

*Enhanced adhesion.* Coatings' adherence to substrates can be strengthened by using nanoparticles. This is essential to guarantee the coating's long-term effectiveness and endurance, particularly in situations where adherence is key.

*Surface properties.* Hydrophobicity, oleophobicity, and antifouling qualities are just a few of the surface qualities that nanocomposite coatings may change [92]. They are therefore appropriate for uses where surface interactions are crucial.

*Electrical and optical qualities.* Coatings that include nanoparticles can acquire special electrical and optical qualities that make them appropriate for use in sensors, optics, and electronics.

#### 12.9.3.2 Limitations

*Cost.* Because of the high cost of nanoparticles and the intricate manufacturing process, producing nanocomposite coatings may be more expensive than producing standard coatings.

*Uniform dispersion.* It might be difficult to get a uniform dispersion of nanoparticles in the coating matrix. It is possible for nanoparticles to aggregate, which might result in irregular characteristics and could lessen the coating's efficacy.

*Compatibility.* It may be problematic if nanoparticles are incompatible with both the substrate and the coating matrix. Poor adhesion, cracking, or other problems that impair performance might result from incompatibility.

*Toxicity.* Certain nanoparticles might be harmful to the environment or human health [93]. It is crucial to carefully consider the selection of nanoparticles to guarantee that they adhere to safety guidelines and laws.

*Increase in size difficulties.* Making the switch from large-scale to lab-scale manufacturing might be difficult [94]. It could need more optimization to keep the same characteristics and consistency on a bigger scale.

*Limited understanding.* Despite developments, much remains unknown regarding the long-term characteristics and possible effects of nanocomposite coatings on the environment [94]. Extensive research on their stability and deterioration over time is necessary.

In conclusion, even though nanocomposite coatings provide a number of performance enhancements, their effective use in a variety of sectors depends on resolving issues with cost, dispersion, compatibility, toxicity, and scale-up. Research and development are still ongoing to get over these restrictions and realize the full potential of nanocomposite coatings.

## 12.10 PHYSICAL VAPOR DEPOSITION (PVD)

The method of depositing thin films on a substrate by condensing a vaporized version of the desired film material is known as physical vapor deposition, or PVD. There is no air around the PVD process. There are various types of physical vapor deposition (PVD) methods, including evaporative deposition, cathode arc deposition, electron beam physical vapor deposition, sputtering, ion plating, and improved sputtering. Using heat or an ion bombardment, the solid coating material is evaporated in the PVD process [95].

A material can be vaporized from a solid or liquid source and then deposited on a substrate using the gaseous-state deposition technique known as PVD. This method can be used to deposit films of one or more elements and is carried out in a vacuum. Vacuum evaporation, or sputtering, is the usual method used in PVD. A heated filament that is always positioned below the substrate takes the place of the magnetron and target in the evaporation configuration, which is remarkably similar to the sputtering arrangement. The thin metallic coatings created by PVD methods are with thicknesses of several nanometers to several micrometers. Using the proper pretreatments, the coatings stick to the substrates decently while maintaining close conformity with the shapes of the coated portions. The coating formation is assumed to occur on the surface that faces the source material, because PVD methods are line-of-sight procedures. Consequently, surfaces that are not in direct contact with the source material may exhibit variations in coating uniformity. Usually, rotating the substrates in front of the source material solves this problem in current PVD equipment. PVD, in general, makes it possible to metalize polymeric components into thin films without exposing any hidden surfaces [96].

The CVD method was first developed by Prof. Karen K. Gleason [97]. CVD occurs in three stages: initiation, propagation, and termination. A polymer film is created by these reactions that occur on the cooled substrate's surface. In this instance, *M* is the monomer, and *I* is the initiator. When the monomer's vinyl bond is attacked by the initiator free radical, initiation results. In the propagation stage, the resultant radical then combines with another vinyl monomer. Primary radical termination and primary radical recombination are two ways in which two of the propagating radicals can react to cause termination.

$$\text{Initiation: I}*\cdot(\text{ad})+\text{M }(\text{ad})\rightarrow\text{M1}*\cdot(\text{ad})\cdot$$

$$\text{Propagation: Mn}*\cdot(\text{ad})+\text{M}\cdot(\text{ad})\rightarrow\text{Mn+1}*\cdot(\text{ad})\cdot$$

$$\text{Termination: Mn}*\cdot(\text{ad})+\text{Mm}*\cdot(\text{ad})\rightarrow\text{Mn + m}\cdot(\text{ad})$$

$$\text{Primary Radical Termination: Mn}*\cdot(\text{ad})+\text{I}*\cdot(\text{ad})\rightarrow\text{Mn}\cdot(\text{ad})$$

$$\text{Primary Radical Recombination: I}*\cdot(\text{ad})+\text{I}*\cdot(\text{ad})\rightarrow\text{I2}\cdot(\text{ad})\cdot$$

Physical vapor deposition is suggested as a new dry-coating technique for creating polymer thin films (PVD). For polymers with poor intermolecular interactions, such as polyethylene or polytetrafluoroethylene, direct vapor deposition can be used. Generally speaking, vapor deposition polymerization is used to create polymer thin films. This process evaporates the monomer ingredients and creates polymer thin films on the substrate surface through a polymerization reaction. Thin films of $\pi$-conjugated polymers can also be prepared using this process. An innovative technique for growing polymer thin films that are chemically bonded to the substrate surface is surface-initiated vapor deposition polymerization, which blends the self-assembled monolayer (SAM) with monomer vapor deposition. PVD's solvent-free characteristic makes it easy to generate multilayer and nanometer-thick films, which are occasionally needed for the construction of devices [98].

Two situations are possible:

1. Adhesion is less than cohesive strength; especially with fragile films, the coating has a tendency to adhere or flake away at the groove's margins. In this case, the measurement of the scratch width will be less precise.
2. Adhesion is greater than cohesive strength when the indenter slides under stress, causing a shear that tends to eliminate any coating depth that may still be present in the scratch channel, instead of the film flaking [99].

## 12.11 CONCLUSION

In conclusion, this chapter has explored a diverse array of coating techniques designed to enhance interfacial adhesion in polymer composites. From traditional chemical treatments to cutting-edge methodologies, such as plasma deposition and electrospinning, the importance of optimizing the interface between reinforcing fillers and the polymer matrix has been underscored. The presented case studies and practical applications highlight the efficacy of these coating strategies in improving the mechanical, thermal, and overall performance of polymer composites. As we look to the future, ongoing research and innovation in coating technologies hold the potential for further advancements, ensuring the continual progress of polymer composites as high-performance materials in various applications.

## ACKNOWLEDGMENTS

The author wishes to thank the Rajshahi University of Engineering and Technology for providing an excellent research opportunity, which helped in writing this book chapter.

## REFERENCE

[1] B. R. Thakur, R. K. Singh, A. K. Handa, and M. A. Rao, "Chemistry and uses of pectin: A review," *Crit. Rev. Food Sci. Nutr.*, vol. 37, no. 1, pp. 47–73, 1997.
[2] T. Darmanin and F. Guittard, "Wettability of conducting polymers: From superhydrophilicity to superoleophobicity," *Prog. Polym. Sci.*, vol. 39, no. 4, pp. 656–682, 2014.
[3] A. Buchman, M. Rotel, and H. Dodiuk-Kenig, "Nd: YAG laser surface treatment of various materials to enhance adhesion," *Laser Surf. Modif. Adhes.*, pp. 1–53, 2014. https://doi.org/10.1002/9781118831670.ch1.
[4] K. L. Mittal, *Adhesion measurement of films and coatings*. CRC Press, Boston, 2014.
[5] D. Briggs and M. P. Seah, *Practical surface analysis: By auger and X-ray photo-electron spectroscopy*, Wiley, Chichester, 1983.
[6] C. Wang, X. Ji, A. Roy, V. V. Silberschmidt, and Z. Chen, "Shear strength and fracture toughness of carbon fibre/epoxy interface: Effect of surface treatment," *Mater. Des.*, vol. 85, pp. 800–807, 2015.
[7] K. L. Mittal and T. Ahsan, *Adhesion in microelectronics*. John Wiley & Sons, Chichester, 2014.
[8] E. Martuscelli, *Polymer blends: Processing, morphology, and properties*. Springer Science & Business Media, New York, 2012.
[9] K. Pielichowski and T. M. Majka, *Polymer composites with functionalized nanoparticles: Synthesis, properties, and applications*. Woodhead Publishing, Duxford, 2018.
[10] N. V. Bhat and D. J. Upadhyay, "Plasma-induced surface modification and adhesion enhancement of polypropylene surface," *J. Appl. Polym. Sci.*, vol. 86, no. 4, pp. 925–936, 2002. https://doi.org/10.1002/app.11024.
[11] S. Yoshida, K. Hagiwara, T. Hasebe, and A. Hotta, "Surface modification of polymers by plasma treatments for the enhancement of biocompatibility and controlled drug release," *Surf. Coat. Technol.*, vol. 233, pp. 99–107, 2013.
[12] P. T. Vu, J. P. Conroy, and A. M. Yousefi, "The effect of argon plasma surface treatment on poly (lactic-co-glycolic acid)/collagen-based biomaterials for bone tissue engineering," *Biomimetics*, vol. 7, no. 4, p. 218, 2022.
[13] V. Luque-Agudo, M. Hierro-Oliva, A. M. Gallardo-Moreno, and M. L. González-Martín, "Effect of plasma treatment on the surface properties of polylactic acid films," *Polym. Test.*, vol. 96, p. 107097, 2021. https://doi.org/10.1016/j.polymertesting.2021.107097.
[14] H. Yu, Y. Zhang, A. Wong, I. M. De Rosa, H. S. Chueh, M. Grigoriev, T. S. Williams, T. Hsu, and R. F. Hicks, "Atmospheric and vacuum plasma treatments of polymer surfaces for enhanced adhesion in microelectronics packaging," in *Adhesion in microelectronics*, Wiley, New York, 2014, pp. 137–172. https://doi.org/10.1002/9781118831373.ch4.
[15] S. B. Idage and S. Badrinarayanan, "Surface modification of polystyrene using nitrogen plasma. An X-ray photoelectron spectroscopy study," *Langmuir*, vol. 14, no. 10, pp. 2780–2785, 1998. https://doi.org/10.1021/la9711286.
[16] N. Hasirci, T. Endogan, E. Vardar, and V. Hasirci, "Effect of oxygen plasma on surface properties and biocompatibility of PLGA films," *Surf. Interface Anal.*, vol. 42, pp. 486–491, 2010. https://doi.org/10.1002/sia.3247.
[17] A. Vesel and M. Mozetic, "Low-pressure plasma-assisted polymer surface modifications," in *Printing on polymers: Fundamentals and applications*, William Andrew, Norwich, 2016, pp. 101–121. https://doi.org/10.1016/B978-0-323-37468-2.00007-5.

[18] J. Kim, R. Mauchauffé, D. Kim, J. Kim, and S. Y. Moon, "Mechanism study of atmospheric-pressure plasma treatment of carbon fiber reinforced polymers for adhesion improvement," *Surf. Coat. Technol.*, vol. 393, p. 125841, 2020. https://doi.org/10.1016/j.surfcoat.2020.125841.

[19] E. M. Liston, L. Martinu, and M. R. Wertheimer, "Plasma surface modification of polymers for improved adhesion: A critical review," *J. Adhes. Sci. Technol.*, vol. 7, no. 10, pp. 1091–1127, 1993. https://doi.org/10.1163/156856193X00600.

[20] J. B. Park, J. S. Oh, E. L. Gil, S. J. Kyoung, J. T. Lim, and G. Y. Yeom, "Polyimide surface treatment by atmospheric pressure plasma for metal adhesion," *J. Electrochem. Soc.*, vol. 157, no. 12, p. D614, 2010. https://doi.org/10.1149/1.3493585.

[21] N. Inagaki, S. Tasaka, H. Kawai, and Y. Yamada, "Surface modification of aromatic polyamide film by remote oxygen plasma," *J. Appl. Polym. Sci.*, vol. 64, no. 5, pp. 831–840, 1997. https://doi.org/10.1002/(SICI)1097–4628(19970502)64:5<831::AID-APP2>3.0.CO;2-Q.

[22] G. O. Mallory and J. B. Hajdu, *Electroless plating: Fundamentals and applications*. William Andrew, Norwich, 1990.

[23] M. Ohring, *Materials science of thin films: Depositon and structure*. Elsevier, Amsterdam, 2001.

[24] K. Dunkel, D. Meyer, T. Lampke, and S. Steinhuser, "Electroless deposition of NiP nanocomposite coatings," *Materwiss. Werksttech.*, vol. 40, no. 12, pp. 888–893, 2009.

[25] C. Deckert, "Electroless copper plating, A review: Part I," *Plat. Surf. Finish.*, vol. 82, no. 2, pp. 48–55, 1995.

[26] M. H. Tusher, *Microbial synthesis of cadmium selenide quantum dots (CdSe QDs), influencing factors and applications*. Springer US, New York, 2023. https://doi.org/10.1007/s11082-023-04632-z.

[27] M. M. H. Tusher, A. Imam, and M. S. I. Shuvo, "Future and challenges of coating materials BT—Coating materials: Computational aspects, applications and challenges," A. Verma, S. K. Sethi, and S. Ogata, Eds., Springer Nature, Singapore, 2023, pp. 229–251. https://doi.org/10.1007/978-981-99-3549-9_11.

[28] G. M. Mirza, M. M. H. Tusher, N. Sakib, and M. D. N. Islam, "Effect of Al-doping on morphology, structure, and optical band gap of ZnO thin films synthesized by sol-gel spin-coating technique," *J. Mater. Sci. Mater. Electron.*, vol. 1542, pp. 1–10, 2023. https://doi.org/10.1007/s10854-023-10958-x.

[29] A. Sehgal, Y. T. Kho, K. Osseo-Asare, and H. W. Pickering, "Comparison of corrosion rate-measuring devices for determining corrosion rate of steel-in-concrete systems," *Corrosion*, vol. 48, no. 10, pp. 871–880, 1992. https://doi.org/10.5006/1.3315888.

[30] A. Jabed, M. H. Tusher, S. I. Shuvo, and A. Imam, "Corrosion of steel rebar in concrete: A review," *Corros. Sci Technol.*, vol. 22, no. 4, pp. 273–286, 2023. https://doi.org/10.14773/cst.2023.22.4.273.

[31] F. C. Walsh, S. Wang, and N. Zhou, "The electrodeposition of composite coatings: Diversity, applications and challenges," *Curr. Opin. Electrochem.*, vol. 20, pp. 8–19, 2020.

[32] C. Kumudinie, "Polymer-ceramic nanocomposites: Interfacial bonding agents," *Encycl. Mater. Sci. Technol.*, pp. 7574–7577, 2001. https://doi.org/10.1016/b0-08-043152-6/01356-5.

[33] P. G. Pape, *25 Adhesion promoters: Silane coupling agents*, 2nd edn., Elsevier Inc., Amsterdam. https://doi.org/10.1016/B978-0-323-39040-8/00026-2.

[34] Y. Xie, C. A. S. Hill, Z. Xiao, H. Militz, and C. Mai, "Silane coupling agents used for natural fiber/polymer composites: A review," *Compos. Part A Appl. Sci. Manuf.*, vol. 41, no. 7, pp. 806–819, 2010. https://doi.org/10.1016/j.compositesa.2010.03.005.

[35] C. Peng, P. Chen, Z. You, S. Lv, R. Zhang, F. Xu, H. Zhang, and H. Chen, "Effect of silane coupling agent on improving the adhesive properties between asphalt binder and aggregates," *Constr. Build. Mater.*, vol. 169, pp. 591–600, 2018. https://doi.org/10.1016/j.conbuildmat.2018.02.186.

[36] F. Ahangaran and A. H. Navarchian, "Recent advances in chemical surface modification of metal oxide nanoparticles with silane coupling agents: A review," *Adv. Colloid Interface Sci.*, vol. 286, p. 102298, 2020. https://doi.org/10.1016/j.cis.2020.102298.

[37] M. C. Brochier Salon and M. N. Belgacem, "Hydrolysis-condensation kinetics of different silane coupling agents," *Phosphorus Sulfur Silicon Relat. Elem.*, vol. 186, no. 2, pp. 240–254, 2011. https://doi.org/10.1080/10426507.2010.494644.

[38] J. Yang, J. Xiao, J. Zeng, L. Bian, C. Peng, and F. Yang, "Matrix modification with silane coupling agent for carbon fiber reinforced epoxy composites," *Fibers Polym.*, vol. 14, no. 5, pp. 759–766, 2013. https://doi.org/10.1007/s12221-013-0759-2.

[39] P. G. Pape and E. P. Plueddemann, "Methods for improving the performance of silane coupling agents," *J. Adhes. Sci. Technol.*, vol. 5, no. 10, pp. 831–842, 1991. https://doi.org/10.1163/156856191X00242.

[40] A. Rezzoug, S. Abdi, N. Bouhelal, and I. Daoud, "Metallic coating for carbon fiber reinforced polymer matrix composite substrate," *Int. J. Mater. Metall. Eng.*, vol. 10, no. 1, pp. 59–64, 2016.

[41] H. F. Wu, D. W. Dwight, and N. T. Huff, "Effects of silane coupling agents on the interphase and performance of glass-fiber-reinforced polymer composites," *Compos. Sci. Technol.*, vol. 57, no. 8, pp. 975–983, 1997. https://doi.org/10.1016/S0266–3538(97)00033-X.

[42] V. Gillet, E. Aubignat, S. Costil, B. Courant, C. Langlade, P. Casari, W. Knapp, and M. P. Planche, "Development of low pressure cold sprayed copper coatings on carbon fiber reinforced polymer (CFRP)," *Surf. Coat. Technol.*, vol. 364, pp. 306–316, 2019.

[43] B. Zhang, L. Jia, M. Tian, N. Ning, L. Zhang, and W. Wang, "Surface and interface modification of aramid fiber and its reinforcement for polymer composites: A review," *Eur. Polym. J.*, vol. 147, p. 110352, 2021.

[44] K. L. Choy, "Chemical vapour deposition of coatings," *Prog. Mater. Sci.*, vol. 48, no. 2, pp. 57–170, 2003.

[45] R. Peelamedu, D. Kumar, and S. Kumar, "Microwave atmospheric pressure plasma for surface treatment and reactive coating on steel surfaces," *Surf. Coat. Technol.*, vol. 201, no. 7, pp. 4008–4013, 2006.

[46] S. R. Culler, H. Ishida, and J. L. Koenig, "FT-IR characterization of the reaction at the silane/matrix resin interphase of composite materials," *J. Colloid Interface Sci.*, vol. 109, no. 1, pp. 1–10, 1986.

[47] R. D. Evans, G. L. Doll, P. W. Morrison Jr, J. Bentley, K. L. More, and J. T. Glass, "The effects of structure, composition, and chemical bonding on the mechanical properties of Si-aC: H thin films," *Surf. Coat. Technol.*, vol. 157, nos. 2–3, pp. 197–206, 2002.

[48] S. Mishra and J. J. Weimer, "Comparative analysis of trimethylmethoxysilane and trimethylchlorosilane bonding on polished copper surfaces," *J. Vac. Sci. Technol. A Vacuum, Surfaces, Film.*, vol. 13, no. 3, pp. 1281–1285, 1995.

[49] T. Kaneko, D. Nemoto, A. Horiguchi, and N. Miyakawa, "FTIR analysis of a-SiC: H films grown by plasma enhanced CVD," *J. Cryst. Growth*, vol. 275, nos. 1–2, pp. e1097–e1101, 2005.

[50] C. R. Taitt, L. C. Shriver-Lake, G. P. Anderson, and F. S. Ligler, "Surface modification and biomolecule immobilization on polymer spheres for biosensing applications," *Biomed. Nanotechnol. Methods Protoc.*, pp. 77–94, 2011.

[51] O. Melnyk, J. Fehrentz, J. Martinez, and H. Gras-Masse, "Functionalization of peptides and proteins by aldehyde or keto groups," *Pept. Sci.*, vol. 55, no. 2, pp. 165–186, 2000.

[52] L. Qiu, Q. Jing, Y. Li, and J. Han, "RNA modification: Mechanisms and therapeutic targets," *Mol. Biomed.*, vol. 4, no. 1, p. 25, 2023.

[53] I. A. A. Ibrahim, A. R. Alzahrani, I. M. Alanazi, N. Shahzad, I. Shahid, A. Hisham Falemban, M. F. N. Azlina, and P. Arulselvan, "Carbohydrate polymers-based surface modified nano delivery systems for enhanced target delivery to colon cancer: A review," *Int. J. Biol. Macromol.*, vol. 253, p. 126581, 2023.

[54] A. A. Homaei, R. Sariri, F. Vianello, and R. Stevanato, "Enzyme immobilization: An update," *J. Chem. Biol.*, vol. 6, pp. 185–205, 2013.

[55] H. Brockman, "Dipole potential of lipid membranes," *Chem. Phys. Lipids*, vol. 73, nos. 1–2, pp. 57–79, 1994.

[56] H. Becker and L. E. Locascio, "Polymer microfluidic devices," *Talanta*, vol. 56, no. 2, pp. 267–287, 2002.

[57] G. Sanità, B. Carrese, and A. Lamberti, "Nanoparticle surface functionalization: How to improve biocompatibility and cellular internalization," *Front. Mol. Biosci.*, vol. 7, p. 587012, 2020.

[58] P. Baumann, P. Tanner, O. Onaca, and C. G. Palivan, "Bio-decorated polymer membranes: A new approach in diagnostics and therapeutics," *Polymers (Basel)*, vol. 3, no. 1, pp. 173–192, 2011.

[59] F. B. Ilhami, Y. S. Birhan, and C.-C. Cheng, "Hydrogen-bonding interactions from nucleobase-decorated supramolecular polymer: Synthesis, self-assembly and biomedical applications," *ACS Biomater. Sci. Eng.*, vol. 10, no. 1, pp. 234–254, 2024. https://doi.org/10.1021/acsbiomaterials.3c01097.

[60] F. Siedenbiedel and J. C. Tiller, "Antimicrobial polymers in solution and on surfaces: Overview and functional principles," *Polymers (Basel)*, vol. 4, no. 1, pp. 46–71, 2012. https://doi.org/10.3390/polym4010046.

[61] Z. Wang, S. Wang, L. Zhang, H. Liu, and X. Xu, "Highly strong, tough and cryogenically adaptive hydrogel ionic conductors via coordination interactions," *Research*, vol. 7, pp. 1–11, 2023. https://doi.org/10.34133/research.0298.

[62] S.-Y. Chen, Z.-H. Zhao, C.-H. Li, and Q. Li, "Responsive materials based on coordination bonds," *Responsive Mater.*, vol. 1, no. 2, p. e20230011, 2023. https://doi.org/10.1002/rpm.20230011.

[63] A. P. Denysenko, A. M. Piddubnyi, I. A. Tkachenko, P. A. Shubin, S. I. Tarabarov, and R. A. Moskalenko, "A comprehensive study of dura mater biomineralization: Morphological, crystallographic, and immunohistochemical aspects," *Rep. Morphol.*, vol. 29, no. 4, pp. 50–57, 2023. https://doi.org/10.31393/morphology-journal-2023-29(4)-07.

[64] J. I. Lim, M. J. Kang, and W.-K. Lee, "Lotus-leaf-like structured chitosan–polyvinyl pyrrolidone films as an anti-adhesion barrier," *Appl. Surf. Sci.*, vol. 320, pp. 614–619, 2014. https://doi.org/10.1016/j.apsusc.2014.09.087.

[65] J. Huang, Y. Jiang, Y. Liu, Y. Ren, Z. Xu, Z. Li, Y. Zhao, X. Wu, and J. Ren, "Marine-inspired molecular mimicry generates a drug-free, but immunogenic hydrogel adhesive protecting surgical anastomosis," *Bioact. Mater.*, vol. 6, no. 3, pp. 770–782, 2021. https://doi.org/10.1016/j.bioactmat.2020.09.010.

[66] A. S. Borisova, T. Virkkala, R. Pylkkänen, M. Kellock, and P. Mohammadi, "Toughening brittle kraft lignin coating on mismatched substrate with spider Silk-Inspired protein as an interfacial modulator," *J. Colloid Interface Sci.*, vol. 655, pp. 789–799, 2024. https://doi.org/10.1016/j.jcis.2023.11.057.

[67] Z. Yu, X. Li, X. Li, B. Zheng, D. Li, D. Xu, and F. Wang, "Nacre-inspired metal-organic framework coatings reinforced by multiscale hierarchical cross-linking for integrated antifouling and antimicrobial corrosion," *Adv. Funct. Mater.*, vol. 33, no. 47, p. 2305995, 2023. https://doi.org/10.1002/adfm.202305995.

[68] Z. Peng, C. Wang, and S. Chen, "The microstructure morphology on ant footpads and its effect on ant adhesion," *Acta Mech.*, vol. 227, no. 7, pp. 2025–2037, 2016. https://doi.org/10.1007/s00707-016-1612-7.

[69] Y. Uyama, K. Kato, and Y. Ikada, "Surface modification of polymers by grafting BT: Grafting/characterization techniques/kinetic modeling," H. Galina, Y. Ikada, K. Kato, R. Kitamaru, J. Lechowicz, Y. Uyama, and C. Wu, Eds., Springer, Berlin, 1998, pp. 1–39. https://doi.org/10.1007/3-540-69685-7_1.

[70] A. C. Jadhav and N. C. Jadhav, "Graft copolymerization of methyl methacrylate on Meizotropis Pellita fibres and their applications in oil absorbency," *Iran. Polym. J.,* English edn., vol. 30, no. 1, pp. 9–24, 2021. https://doi.org/10.1007/s13726-020-00869-7.

[71] V. Sethi, C. Verma, S. Mukhopadhyay, A. Gupta, and B. Gupta, "Functional surfaces by plasma grafting of itaconic acid onto polypropylene mesh: Synthesis and structural investigations," *Polym. Int.*, vol. 73, no. 3, p. 167, 2023. https://doi.org/10.1002/pi.6603.

[72] G. Gatseva-Bogoeva and G. Petrov, "Photoinitiated graft polymerization of acrylamide to polyethylene terephthalate and properties of the modified fibres," *Fibre Chem.*, vol. 17, no. 4, pp. 300–304, 1986. https://doi.org/10.1007/BF00545655.

[73] K. P. Santo, A. Vishnyakov, Y. Brun, and A. V. Neimark, "Adhesion and separation of nanoparticles on polymer-grafted porous substrates," *Langmuir*, vol. 34, no. 4, pp. 1481–1496, 2018. https://doi.org/10.1021/acs.langmuir.7b02914.

[74] A. C. Genix, V. Bocharova, B. Carroll, P. Dieudonné-George, E. Chauveau, A. P. Sokolov, and J. Oberdisse, "Influence of the graft length on nanocomposite structure and interfacial dynamics," *Nanomaterials*, vol. 13, no. 4, pp. 1–18, 2023. https://doi.org/10.3390/nano13040748.

[75] K. Nayak, A. Kumar, P. Das, and B. P. Tripathi, "Amphiphilic antifouling membranes by polydopamine mediated molecular grafting for water purification and oil/water separation," *J. Membr. Sci.*, vol. 630, p. 119306, 2021. https://doi.org/10.1016/j.memsci.2021.119306.

[76] Q. Zhang, J. Chen, H. Wang, D. Xie, Z. Yang, J. Hu, H. Luo, and Y. Wan, "Water-induced expanded bilayer vascular graft with good hemocompatibility and biocompatibility," *Macromol. Biosci.*, vol. 24, no. 4, p. 2300401, 2023. https://doi.org/10.1002/mabi.202300401.

[77] Y. Zhu, Y. Ma, C. Yan, H. Xu, D. Liu, G. Chen, P. Shi, J. Hu, and C. Gao, "Improved interfacial shear strength of CF/PA6 and CF/epoxy composites by grafting graphene oxide onto carbon fiber surface with hyperbranched polyglycerol," *Surf. Interface Anal.*, vol. 53, no. 10, pp. 831–843, 2021. https://doi.org/10.1002/sia.6984.

[78] J. Nasser, "Nanostructured interphases for improved interfacial adhesion in structural and ballistic composites," *Deep Blue Data*, 2021. https://doi.org/10.7302/1370.

[79] S. Temurlu, M. A. A. Abureesh, A. Abourajab, P. Karsili, M. Dinleyici, S. Altınışık, S. Koyuncu, and H. Icil, "Grafting of perylene and naphthalene fluorophores onto chitosan for improved thermal, optical and electrical properties," *Macromol. Res.*, 2024. https://doi.org/10.1007/s13233-023-00233-7.

[80] M. Zhang, M. Liu, S. Bewick, and Z. Suo, "Nanoparticles to increase adhesive properties of biologically secreted materials for surface affixing," *J. Biomed. Nanotechnol.*, vol. 5, no. 3, pp. 294–299, 2009. https://doi.org/10.1166/jbn.2009.1034.

[81] H. Yang, H. G. Wang, X. P. Cao, and L. Wang, "Influence of substrate surface roughness on adhesive strength," *ITSC2006*, pp. 1277–1282, 2006. https://doi.org/10.31399/asm.cp.itsc2006p1277.

[82] H. Kang and Y. Nam, "The effect of electrostatic force between the nanoparticles and the substrate on the uniform assembly of inkjet-printed nanoparticles," *J. Flex. Print. Electron.*, vol. 1, no. 1, pp. 79–89, 2022. https://doi.org/10.56767/jfpe.2022.1.1.79.

[83] R. Rathore, D. Suhag, F. Wan, A. Thakur, and P. Thakur, "Toxicology and toxicity studies of nanobiomaterials BT: Integrated nanomaterials and their applications," D. Suhag, A. Thakur, and P. Thakur, Eds., Springer Nature, Singapore, 2023, pp. 433–456. https://doi.org/10.1007/978-981-99-6105-4_20.

[84] M. Ahmadinejad and J. S. Marshall, "Magnetic nanoparticle interaction with a hydrogel in an oscillating magnetic field," *Phys. Fluids*, vol. 36, no. 1, p. 13104, 2024. https://doi.org/10.1063/5.0177375.

[85] Y.-J. Park, D.-H. Lim, H.-J. Kim, D.-S. Park, and I.-K. Sung, "UV-and thermal-curing behaviors of dual-curable adhesives based on epoxy acrylate oligomers," *Int. J. Adhes. Adhes.*, vol. 29, no. 7, pp. 710–717, 2009. https://doi.org/10.1016/j.ijadhadh.2009.02.001.

[86] H. Ren, Z. Zhang, X. Cheng, Z. Zou, X. Chen, and C. He, "Injectable, self-healing hydrogel adhesives with firm tissue adhesion and on-demand biodegradation for sutureless wound closure," *Sci. Adv.*, vol. 9, no. 33, p. eadh4327, 2024. https://doi.org/10.1126/sciadv.adh4327.

[87] D. Dhakal, P. Lamichhane, L. Baxter, B. Sedai, and R. Vaidyanathan, "Nanosized hybrid polymer modifiers (HPM) for improved mechanical and thermal behavior of carbon fiber reinforced composites BT: Proceedings of the international conference on metallurgical engineering and centenary celebration," S. Patra, S. Sinha, G. S. Mahobia, and D. Kamble, Eds., Springer Nature, Singapore, 2024, pp. 157–165.

[88] H. Li, X. Wang, D. Qiao, J. Li, W. Zhu, and H. Li, "Mechanism of nanoparticle aggregation in gas-liquid microfluidic mixing," *Chin. Chem. Lett.*, p. 108747, 2023. https://doi.org/10.1016/j.cclet.2023.108747.

[89] J. Mahalingam, S. Rajendran, N. Shanmugasundram, H. Kuppuswamy, and R. Thirumurugan, "Investigation of acoustic, mechanical, thermal, and moisture properties of milled alkali-treated coconut tree secondary flower leaf stalk fiber-reinforced polymer composite," *Polym. Compos.*, vol. 45, no. 2, 2023. https://doi.org/10.1002/pc.28054.

[90] S. Pourhashem, F. Saba, J. Duan, A. Rashidi, F. Guan, E. G. Nezhad, and B. Hou, "Polymer/Inorganic nanocomposite coatings with superior corrosion protection performance: A review," *J. Ind. Eng. Chem.*, vol. 88, pp. 29–57, 2020. https://doi.org/10.1016/j.jiec.2020.04.029.

[91] Q. Fu, P. Zhang, L. Zhuang, L. Zhou, J. Zhang, J. Wang, X. Hou, R. Riedel, and H. Li, "Micro/nano multiscale reinforcing strategies toward extreme high-temperature applications: Take carbon/carbon composites and their coatings as the examples," *J. Mater. Sci. Technol.*, vol. 96, pp. 31–68, 2022. https://doi.org/10.1016/j.jmst.2021.03.076.

[92] Y. Abebe, P. Sivaprakasam, M. Desta, J. Udayaprakash, and M. P. Saravanan, "Tribological behavior of physical vapor deposition coating for punch and dies: An overview," *Int. J. Veh. Struct. Syst.*, vol. 15, no. 3, pp. 319–325, 2023. https://doi.org/10.4273/ijvss.15.3.08.

[93] G. Singh, N. Yadav, P. Yadav, S. K. Singh, and A. Kumar, "Incorporation of nanotechnology in wastewater remediation: Advancement and challenges," *Adv. Chem. Pollution, Environ. Manag. Prot.*, 2023. https://doi.org/10.1016/bs.apmp.2023.08.006.

[94] C. Fernandes, M. Jathar, B. K. S. Sawant, and T. Warde, "Scale-up of nanoparticle manufacturing process BT: Pharmaceutical process engineering and scale-up principles," A. B. Jindal, Ed., Springer Nature, Cham, 2023, pp. 173–203. https://doi.org/10.1007/978-3-031-31380-6_12.

[95] S. Shahidi, B. Moazzenchi, and M. Ghoranneviss, "A review-application of physical vapor deposition (PVD) and related methods in the textile industry," *Eur. Phys. J. Appl. Phys.*, vol. 71, no. 3, p. 31302, 2015. https://doi.org/10.1051/epjap/2015140439.

[96] R. Melentiev, A. Yudhanto, R. Tao, T. Vuchkov, and G. Lubineau, "Metallization of polymers and composites: State-of-the-art approaches," *Mater. Des.*, vol. 221, p. 110958, 2022. https://doi.org/10.1016/j.matdes.2022.110958.

[97] K. K. Gleason, *CVD polymers: Fabrication of organic surfaces and devices*, Wiley, New York, 2015. https://doi.org/10.1002/9783527690275.

[98] H. Usui, "Preparation of polymer thin films by physical vapor deposition," *Funct. Polym. Film.*, vol. 1, pp. 287–318, 2011. https://doi.org/10.1002/9783527638482.ch9.

[99] F. Ashrafizadeh, "Adhesion evaluation of PVD coatings to aluminium substrate," *Surf. Coat. Technol.*, vol. 130, nos. 2–3, pp. 186–194, 2000. https://doi.org/10.1016/S0257–8972(00)00670–8.

# 13 Experimental Evaluation of Nanocomposites in Polymer Matrix

*Priya Goel, Arti Hansda, Yesha Master, Priyabrata Mandal, and Swati Mohapatra*

## 13.1 INTRODUCTION

*Nanocomposites*, a new class of materials with significant applications across various scientific, industrial, and technical fields, are the result of the union of nanotechnology and polymer matrices. These compounds comprise nanoscale filler substances, often known as nanoparticles or nanofillers, disseminated within a polymer matrix. Especially compared to conventional polymers, the resulting nanocomposite materials exhibit distinctive and improved features, making them a focus of investigation and advancement in the fields of materials science and engineering. The importance of nanocomposites in polymer matrices can be attributed to their remarkable ability to provide an array of enhanced mechanical, thermal, electrical, and barrier properties. Engineers and scientists can modify the properties of polymer matrices by adding nanoparticles to satisfy the needs of certain applications. This exceptionally fine control over material characteristics has created a wide range of opportunities in plenty of industries, including aerospace engineering, electronics, therapeutic devices, and preservation of the environment. The reinforcement of polymer materials with nanofillers, often nanoparticles like carbon nanotubes, graphene, clay nanoparticles, or metal oxides, is one of the most acclaimed achievements in the field of nanocomposites. The mechanical strength, stiffness, and toughness of polymers can be significantly improved by introducing these nanoscale reinforcements, making them appropriate for high-performance applications. Some of the most important and distinctive characteristics of nanoparticles and their potential uses in a variety of industries are high surface area, quantum confinement, tunable optical properties, enhanced chemical and mechanical reactivity, superparamagnetic, heat conductivity, selective adsorption and filtration, ultimate targeted drug delivery, and ultralubrication.

Nanocomposites, for instance, have the potential to supplant conventional materials in lightweight structural components, resulting in decreased energy consumption and greenhouse gas emissions in the transportation industry. These properties of nanocomposites have led to their adoption in the automotive and aircraft industries, which not only lower the environmental imprint but also improve performance and safety. The creation of conductive and dielectric nanocomposites is propelling the development of highly efficient, miniature electronics. Nanocomposites are being used to increase shelf life and decrease waste in the world of food packaging and barrier materials. In addition, nanocomposite-based drug delivery systems, tissue engineering scaffolds, and diagnostic tools have emerged in the medical field, all of which promise revolutionary improvements in health care. These developments provide potential for regenerative therapeutics and personalized medicine. It is essential to explore the most contemporary advances and difficulties of this reprimanding which are continuously reshaping our world to fully comprehend the relevance of nanocomposites in polymer matrices. Additionally, nanoparticles frequently exhibit distinct optical properties due to their size and chemical composition. For instance, quantum dots have size-dependent emission colors and are utilized in display technologies and bioimaging. Similar to silver nanoparticles, gold nanoparticles are renowned for their plasmonic properties, which enable sensing, imaging, and

DOI: 10.1201/9781003476337-13

cancer therapy applications. Superparamagnetic is a property that exists in certain nanoparticles, such as iron oxide nanoparticles. They are utilized in medicinal applications, specifically in medication delivery systems and magnetic resonance imaging (MRI).

One notable research gap in the experimental evaluation of nanocomposites in polymer matrices is the need for comprehensive studies focused on the long-term durability and aging behavior of these materials. While there is extensive research on the initial properties and short-term performance of nanocomposites, there is a relative lack of data regarding how these materials perform over extended periods, especially in real-world conditions. The properties and behavior of materials must be thoroughly investigated through systematic experimental research because this is the only way to build knowledge in science, progress technology, and create new applications in a variety of disciplines. For instance, researchers have used nanoparticle-based adsorbents in environmental science to remove heavy metals from wastewater [1]. A recent study by Zhu et al. demonstrated the use of magnetic iron oxide nanoparticles for efficient lead removal from water [2]. In the electronics industry, quantum dots have become integral to next-generation displays [3]. Semiconductor nanoparticles, also known as quantum dots, are utilized in the electronics industry to take advantage of a phenomenon called quantum confinement. By adjusting the size of these nanoparticles, it is possible to manipulate their electronic properties [4, 5]. Huang et al. explored the use of quantum dots for high-quality displays in electronic devices [6]. In the solar industry, researchers are using quantum dots to enhance the efficiency of solar cells [7]. A recent article by Zhou et al. delves into the potential of quantum dots for solar energy applications [8]. Iron oxide nanoparticles are widely used in MRI contrast agents, and Lodhia et al. and Smith et al. discuss the development of novel iron oxide nanoparticles for improved MRI imaging [9]. The medical industry is advancing targeted therapies using nanoparticles. A recent publication by Fan et al. deliberated the potential of nanoparticle-based drug delivery systems for cancer treatment [10]. Moreover, a study by Lu et al. explores the use of nanoparticles as lubricant additives in the manufacturing industry to enhance wear resistance and reduce energy consumption [11]. Nanotechnology continues to inspire innovation, offering solutions to complex challenges and improving the performance of existing technologies. The key considerations for using NMs are affordability, regulatory compliance, and product development.

## 13.2 OVERVIEW OF NMS IN POLYMER COMPOSITES

The integration of diverse NMs into polymer matrices has revolutionized the field of composite materials, providing enhanced properties and functionalities. Various NMs, each with unique characteristics, are employed to tailor the performance of polymer composites for specific applications (Table 13.1). Nanoparticles are scattered across a polymer matrix to create composite materials known as nanocomposites. Techniques like solution blending, melt mixing, in situ polymerization, and electrospinning are used to incorporate NMs into polymers [12]. These 1 to 100 nm sized nanoparticles are available in a variety of shapes and sizes, each with unique characteristics and uses. The diverse range of NMs allows for tailoring composite materials to meet specific requirements, such as mechanical strength, electrical conductivity, optical properties, biocompatibility, etc. The resultant nanocomposites have remarkable qualities that make them useful in a variety of industries, including electronics, health care, aerospace, and environmental engineering. The careful selection and incorporation of these NMs contribute to the development of advanced polymer composites with superior performance characteristics.

From the last decades, carbon nanotubes (CNTs) are one of the nanoparticles which have drawn a considerable amount of attention. CNTs are 1D hollow cylindrical tube made up of atom-thick layers of carbon atoms, having a diameter measuring on the nanometer scale. CNTs are either single-walled carbon nanotubes (SWCNTs; seamless cylinders, comprising a single graphene sheet, with diameters ranging from 0.4 to 3 nm) or multiwalled carbon nanotubes (MWCNTs; concentric rolls of graphene sheets). Since the discovery of CNTs, its fascinating properties, such as mechanical strength and excellent electrical conductivity, high aspect ratio, and high surface area, and its

**TABLE 13.1**
**Overview of the Diverse NMs Utilized in Polymer Composites and Its Applications**

| Nanomaterial | Properties | Applications | Ref. |
|---|---|---|---|
| **Carbon nanotubes (CNTs)** | Exceptional mechanical strength, high thermal conductivity, excellent electrical conductivity due to their unique cylindrical nanostructure | Reinforcement in polymers for aerospace components, conductive materials in electronics, and drug delivery systems in biomedicine | [27] |
| **Graphene/graphene oxide** | Single layer of carbon atoms arranged in a hexagonal lattice, high mechanical strength, thermal conductivity, and electrical conductivity | Flexible electronics, energy storage devices (batteries and supercapacitors), and high-performance composites for automotive and aerospace industries | [28] |
| **Clay nanoparticles (e.g. montmorillonite, kaolinite)** | High aspect ratio, large surface area, and excellent mechanical strength, improved barrier properties, and flame resistance | Packaging materials, automotive parts, and nanocomposite coatings for corrosion protection | [16, 29] |
| **Metal (e.g., silver, gold)** | Unique optical and antimicrobial properties, good electrical conductivity | Antimicrobial coatings, conductive polymers for electronics | [30] |
| **Metal oxides (e.g., $TiO_2$, ZnO, $Al_2O_3$)** | Photocatalytic activity, UV absorption, antimicrobial behavior, excellent mechanical strength | UV-blocking coatings in sunscreens, antimicrobial films in packaging, and reinforcement in polymers for mechanical strength | [24] |

cylindrical structures made up of carbon have been used extensively in electronics and aerospace industries [13]. Despite the excellent properties of CNTs, it is challenging to attain a homogeneous dispersion of CNTs in many solvents. The fabrication of polymer nanocomposites with homogeneously dispersed CNTs is contingent upon strong interfacial bonding between the CNTs and the polymer matrix. So the functionalization of CNTs is sometimes performed in order to ensure uniform dispersion in the polymer matrix and improved adhesion of CNTs to the polymer.

Recently, graphene and graphene-based materials have diversified applications because of their outstanding adaptability and stable functional properties. Graphene-based materials have come out as excellent nanomaterial in various applications, such as energy storage, fuel cells, catalysis, vapor transport, hydrogen separation, proton exchange, desalination, bacterial inactivation and removal, water purification, energy storage, supercapacitors, biomedical, environmental, and many more. GO, a 2D nanomaterial derived from graphite, is comprised of carbon sheets decorated with oxygen-containing functionalities on the surface (hydroxyl and epoxide groups) and on the edges (hydroxyl, carbonyl, and carboxyl groups). Its ease of dispersibility and reactivity in organic and aqueous media make GO a versatile starting material with scalable production. Since the discovery of GO, it has drawn a considerable amount of attention due to its fascinating electrical and mechanical an selective barrier properties and excellent chemical, mechanical, and thermal stability, combined with a high surface area. In 2012, Geim et al. firstly revealed from his research that the GO membrane allows unimpeded permeation of water, whereas it blocks everything else in the vapor form [14]. Synthesis of GO from graphite includes mainly five methods (Staudenmaier, Brodie, Hofmann, Hummer's, Tour), in which the modified Hummer's method is the most preferable. The oxygenic groups (-OH, -COOH) in the carbon lattice allow GO to be modified by chemical reactions to introduce additional functional groups, which assist in boosting the nanocomposite properties even further. As mentioned earlier, functionalization serves to improve the dispersion of NMs in solvents and casting solutions. Except GO-functionalized analogues of graphene-based materials, for example, reduced GO is another strategy for enhancing nanocomposites with carbonaceous nanofillers. The chemical structure of rGO is characterized by carbon vacancies, residual oxygen functional groups, and clustered pentagons and heptagons carbon structures prepared through chemical reducing agents, photoreduction, thermal annealing, or microwave-assisted reduction. Unlike GO, rGO is hydrophobic due to the reduction of polar functional groups on its surface and possesses properties that approach that of pristine graphene. Similarly, with its exceptional mechanical strength and thermal conductivity, graphene—a single sheet of carbon atoms organized in a hexagonal lattice—finds applications in flexible electronics and energy storage devices [15].

The presence of inorganic fillers in the nanocomposites gives them greater potential to compete with the pristine analogues in terms of thermal and chemical stabilities at intermediate temperature as they take on the characteristics of inorganic materials that are stable at high temperatures. Inorganic fillers such as clay, silica, titanium (IV) oxide ($TiO_2$), aluminum oxide ($Al_2O_3$), iron (II, III) oxide ($Fe_3O_4$), zirconium phosphate, silver (Ag), zeolite, and zinc oxide (ZnO) are popularly used as nanofillers in nanocomposite fabrication. As these nanoparticles are incorporated into polymer matrices, they produce nanocomposites with enhanced mechanical, thermal, and electrical properties. This pushes the boundaries of materials science forward and opens up a new era for the applications of electronics to medicine, everywhere.

Clay nanoparticles with a high aspect ratio, such as kaolinite and montmorillonite, are well-known for improving the barrier qualities of packing materials and the flame resistance of coatings [16]. Silica NPs are one of the widely investigated oxide nanoparticle types. By virtue of their simple fabrication, chemical inertness, and relatively affordable precursors, silica NPs have received substantial considerations for various applications, such as in biomedical applications, catalytic applications, and fillers in polymer nanocomposites. This is motivated by the increased desire for materials with improved thermal, mechanical, physical, and chemical properties. A number of methods/techniques are available for preparing silica NMs, but the sol-gel technique is commonly used due to a high level of control over the particle size, size distribution, and morphology. Due to the advantage of easy surface modification, silica NMs are functionalized with a variety of functional groups via physical adsorption or covalent conjugation using silane chemistry. Depending upon the specific application, the surface property of silica NMs could be easily tuned with positive, negative, or zwitterionic surface charge.

ZnO NPs are another important NMs that are well applied in photocatalysis [17], sensors or biosensors [18], antibacterial materials [19], water disinfection [20, 21], biomedical applications [22], dye-sensitized solar cells [23], and many more [24]. Additionally, the low cost and high surface area of ZnO NPs make them attractive alternatives to other NMs. Recently, various approaches, including physical, chemical, and biological ("green chemistry"), have been used to prepare ZnO nanocomposites with different morphologies [25]. ZnO NPs can be prepared using physical methods (physical vapor deposition, arc plasma, thermal evaporation, ultrasonic irradiation), chemical methods (microemulsion, sol-gel, precipitation, reduction, hydrothermal, solvothermal, chemical vapor deposition), and biological methods (plant extracts, microorganisms, biochemistry). The impressive magnetic and selective adsorption properties of $Fe_3O_4$ NPs make them attractive candidates for water treatment applications.

Selecting the appropriate polymer matrix is essential for creating efficient nanocomposites, since it affects the dispersion, compatibility, and overall functionality of the nanoparticles in the composite material [26]. Different types of nanoparticles have varying interactions with polymers, and selecting a suitable polymer matrix is critical for maximizing the benefits of nanocomposites, for example, polyethylene (PE) can hold various types of nanoparticles because of its adaptability and low processing temperature, making it a good choice for reinforcing agents like graphene and carbon nanotubes (CNTs). Similarly, metal oxide nanoparticles can be added to polyvinyl chloride (PVC) to improve its UV resistance and mechanical characteristics, which makes it appropriate for outdoor applications. When combined with particular nanoparticles, each of these carefully chosen polymers offers special benefits, which guarantee a broad range of applications across many industries.

## 13.3 CHARACTERIZATION TECHNIQUES

Characterizing the nanocomposite components is crucial for a fuller knowledge of polymer nanocomposites. Among these are the following: (1) filler dispersion in the polymer matrix; (2) filler surface modification effects on dispersion and composite properties; (3) interactions between the filler modification and polymer chains, including chemical reactions between them; (4) orientation or alignment related to the processing method used; (5) morphology and properties of the nanocomposites changed by changes in process parameters; as well as (6) assessments of a wide range of properties to determine a nanocomposite's application potential.

A precise description of *nanocomposite* may need the use of multiple methodologies, as there are numerous approaches available to characterize nanocomposites based on their structures and applications. After sample preparation, which entails dispersing nanoparticles uniformly in the polymer matrix, a systematic evaluation and a comprehensive grasp of the morphological, dispersion, and structural characteristics of nanocomposites are required (Figure 13.1). SEM and TEM are particularly useful for morphological examination because they produce high-resolution pictures that are essential for evaluating the morphology and dispersion of nanoparticles inside the polymer matrix. XRD and FTIR spectroscopy are two methods that are used to carefully examine structural qualities in order to reveal information about crystalline structures and chemical interactions. Moreover, NMR spectroscopy enhances our knowledge of nanocomposite compositions by helping us comprehend atomic-level features. Techniques like DLS and AFM are used to comprehensively study the dispersion of nanoparticles within the polymer matrix, a significant aspect impacting performance. AFM provides a three-dimensional profile of the nanoscale surface, while DLS provides

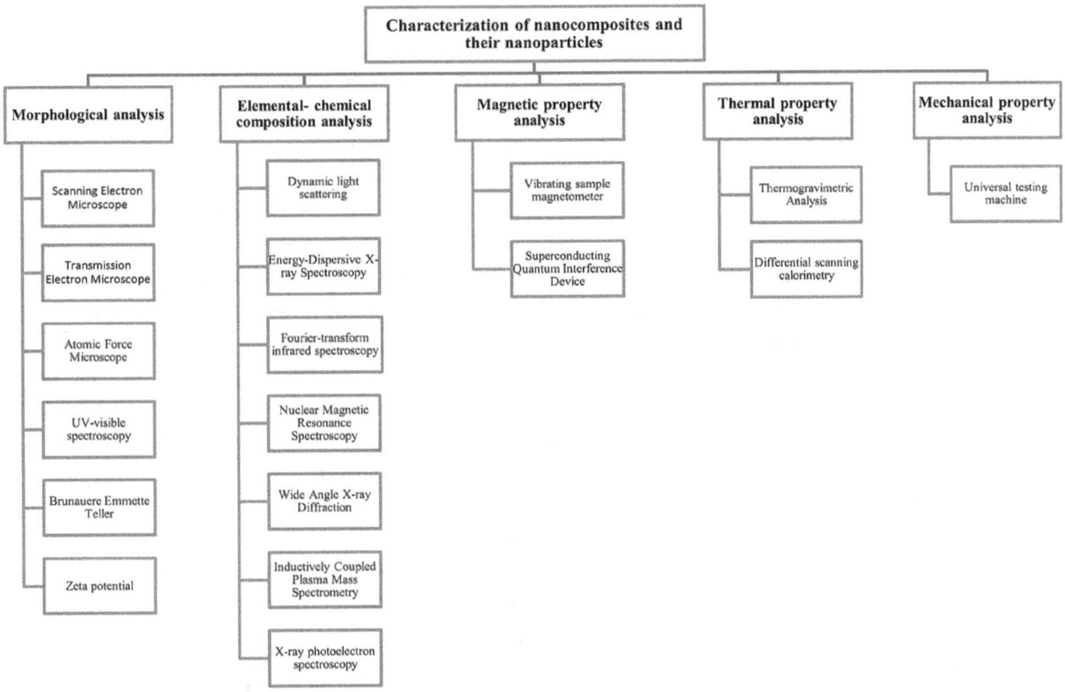


**FIGURE 13.1** Various characterizations of nanoparticles/nanocomposites to investigate their morphological, mechanical, magnetic, and thermal analysis.

information on the particle size distribution in solution. Tensile testing, impact testing, hardness testing, and other methods are used to thoroughly evaluate mechanical properties, which are essential for assessing the performance of nanocomposite materials. Simultaneously, methods like DSC and TGA are employed to examine thermal properties, offering vital information about thermal behavior and stability. Essentially, the careful incorporation of these methods into the experimental framework guarantees a comprehensive comprehension of nanocomposites, including their morphological, structural, and functional features. For nanocomposite materials to be developed and used in a variety of sectors, thorough characterization is essential.

### 13.3.1 Morphological Analysis

The study and characterization of the physical structure, shape, and distribution of nanoparticles inside the polymer matrix are all part of the morphological investigation of polymer nanocomposites. Understanding the various elements' distribution within the polymer matrix, as well as their interactions and general layout, is critical. Examples of these elements include fibers and nanoparticles. SEM helps evaluate the homogeneity and dispersion of nanoparticles by providing detailed surface morphology information that shows the distribution and arrangement of nanoparticles inside the polymer matrix. It is employed to assess the extent to which nanoparticles inside the polymer matrix have aggregated, clustered, or segregated. On the composite surface, well-dispersed nanoparticles manifest as single particles or clusters that are uniformly distributed. Agglomerated nanoparticles may be a sign of poor dispersion and have an impact on the characteristics of the nanocomposite. It is evident that a common problem following the dispersion of nanoparticles into liquid polymer suspension is the creation of agglomerates of particles (instead of individual particles); hence, SEM examination is required for further use. Goel et al. showed a $TiO_2$-incorporated nanocomposite membrane in which agglomeration at a higher concentration loading of nanoparticle within the polymer matrix reduced the overall performance of the membrane [31]. SEM images clearly showed the agglomerated sites that can be improved via utrasonication/probe sonication to break the weak van der Waals forces between particles to segregate the particles and enhance the overall performance of any nanocomposite. Similarly, Liu et al. show a drop in the water flux due to the agglomeration of graphene oxide (GO)–tailored polyamide composite reverse osmosis (RO) membranes [32].

TEM offers atomic-level imaging and precise structural information which deviates from the SEM technique and fosters insights into the microstructure, flaws, and composition of nanocomposite/nanomaterial. It provides information about the internal structure of the nanocomposite and the size, shape, and dispersion of the individual nanoparticles. It facilitates comprehension of the interactions at the interface between the polymer matrix and nanoparticles. Monticelli et al. used TEM to analyze the internal structure and spatial distribution of the various phases of MMT-incorporated nanocomposites. Kaveh et al. investigated the dispersion of modified GO into the aramid layer [33]. Li et al. explored TEM images to verify the expected hydrophilic/hydrophobic micro–phase separation structure of prepared PSf-BDMImOH nanocomposite [34]. Tang et al.'s study used TEM to characterize the morphology of sulfonated polyarylene ether nitrile (SPEN)/graphene oxide (GO) hybrid membranes prepared in this study, which have nanosheets [35].

The surface topography of the polymer nanocomposite, including its texture, roughness, and nanoparticle presence, may be seen at the nanoscale in AFM pictures. The ability to see individual or aggregating nanoparticles on the surface of the nanocomposite provides insight about the dispersion quality of the nanoparticles. Phase imaging gives contrast based on variations in material properties, such as stiffness and composition, while thickness and surface forces are measured to evaluate the elasticity and stickiness of the nanocomposite material. Zucker et al. used AFM to find out the thickness of various nanosheets (e.g., GO, rGO, MoS2, CuO etc.) [36]. Tian et al. quantitatively characterize the thickness of interphase and directly identify the interfacial double-layer structure of SiO2/cured rubber nanocomposites by using AFM [37].

A popular method for analyzing nanocomposites is UV-Vis spectroscopy, which offers details on the materials' optical characteristics, electronic transitions, and absorbing species concentration [38–40]. By connecting absorbance with known nanoparticle concentrations, it can be used to calculate the concentration of nanoparticles in a nanocomposite. Surface plasmon resonance is exhibited by metallic nanoparticles, such gold and silver, and results in characteristic absorption peaks in the UV-Vis spectrum [29, 30]. Information about the size, shape, and dispersion of nanoparticles can be obtained by analyzing these peaks. By tracking variations in absorption spectra, it can be helpful in determining coatings on nanoparticles [41]. UV-Vis spectroscopy can monitor changes in the absorption spectra that reflect the dispersion of nanoparticles or the production of new materials during the synthesis of nanocomposites. Yu et al. reported an improved $NaNO_3$-free Hummer's method. Yu compared both the GOs with various instrumental techniques, and GO1 and GO2 exhibit similar characteristics in UV-Vis spectra and indicate that they are similar in domain structures with similar oxygen groups [38].

To measure the specific surface area of materials used in nanocomposite construction, BET analysis is utilized [42]. In order to comprehend the reactivity and adsorption capability of nanocomposites, this is especially significant [43]. BET analysis can help characterize the internal structure of nanocomposites by supplying details about the pore size distribution in such materials [44, 45]. The effectiveness of surface changes on nanoparticles inside nanocomposites can be determined by alterations in the specific surface area detected by BET analysis. BET analysis can be used to ensure consistent surface features during the synthesis or production of nanocomposites by providing quality control.

When characterizing nanocomposites, zeta potential is an important factor, particularly when working with colloidal systems [46]. The measured potential is influenced by the presence of ions at the shear plane, which is the plane inside the electrical double layer. Nanoparticles scattered in a liquid media are a common ingredient in nanocomposites. The surface charge of these nanoparticles can be determined using the zeta potential [47, 48]. It is an essential parameter in determining how stable colloidal systems are. Higher zeta potential values for nanoparticles often translate into better stability because they avert flocculation and agglomeration through enhanced electrostatic repulsion. The effectiveness of surface modifications or functionalization of nanoparticles inside nanocomposites can be determined by changes in zeta potential [49]. Functional groups have the ability to change the zeta potential by changing the surface charge. It also affects how particles interact with one another within a colloidal system. Optimized zeta potential values can lead to more uniform and improved dispersion of nanocomposites. It is connected to the nanocomposites' rheological characteristics. Improved dispersion can result in a higher zeta potential, which might affect the behavior and overall performance of the material.

### 13.3.2 Elemental-Chemical Composition Analysis

The elemental-chemical composition analysis of nanocomposites involves determining the presence and quantity of different elements in the material. Several analytical techniques are employed for this purpose, providing insights into the composition of both the polymer matrix and the nanoparticles.

DLS, also known as photon correlation spectroscopy, is a sensitive technique used to measure particle size and size distribution in a solution, including nanoparticles in polymer composites. It measures the hydrodynamic size of particles in suspension, providing information about particle size distribution [50, 51], analyzing the size distribution of nanoparticles in solution and assessing their aggregation state. It calculates the polydispersity index, representing the width of the particle size distribution curve [52]. Its advantages lie in rapid analysis, non-invasiveness, and the ability to analyze a wide size range.

EDS is a powerful analytical technique commonly used for the elemental analysis of nanocomposites. It is often coupled with scanning electron microscopy (SEM) to provide detailed information about the distribution and concentration of elements within the nanomaterial. EDS can provide spatial information about the distribution of elements within the nanocomposite [53, 54]. Elemental maps generated by scanning the electron beam across the sample surface reveal the spatial arrangement of different elements. It enables the quantification of elemental concentrations in nanocomposites. By comparing X-ray intensities to standard references, the relative abundance of each element can be determined. EDS aids in identifying different phases or components in nanocomposites. It can distinguish between the elemental composition of nanoparticles and the surrounding matrix. It is valuable for studying the elemental composition of nanoparticles embedded in the nanocomposite matrix, providing insights into their size, shape, and dispersion [55].

FTIR spectroscopy serves as a robust analytical tool for delving into the study of the molecular composition, chemical bonding, and interactions of chemical analysis for identifying functional groups and molecular structures, material characterization, providing insights into the composition and structure of materials, within polymer and nanomaterial [56].

NMR spectroscopy is a powerful analytical technique to investigate the molecular structure, dynamics, and interactions within nanocomposites. It provides information about the local environment of atoms (1H or C-13), revealing interactions between nanoparticles and polymer chains. Also, NMR spectra provide information about the chemical structure of the polymers, nanoparticles, and any functional groups in the nanocomposite [57, 58]. NMR elucidates the three-dimensional structure of compounds, quantitative analysis of complex mixtures, and in studying molecular dynamics, to playing a pivotal role in protein and biomolecular research, monitoring chemical reactions in real time, and providing insights into reaction kinetics and intermediates.

XRD is a technique used to analyze the crystallographic structure, composition, and crystallinity of NMs and nanocomposites. It measures the diffraction pattern when the crystal lattice of the material scatters X-rays [59]. XRD identifies crystalline phases and quantitative analysis of their abundance to determine crystal structure, analyze crystallite size and strain, and study phase transformations. Its high sensitivity to atomic arrangements and nondestructive nature infer applicability to a wide array of materials. However, XRD is most effective for crystalline materials. It identifies the crystalline structure of nanoparticles and their orientation within the nanocomposite and helps determine the extent of nanoparticle exfoliation in layered nanocomposites [60].

ICP-MS is typically used for elemental analysis in a wide range of materials; its application to nanocomposites allows researchers to assess the elemental composition and concentration of nanoparticles or nanofillers within the composite matrix [61]. It can determine the elemental composition of nanocomposites by ionizing the elements present in the sample. This is particularly useful for nanocomposites containing metallic nanoparticles or other elements with distinctive mass-to-charge ratios. It allows for the quantification of the concentration of specific elements, providing information about the amount of nanoparticles incorporated into the nanocomposite [62, 63]. This quantitative data is crucial for understanding the loading of nanoparticles and optimizing composite properties. This is highly sensitive and capable of detecting trace elements, which is essential for identifying impurities or unintended elements in the nanocomposite. While ICP-MS itself does not provide information on the size or distribution of nanoparticles, it can be complemented with other techniques, such as dynamic light scattering or electron microscopy, for a comprehensive characterization of nanocomposite properties.

XPS, also known as electron spectroscopy for chemical analysis (ESCA), is a surface-sensitive technique that can be applied to analyze nanocomposite. It provides information about the elemental composition, chemical states, and electronic environments of the surface elements in a material or on nanocomposite surface [64]. This is particularly valuable for understanding the presence and distribution of elements, including nanoparticles or surface coatings. It can distinguish between different oxidation states and chemical environments, aiding in the analysis of functional groups and surface modifications. It can be used to identify the presence of nanoparticles by detecting characteristic peaks associated with specific elements [49]. It can differentiate between different forms of the same element based on their binding energies. Though XPS is primarily a surface analysis technique, depth profiling methods can be employed to investigate the elemental composition as a function of depth. This can provide insights into the distribution of nanoparticles within the composite, interactions between nanoparticles and the polymer matrix, as well as any chemical modifications that may occur during the synthesis or processing of the nanocomposite. It allows for quantitative analysis of elemental composition, providing information about the concentration of specific elements on the nanocomposite surface. XPS is capable of detecting chemical changes induced by interactions between nanoparticles and the polymer matrix.

By employing these advanced characterization techniques, researchers can gain a comprehensive understanding of nanocomposites, aiding in the design and optimization of materials for specific applications. These methods provide essential information on the morphology, dispersion, and structural properties of nanocomposites, contributing to the ongoing advancements in the field of materials science and engineering.

### 13.3.3 Magnetic Properties Analysis

Analyzing the magnetic properties of nanocomposites is crucial for understanding their behavior and potential applications. Magnetic properties are often influenced by the presence of magnetic nanoparticles within the polymer matrix. The choice of magnetic nanoparticles, such as iron oxides (e.g., magnetite, maghemite) or other magnetic materials, impacts the overall magnetic behavior. The size and shape of magnetic nanoparticles influence their superparamagnetic or ferromagnetic behavior. Smaller nanoparticles may exhibit superparamagnetism, while larger ones may show ferromagnetic characteristics. Magnetic saturation, the point at which a material cannot be magnetized further, is an important parameter. It depends on the saturation magnetization of the magnetic nanoparticles and their concentration within the nanocomposite. The hysteresis loop is a crucial graph that illustrates the relationship between magnetic field strength ($H$) and magnetization ($M$). It provides information about the coercivity, remanence, and overall magnetic behavior of the nanocomposite.

Understanding the magnetic properties is crucial for developing magnetic sensors and devices based on nanocomposite materials. VSM is a common technique for measuring the magnetic properties of nanocomposites which involves subjecting the sample to a varying magnetic field while measuring its magnetization. It provides valuable information about the magnetization behavior, magnetic hysteresis, and other key parameters of NMs and nanoparticles embedded in nanocomposites [65]. It is valuable for studying the magnetic behavior or providing information about their superparamagnetic or ferromagnetic characteristics.

SQUID magnetometry is another highly sensitive method used to measure the magnetic properties of NMs, including the magnetization and susceptibility of NMs/nanocomposites [65]. It is renowned for its exceptional sensitivity, making it suitable for studying the magnetic properties of nanocomposites, including those with superparamagnetic or weak magnetic responses. It can be used to identify the critical temperature (transition temperature) at which a material undergoes a transition from superconducting to normal state. It is also valuable for characterizing the magnetic behavior of nanoparticles embedded in nanocomposites, providing insights into their superparamagnetic or ferromagnetic properties.

### 13.3.4 Thermal Properties Analysis

Thermal property analysis of nanocomposites is crucial for understanding how the addition of nanoparticles influences the thermal behavior of the material. Understanding how the inclusion of nanoparticles affects overall performance and durability is crucial. Parameters such as tensile strength, compressive strength, and flexural strength are essential for evaluating the properties of nanocomposite-integrated materials. The inclusion of nanoparticles in polymer matrices often leads to improvements in various properties compared to traditional polymers. Various techniques are employed to investigate thermal properties, including thermal stability, conductivity, and expansion.

Thermogravimetric analysis (TGA) is a widely used technique for studying the thermal properties of nanocomposites. TGA measures changes in a material's weight as a function of temperature or time, providing valuable information about thermal stability, decomposition temperatures, and mass loss. TGA helps in the selection of suitable nanofillers for enhancing the thermal stability of polymers. It aids in identifying materials that can withstand higher temperatures without significant degradation. TGA is employed for quality control in the production of nanocomposites, ensuring consistency in thermal properties and confirming the effectiveness of thermal stabilizers. For nanocomposites designed for flame-retardant applications, TGA provides insights into the temperature range at which flame-retardant mechanisms are activated. TGA helps in understanding the behavior of nanoparticles during thermal events, including their influence on decomposition kinetics and the formation of residues.

Differential scanning calorimetry (DSC) is a widely used technique for studying the thermal properties of nanocomposites. DSC measures the heat flow associated with thermal transitions in a material as a function of temperature. DSC is fundamental for characterizing the thermal behavior of polymers and understanding how the incorporation of nanoparticles influences polymer properties.

## 13.4 CHALLENGES AND LIMITATIONS

The investigation of experimental design in assessing nanocomposites inside polymer matrices comes to a close with a critical analysis of the difficulties and potential directions that shape the field of hybrid approaches. One of the major challenges that affect the entire performance of nanocomposite materials is addressing the complex problem of dispersion, namely, achieving uniform distribution of nanoparticles inside the polymer matrix. Important topics for continued research and development include methods for improving dispersion, such as surface functionalization, the use of stabilizers, the choice of organic solvents, and production processes. Hybrid fillers are another recommended strategy, since their combined effects may improve the dispersion situation. Because of their special qualities, NMs—found in items like high-definition electronics and medicine delivery systems—are essential to study. Another obstacle to the advancement of nanocomposites is the switch from lab-scale to large-scale manufacturing. The need for creative production methods and optimized processing parameters to scale up operations without sacrificing quality and economy represents a crucial area for further research and technical advancement. Because of these unique properties, the application of nanocomposites in polymer matrices has shown great promise in a number of industrial domains. Nonetheless, a critical component remains the large-scale production processes' cost-effectiveness. The raw materials—more especially, the nanoparticles—that go into making composites have a big impact on their price. The high cost of creating nanoparticles is a significant factor that is often linked to their synthesis from rare or expensive components. The cost of acquiring and synthesizing nanoparticles is influenced by the need for them to be of the highest caliber and purity. Cost-effectiveness is still a major barrier to the widespread use of nanocomposites; in order to achieve wider use, costs related to NMs and sophisticated filler coatings must be lowered. Overcoming financial obstacles requires research efforts focused on sustainable nanomaterial sources and cost-effective synthesis techniques. Energy use and equipment expenses vary throughout procedures, including solution mixing, melt mixing, and in situ polymerization.

Sustainable development demands that nanoparticles and coatings be carefully considered due to safety and environmental issues. The conversation emphasizes how critical it is to address these issues through waste management, eco-friendly nanomaterial research, and responsible manufacturing methods. It becomes clear that regulation and standardization are essential to the future of nanocomposites. Regulatory frameworks and industry standards are put in place to guarantee the safe and reliable application of hybrid materials in a variety of settings. To create rules and protocols for the synthesis, characterization, and application of nanocomposites, researchers, companies, and regulatory agencies must work together. The prospects for nanocomposite research in the future are addressed. Future efforts are anticipated to concentrate on multifunctional materials that are suited for particular purposes. Sustainability is the main focus, and eco-friendly NMs that support international environmental goals are becoming more and more important. The nanocomposite environment offers a multitude of innovative avenues, as seen by advances in biomedical applications, advanced production techniques, and the integration of artificial intelligence. To summarize, the conversation on obstacles and potential directions provides guidance for scientists, engineers, and business experts negotiating the ever-changing landscape of nanocomposites and filler coatings–enhanced polymer composites. It offers insightful information that directs the development of technology, production, and application, promoting the ongoing growth of hybrid materials for a range of sustainable applications.

## 13.5 FUTURE PROSPECTS AND EMERGING TECHNOLOGIES

Advanced nanoparticles, including quantum dots and metal–organic frameworks (MOFs), have shown tremendous potential in revolutionizing various fields, particularly in the development of nanocomposites. These materials offer unique properties and functionalities that can be harnessed for a wide range of applications. *Quantum dots* are semiconductor nanocrystals with size-dependent electronic and optical properties. They exhibit quantum confinement effects, resulting in tunable bandgaps and emission wavelengths. QDs have found applications in several areas, such as electronics, photonics, and medicine. In nanocomposites, it can be incorporated to enhance the optical and electronic properties of the material. Researchers can create materials with improved conductivity, luminescence, and sensing capabilities by integrating quantum dots into nanocomposites. This is particularly relevant in the development of advanced sensors, optoelectronic devices, and quantum dot–based nanocomposite materials. Additionally, MOFs are porous materials composed of metal ions or clusters connected by organic linkers. They exhibit high surface areas, tunable pore sizes, and customizable functionalities. MOFs have shown promise in gas storage, catalysis, drug delivery, and sensing applications. Their versatility makes them suitable for incorporation into nanocomposites for various purposes. When integrated into nanocomposites, MOFs can provide enhanced surface area, selective adsorption, and improved mechanical properties. This can be exploited in gas sensing, drug release systems, and composite materials with tailored properties.

The combination of quantum dots and MOFs in nanocomposites can result in multifunctional materials with diverse properties, ranging from optical and electronic to catalytic and adsorption capabilities. Quantum dots and MOFs can introduce “smart” characteristics to nanocomposites, responding to external stimuli such as light, temperature, or chemical signals. This can lead to the development of advanced materials with adaptive and responsive features. Nanocomposites incorporating these advanced nanoparticles have potential energy storage, conversion, and harvesting applications. For instance, improved catalytic activity of MOFs and enhanced electronic properties of quantum dots can contribute to more efficient energy-related devices. Quantum dots may contain heavy metals, raising concerns about their potential toxicity. Proper encapsulation or surface modification is crucial to mitigate these risks. Ensuring the stability of both quantum dots and MOFs within nanocomposites over time is essential for practical applications. The scalability of production processes for these advanced nanoparticles needs to be addressed to facilitate their widespread adoption in industrial applications.

In summary, integrating quantum dots and metal–organic frameworks into nanocomposites holds great promise for advancing materials science and technology. Their unique properties can be tailored for specific applications, ranging from electronics and optics to medicine and energy. As research in this field progresses, addressing challenges and optimizing fabrication processes will be crucial for realizing the full potential of these advanced nanoparticles in the future of nanocomposites.

## 13.6 CONCLUSION

Future research will focus on real-time characterization of nanocomposites during processes using multiscale techniques. Machine learning and data analytics will process large datasets to reveal hidden patterns. Nondestructive testing and correlative microscopy will be used to assess nanocomposite integrity and structure. Experimental assessments are important in the industrial use of nanocomposites, with major consequences for various industry segments. These consequences comprise systematic tests that assist in refining nanocomposite compositions, improving their performance, and customizing them to meet particular industrial needs. Thorough testing guarantees that nanocomposites adhere to industry benchmarks for dependability and security. Experimental assessments can determine cost-efficient production techniques and materials, potentially decreasing manufacturing costs and improving commercial feasibility. It is important to explore potential areas for future study in the realm of advanced nanocomposites to expand the frontiers of materials science and technology. This involves the creation of sophisticated methods for understanding and describing things, as well as the invention of creative uses. The development of methodologies for in situ characterization enables real-time monitoring of nanocomposite behavior throughout several processes, including synthesis, deformation, and exposure to the environment, to explore nondestructive testing techniques, like ultrasound and terahertz spectroscopy, and to evaluate the quality and functionality of nanocomposite materials without causing harm, which is crucial for their use in aerospace, civil engineering, and infrastructure applications. Innovative application of nanocomposite materials includes their use in various fields, such as in biomedical and health-care services; in flexible electronic devices and wearables, energy storage, and conversion; in the aerospace and automotive industries; in research on smart materials; and in exploring their potential in construction for enhancing sustainability, durability, and thermal efficiency of buildings and infrastructure.

## ACKNOWLEDGMENTS

I am grateful to GSFC University, Department of Chemical Sciences, for providing the necessary resources and facilities that facilitated the research and writing of this chapter. I would like to express my sincere gratitude to all co-authors who collaborated with me on this chapter to enrich the content and strengthen the overall quality of the chapter.

## REFERENCES

[1] J. Yang, B. Hou, J. Wang, B. Tian, J. Bi, N. Wang, X. Li, and X. Huang, "Nanomaterials for the removal of heavy metals from wastewater," *Nanomaterials*, vol. 9, no. 3, p. 424, 2019.

[2] F. Zhu, Y. M. Zheng, B. G. Zhang, and Y. R. Dai, "A critical review on the electrospun nanofibrous membranes for the adsorption of heavy metals in water treatment," *J. Hazard. Mater.,* vol. 401, p. 123608, 2021.

[3] Y. Shu, X. Lin, H. Qin, Z. Hu, Y. Jin, and X. Peng, "Quantum dots for display applications," *Angew. Chem.*, vol. 132, no. 50, pp. 22496–22507, 2020.

[4] M. Jha, *Current trends in industrial scale synthesis of quantum dots and its application in electronics.* Elsevier Inc., 2018.

[5] M. Liu, N. Yazdani, M. Yarema, M. Jansen, V. Wood, and E. H. Sargent, "Colloidal quantum dot electronics," *Nat. Electron.*, vol. 4, no. 8, pp. 548–558, 2021.

[6] Y. M. Huang, K. J. Singh, A. C. Liu, C. C. Lin, Z. Chen, K. Wang, Y. Lin, Z. Liu, T. Wu, and H. C. Kuo, "Advances in quantum-dot-based displays," *Nanomaterials*, vol. 10, no. 7, pp. 1–29, 2020.

[7] K. A. Sablon, J. W. Little, V. Mitin, A. Sergeev, N. Vagidov, and K. Reinhardt, "Strong enhancement of solar cell efficiency due to quantum dots with built-in charge," *Nano Lett.*, vol. 11, no. 6, pp. 2311–2317, 2011.

[8] Y. Zhou, J. Yang, X. Luo, Y. Li, Q. Qiu, and T. Xie, "Selection, preparation and application of quantum dots in perovskite solar cells," *Int. J. Mol. Sci.*, vol. 23, no. 16, p. 9482, 2022.
[9] L. Smith, Z. Kuncic, H. L. Byrne, and D. Waddington, "Nanoparticles for MRI-guided radiation therapy: A review," *Cancer Nanotechnol.*, vol. 13, no. 1, pp. 1–28, 2022.
[10] D. Fan, Y. Cao, M. Cao, Y. Wang, Y. Cao, and T. Gong, "Nanomedicine in cancer therapy," *Signal Transduct. Target. Ther.*, vol. 8, no. 1, p. 293, 2023.
[11] Z. Lu, Q. Lin, Z. Cao, W. Li, J. Gong, Y. Wang, K. Hu, and X. Hu, "$MoS_2$ nanomaterials as lubricant additives: A review," *Lubricants*, vol. 11, no. 12, 2023.
[12] A. Tarhini and A. R. Tehrani-Bagha, "Advances in preparation methods and conductivity properties of graphene-based polymer composites," *Appl. Compos. Mater.*, vol. 30, no. 6, pp. 1737–1762, 2023.
[13] S. K. Soni, B. Thomas, and V. R. Kar, "A comprehensive review on CNTs and CNT-reinforced composites: Syntheses, characteristics and applications," *Mater. Today Commun.*, vol. 25, p. 101546, 2020.
[14] A. K. Geim, R. R. Nair, H. A. Wu, P. N. Jayaram, and I. V. Grigorieva, "Unimpeded permeation of water," *Science*, vol. 335, pp. 442–444, 2012.
[15] H. Kim, A. A. Abdala, and C. W. MacOsko, "Graphene/polymer nanocomposites," *Macromolecules*, vol. 43, no. 16, pp. 6515–6530, 2010.
[16] S. Sen Gupta and K. G. Bhattacharyya, "Adsorption of heavy metals on kaolinite and montmorillonite: A review," *Phys. Chem. Chem. Phys.*, vol. 14, no. 19, pp. 6698–6723, 2012.
[17] C. B. Ong, L. Y. Ng, and A. W. Mohammad, "A review of ZnO nanoparticles as solar photocatalysts: Synthesis, mechanisms and applications," *Renew. Sustain. Energy Rev.*, vol. 81, pp. 536–551, 2018.
[18] N. P. Shetti, S. D. Bukkitgar, K. R. Reddy, C. V. Reddy, and T. M. Aminabhavi, "ZnO-based nanostructured electrodes for electrochemical sensors and biosensors in biomedical applications," *Biosens. Bioelectron.*, vol. 141, p. 111417, 2019.
[19] R. Kumar, A. Umar, G. Kumar, and H. S. Nalwa, "Antimicrobial properties of ZnO nanomaterials: A review," *Ceram. Int.*, vol. 43, no. 5, pp. 3940–3961, 2017.
[20] E. A. S. Dimapilis, C. S. Hsu, R. M. O. Mendoza, and M. C. Lu, "Zinc oxide nanoparticles for water disinfection," *Sustain. Environ. Res.*, vol. 28, no. 2, pp. 47–56, 2018.
[21] T. Naseem and T. Durrani, "The role of some important metal oxide nanoparticles for wastewater and antibacterial applications: A review," *Environ. Chem. Ecotoxicol.*, vol. 3, pp. 59–75, 2021.
[22] P. K. Mishra, H. Mishra, A. Ekielski, S. Talegaonkar, and B. Vaidya, "Zinc oxide nanoparticles: A promising nanomaterial for biomedical applications," *Drug Discov. Today*, vol. 22, no. 12, pp. 1825–1834, 2017.
[23] R. Shashanka, H. Esgin, V. M. Yilmaz, and Y. Caglar, "Fabrication and characterization of green synthesized ZnO nanoparticle based dye-sensitized solar cells," *J. Sci. Adv. Mater. Dev.*, vol. 5, no. 2, pp. 185–191, 2020.
[24] A. R. Prasad, L. Williams, J. Garvasis, K. O. Shamsheera, S. M. Basheer, M. Kuruvilla, and A. Joseph, "Applications of phytogenic ZnO nanoparticles: A review on recent advancements," *J. Mol. Liq.*, vol. 331, p. 115805, 2021.
[25] A. Król, P. Pomastowski, K. Rafińska, V. Railean-Plugaru, and B. Buszewski, "Zinc oxide nanoparticles: Synthesis, antiseptic activity and toxicity mechanism," *Adv. Colloid Interface Sci.*, vol. 249, pp. 37–52, 2017.
[26] R. Hsissou, R. Seghiri, Z. Benzekri, M. Hilali, M. Rafik, and A. Elharfi, "Polymer composite materials: A comprehensive review," *Compos. Struct.*, vol. 262, pp. 1–3, 2021.
[27] S. Rathinavel, K. Priyadharshini, and D. Panda, "A review on carbon nanotube: An overview of synthesis, properties, functionalization, characterization, and the application," *Mater. Sci. Eng. B Solid-State Mater. Adv. Technol.*, vol. 268, p. 115095, 2021.
[28] F. Perreault, A. F. De Faria, and M. Elimelech, "Environmental applications of graphene-based nanomaterials," *Chem. Soc. Rev.*, vol. 44, no. 16, pp. 5861–5896, 2015.
[29] I. Fatimah, G. Fadillah, I. Yanti, and R. A. Doong, "Clay-supported metal oxide nanoparticles in catalytic advanced oxidation processes: A review," *Nanomaterials*, vol. 12, no. 5, p. 825, 2022.
[30] X. Huang and M. A. El-Sayed, "Gold nanoparticles: Optical properties and implementations in cancer diagnosis and photothermal therapy," *J. Adv. Res.*, vol. 1, no. 1, pp. 13–28, 2010.
[31] P. Goel, P. Mandal, E. Bhuvanesh, V. K. Shahi, and S. Chattopadhyay, "Sulfonated poly (ether ether ketone) composite cation exchange membrane for NaOH production by electro-electrodialysis using agro-based paper mill green liquor," *J. Environ. Chem. Eng.*, vol. 9, no. 6, p. 106409, 2021.
[32] Q. Liu and G. R. Xu, "Graphene oxide (GO) as functional material in tailoring polyamide thin film composite (PA-TFC) reverse osmosis (RO) membranes," *Desalination*, vol. 394, pp. 162–175, 2016.

[33] O. Monticelli, Z. Musina, S. Russo, and S. Bals, "On the use of TEM in the characterization of nanocomposites," *Mater. Lett.*, vol. 61, no. 16, pp. 3446–3450, 2007.
[34] T. Li, X. Yan, J. Liu, X. Wu, X. Gong, D. Zhen, S. Sun, W. Chen, and G. He, "Friedel-Crafts alkylation route for preparation of pendent side chain imidazolium-functionalized polysulfone anion exchange membranes for fuel cells," *J. Membr. Sci.*, vol. 573, pp. 157–166, 2019.
[35] X. Tang, Y. You, H. Mao, K. Li, R. Wei, and X. Liu, "Improved energy storage density of composite films based on poly(arylene ether nitrile) and sulfonated poly(arylene ether nitrile) functionalized graphene," *Mater. Today Commun.*, vol. 17, pp. 355–361, 2018.
[36] I. Zucker, J. R. Werber, Z. S. Fishman, S. M. Hashmi, U. R. Gabinet, X. Lu, C. O. Osuji, L. D. Pfefferle, and M. Elimelech, "Loss of phospholipid membrane integrity induced by two-dimensional nanomaterials," *Environ. Sci. Technol. Lett.*, vol. 4, no. 10, pp. 404–409, 2017.
[37] C. Tian, G. Chu, Y. Feng, Y. Lu, C. Miao, N. Ning, L. Zhang, and M. Tian, "Quantitatively identify and understand the interphase of $SiO_2$/rubber nanocomposites by using nanomechanical mapping technique of AFM," *Compos. Sci. Technol.*, vol. 170, pp. 1–6, 2019.
[38] H. Yu, B. Zhang, C. Bulin, R. Li, and R. Xing, "High-efficient synthesis of graphene oxide based on improved hummers method," *Sci. Rep.*, vol. 6, pp. 1–7, 2016.
[39] X. Gao, H. Liu, D. Wang, and J. Zhang, "Graphdiyne: Synthesis, properties, and applications," *Chem. Soc. Rev.*, vol. 48, no. 3, pp. 908–936, 2019.
[40] P. Narayanaswamy Venkatesan and S. Dharmalingam, "Effect of cation transport of SPEEK: Rutile $TiO_2$ electrolyte on microbial fuel cell performance," *J. Membr. Sci.*, vol. 492, pp. 518–527, 2015.
[41] S. Pilotek and F. Tabellion, "Nanoparticles in coatings: Tailoring properties to applications," *Eur. Coat. J.*, vol. 49, no. 4, pp. 170–176, 2005.
[42] R. Bardestani, G. S. Patience, and S. Kaliaguine, "Experimental methods in chemical engineering: Specific surface area and pore size distribution measurements—BET, BJH, and DFT," *Can. J. Chem. Eng.*, vol. 97, no. 11, pp. 2781–2791, 2019.
[43] H. Zhu, Y. Ju, C. Huang, F. Chen, B. Chen, and K. Yu, "Microcosmic gas adsorption mechanism on clay-organic nanocomposites in a marine shale," *Energy*, vol. 197, p. 117256, 2020.
[44] R. Saravanan, N. Karthikeyan, V. K. Gupta, E. Thirumal, P. Thangadurai, V. Narayanan, and A. Stephen, "ZnO/Ag nanocomposite: An efficient catalyst for degradation studies of textile effluents under visible light," *Mater. Sci. Eng. C*, vol. 33, no. 4, pp. 2235–2244, 2013.
[45] Z. F. Li, H. Zhang, Q. Liu, L. Sun, L. Stanciu, and J. Xie, "Fabrication of high-surface-area graphene/polyaniline nanocomposites and their application in supercapacitors," *ACS Appl. Mater. Interfaces*, vol. 5, no. 7, pp. 2685–2691, 2013.
[46] C. N. Lunardi, A. J. Gomes, F. S. Rocha, J. De Tommaso, and G. S. Patience, "Experimental methods in chemical engineering: Zeta potential," *Can. J. Chem. Eng.*, vol. 99, no. 3, pp. 627–639, 2021.
[47] P. Leroy, C. Tournassat, and M. Bizi, "Influence of surface conductivity on the apparent zeta potential of $TiO_2$ nanoparticles," *J. Colloid Interface Sci.*, vol. 356, no. 2, pp. 442–453, 2011.
[48] M. D. Butterworth, R. Corradi, J. Johal, S. F. Lascelles, S. Maeda, and S. P. Armes, "Zeta potential measurements on conducting polymer-inorganic oxide nanocomposite particles," *J. Colloid Interface Sci.*, vol. 174, no. 2. pp. 510–517, 1995.
[49] P. Goel, E. Bhuvanesh, P. Mandal, V. K. Shahi, A. Bandyopadhyay, and S. Chattopadhyay, "Di-quaternized graphene oxide based multi-cationic cross-linked monovalent selective anion exchange membrane for electrodialysis," *Sep. Purif. Technol.*, vol. 276, p. 119361, 2021.
[50] S. Bhattacharjee, "DLS and zeta potential: What they are and what they are not?" *J. Control. Release*, vol. 235, pp. 337–351, 2016.
[51] T. Provder, "Challenges in particle size distribution measurement past, present and for the 21st century," *Prog. Org. Coat.*, vol. 32, nos. 1–4, pp. 143–153, 1997.
[52] H. Nolte, C. Schilde, and A. Kwade, "Determination of particle size distributions and the degree of dispersion in nanocomposites," *Compos. Sci. Technol.*, vol. 72, no. 9, pp. 948–958, 2012.
[53] B. Eswaraswamy, A. Suhag, P. Goel, P. Mandal, and S. Chattopadhyay, "Potential of montmorillonite nanoclay as water dissociation catalyst at the interface of bipolar membrane," *Sep. Purif. Technol.*, vol. 295, p. 121257, 2022.
[54] P. Goel, P. Mandal, E. Bhuvanesh, V. K. Shahi, and S. Chattopadhyay, "Temperature resistant cross-linked brominated poly phenylene oxide-functionalized graphene oxide nanocomposite anion exchange membrane for desalination," *Sep. Purif. Technol.*, vol. 255, p. 117730, 2021.

[55] S. Egodawatte, A. Datt, E. A. Burns, and S. C. Larsen, "Chemical insight into the adsorption of chromium(III) on iron oxide/mesoporous silica nanocomposites," *Langmuir*, vol. 31, no. 27, pp. 7553–7562, 2015.

[56] P. Goel, P. Mandal, E. Bhuvanesh, V. K. Shahi, and S. Chattopadhyay, "Surface smoothening and formation of nano-channels improved mono-selectivity and antifouling property in $TiO_2$ incorporated cation exchange membrane," *Chem. Eng. Res. Des.*, vol. 184, pp. 378–391, 2022.

[57] Y. Zhao, Y. Fu, B. Hu, and C. Lü, "Quaternized graphene oxide modified ionic cross-linked sulfonated polymer electrolyte composite proton exchange membranes with enhanced properties," *Solid State Ionics*, vol. 294, pp. 43–53, 2016.

[58] H. M. Saif, R. M. Huertas, S. Pawlowski, J. G. Crespo, and S. Velizarov, "Development of highly selective composite polymeric membranes for $Li^+/Mg^{2+}$ separation," *J. Membr. Sci.*, vol. 620, p. 118891, 2020.

[59] H. Khan, A. S. Yerramilli, A. D'Oliveira, T. L. Alford, D. C. Boffito, and G. S. Patience, "Experimental methods in chemical engineering: X-ray diffraction spectroscopy—XRD," *Can. J. Chem. Eng.*, vol. 98, no. 6, pp. 1255–1266, 2020.

[60] G. O. Noonan, A. J. Whelton, D. Carlander, and T. V. Duncan, "Measurement methods to evaluate engineered nanomaterial release from food contact materials," *Compr. Rev. Food Sci. Food Saf.*, vol. 13, no. 4, pp. 679–692, 2014.

[61] I. Abad-Alvaro, E. Bolea, F. Laborda, and J. R. Castillo, "An ICP-MS-based platform for release studies on silver-based nanomaterials," *J. Anal. At. Spectrom.*, vol. 32, no. 6, pp. 1101–1108, 2017.

[62] M. Azizi-Lalabadi, F. Garavand, and S. M. Jafari, "Incorporation of silver nanoparticles into active antimicrobial nanocomposites: Release behavior, analyzing techniques, applications and safety issues," *Adv. Colloid Interface Sci.*, vol. 293, p. 102440, 2021.

[63] X. Qian, N. Li, Q. Wang, and S. Ji, "Chitosan/graphene oxide mixed matrix membrane with enhanced water permeability for high-salinity water desalination by pervaporation," *Desalination*, vol. 438, pp. 83–96, 2018.

[64] P. Chitra, A. Muthusamy, R. Jayaprakash, and E. Ranjith Kumar, "Effect of ultrasonication on particle size and magnetic properties of polyaniline $NiCoFe_2O_4$ nanocomposites," *J. Magn. Magn. Mater.*, vol. 366, pp. 55–63, 2014.

[65] M. Škrátek, A. Dvurečenskij, M. Kluknavský, A. Barta, P. Bališ, A. Mičurová, A. Cigáň, A. Eckstein-Andicsová, J. Maňka, and I. Bernátová, "Sensitive SQUID bio-magnetometry for determination and differentiation of biogenic iron and iron oxide nanoparticles in the biological samples," *Nanomaterials*, vol. 10, no. 10, pp. 1–20, 2020.

# 14 Computational Modeling of Nanocomposites and Interfacial Adhesion

*Tushar Kanti Maiti, Manjinder Singh, Subrata Kumar Maiti, Sakchi Bhushan, Sujay Chattopadhyay, and Gaurav Manik*

## 14.1 INTRODUCTION

A *nanocomposite* is a material comprising generally two phases: reinforcement (i.e., nanofillers) and matrix, where the nanoscale reinforcement phase, typically ranging from 1 to 100 nm, is dispersed into the matrix phase. The matrix functions as a protector, binding agent for the nanofillers, and transmitter of external forces to the nanomaterial. Polymer nanocomposites are widely used in the automotive, coating, rubber, adhesive, and construction industries due to their excellent mechanical properties, light weight, suitable thermal properties, and better chemical resistance [1–3]. Various nanofillers, like carbon nanotubes (CNTs), nanoclays, titanium dioxide nanopowder, zinc oxide nanoparticles, graphene, silica nanoparticles, and carbon black, have displayed promise in advanced polymer nanocomposites. This is attributed to their elevated surface area and superior mechanical, electrical, optical, and thermal properties. These nanoparticles have substantial potential for altering material properties, offering opportunities for a diverse range of applications. Therefore, several new materials have been created by incorporating graphene, CNTs, GOs, carbon black, and metal oxide nanoparticles as reinforcing agents within the polymer matrix. The performance of polymer nanocomposites depends upon a lot of factors, such as the dispersion of nanoparticles; the loading of nanoparticles into polymer matrix; the structure of the nanoparticles; the types, length, purity, and aspect ratio of nanoparticles; the alignment of nanoparticles; and the interaction of nanoparticles with the considered polymer matrix [4–7].

Over the past decade, polymer nanocomposites have received substantial industrial applications, such as in automotive, packaging, construction, coating, rubber, and pigments. The progress of these innovative materials depends primarily on empirical approaches, which are costly, time-consuming, and complicated processes that significantly optimize formulation and process. On the contrary, computer modeling and simulation offer a valuable means to tackle the challenges [6–9]. Developing cutting-edge polymer nanocomposites requires in-depth knowledge of fundamental principles in chemistry, polymer physics, surface science, and nanoscience. Employing computational modeling furnishes essential insights for their development and enriches the understanding of the pivotal influence of interfacial adhesion on their properties. The goal of computational materials science is to rapidly and precisely predict the properties of new materials, overcoming challenges posed by the limitations of traditional empirical methods [10–12]. As a result, several types of computational modeling and simulation strategies have been adopted to resolve the complex issues of polymer nanocomposites. This chapter presents various types of computational modeling and simulation techniques for polymer nanocomposites, along with studies on interfacial adhesion.

 DOI: 10.1201/9781003476337-14

## 14.2 VARIOUS APPROACHES OF COMPUTATIONAL MODELING AND SIMULATION METHODS OF POLYMER NANOCOMPOSITES

### 14.2.1 Molecular Dynamics Simulation of Polymer Nanocomposites

Polymer nanocomposites represent rapidly growing industrial and research areas that consist of polymer matrices incorporating organic or inorganic fillers, characterized by at least one dimension at the nanoscale, typically considered to be up to 100 nm. The exponential growth of nanoscience and nanotechnology helps develop many potential nanomaterials that can be used to prepare polymer nanocomposites for high-performance applications. Many experimental and computational techniques have been developed to explore the interaction between nanoscale materials and polymer matrices. Molecular dynamics (MD) simulation is a computational technique that can determine the properties of materials. Molecular mechanics has successfully determined polymer composites' properties, such as elastic modulus, elongation at break, density, tensile strength, etc. [12–14]. MD simulations explicitly consider the particulate nature of matter, representing each constituent atom as individual point-like sites, engaging in pairwise unbound interactions and bound interactions [15]. To reduce the experimental burden, an MD simulation provides useful information on composite materials, including mechanical, thermal, and electrical properties. In the last decade, MD simulations have been used to investigate the complicated properties of polymer nanocomposites, such as crystallization temperature, glass transition temperature, cyclic loading response, creep response, and stress–strain behavior, which help develop novel composite materials [16–18]. Monte Carlo (MC) and MD simulations are the two major simulation types under molecular mechanics. Among these two simulation techniques, MD is more widely used than the MC simulation technique to determine material properties. In MD simulations, particles in a system obey Newton's second law of motion, and the position of the atoms or particles follows as a function of time. For carrying out MD simulation, the atomic system must be defined by boundary conditions, thermodynamic constraints, specific forms of interparticle interactions, and proper initial state. The applications of MD simulation are diverse; it is used in materials science to determine mechanical, thermal, and rheological properties of polymers; in biochemistry to assess the performance of drugs; in civil engineering to determine the properties of various materials used for construction; in nanotechnology to determine the surface properties of nanomaterials, etc. In MD simulations, the behavior of particles in a system follows Newton's second law of motion, where the positions of atoms/particles change over time. Establishing well-defined boundary conditions that ensure a suitable initial state for the atomic system is essential. In MD simulations, the movement of particles is monitored over time through the numerical integration of Newton's equations of motion. The energy of an atomistic system is articulated as a sequence of interactions among the particles. The determination of forces acting on each particle is achieved by calculating the derivative of the system energy concerning the position vector of a particle pair. In its basic form, MD tackles problems characterized by a constant number of volumes, atoms, and energy, denoted as NVE [15, 19–21]. Various methods have been devised to broaden the scope of MD simulations to enable pressure and temperature control within the system, such as NPT (pressure, constant number of atoms, and temperature) and NVT (constant number of volumes, atoms, and temperature). Understanding how materials behave on the nanoscale allows for the analysis of experimental data and opens avenues for developing novel designs for nanocomposites. For instance, consider thermosetting polymers, such as epoxy resins, which belong to a class of polymers characterized by their irreversible curing process. This transformation changes the material from a liquid-soluble prepolymer to a solid form. The resultant physical characteristics of the extensively interconnected network are predominantly influenced by the molecular structure of the underlying network, a feature strongly affected by the constituent monomers and the polymerization reaction. As there is a diverse range of constituent monomers that allow for numerous combinations, employing computer simulations becomes beneficial in reducing both costs and time when investigating new polymers. MD simulations provide an efficient approach to simulate the curing process and predict the structural evolution of polymers. A recent study explored diverse mine curing agents and epoxy resin combinations through MD simulations and experimental techniques [22]. The results of molecular modeling exhibited satisfactory agreement with experimental findings, particularly in terms of the general trends observed in the curing curves and the reactivity of the epoxy groups. Sharp and his team [23] investigated how water affects the cross-linking process of epoxy resins through MD simulations. The findings are consistent with experimental data, suggesting that the introduction of water boosts molecular diffusion, leading to a faster curing rate for less-extensive cross-linking.

The COMPASS (condensed-phase optimized molecular potentials for atomistic simulation studies) force field was used to determine the potential energy of the several proposed nanocomposite systems, which several researchers have extensively used for the estimation of the interfacial and bulk properties of the composite materials [11, 24, 25]. The non-bonded electrostatic and van der Waals interactions were calculated using Ewald summation and atom-based methods. A cut-off distance of 15 Å was considered to calculate the electrostatic and van der Waals interactions [26]. The quantum mechanical techniques can precisely determine the behavior of the subatomic particles. However, it is challenging to determine the bulk properties of a system by using quantum mechanical techniques because it involves the interaction with a large number of particles. The MD simulation is a suitable technique to determine the bulk properties of a system that considers the interaction between many particles [27]. In MD simulation, the properties of a system are determined based on the location of atomic nuclei within a system. The determination of the potential (force field) energy of a system is essential, helping determine the location and the set confirmation of the system's particles. The choice of force fields is necessary for multiparticle methods to assess a system's properties. Therefore, a brief overview of a molecular system's potential energy is given in this section, which includes the following term [27].

$$E_{total} = E_{valence} + E_{non-bond} + E_{cross-term} \tag{14.1}$$

$$E_{valence} = E_{bond} + E_{angle} + E_{torsion} + E_{opp} + E_{UB} \tag{14.2}$$

$$E_{non-bond} = E_{vdW} + E_{Columb} + E_{H-bond} \tag{14.3}$$

$$\begin{aligned} E_{Cross-term} = {} & E_{bond-bond} + E_{angle-angle} + E_{bond-angle} + E_{end-bond-torsion} \\ & + E_{middle-bond-torsion} + E_{angle-torsion} + E_{angle-angle-torsion} \end{aligned} \tag{14.4}$$

In general, the valence energy considers the following terms: a bond stretching term, $E_{bond}$; a two-bond angle term, $E_{angle}$; a dihedral bond–torsion term, $E_{torsion}$; an out-of-plane interaction term, $E_{oop}$; and a Urey–Bradlay term (involves interactions between two atoms bonded to a common atom), $E_{UB}$. The non-bonded interaction term, $E_{non\text{-}bond}$, considers the van der Waals energy, $E_{vdW}$; the hydrogen bond energy, $E_{H\text{-}bond}$; and the Coulomb's electrostatic energy, $E_{Coulomb}$.

The cross-term interacting energy, $E_{cross\text{-}term}$, considers the following interaction energy term: stretch–stretch interactions between two adjacent bonds, $E_{bond\text{-}bond}$; stretch–bend interactions between a two-bond angle and one of its bonds, $E_{bond\text{-}angle}$; bend–bend interactions between two valence angles associated with a common vertex atom, $E_{angle\text{-}angle}$; stretch–torsion interactions between a dihedral angle and one of its end bonds, $E_{end\text{-}bond\text{-}torsion}$; bend–bend–torsion interactions between a dihedral angle and its two valence angles, $E_{angle\text{-}angle\text{-}torsion}$; bend–torsion interactions between a dihedral angle and one of its valence angles, $E_{angle\text{-}torsion}$; and stretch–torsion interactions between a dihedral angle and its middle bond, $E_{middle\text{-}bond\text{-}torsion}$.

The COMPASS force field was used in several research studies to calculate and determine the potential energy of the proposed simulated system, which several researchers have extensively used to estimate the composite materials' interfacial and bulk properties. The COMPASS force field uses the following expressions for the various components of the potential energies [27–29].

$$E_{bond} = \sum_{b} K_2 (b - b_0)^2 + K_3 (b - b_0)^3 + K_4 (b - b_0)^4 \quad (14.5)$$

$$E_{angle} = \sum_{\theta} H_2 (\theta - \theta_0)^2 + H_3 (\theta - \theta_0)^3 + H_4 (\theta - \theta_0)^4 \quad (14.6)$$

$$E_{torsion} = \sum_{\varnothing} \left\{V_1 \left[1 - cos\ cos\left(\varnothing - \varnothing_1^0\right)\right]\right\} + \left\{V_2 \left[1 - cos\ cos\left(\varnothing - \varnothing_2^0\right)\right]\right\} + \left\{V_3 \left[1 - cos\ cos\left(\varnothing - \varnothing_3^0\right)\right]\right\} \quad (14.7)$$

$$E_{opp} = \sum_{x} K_x^{\ 2} \quad (14.8)$$

$$E_{Coulomb} = \sum_{i>j} \frac{q_i q_j}{\in r_{ij}} \quad (14.9)$$

$$E_{vdW} = \sum_{i>j} \left[\frac{A_{ij}}{r_{ij}^9} - \frac{B_{ij}}{r_{ij}^6}\right] \quad (14.10)$$

$$E_{bond-bond} = \sum_{b} \sum_{b'} F_{bb'} (b - b_0)(b' - b_0') \quad (14.11)$$

$$E_{angle-angle} = \sum_{b} \sum_{b'} F_{\theta\theta'} (\theta - \theta_0)(\theta' - \theta_0') \quad (14.12)$$

$$E_{bond-angle} = \sum_{b} \sum_{\theta} F_{b\theta} (b - b_0)(\theta - \theta_0) \quad (14.13)$$

$$E_{end-bond-torsion} = \sum_{b} \sum_{\theta} F_{b\theta} (b - b_0)[V_1 cos\varnothing + V_2 cos2\varnothing + V_3 cos3\varnothing] \quad (14.14)$$

$$E_{middle-bond-torsion} = \sum_{b'} \sum_{\theta} F_{b'\theta} (b' - b_0')(b' - b_0')[F_1 cos\varnothing + F_2 cos2\varnothing + F_3 cos3\varnothing] \quad (14.15)$$

$$E_{angle-angle-torsion} = \sum_{\phi} \sum_{\theta} \sum_{\theta} K_{\phi\theta\theta'} cos\varnothing (\theta - \theta_0)(\theta' - \theta_0') \quad (14.16)$$

$$E_{angle-torsion} = \sum_{\theta} \sum_{\varnothing} F_{\theta\varnothing} (\theta - \theta_0)[V_1 cos\varnothing + V_2 cos2\varnothing + V_3 cos3\varnothing] \quad (14.17)$$

Where $b$ and $b'$ are the lengths of two adjacent bonds, $\varphi$ is the dihedral torsion angle, $\theta$ is the two-bond angle, $\phi$ is the out-of-plane angle, $q$ is the atomic charge, and $\in$ is the dielectric constant. $A_{ij}$, $B_{ij}$, $K_{\varphi\theta\theta'}$, $F_{\theta\varphi}$, $b_0$, $k_i(i = 2 - 4)$, $\theta_0$, $H_i(i = 2 - 4)$, $\varphi_i^0(i = 1 - 3)$, $V_i(i = 1 - 3)$, $F_{bb'}$, $b_{0'}$, $F_{\theta\theta'}$, $\theta_{0'}$, $F_{\theta\theta'}$, $\theta_{0'}$, $F_{b\theta}$, $F_{b\varphi}$, $F_{b'\theta}$, $F_i(i = 1 - 3)$ are fitted from quantum mechanics calculations and are implemented into the Discover module of Materials Studio [27], a powerful commercial atomic simulation program used in this chapter.

### 14.2.2 Coarse-Grained Modeling of Polymer Nanocomposites

Another popular computer modeling and simulation technique is the coarse-grained modeling of polymer nanocomposites, where basic features of systems are captured at a computational cost much lower than that of atomistic simulations. Representing several atoms by a single interaction site/bead in coarse-grained models decreases the number of degrees of freedom in the system, which allows for faster computations. CG models can be categorized into two primary groups: (1) generic models, intended for qualitative assessment of properties, and (2) specific models, designed to predict properties quantitatively. The widespread use of molecular simulations in nanocomposite modeling is hindered by significant computational requirements, limiting their applicability to small molecular systems and constrained timescales; additionally, there is a notable tendency for nanoparticles to cluster within a polymer matrix [20]. When boehmite nanoparticles are added to an epoxy matrix, they often form clusters (100–200 nm), and single particles are not very common. Given that current and future computer capabilities effectively allow simulations for systems only two orders of magnitude smaller, it is essential to include several tens of filler particles for a statistically reliable representation, demanding an unfeasible number of atoms—around 100 million—to accurately simulate a representative volume element of a nanoparticle/polymer system [20, 30, 31]. These methods facilitate the exploration of higher-length scales by simplifying the intricate details at the microscopic level in the underlying atomistic system. Coarse-grained models have been systematically developed for a variety of polymer nanocomposites. These models entail replacing a small number of atoms with a single bead, which allows the establishment of cohesive, effective interaction potentials among these coarse-grained beads. Within this context, a particular category of coarse-grained force fields has been employed to depict polymer chains through a bead-spring model [32, 33]. In an alternative class of coarse-grained force fields, more intricate information about the precise chemical composition of the effective bead is integrated, which facilitates an in-depth examination of the influences arising from monomer–solvent, monomer–monomer, and monomer–particle interactions [34, 35]. Extensive research in this field aims to comprehend how polymer coarse-graining influences the rheological, structural, and thermal properties across various bulk polymers, including polystyrene, polyethylene, poly(methyl methacrylate) (PMMA), polyisoprene, sulfonated polyetherketone, poly(3-hexylthiophene), polyvinyl chloride, and aromatic polyimides. Coarse-grained models for polymer nanocomposites are performed to precisely determine quantitative features of dynamic, structural, mechanical, and thermodynamic properties.

### 14.2.3 Multiscale Modeling in Polymer Nanocomposites

Computational materials science aims to rapidly and precisely predict novel materials and their properties, facing challenges posed by the limitations of traditional modeling and simulation methods, which are constrained to a single length and timescale [36, 37]. Therefore, there is an expectation of employing multiscale simulation strategies to seamlessly integrate models and simulation techniques across various lengths and timescales, which aim to reveal detailed molecular descriptions of macroscopic or mesoscopic material behaviors. Multiscale modeling in polymer nanocomposites is a computational approach essential for simulating and comprehending phenomena across diverse length scales, ranging from macroscopic to microscopic levels, as it addresses the complex interplay of polymers and nanoscale fillers, shaping the overall material behavior [37, 38]. Successfully navigating multiscale simulations requires smooth transitions between various scales, ensuring efficient transfer of computed parameters. In the case of dispersing a particle in a polymer matrix to form polymer nanocomposites containing 5% clay by weight, the resulting system will consist of around 3 billion atoms. Consequently, it becomes impractical for classical molecular dynamics and significantly exceeds the computational limits of quantum mechanics. Furthermore, the observable characteristics of materials are influenced by a hierarchical structure encompassing multiple levels, ranging from atomic-level chemical details and individual chains to microscopic features involving chain aggregates and clusters of clay platelets, extending up to macroscopic continuum phenomena. This emphasizes the importance of providing a comprehensive description of a polymer nanocomposite that considers a wide range of length scales, from the atomic scale to chain aggregates [39, 40].

## 14.3 NANOFILLERS USED IN POLYMER NANOCOMPOSITES

Different kinds of nanomaterials, including graphene, graphene oxide (GO), silica nanoparticles, carbon nanotubes, nanoclays, and titanium dioxide nanoparticles, are extensively employed to enhance the properties of polymer composite materials due to their excellent mechanical properties, reliable thermal stability, Young's modulus, and exceptional electrical characteristics. Over the past decade, numerous researchers have dedicated efforts to enhance the crucial factor of the interaction force between nanofillers and the polymer matrix, whose objective is to improve diverse properties in polymer composite materials [41–43]. In recent years, cellulose has garnered significant interest in creating advanced nanocomposite materials that are both lightweight and high-performing. Interest in developing cellulose-based advanced materials arises due to their renewability, cost-effectiveness, and compatibility with living organisms. The most common examples of cellulose-based nanomaterials are cellulose nanofibrils and cellulose nanocrystals. Nanocellulose demonstrates remarkable mechanical and thermal properties and low coefficient of thermal expansion due to the elaborate hydrogen bonding network between closely packed linear cellulose chains in the crystals [44, 45]. Active computational efforts are underway to develop surface chemistries for nanoparticles, to measure surface energy and dispersion properties, to inform the design of nanocomposites. Gaining insights into nanoscale interfaces is crucial for improving polymer nanocomposites' thermal, mechanical, and chemical characteristics reinforced with nanocellulose. A recent study utilized MD simulations to investigate the interfacial bonding between PMMA and nanocellulose [46]. The higher adhesion is primarily created from more hydrogen bonds between the polymer chains and cellulose. The higher adhesion energy between fillers in the amorphous cellulose contributes to notably better dispersibility of the nanofiller within polymer matrices when contrasted with cellulose nanocrystals. In Figure 14.1a, a depiction of a standard polymer nanocomposite reveals the integration of cellulose-based nanofillers into the polymer matrix. On the other hand, Figure 14.1b presents atomistic simulation snapshots illustrating a PMMA top layer supported by amorphous cellulose and cellulose nanocrystals. Figure 14.1c presents illustrations of cellulose–polymer adhesion energy evaluations, Figure 14.1d presents interfacial traction force between cellulose and PMMA, Figure 14.1e presents the slopes of the dashed lines indicating the stiffness of the interfaces, and Figure 14.1f is an illustration depicting the assessment of cellulose–polymer adhesion energy. The research provided valuable insights into the molecular-level interfacial processes occurring in nanocellulose/polymer interfaces. These findings suggest that the intentionally induced surface disorder during production can potentially improve interfacial adhesion.

Several studies suggested that achieving uniform dispersion of SWCNTs/MWCNTs/graphene is crucial for minimizing stress concentration and ensuring effective load transfer between the polymer matrix and the nanofiller. Investigations have revealed that inadequate dispersion, particularly at higher CNT content, results in lower hardness, yield strength, and indentation modulus [47]. Hence, ongoing efforts have been directed toward enhancing the dispersion of CNTs, employing methods such as electrostatic treatment, physical treatment, and surfactants. A reasonably uniform dispersion can result in a noticeable increase in Young's modulus for the CNT/polypropylene composite. While low-dimensional nanostructures are widely used as reinforcements in polymer nanocomposites, most studies have concentrated on samples containing a single nanofiller or perfectly aligned nanofillers. In a recent study, molecular dynamics simulations are utilized to methodically

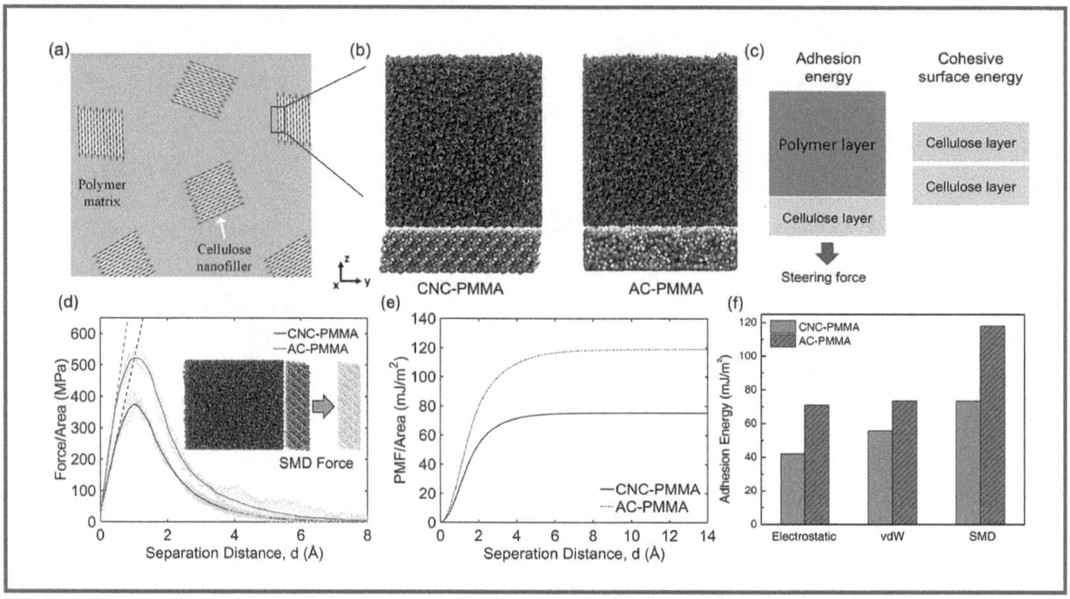


**FIGURE 14.1** (a) Illustration of a typical polymer nanocomposite containing cellulose-based nanofillers. (b) Atomistic simulation snapshots displaying a PMMA/cellulose nanocrystal and PMMA/amorphous cellulose. (c) Illustrations of cellulose–polymer adhesion energy evaluations. (d) The interfacial traction force between cellulose and PMMA. (e) Slopes of the dashed lines indicating the stiffness of the interfaces. (f) Illustration of the assessment of cellulose–polymer adhesion energy.

**Source: Reproduced from ref. [46].**

assess the tensile properties of PMMA nanocomposites that incorporate randomly distributed carbon nanothreads as a one-dimensional nanofiller [47]. Figure 14.2 illustrates a schematic representation of the PMMA nanocomposite: (a) the structure of the original achiral carbon nanothreads and PMMA, and (b, c) the arrangement of achiral carbon nanothreads fillers within the PMMA matrix. The results indicate that the presence of carbon nanothreads has an adverse effect on tensile properties by creating continuous interphases with lower density around the filler. However, functionalization can mitigate these negative impacts and lead to positive improvements, especially through cross-linking at the carbon nanothreads/PMMA interface.

The aspect ratio of the nanofillers used in polymer nanocomposites is a vital factor that can greatly influence the properties of the composite material. A broad spectrum of aspect ratios is produced when working with nanoscale fillers/particles, such as carbon nanotubes and graphene nanoplatelets. Findings from several studies demonstrate an improvement in the engineering properties of composites as the aspect ratios of fillers increase. Unfortunately, due to impractical experimental methodologies for investigating the impact of the aspect ratio of nanoscale fillers, there is a pressing need to thoroughly understand the consequences of altering the aspect ratio on the properties of the polymer matrix [48–50]. Although various experimental characterization tools provide insights into the structure and properties of nanocomposites, some information is difficult to obtain through these methods. Challenges arise with aspects of understanding interphase region characteristics, intermolecular interaction energy, and molecular-level mechanical failure behavior, which exceed the capabilities of current experimental tools. Consequently, incorporating computational simulation techniques has become essential to achieve a more thorough understanding of material structure and behavior at the atomic scale [48, 49, 51, 52].

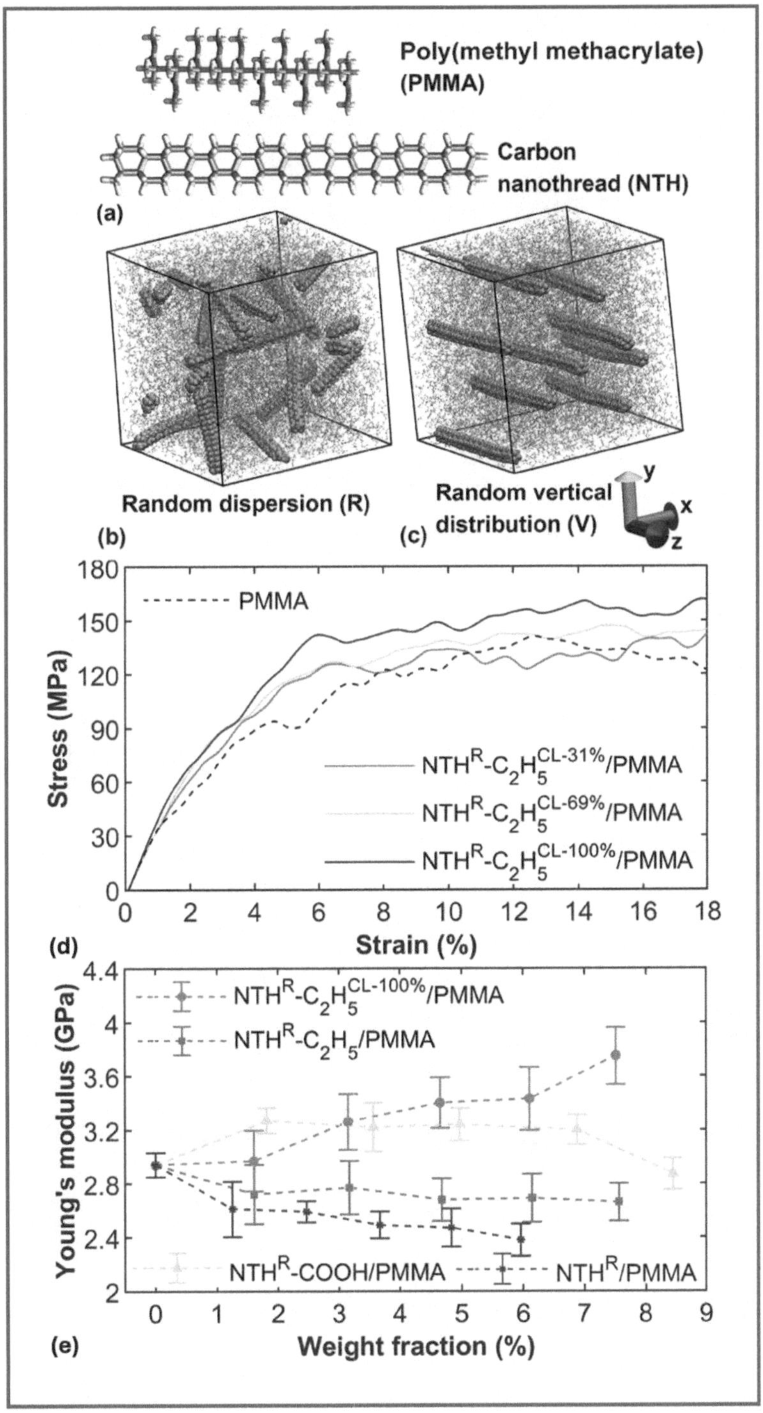


**FIGURE 14.2** (a) Schematic illustration of the PMMA, carbon nanothreads (NTH). (b) Random dispersion NTH in the composite in a simulation box. (c) Vertically aligned distribution NTH in composite in a simulation box. (d) Stress–strain curves of PMMA nanocomposites incorporating randomly dispersed NTHs with cross-linking. (e) Young's modulus of the PMMA nanocomposite with varying weight fractions of NTH fillers.

**Source: Reproduced from ref. [47].**

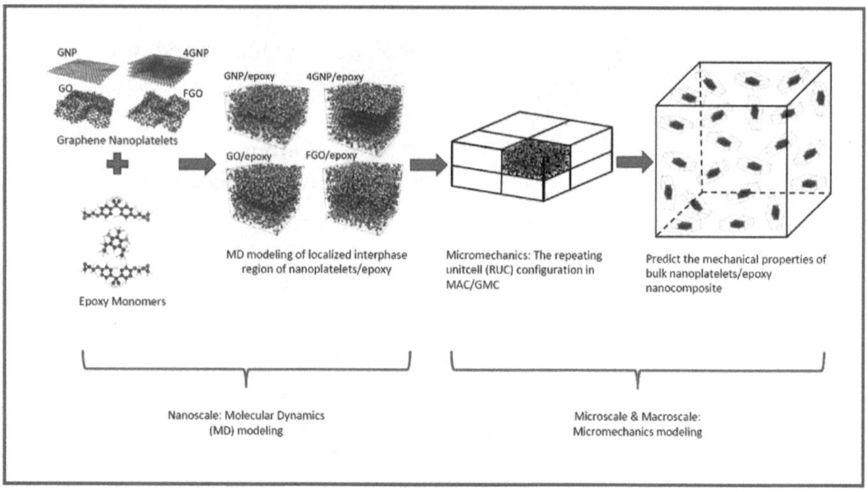


**FIGURE 14.3** Illustration of a multiscale modeling approach for epoxy-based nanocomposites reinforced by both nonfunctionalized and functionalized reinforcements [49].

**Source: Reproduced from ref. [49].**

MD simulation techniques are recognized as a potent computational technique for anticipating characteristics like intermolecular interaction energy and the molecular-level mechanical failure behavior across diverse materials. In computational investigations, MD modeling and simulations have been employed to forecast the mechanical characteristics of epoxy polymers containing randomly dispersed graphene nanoplatelets. Various considerations, such as the levels of dispersion, the content of graphene nanoplatelets, and modifications to the nanoplatelet surface, can be considered. Incorporating a variety of functional groups onto the surface of graphene nanoplatelets provides possibilities for improving the engineering properties of the composite material. The modified graphene nanoplatelets demonstrate improved dispersion levels and efficient interaction with the polymer matrix [52–54]. A recent study has explored the impact on the mechanical properties of an epoxy resin reinforced with pristine graphene nanoplatelets, functionalized graphene oxide (FGO), and GO-utilizing MD employing a reactive force field [49], as illustrated in Figure 14.3. A computational simulation strategy was implemented to determine the impact on the mechanical properties of the epoxy matrix. The modeling outcomes reveal a substantial reduction in the nanocomposite's elastic and shear modulus when introducing significant quantities of oxygen functional groups to the graphene nanoplatelets. Nevertheless, the wrinkled structure of GO and FGO enhances the interlocking mechanism within the polymer matrix, which leads to a substantial enhancement in the shear modulus.

## 14.4 CNT-BASED POLYMER NANOCOMPOSITES

CNTs are considered suitable materials for enhancing the performance of polymer composite materials because of their impressive characteristics, such as a higher surface area, excellent mechanical characteristics, chemical durability, and excellent electrical conductivity. Carbon nanomaterials such as SWCNTs, MWCNTs, and graphene have high conductivity; large surface areas, aspect ratios, and electrochemical stability; and low mass densities, making them attractive for use in high-performance polymer composites [55]. Notably, synthesis processes may be fine-tuned to optimize these physical features, making carbon nanomaterials excellent for composite applications. The huge surface area and higher mechanical properties of the carbon nanomaterials, combined with outstanding chemical stability, allow for the development of high-performance polymer composites, which is an interesting application of these materials [56–58]. CNT-based polymer composites have shown substantially higher specific strength than conventional polymer composite materials.

The significance of the interaction between the polymer matrix and carbon nanotubes in nanocomposites is well-known to improve mechanical characteristics [59]. To fully harness the impressive strength of these nanocomposites, establishing a strong interfacial connection becomes crucial for effective load transfer from the polymer matrix to the CNTs. The fracture surface morphology in CNT–epoxy composites was examined by the fracture test, where researchers observed unbroken CNTs, which indicated insufficient interfacial binding for effective load transfer [59]. A clear approach to enhancing this binding involves chemical functionalization, which is achieved by attaching side chains to the CNTs. A straightforward approach to enhance interfacial binding involves chemically functionalizing carbon nanotubes.

MD simulations were applied to investigate the molecular interactions and distribution of loads during the extraction process of SWCNT and nanotube ropes [60]. The study involved pulling out a single nanotube from the epoxy matrix, as illustrated in Figure 14.4. The interfacial shear strength between the nanotube and the epoxy resin was determined to be as high as 75 MPa, which indicates effective stress transfer from the epoxy resin to the nanotube. The simulation findings indicated that individual nanotubes exhibit stronger interactions with epoxy resins, leading to superior load transfer compared to nanotube ropes.

Carbon nanotube (CNT)–asphalt nanocomposites face a challenge in the comprehension of the interaction between asphalt and CNT at the interface. Further, the mechanisms responsible for property enhancement remain incomplete. The widespread application of CNT–asphalt nanocomposites is due to a lack of understanding, which has led to the utilization of CNT as a representative in a research study exploring one-dimensional nanomaterials. A recent effort established a theoretical foundation for the groundwork for the application of CNT–asphalt nanocomposites in highway settings. A distinctive face-to-face stacking is evident between CNT and aromatic, where the benzene rings of aromatic align in parallel with those of the CNT at a spacing of 3.816 Å (as illustrated in Figure 14.5).

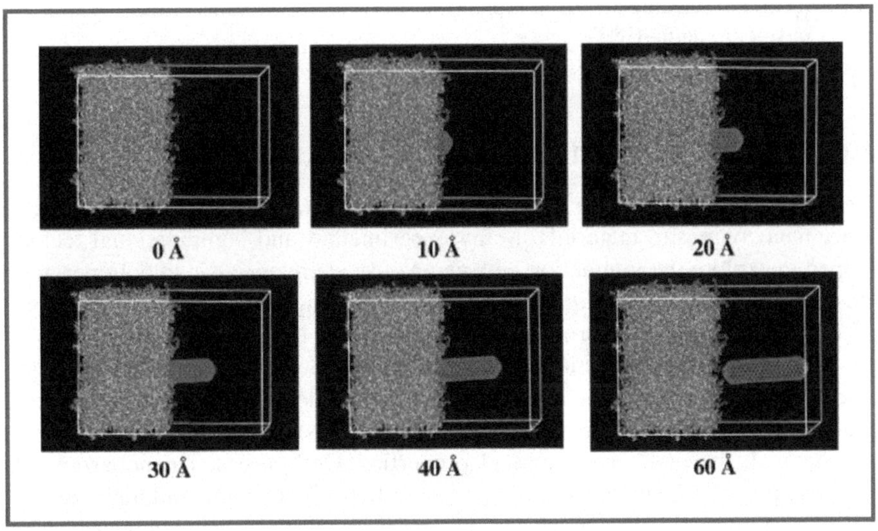


**FIGURE 14.4** Snapshots of the pullout simulations of AWCNT/polymer nanocomposite systems.

Source: [60].

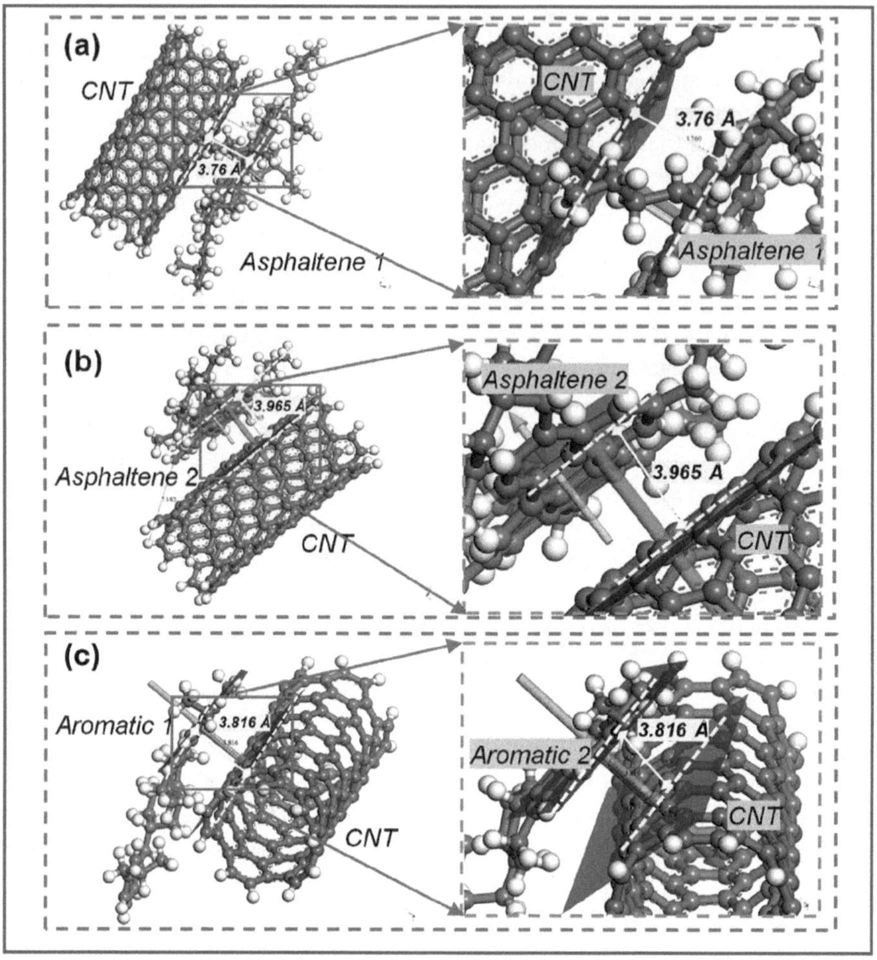


**FIGURE 14.5** (a, b) π–π stacking details of asphaltene–CNT. (c) Illustrations of the π–π stacking morphology of aromatic–CNT.

**Source:** [44].

## 14.5 INTERFACIAL PROPERTIES OF POLYMER NANOCOMPOSITES

The rapid expansion of nanoscience contributes significantly to the creation of numerous promising polymer nanocomposite materials. Many experimental and computational techniques have been developed to explore the interaction between nanoscale materials and polymer matrices. MD simulation is a computational technique that deals with the length scale of a few angstroms to nanometers to determine the materials' properties. MD simulation has successfully determined polymer composites' properties, such as elastic modulus, elongation at break, density, tensile strength, etc. Carbon-based nanoparticles, various metal oxide nanoparticles, and clay nanoparticles are extensively employed in the formulation of polymer nanocomposites due to their excellent ability to improve mechanical, thermal, and chemical properties. Carbon-based nanoparticles are highly used to formulate polymer nanocomposites because of their light weight and high specific strength. A lot of research on improving the interfacial interaction of carbon-based nanoparticles with different polymer matrixes has been carried out for high-performance applications. There is a growing demand for high-performance polymer nanocomposites, driven by their high specific strength and light weight, with the interfacial properties between the polymer matrix and nanofiller playing a pivotal role in shaping the mechanical properties and determining the efficiency of load transfer within these materials. Table 14.1 shows the interfacial strength of various polymer nanocomposites reported in the literature. Improving the interfacial interaction between the polymer and nanofiller can be achieved through both covalent and noncovalent bonding. The established method for evaluating the mechanical characteristics of the polymer nanocomposite interface is through pullout tests, providing a direct means of assessing interfacial strength, which is crucial for ensuring effective load transfer.

**TABLE 14.1**
**Some Experimental and Theoretical Results Obtained for the Shear Strength of the Interface/Interfacial Strength Reported in the Open Literature and Present Work**

| Authors | Filler Type | Interfacial Strength (MPa) | Interlinker Type | Matrix Type | Analysis Type |
|---|---|---|---|---|---|
| Barber et al. [61] | MWCNT | 47 MPa | – | Polyethylene-butene | Experimental AFM |
| Barber et al. [62] | MWCNT | 23–37 MPa | – | Epoxy | Experimental AFM |
| Manoharan et al. [63] | Carbon nanofiber | 170 MPa | – | Epoxy | Experimental AFM |
| Tsuda et al. [64] | MWCNT | 3.5–14 MPa | – | PEEK | Experimental AFM |
| Siddiqui et al. [65] | Carbon fiber | 59.4 ± 4.1 MPa | | Epoxy | Experimental |
| Rahman et al. [66] | GNP | 88 | – | Epoxy | MD |
| Zhang et al. [67] | SWCNT | 46.48 | – | Polyethylene | MD |
| Zhang et al. [67] | SWCNT | 100.85 | $(CH_2)_{10}N_2H_4$ | Polyethylene | MD |
| Zhang et al. [67] | SWCNT | 63.45 | PDA | Polyethylene | MD |
| Zhang et al. [67] | SWCNT | 90.26 | $(CH_2)_3N_2H_4$ | Polyethylene | MD |
| Chandra et al. [68] | SWCNT | 88.42 | – | Epoxy | Atomistic |
| Chandra et al. [68] | SWCNT | 88.42 | PDA | Epoxy | Atomistic |
| Chandra et al. [68] | SWCNT | 88.42 | $(CH_2)_3N_2H_4$ | Epoxy | Atomistic |
| Chandra et al. [68] | SWCNT | 98.13 | $(CH_2)_{10}N_2H_4$ | Epoxy | Atomistic |
| Wagner et al. [69] | MWCNT | 500.00 | – | Urethane-based | Experimental |
| Gou et al. [70] | SWCNT | 75.00 | – | Epoxy | MD |
| Liao et al. [71] | SWCNT | 160 | – | Polystyrene | MD |
| Rahman et al. [66] | GNP | 88 | – | Epoxy | MD |
| Zhang et al. [67] | SWCNT | 46.48 | – | Polyethylene | MD |
| Zhang et al. [67] | SWCNT | 100.85 | $(CH_2)_{10}N_2H_4$ | Polyethylene | MD |
| Zhang et al. [67] | SWCNT | 63.45 | PDA | Polyethylene | MD |
| Zhang et al. [67] | SWCNT | 90.26 | $(CH_2)_3N_2H_4$ | Polyethylene | MD |
| Chandra et al. [68] | SWCNT | 88.42 | – | Epoxy | Atomistic |
| Chandra et al. [68] | SWCNT | 88.42 | PDA | Epoxy | Atomistic |
| Chandra et al. [68] | SWCNT | 88.42 | $(CH_2)_3N_2H_4$ | Epoxy | Atomistic |
| Chandra et al. [68] | SWCNT | 98.13 | $(CH_2)_{10}N_2H_4$ | Epoxy | Atomistic |
| Sager et al. [72] | MWCNT | 86.6 | – | T650 carbon–Epoxy | Kelly–Tyson model |
| Wagner et al. [69] | MWCNT | 500.00 | – | Urethane-based | Experimental |
| Gou et al. [70] | SWCNT | 75.00 | – | Epoxy | MD |
| Liao et al. [71] | SWCNT | 160 | – | Polystyrene | MD |

Molecular dynamics simulation studies have been broadly used for examining the interaction between nanofillers and polymer matrices. These simulations can support understanding of the possible interaction between nanotubes and polymer matrix and suggest possible methods to increase the interaction between nanofillers and polymer matrix. For example, using MD simulations, Zheng et al. [27] applied MD simulations to study the interaction between carbon nanotubes and various types of polymer matrix, including polyethylene, polyaniline, polystyrene, and polypropylene. Based on their research findings, the interaction between the carbon nanotubes and the polymer is strongly influenced by both the structure of the polymer's repeat unit and the radius of the carbon nanotubes. Foroutan et al. [59] performed MD simulations to understand the interaction between carbon nanotubes and conjugated polymers, including poly-(2,6-pyridinylenevinylene-co-2,5-dioctyloxy-p-phenylenevinylene) (PPyPV), polypyrrole (PPy), polythiophenes (PT), and poly(m-phenylenevinylene-co-2,5-dioctyloxyp-phenylenevinylene) (PmPV). Utilizing the intermolecular interaction energies derived from molecular dynamics simulations, they concluded that polythiophene (PT) and polypyrrole (PPy) show better interaction energy with nanotubes than poly-(2,6-pyridinylenevinylene-co-2,5-dioctyloxy-p-phenylenevinylene) (PPyPV), poly(m-phenylenevinylene-co-2,5-dioctyloxyp-phenylenevinylene) (PmPV). In addition, the results have shown that the aromatic rings of the conjugated polymers are positioned parallel to the nanotube surface. The mechanical properties of the polymer nanocomposites can be demonstrated by evaluating the interaction energy between the polymer molecules and the nanofillers. Generally, the interaction energy of polymer composite systems is assessed from the difference between the potential energy of the composite systems and the sum of the potential energy of the individual components from which the composite system is created. The interaction energy is estimated according to the following equation:

$$E_{interaction} = E_{composite} - \sum E_{individual} \tag{14.18}$$

Molecular dynamics (MD) simulation is a powerful technique that offers essential information to develop novel composite materials. The interaction energy between the polymer matrix and reinforcing filler is the critical parameter that indicates the extent of improving the polymer composite properties, including the polymer matrix's mechanical, electrical, and thermal properties. The MD simulation technique has been widely used to evaluate the various properties of CNTs-based polymer composites [27, 73, 74]. Xue et al. employed the MD simulation technique to determine the interaction energy between CNTs and various polymer matrices, including polyethylene [74], polyphenylacetylene [73], polystyrene, polypropylene, and polyaniline [27]. According to their research study, the interfacial property between SWCNTs strongly depends on their radius and chirality. In recent years, significant efforts have been made to improve the polymer matrix's mechanical properties by incorporating nanomaterials. Among the various nanomaterials, graphene (which is a two-dimensional layer structure of elemental carbon) has been considered a potential reinforcing agent for the polymer matrix because of its excellent mechanical, electrical, and chemical properties and good thermal stability [75–77]. In recent years, polymer nanocomposites based on CNTs have been the focus of research areas because of their outstanding properties, including high specific strength, light weight, and attractive thermal, optical, and electrical properties. After significant progress in the field of nanotechnology, several researchers in the field of polymer science have realized the possibility of CNTs as a reinforcing agent in polymer nanocomposites. The interfacial characteristics and the efficiency of the load transfer through the interface significantly influence the improvement of mechanical properties of CNTs-based polymer nanocomposites. Significant focus is placed on understanding the interfacial characteristics of polymer nanocomposites, as these interface characteristics significantly influence the effectiveness of load transfer. Although considerable advancement has been made in polymer nanocomposites using CNTs as reinforcing agents, further development in this field is hindered because of a lack of understanding of characteristics at the interface. The fundamental knowledge of interfacial interaction between CNTs and polymer matrix is essential to understanding the possibility of

CNTs for usage in polymer nanocomposites. Both direct measurement techniques and molecular dynamics simulations have been applied to measure the interfacial interactions of the polymer nanocomposites, which reveal that interactions at the interface play a crucial role in both the load transfer across the interface and the dispersion of nanofillers (e.g., CNTs/GOs/graphenes) in polymer matrices. Various approaches have been made to improve the interaction at the interface between CNTs and polymer matrixes. Grafting functional groups on the surface of CNTs is one of the effective approaches to improve the interactions at the interface. Generally, the interactions at the interface can be classified into two types: covalent and noncovalent interactions. The chemical structure of the polymer strongly influences the strength of interactions at the interface of CNT-based polymer nanocomposites. For example, the strength of interactions at the interface can be significantly enhanced by utilizing conjugated polymers holding aromatic groups. In the case of chemical functionalization, chemical bonds are formed at the interface between CNTs and polymer chains. The chemical functionalization is more effective in improving the properties of the formulated polymer nanocomposites. Covalent functionalization involves the formation of covalent bonds onto either the surface of the sidewalls or the ends of CNTs. CNTs can be modified by various functional groups, such as hydroxide, carboxylic acid, epoxy, and halogen-containing groups. The covalent functionalization can significantly enhance the interactions at the interface and consequently improve the properties of the polymer nanocomposites.

Among various fiber-reinforced polymer nanocomposites, carbon fiber–reinforced polymer composites show superior mechanical properties. The distinctive characteristics of carbon fiber–reinforced polymer nanocomposites have allowed their widespread application across diverse sectors, such as aerospace, wind turbines, gas storage, automotive, energy systems, etc. Various surface modification techniques, such as nanoparticle coatings, chemical grafting, sizing, and plasma treatment, have been applied to carbon fibers to enhance their interfacial adhesion with various polymer matrices [42, 78, 79]. In addition to several experimental efforts, several recent efforts have focused on predicting the interphase properties of surface-modified carbon fibers (CF) through MD simulations. The information obtained at the molecular level via MD simulations can provide a basis for optimally designing effective strategies for surface functionalization of carbon fibers in experimental settings. Molecular simulations were utilized to methodically construct a conceptual framework for investigating the interfacial shear behavior of carbon fiber/epoxy composites, which demonstrated that the interfacial shear strength (IFSS) of carbon fiber with surface-grafted molecules experienced a substantial increase, approximately 51- to 76-fold [79]. A fully atomistic molecular dynamics simulation thoroughly examined interfacial interactions between carbon fiber and a polypropylene matrix [78]. Among the simulated composites, the one reinforced with amine-grafted carbon fiber exhibited a substantial ~126% increase in interfacial shear strength compared to the composite with pristine carbon fiber. Figure 14.6a shows the molecular configuration of the base matrix incorporating polypropylene (PP)– and maleic anhydride–grafted PP (MA-g-PP) chains. Further, Figure 14.6b visualizes potential van der Waals interactions involving maleic anhydride–grafted polypropylene and CNT-coated 4 wt.% functionalized carbon fibers (CF). Additionally, the mechanical properties of the composites showed significant enhancement through surface functionalization and CNT coating. Therefore, incorporating functionalized carbon nanotubes is a logical approach to improving interfacial interactions.

The importance of the interaction between the polymer matrix and carbon nanotubes in nanocomposites is well-established for enhancing mechanical, thermal, and electrical properties. To fully exploit the impressive strength of these nanocomposites, establishing a robust interfacial connection is essential for the efficient transfer of loads from the polymer matrix to the CNTs [59]. A clear approach to enhancing this binding involves chemical functionalization, which is achieved by attaching side chains to the CNTs. A straightforward approach to enhance interfacial binding involves chemically functionalizing carbon nanotubes. For example, various functional groups, such as -OH, -F, and -COOH, have been directly attached to CNTs, which were subsequently altered through substitution reactions [80].

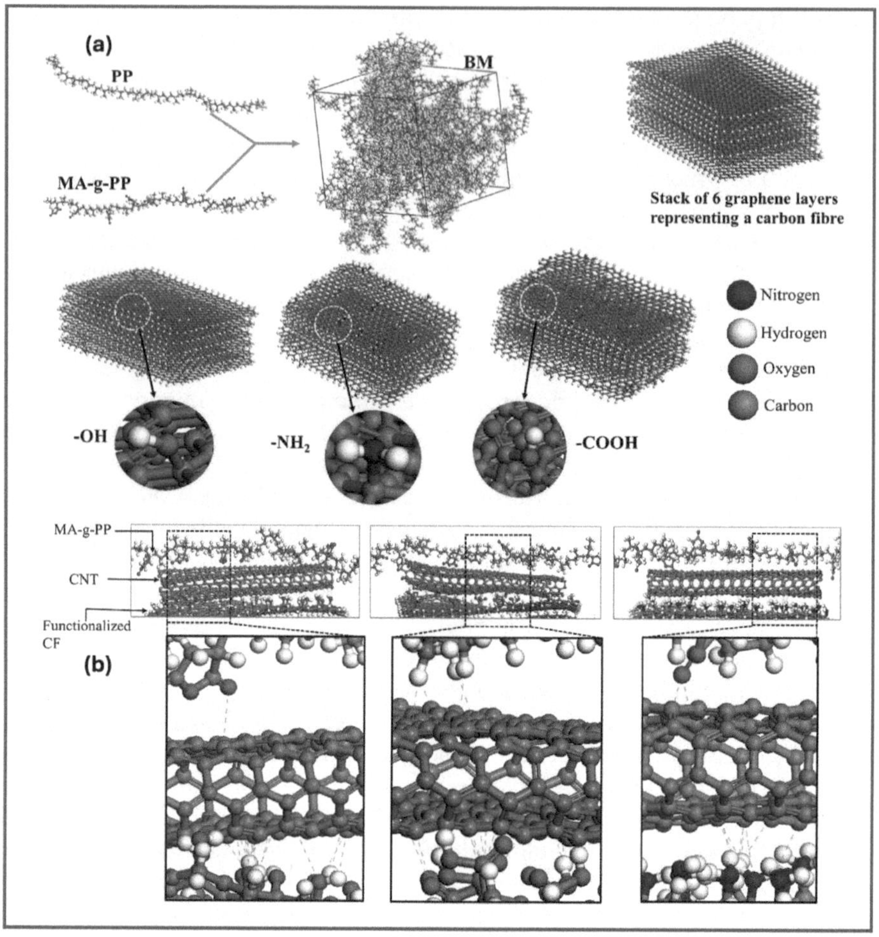


**FIGURE 14.6** (a) Illustration of the atomistic structure of polypropylene (PP)–and maleic anhydride–grafted PP and CF as a stack of graphene layers. (b) Schematic presentation of possible vdW interactions of functionalized among CNT and MA-g-PP.

**Source: Reproduced from ref. [78].**

Nonetheless, employing chemical functionalization as a processing step presents a significant drawback because it introduces atomic defects into carbon nanotubes. MD simulations were conducted to investigate the interfacial binding between SWCNTs and conjugated polymers like polypyrrole (PP), polythiophene (PT), and other analogous polymers, as depicted in Figure 14.7 [59]. The simulation results revealed the strongest interaction between SWCNTs and conjugated polymers. The simulation study suggested that there is an effective load transfer at the interface between the heterocyclic conjugated polymer and the nanotube.

## 14.6 APPROACHES TO IMPROVE INTERFACIAL INTERACTIONS OF NANOCOMPOSITES

Several research efforts have found that the inadequate interfacial interactions between the nanofillers (e.g., CNT, graphene, GO) and the polymer matrix play a crucial role in causing lower mechanical properties of polymer nanocomposites [81–83]. Another vital reason for the lack of the expected enhancement in the mechanical properties of the CNT-, graphene-, and GO-based polymer

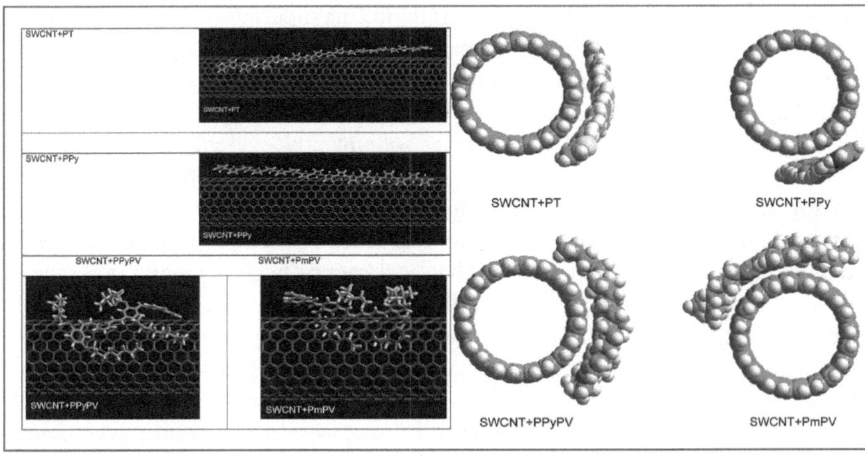


**FIGURE 14.7** MD simulation snapshots of the wrapping of an SWCNT by different polymers.

**Source: Reproduced from ref. [59].**

nanocomposites is an agglomeration of nanoparticles because of the van der Waals attractive forces. Since graphene, GOs, and CNTs are comprised of the benzene moieties in which carbon atoms are $sp^2$-hybridized, the accumulation of neighboring graphene nanoparticles and CNTs by the van der Waals attractive forces certainly occurs in the polymer nanocomposite. Moreover, the occurrence of mechanical interlocking, widely recognized as the primary mechanism for interfacial bonding between the polymer matrix and nanofillers, is scarcely observed at the interface between the polymer matrix and CNT/graphene. Since their discovery, several research groups have constantly tried to develop high-performance polymer nanocomposites, introducing CNTs and graphene into the polymer matrices. Specifically, the demand for lightweight, high-strength polymer nanocomposites has been growing, utilizing excellent mechanical properties of CNTs and graphene, such as their tensile modulus, tensile strength, good processability, and low product cost. However, experimental results reported in the literature have shown lower than theoretical expectations. Therefore, several researchers have investigated to identify the reasons for the unsatisfactory outcomes [81–83]. Despite their excellent physical properties, several research teams have pinpointed that the subpar interfacial interactions between the nanofillers (such as CNTs, graphene, and GOs) and the polymer matrix, along with the agglomeration of nanoparticles caused by van der Waals attractive forces, may be the primary causes for the unsatisfactory outcomes.

While covalent functionalization improves the dispersal of CNTs, GOs, and graphene in the polymer matrix and improves their interaction with the polymer matrix, this functionalization has certain drawbacks, which include alterations in the structural properties of graphene, GOs, and CNTs, degradation caused by strong oxidants, and damage occurring during the processing of the functionalized groups. In addition, the synthesis techniques of covalently functionalized CNTs and graphene are sometimes complicated, involving multistep, time-consuming, and industrially unsuitable processes. Noncovalent functionalization has been employed to increase the dispersion of graphene and CNTs into polymer matrices, to reduce the complexity involved with covalent functionalization, and to avoid disrupting the structure of carbon nanotubes. The noncovalent functionalization approach aids in connecting the molecules without creating chemical bonds. This can be achieved by Π–Π interactions, electrostatic interactions, using surfactants, and wrapping molecules around the nanofillers. Several research teams have shown interest in the noncovalent functionalization of nanofillers, such as graphene, CNTs, GOs, nanoclays, and silica nanoparticles [81, 84–86]. This approach is appealing because it preserves the fundamental properties of nanoparticles while enhancing their dispersion in the polymer matrix, processability, interaction at the interface of nanocomposites, solubility in a solvent, and prevents the agglomeration of nanoparticles. In the case of noncovalent functionalization, biomacromolecules and aromatic groups containing polymers are used to prevent agglomeration of graphene nanoparticles and CNTs to enhance the interaction of these nanoparticles with polymer. Generally, cationic, anionic, cationic, and non-ionic surfactants are used to disperse CNTs in water. Anionic surfactants, for example, sodium dodecylbenzene sulfonate and sodium dodecyl sulfate, are commonly used to prevent the agglomeration of CNTs in water. The interaction between surfactants and CNTs is significantly influenced by the attributes of the surfactants, such as their alkyl chain length, head group size, and charge [4, 86, 87]. CNTs and graphene have the ability to covalently connect with various chemical groups because of the presence of Π bonds between $sp^2$-hybridized C atoms. Because of covalent functionalization, $sp^2$ carbon atoms of CNTs and graphene are changed to sp3 carbon atoms, and some intrinsic properties of CNTs and graphene are changed. However, the covalent functionalization of CNTs and graphene can increase dispersion into the polymer matrix and interfacial interaction of polymer nanocomposites. Typically, the oxidation process on carbon-based nanofillers involves sulfuric acid, nitric acid, oxygen, and hydrogen peroxide. This results in the formation of various functional groups, such as hydroxyl and carboxylic

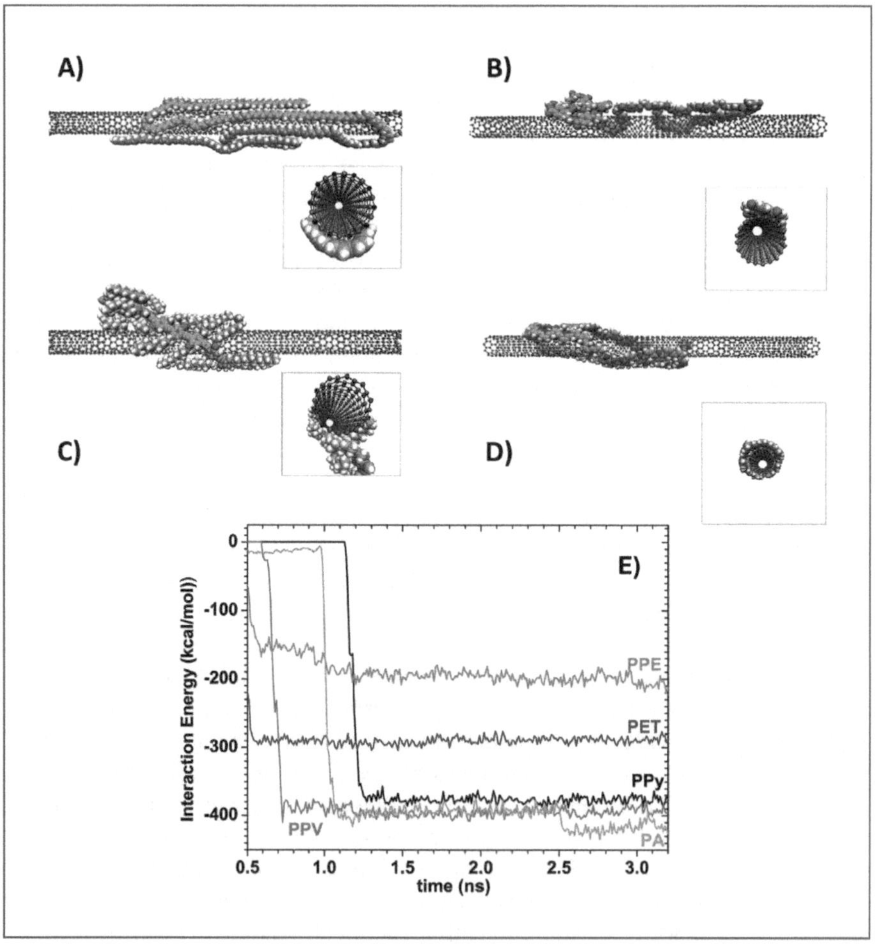


**FIGURE 14.8** MD simulation snapshots of polymers interacting with the SWCNT: (a) poly(acetylene), (b) poly(ethylene teraphthalate), (c) poly(aryleneethynylene), and (d) poly(pyrrole). (e) Illustration of interaction energies between SWCNT and polymer molecules.

**Source: Reproduced from ref. [88].**

groups. Carboxylic acid and hydroxyl groups on the nanofillers play a crucial role in facilitating the bonding with organic materials, which is essential for effectively dispersing nanofillers within polymer matrices. Click chemistry is an ideal approach for synthesizing and modifying novel materials and for self-assembling nanomaterials. Click chemistry is very suitable for creating biologically originated macromolecular structures and dendrimers. Click chemistry is widely accepted because of its high yield, short reaction time, regiospecificity, and suitability under an aqueous medium. Azides and acetylenes used in click chemistry are highly stable in biological environments and in a wide range of organic reaction conditions, yet they are highly reactive functional groups, and the products are easily separated from the reaction mixture since no protecting groups are utilized. Utilizing click chemistry presents an optimal method for integrating a diverse array of organic molecules onto the surface of CNTs/graphene. Employing this technique allows for the grafting of desired molecules onto the CNT surface, enhancing interfacial interaction and load transfer efficiency within polymer/CNT nanocomposites. Click chemistry could be an ideal approach for the incorporation of a wide range of organic molecules onto the surface of CNTs. By using this approach, the desired molecules can be grafted onto the surface of CNTs, which improves the interfacial interaction and efficiency of load transfer across the interface of polymer/CNT nanocomposites.

Implementing covalent bonding to improve the interfacial bonding between the polymer matrix and CNTs is a potential strategy, but it has several drawbacks, such as the overall changes in CNT properties and adverse effects on the overall material characteristics. Consequently, an alternative technique involves noncovalent engineering of the interface between SWCNTs and the polymer. The noncovalent strategy has proven to bring about significant enhancements in various aspects of nanocomposite materials. A recent study aims to explore the interaction between SWCNTs and polymer chains with rigid backbones using MD simulations [88]. **Figure** 14.8 showcases snapshots derived from MD simulations which depict the interplay between SWCNT and polymer molecules. The impact of aromatic moieties along the polymer backbone on adsorption conformation is evident, likely linked to a preference for optimizing pi–pi interactions [88]. Results from the simulations indicate that polymers with stiff backbones exhibit distinct conformations as they wrap around the SWCNT, unlike those with flexible backbones.

## 14.7 CONCLUSION

The development of innovative polymer nanocomposites has significant potential thanks to nanomaterials like carbon nanotubes, graphene, graphene oxides, nanoclays, nanocellulose, nanosilica, and titanium dioxide nanoparticles. Polymer nanocomposites have a broad spectrum of applications, covering industries such as construction, packaging, rubber, electronics, paper, automotive, cosmetics, coatings, adhesives, textiles, tissue engineering, and water purification. Developing innovative polymer nanocomposites necessitates a profound grasp of polymer physics, surface science, polymer chemistry, and nanoscience, and employing computational modeling offers crucial insights into their development by elucidating the pivotal role of interfacial adhesion in shaping their characteristics. Integrating various computational modeling and simulation methods can enhance the overall understanding of these materials, aiding in designing and improving advanced nanocomposites customized for various applications. Ensuring a strong interfacial connection becomes crucial for effectively transferring loads from the polymer matrix to the nanofillers, maximizing the impressive strength of these nanocomposites. MD simulations have been widely employed to investigate the interplay between nanofillers and polymer matrices. These simulations aid in comprehending potential interactions between nanotubes and polymer matrices, offering insights and proposing potential strategies to enhance the connection between nanofillers and polymer matrices.

## ACKNOWLEDGMENTS

We gratefully acknowledge IIT Roorkee for offering all literature.

## DECLARATION OF COMPETING INTEREST

The authors state that they have no recognized competing financial interests.

## REFERENCES

[1] Zhaoying Z, Yuhui H, Bing L, Guangming C. Studies on particle size of waterborne emulsions derived from epoxy resin. Eur Polym J 2001;37:1207–11. https://doi.org/10.1016/S0014-3057(00)00192-0.
[2] Marouf BT, Mai YW, Bagheri R, Pearson RA. Toughening of epoxy nanocomposites: Nano and hybrid effects. Polym Rev 2016;56:70–112. https://doi.org/10.1080/15583724.2015.1086368.
[3] Gu J, Liang C, Zhao X, Gan B, Qiu H, Guo Y, Yang X, Zhang Q, Wang D-Y. Highly thermally conductive flame-retardant epoxy nanocomposites with reduced ignitability and excellent electrical conductivities. Compos Sci Technol 2017;139:83–89. https://doi.org/10.1016/j.compscitech.2016.12.015.
[4] Sáfar GAM, Ribeiro HB, Malard LM, Plentz FO, Fantini C, Santos AP, Idemori YM. Optical study of porphyrin-doped carbon nanotubes. Chem Phys Lett 2008;462:109–11. https://doi.org/10.1016/j.cplett.2008.07.093.
[5] Shen XJ, Liu Y, Xiao HM, Feng QP, Yu ZZ, Fu SY. The reinforcing effect of graphene nanosheets on the cryogenic mechanical properties of epoxy resins. Compos Sci Technol 2012;72:1581–87. https://doi.org/10.1016/j.compscitech.2012.06.021.
[6] Barkoula NM, Alcock B, Cabrera NO, Peijs T. Flame-retardancy properties of intumescent ammonium poly(phosphate) and mineral filler magnesium hydroxide in combination with graphene. Polym Polym Compos 2008;16:101–13. https://doi.org/10.1002/pc.
[7] Wang S, Tambraparni M, Qiu J, Tipton J, Dean D. Thermal expansion of graphene composites. Macromolecules 2009;42:5251–55. https://doi.org/10.1021/ma900631c.
[8] Vishwakarma A, Singh M, Weclawski B, Reddy VJ, Kandola BK, Manik G, Dasari A, Chattopadhyay S. Construction of hydrophobic fire retardant coating on cotton fabric using a layer-by-layer spray coating method. Int J Biol Macromol 2022;223:1653–66. https://doi.org/10.1016/j.ijbiomac.2022.10.231.
[9] Singh M, Sethi SK, Manik G. Pressure-sensitive adhesives based on acrylated epoxidized linseed oil: A computational approach. Int J Adhes Adhes 2022;112:103031. https://doi.org/10.1016/j.ijadhadh.2021.103031.
[10] Kadian S, Manik G, Kalkal A, Singh M, Chauhan RP. Effect of sulfur doping on fluorescence and quantum yield of graphene quantum dots: An experimental and theoretical investigation. Nanotechnology 2019;30. https://doi.org/10.1088/ 1361–6528 /ab3566.
[11] Sethi SK, Singh M, Manik G. A multi-scale modeling and simulation study to investigate the effect of roughness of a surface on its self-cleaning performance. Mol Syst Des Eng 2020;5:1277–89. https://doi.org/10.1039/d0me00068j.
[12] Singh M, Sahoo SK, Manik G. Combined computational and experimental approach for bio-sourced monomers to design green pressure-sensitive adhesives. Mol Syst Des Eng 2023;8:1498–511. https://doi.org/10.1039/d3me00080j.
[13] Sindu BS, Sasmal S. Evaluation of mechanical characteristics of nano modified epoxy based polymers using molecular dynamics. Comput Mater Sci 2015;96:146–58. https://doi.org/10.1016/j.commatsci.2014.09.003.
[14] Singh M, Sethi SK, Manik G. Pressure-sensitive adhesives based on acrylated epoxidized linseed oil: A computational approach. Int J Adhes Adhes 2022;112:103031. https://doi.org/10.1016/j.ijadhadh.2021.103031.
[15] Kalra A, Parks DM, Rutledge GC. Molecular simulation of strain dependence of vibrational frequencies for montmorillonite clay and analysis of strain transfer in a polymer-clay nanocomposite. Macromolecules 2007;40:140–44. https://doi.org/10.1021/ma062144p.
[16] Koyanagi J, Itano N, Yamamoto M, Mori K, Ishida Y, Bazhirov T. Evaluation of the mechanical properties of carbon fiber/polymer resin interfaces by molecular simulation. Adv Compos Mater 2019;28:639–52. https://doi.org/10.1080/09243046.2019.1630069.
[17] Cheng X, Yokozeki T, Yamamoto M, Wang H, Wu L, Koyanagi J, Sun Q. The decoupling electrical and thermal conductivity of fullerene/polyaniline hybrids reinforced polymer composites. Compos Sci Technol 2017;144:160–68. https://doi.org/10.1016/j.compscitech.2017.03.030.
[18] Niuchi T, Koyanagi J, Inoue R, Kogo Y. Molecular dynamics study of the interfacial strength between carbon fiber and phenolic resin. Adv Compos Mater 2017;26:569–81. https://doi.org/10.1080/09243046.2017.1286543.

[19] Singh M, Manik G. Computationally developed acrylated epoxidized methyl ester based pressure-sensitive adhesives. Comput Mater Sci 2023;228:112329. https://doi.org/10.1016/j.commatsci.2023.112329.

[20] Mazumdar P, Chockalingam S, Rattan S, Gupta BK. Tunable mechanical, electrical, and thermal properties of polymer nanocomposites through GMA bridging at interface. ACS Omega 2018;3:3675–87. https://doi.org/10.1021/acsomega.8b00194.

[21] Khan P, Kaushik R, Jayaraj A. Approaches and perspective of coarse-grained modeling and simulation for polymer-nanoparticle hybrid systems. ACS Omega 2022;7:47567–86. https://doi.org/10.1021/acsomega.2c06248.

[22] Okabe T, Takehara T, Inose K, Hirano N, Nishikawa M, Uehara T. Curing reaction of epoxy resin composed of mixed base resin and curing agent: Experiments and molecular simulation. Polymer (Guildf) 2013;54:4660–68. https://doi.org/10.1016/j.polymer.2013.06.026.

[23] Sharp N, Li C, Strachan A, Adams D, Pipes RB. Effects of water on epoxy cure kinetics and glass transition temperature utilizing molecular dynamics simulations. J Polym Sci Part B Polym Phys 2017;55:1150–59. https://doi.org/10.1002/polb.24357.

[24] Sethi SK, Shankar U, Manik G. Fabrication and characterization of non-fluoro based transparent easy-clean coating formulations optimized from molecular dynamics simulation. Prog Org Coat 2019;136:105306. https://doi.org/10.1016/j.porgcoat.2019.105306.

[25] Agrawal G, Samal SK, Sethi SK, Manik G, Agrawal R. Microgel/silica hybrid colloids: Bioinspired synthesis and controlled release application. Polymer (Guildf) 2019;178:121599. https://doi.org/10.1016/j.polymer.2019.121599.

[26] Olivei A. Surface imaging by scanning-laser-induced voltage. J Phys D Appl Phys 1976;9:183–95. https://doi.org/10.1088/ 0022–3727 /9/2/008.

[27] Zheng Q, Xue Q, Yan K, Hao L, Li Q, Gao X. Investigation of molecular interactions between SWNT and polyethylene/polypropylene/polystyrene/polyaniline molecules. J Phys Chem C 2007;111:4628–35. https://doi.org/10.1021/jp066077c.

[28] Sun H. The COMPASS force field: Parameterization and validation for phosphazenes. Comput Theor Polym Sci 1998;8:229–46. https://doi.org/10.1016/S 1089–3156 (98) 00042–7.

[29] Sun H, Jin Z, Yang C, Akkermans RLC, Robertson SH, Spenley NA, Miller S, Todd SM. COMPASS II: Extended coverage for polymer and drug-like molecule databases. J Mol Model 2016;22:1–10. https://doi.org/10.1007/s00894-016-2909-0.

[30] Volgin IV, Larin SV, Lyulin AV, Lyulin SV. Coarse-grained molecular-dynamics simulations of nanoparticle diffusion in polymer nanocomposites. Polymer (Guildf) 2018;145:80–87. https://doi.org/10.1016/j.polymer.2018.04.058.

[31] Khan P, Goel G. Martini coarse-grained model for clay-polymer nanocomposites. J Phys Chem B 2019;123:9011–23. https://doi.org/10.1021/acs.jpcb.9b06708.

[32] Dhamankar S, Webb MA. Chemically specific coarse-graining of polymers: Methods and prospects. J Polym Sci 2021;59:2613–43. https://doi.org/10.1002/pol.20210555.

[33] Mehandzhiyski AY, Zozoulenko I. A review of cellulose coarse-grained models and their applications. Polysaccharides 2021;2:257–70. https://doi.org/10.3390/polysaccharides2020018.

[34] Kmiecik S, Gront D, Kolinski M, Wieteska L, Dawid AE, Kolinski A. Coarse-grained protein models and their applications. Chem Rev 2016;116:7898–936. https://doi.org/10.1021/acs.chemrev.6b00163.

[35] De Jong DH, Singh G, Bennett WFD, Arnarez C, Wassenaar TA, Schäfer LV, Periole X, Tieleman DP, Marrink SJ. Improved parameters for the martini coarse-grained protein force field. J Chem Theory Comput 2013;9:687–97. https://doi.org/10.1021/ct300646g.

[36] Araújo MC, Martins JP, Mirkhalaf SM, Lanceros-Mendez S, Pires FMA, Simoes R. Predicting the mechanical behavior of amorphous polymeric materials under strain through multi-scale simulation. Appl Surf Sci 2014;306:37–46. https://doi.org/10.1016/j.apsusc.2014.03.072.

[37] Çağin T, Che J, Qi Y, Zhou Y, Demiralp E, Gao G, Goddard WA. Computational materials chemistry at the nanoscale. J Nanopart Res 1999;1:51–69. https://doi.org/10.1023/A:1010009630519.

[38] Sahputra IH, Echtermeyer AT. Molecular dynamics simulations of strain-controlled fatigue behaviour of amorphous polyethylene. J Polym Res 2014;21. https://doi.org/10.1007/s10965-014-0577-2.

[39] Zeng QH, Yu AB, Lu GQ. Multiscale modeling and simulation of polymer nanocomposites. Prog Polym Sci 2008;33:191–269. https://doi.org/10.1016/j.progpolymsci.2007.09.002.

[40] Lee J, Wang H, Kim JH, Shin H. Recent studies on the multiscale modeling and simulation of polymer nanocomposites. JMST Adv 2023;5:85–91. https://doi.org/10.1007/s42791-023-00051-z.

[41] Bekyarova E, Thostenson ET, Yu A, Kim H, Gao J, Tang J, Hahn HT, Chou T-W, Itkis ME, Haddon RC. Multiscale carbon nanotube-carbon fiber reinforcement for advanced epoxy composites. Langmuir 2007;23:3970–74. https://doi.org/10.1021/la062743p.

[42] Lv P, Feng YY, Zhang P, Chen HM, Zhao N, Feng W. Increasing the interfacial strength in carbon fiber/epoxy composites by controlling the orientation and length of carbon nanotubes grown on the fibers. Carbon NY 2011;49:4665–73. https://doi.org/10.1016/j.carbon.2011.06.064.

[43] Sharma M, Gao S, Mäder E, Sharma H, Wei LY, Bijwe J. Carbon fiber surfaces and composite interphases. Compos Sci Technol 2014;102:35–50. https://doi.org/10.1016/j.compscitech.2014.07.005.

[44] Yu C, Yang Q. Multiscale interfacial interactions in carbon nanotube: Asphalt nanocomposites: Implications for asphalt pavement materials. ACS Appl Nano Mater 2023;6:7715–30. https://doi.org/10.1021/acsanm.3c00856.

[45] Kargarzadeh H, Mariano M, Huang J, Lin N, Ahmad I, Dufresne A, Thomas S. Recent developments on nanocellulose reinforced polymer nanocomposites: A review. Polymer (Guildf) 2017;132:368–93. https://doi.org/10.1016/j.polymer.2017.09.043.

[46] Xia W, Qin X, Zhang Y, Sinko R, Keten S. Achieving enhanced interfacial adhesion and dispersion in cellulose nanocomposites via amorphous interfaces. Macromolecules 2018;51:10304–11. https://doi.org/10.1021/acs.macromol.8b02243.

[47] Li C, Wei H, Zhan H, Bai J, Kou L, Gu Y. Tensile performance of polymer nanocomposites with randomly dispersed carbon nanothreads. Macromolecules 2021;54:11486–96. https://doi.org/10.1021/acs.macromol.1c01711.

[48] Ding P, Wu J, Zhang J, Shao J, Tang W, Hou G, Xang L, Chen X. Role of geometric shapes on the load transfer in graphene-PMMA nanocomposites. Comput Mater Sci 2020;184:109863. https://doi.org/10.1016/j.commatsci.2020.109863.

[49] Al Mahmud H, Radue MS, Chinkanjanarot S, Odegard GM. Multiscale modeling of epoxy-based nanocomposites reinforced with functionalized and non-functionalized graphene nanoplatelets. Polymers (Basel) 2021;13. https://doi.org/10.3390/polym13121958.

[50] Ma PC, Siddiqui NA, Marom G, Kim JK. Dispersion and functionalization of carbon nanotubes for polymer-based nanocomposites: A review. Compos Part A Appl Sci Manuf 2010;41:1345–67. https://doi.org/10.1016/j.compositesa.2010.07.003.

[51] Ramanathan T, Abdala AA, Stankovich S, Dikin DA, Herrera-Alonso M, Piner RD, Adamson DH, Schniepp HC, Chen X, Ruoff RS, Nguyen ST, Aksay IA, Prud'Homme RK, Brinson LC. Functionalized graphene sheets for polymer nanocomposites. Nat Nanotechnol 2008;3:327–31. https://doi.org/10.1038/nnano.2008.96.

[52] King JA, Klimek DR, Miskioglu I, Odegard GM. Mechanical properties of graphene nanoplatelet/epoxy composites. J Appl Polym Sci 2013;128:4217–23. https://doi.org/10.1002/app.38645.

[53] Geng Y, Wang SJ, Kim JK. Preparation of graphite nanoplatelets and graphene sheets. J Colloid Interface Sci 2009;336:592–98. https://doi.org/10.1016/j.jcis.2009.04.005.

[54] Shen B, Zhai W, Chen C, Lu D, Wang J, Zheng W. Melt blending in situ enhances the interaction between polystyrene and graphene through π-π stacking. ACS Appl Mater Interfaces 2011;3:3103–9. https://doi.org/10.1021/am200612z.

[55] Simon P, Gogotsi Y. Carbon à electrolyte systems. Acc Chem Res 2013;46:1094.

[56] Kaempgen M, Chan CK, Ma J, Cui Y, Gruner G. Printable thin film supercapacitors using single-walled carbon nanotubes. Nano Lett 2009;9:1872–76. https://doi.org/10.1021/nl8038579.

[57] Stoller MD, Park S, Zhu Y, An J, Ruoff RS. Graphene-based ultracapacitors. Nano Lett 2008:8;3498–502. https://doi.org/10.1021/nl802558y.

[58] Wang Y, Shi Z, Huang Y, Ma Y, Wang C, Chen M, Chen Y. Supercapacitor devices based on graphene materials. J Phys Chem C 2009;113:13103–7. https://doi.org/10.1021/jp902214f.

[59] Foroutan M, Nasrabadi AT. Investigation of the interfacial binding between single-walled carbon nanotubes and heterocyclic conjugated polymers. J Phys Chem B 2010;114:5320–26. https://doi.org/10.1021/jp100960u.

[60] Gou J, Liang Z, Zhang C, Wang B. Computational analysis of effect of single-walled carbon nanotube rope on molecular interaction and load transfer of nanocomposites 2005;36:524–33. https://doi.org/10.1016/j.compositesb.2005.02.004.

[61] Barber AH, Cohen SR, Wagner HD. Measurement of carbon nanotube-polymer interfacial strength. Appl Phys Lett 2003;82:4140–42. https://doi.org/10.1063/1.1579568.

[62] Barber AH, Cohen SR, Eitan A, Schadler LS, Wagner HD. Fracture transitions at a carbon-nanotube/polymer interface. Adv Mater 2006;18:83–87. https://doi.org/10.1002/adma.200501033.

[63] Manoharan MP, Sharma A, Desai AV, Haque MA, Bakis CE, Wang KW. The interfacial strength of carbon nanofiber epoxy composite using single fiber pullout experiments. Nanotechnology 2009;20. https://doi.org/10.1088/ 0957–4484 /20/29/295701.

[64] Tsuda T, Ogasawara T, Deng F, Takeda N. Direct measurements of interfacial shear strength of multi-walled carbon nanotube/PEEK composite using a nano-pullout method. Compos Sci Technol 2011;71:1295–300. https://doi.org/10.1016/j.compscitech.2011.04.014.

[65] Siddiqui NA, Khan SU, Kim JK. Experimental torsional shear properties of carbon fiber reinforced epoxy composites containing carbon nanotubes. Compos Struct 2013;104:230–38. https://doi.org/10.1016/j.compstruct.2013.04.033.

[66] Rahman R, Haque A. Molecular modeling of crosslinked graphene-epoxy nanocomposites for characterization of elastic constants and interfacial properties. Compos Part B Eng 2013;54:353–64. https://doi.org/10.1016/j.compositesb.2013.05.034.

[67] Zhang J, Jiang D, Scarpa F, Peng HX. Enhancement of pullout energy in a single-walled carbon nanotube-polyethylene composite system via auxetic effect. Compos Part A Appl Sci Manuf 2013;55:188–94. https://doi.org/10.1016/j.compositesa.2013.09.006.

[68] Chandra Y, Scarpa F, Adhikari S, Zhang J, Saavedra Flores EI, Peng HX. Pullout strength of graphene and carbon nanotube/epoxy composites. Compos Part B Eng 2016;102:1–8. https://doi.org/10.1016/j.compositesb.2016.06.070.

[69] Wagner HD, Lourie O, Feldman Y, Tenne R. Stress-induced fragmentation of multiwall carbon nanotubes in a polymer matrix. Appl Phys Lett 1998;72:188–90. https://doi.org/10.1063/1.120680.

[70] Gou J, Minaie B, Wang B, Liang Z, Zhang C. Computational and experimental study of interfacial bonding of single-walled nanotube reinforced composites. Comput Mater Sci 2004;31:225–36. https://doi.org/10.1016/j.commatsci.2004.03.002.

[71] Liao K, Li S. Interfacial characteristics of a carbon nanotube-polystyrene composite system. Appl Phys Lett 2001;79:4225–27. https://doi.org/10.1063/1.1428116.

[72] Sager RJ, Klein PJ, Lagoudas DC, Zhang Q, Liu J, Dai L, Baur JW. Effect of carbon nanotubes on the interfacial shear strength of T650 carbon fiber in an epoxy matrix. Compos Sci Technol 2009;69:898–904. https://doi.org/10.1016/j.compscitech.2008.12.021.

[73] Chen H, Xue Q, Zheng Q, Xie J, Yan K. Influence of nanotube chirality, temperature, and chemical modification on the interfacial bonding between carbon nanotubes and polyphenylacetylene. J Phys Chem C 2008;112:16514–20. https://doi.org/10.1021/jp803615v.

[74] Zheng Q, Xia D, Xue Q, Yan K, Gao X, Li Q. Computational analysis of effect of modification on the interfacial characteristics of a carbon nanotube-polyethylene composite system. Appl Surf Sci 2009;255:3534–43. https://doi.org/10.1016/j.apsusc.2008.09.077.

[75] Naebe M, Wang J, Amini A, Khayyam H, Hameed N, Li LH, Chen Y, Fox B. Mechanical property and structure of covalent functionalised graphene/epoxy nanocomposites. Sci Rep 2014;4:1–7. https://doi.org/10.1038/srep04375.

[76] Rafiee MA, Rafiee J, Wang Z, Song H, Yu ZZ, Koratkar N. Enhanced mechanical properties of nanocomposites at low graphene content. ACS Nano 2009;3:3884–90. https://doi.org/10.1021/nn9010472.

[77] Tang LC, Wan YJ, Yan D, Pei YB, Zhao L, Li YB, Wu LB, Jiang JW, Lai GQ. The effect of graphene dispersion on the mechanical properties of graphene/epoxy composites. Carbon NY 2013;60:16–27. https://doi.org/10.1016/j.carbon.2013.03.050.

[78] Gogoi R, Sethi SK, Manik G. Surface functionalization and CNT coating induced improved interfacial interactions of carbon fiber with polypropylene matrix: A molecular dynamics study. Appl Surf Sci 2021;539:148162. https://doi.org/10.1016/j.apsusc.2020.148162.

[79] Demir B, Henderson LC, Walsh R. Design rules for enhanced interfacial shear response in functionalized carbon fiber epoxy composites. ACS Appl Mater Interface 2017. https://doi.org/10.1021/acsami.6b16041.

[80] Kuzmany H, Kukovecz A, Simon F, Holzweber M, Kramberger C, Pichler T. Functionalization of carbon nanotubes. Synthetic Metals 2004;141:113–22. https://doi.org/10.1016/j.synthmet.2003.08.018.

[81] Du JH, Bai J, Cheng HM. The present status and key problems of carbon nanotube based polymer composites. Express Polym Lett 2007;1:253–73. https://doi.org/10.3144/expresspolymlett.2007.39.

[82] Zhbanov AI, Pogorelov EG, Chang YC. Van der Waals interaction between two crossed carbon nanotubes. ACS Nano 2010;4:5937–45. https://doi.org/10.1021/nn100731u.

[83] Spitalsky Z, Tasis D, Papagelis K, Galiotis C. Carbon nanotube-polymer composites: Chemistry, processing, mechanical and electrical properties. Prog Polym Sci 2010;35:357–401. https://doi.org/10.1016/j.progpolymsci.2009.09.003.

[84] Paredes JI, Burghard M. Dispersions of individual single-walled carbon nanotubes of high length owing to their unique physical and chemical properties, 1, 2 carbon nanotubes and particularly single-walled carbon nanotubes (SWCNTs) are very promising for a wide range of app. Langmuir 2004;20:5149–52. https://doi.org/10.1021/la049831z.
[85] Moore VC, Strano MS, Haroz EH, Hauge RH, Smalley RE, Schmidt J, Talmon Y. Individually suspended single-walled carbon nanotubes in various surfactants. Nano Lett 2003;3:1379–82. https://doi.org/10.1021/nl034524j.
[86] Islam MF, Rojas E, Bergey DM, Johnson AT, Yodh AG. High weight fraction surfactant solubilization of single-wall carbon nanotubes in water. Nano Lett 2003;3:269–73. https://doi.org/10.1021/nl025924u.
[87] Tan Y, Resasco DE. Dispersion of single-walled carbon nanotubes of narrow diameter distribution. J Phys Chem B 2005;109:14454–60. https://doi.org/10.1021/jp052217r.
[88] Tallury SS, Pasquinelli MA. Molecular dynamics simulations of polymers with stiff backbones interacting with single-walled carbon nanotubes. J Phys Chem B 2010;114:9349–55.

# 15 Hybrid Approaches
## *Nanocomposites and Filler Coating for Enhanced Performance in Polymer Composites*

*Vishwa Dave, Debajyoti Palai, Vishwa Patel, Jolly Mistry, Zarana Patel, Priyabrata Mandal, and Priya Goel*

### 15.1 INTRODUCTION

Combining two or more different types of materials—one of which is a polymer matrix and the other is a reinforcing material, which is frequently in the form of fibers, particles, or other structural elements—is known as polymer composites [1–9]. These composites are made to benefit from the unique characteristics of the reinforcing material and the polymer matrix. The resultant composite material shows enhanced overall performance and customized features compared to the individual components. These composites combine the desirable properties of the polymer matrix and the reinforcing materials, resulting in improved materials characteristics compared to the individual components. Polymer nanocomposites represent a significant advancement in materials science, combining the versatility of polymers with the unique properties of nanomaterials. A recent report estimated that the expected value of the global market for advanced polymer composites as of 2022 was US$ 11.12 billion. Future Market Insights analysts see significant growth, estimating that the market would surpass US$ 19.73 billion by 2032, exhibiting a 5.9% compound annual growth rate (CAGR) [10]. Due to their exceptional qualities and the growing needs of different industries, advanced polymer composites are becoming more and more important across a wide range of industries. This is demonstrated by their strong growth trajectory [11].

The concept of incorporating nanomaterials into polymers dates back to the early 20th century, when researchers began experimenting with the addition of various fillers and reinforcements to enhance polymer properties [12]. However, the true potential of polymer nanocomposites was realized with the advent of nanotechnology and the synthesis of engineered nanomaterials in the late 20th century. The development of nanotechnology allowed for the precise control and manipulation of materials at the nanoscale. This breakthrough opened up new possibilities for creating nanoscale reinforcements, such as carbon nanotubes, graphene, and nanoclays, which could be incorporated into polymers to enhance their performance [13]. Throughout the late 20th and early 21st centuries, researchers achieved several key milestones in polymer nanocomposites, which included the successful dispersion of nanomaterials within polymer matrices, the discovery of significant property enhancements, and the development of various processing techniques for nanocomposite fabrication [14]. The field of polymer nanocomposites continues to evolve with ongoing research to further enhance performance, scalability, and cost-effectiveness. Emerging trends include the development of sustainable nanomaterials, smart nanocomposites, and the integration of nanocomposites. Polymer nanocomposites represent a convergence of materials science, nanotechnology, and engineering, offering a promising avenue for addressing the demands of modern industries for high-performance materials with tailored properties. Enhancing the performance of polymer composites is a complex and multidisciplinary task, and it involves overcoming various challenges [15]. Choosing the right combination of polymers and reinforcing materials (e.g., fibers, nanoparticles) is also important.

DOI: 10.1201/9781003476337-15

The objective of using nanocomposites and filler coatings in polymer composites is to enhance the performance and properties of the resulting material. These materials are widely used in various industries, such as aerospace, automotive, construction, electronics, and renewable energy, due to their unique combination of properties, such as improved mechanical behavior, thermal stability, barrier effect, electrical conductivity, flame retardancy, tribological properties, light weight, etc. [16–18].

## 15.2 NANOCOMPOSITES: TRANSFORMING MATERIALS AT THE NANOSCALE

A groundbreaking development in materials science, nanocomposites are a revolutionary material that leverages nanomaterials' unique properties to improve conventional composites' performance. The integration of nanoparticles, such as graphene, nanotubes, and nanofibers, into a polymer matrix is the fundamental process of creating nanocomposites [19, 20]. The extraordinary features of polymer composites, which range from improved electrical conductivity and thermal stability to higher mechanical strength, are imparted by this nanoscale blending [21]. Typically, nanocomposites are made of a polymer matrix mixed with nanoscale reinforcing components. The selection of the nanomaterials and polymer matrix determines the ultimate properties of the composite.

Fullerenes, carbon nanotubes, graphene and its derivatives, and quantum dots are examples of carbon-based nanomaterials widely used because of their exceptional strength, electrical conductivity, and thermal qualities [22]. Furthermore, various nanoparticles, such as metal oxides, nanoclays, and nanofibers, are used in accordance with certain performance standards [23]. The potential of nanocomposites to greatly improve mechanical characteristics is one of their most remarkable qualities.

Composites with enhanced tensile strength, modulus, and impact resistance are produced through the polymer matrix's reinforcement by adding nanoparticles, which frequently have high aspect ratios and remarkable strength. For example, carbon nanotubes can be used in structural applications in the automotive and aerospace sectors because of their remarkable tensile strength, which allows them to function as effective reinforcements in polymers [24]. Applications that demand exceptional thermal characteristics are well-suited for nanocomposites. Certain nanomaterials, such as graphene, have a high thermal conductivity, which guarantees effective heat dissipation and makes them indispensable in electrical systems, where thermal management is crucial [25, 26]. Adding these nanoscale reinforcements improves the composite's thermal resilience, increasing its applicability in hot conditions. Electrical conductivity undergoes a fundamental change thanks to nanocomposites. Applications in electronics, sensors, and conductive coatings are made possible by the introduction of conductive pathways into the composite by carbon-based nanomaterials, particularly carbon nanotubes and graphene. This creates opportunities for the creation of cutting-edge, high-performing electronics.

Despite the remarkable properties of nanocomposites, challenges persist. Research is still being done in the areas of cost-effectiveness, scalability of production, and homogeneous dispersion of nanoparticles inside the polymer matrix. To overcome these obstacles and realize the full potential of nanocomposites, cutting-edge manufacturing approaches, such as in situ polymerization and functionalization of nanoparticles, are being investigated. Nanocomposites are finding applications in various industries, surpassing previous boundaries, for example, the aerospace industry, contributing to lightweight yet robust components [27]. Their improved mechanical qualities are advantageous for automotive parts, and the electronics sector uses their electrical and thermal conductivity [28]. Energy storage systems, construction materials, and even medical gadgets have embraced the adaptability of nanocomposites [29].

### 15.2.1 Organic Nanoparticles

*Organic nanoparticles* refer to nanoscale particles primarily composed of organic (carbon-based) materials. They exhibit diverse applications in various fields due to their biocompatibility, tunability, and versatility. Overall, organic nanoparticles represent a versatile and promising class of nanomaterials with a wide range of applications in medicine, diagnostics, sensing, catalysis, and electronics. These nanoparticles' unique characteristics and wide range of applications have drawn much interest from researchers in several disciplines. Numerous organic substances, including polymers, lipids, dendrimers, and tiny organic molecules, can synthesize organic nanoparticles. The following are some typical forms and applications of organic nanoparticles:

#### 15.2.1.1 Polymeric Nanoparticles

*Polymeric nanoparticles* are small particles composed of polymers, long chains of repeating molecular units. These nanoparticles are generally in the size range of 1 to 1,000 nm and can be precisely created with specified features for a variety of applications [30]. These are loaded with active compounds within surface adsorbed (nanospheres) or entrapped within polymeric cores (nanocapsules) [31]. Polymeric nanoparticles are frequently employed in various disciplines, such as medicine, drug transport, imaging, and diagnostics, because of their versatility, biocompatibility, and adjustable characteristics. Different polymeric nanoparticles and their applications are listed here.

- *Polymeric micelles.* Mainly used for drug delivery to improve therapeutic efficacy due to their ability to encapsulate hydrophobic drugs [32].
- *Polymeric nanoparticles.* Utilized as carriers for drugs, peptides, and vaccines. It is also used in bioimaging and diagnostics [33].

#### 15.2.1.2 Lipid-Based Nanoparticles

Numerous bioactive molecules that have therapeutic effects are either chemically and biologically toxic or fragile or insoluble in water. Lipid-based nanoparticles act as nanoscale carriers primarily composed of lipids, naturally occurring molecules with hydrophobic and hydrophilic regions. These nanoparticles are designed as a colloidal carrier to encapsulate and deliver various therapeutic agents. Furthermore, the combination of lipid nanoparticles and chemotherapeutic drugs lowers the toxicity and active therapeutic dose, lowers drug resistance, and raises drug levels in tumor tissue by lowering levels in healthy tissue. Lipid-based nanoparticles offer several advantages for drug delivery and other biomedical applications, including biocompatibility, versatility, and the ability to overcome biological barriers. Different applications of the lipid-based nanoparticles are listed here.

*Liposomes.* Commonly used for oral-based drug delivery, gene therapy, and diagnostics. It can encapsulate various therapeutic agents, such as small molecules, nucleic acids, and monoclonal antibodies [34].

*Solid lipid nanoparticles (SLNs).* Often used for controlled gene delivery and anticancer drug release. It has the ability to encapsulate all types of xenobiotics and skin delivery [35].

#### 15.2.1.3 Dendrimers

A small atom or collection of atoms is encircled by symmetric branches, or dendrons, known as dendrimers, which are symmetrical molecules ranging from 1 to 10 nm [36]. The primary impact on the chemical and physical characteristics of dendrimers comes from their structure. They proliferate through the core–shell structure, which then interacts with monomers that contain either two dormant or one reactive molecule. Due to their distinct characteristics, including biocompatibility, well-defined spherical structure, and hyperbranching, dendrimers find extensive use in various fields, including medicine and biomedicine. It is primarily known as artificial macromolecules. It has remarkable applications in anticancer therapies, nanoscale delivery devices, and diagnostic imaging [37].

#### 15.2.1.4 Carbon-Based Nanoparticles

*Carbon-based nanoparticles* refer to nanoscale structures composed primarily of carbon atoms. These nanoparticles exhibit unique properties and have a wide range of applications across various fields due to the versatility and abundance of carbon. They offer several advantages, including high surface area, tunable properties, biocompatibility, and chemical stability. It plays a significant role in advancing technology and addressing various challenges in electronics, health care, energy, and environmental sustainability. Different carbon-based nanoparticles and their application are listed here:

- *Fullerenes.* Used in gene and drug delivery system applications, also known for biomarker discovery and as a biocontrasting agent during MRI and X-ray imaging [37].
- *Carbon nanotubes.* In this modern era, the application of carbon nanotubes is everywhere, such as the energy sector, agriculture, pharmacy, electronics, coatings, etc. [37].

#### 15.2.1.5 Biodegradable Nanoparticles

*Biodegradable nanoparticles* are nanoscale particles designed to degrade or break down into non-toxic components in biological environments. Their ability to degrade into nontoxic components, controlled release capabilities, and biocompatibility make them valuable tools for advancing drug delivery, imaging, tissue engineering, and environmental remediation technologies. It is widely used for the controlled release of drugs or other therapeutic agents and for reducing potential long-term toxicity [38].

Organic nanoparticles are used in many sectors, such as materials research, imaging, medicine, and agriculture. Their distinct characteristics, adaptability, and biocompatibility render them advantageous instruments for tackling obstacles in medical science, diagnosis, and several other technology-related uses. Novel organic nanoparticles are still being developed and explored by researchers for various applications, and there is still more work to be done on their potential in nanotechnology.

### 15.2.2 Inorganic Nanoparticles

Small particles, typically with dimensions in the nanometer range (1–100 nm), are made of inorganic materials, known as inorganic nanoparticles. Large material chunks can be divided into inorganic nanoparticles using the top-down method, or individual atoms and molecules can be integrated into larger nanostructures using the bottom-up method. The top-down strategy is typically centered on mechanical/physical approaches, while the bottom-up approach mainly concerns chemical methods of creating nanosized particles. Due to their small size and larger surface area than their bulk counterparts, these nanoparticles exhibit unique features. Many materials, such as metals, metal oxides, metal sulfides, and other inorganic compounds, are considered inorganic nanoparticles.

The following list includes typical inorganic nanoparticles and their uses:

#### 15.2.2.1 Metal Nanoparticles

*Metal nanoparticles* are nanoscale particles composed primarily of metal atoms. These nanoparticles' small size, high surface-area-to-volume ratio, and quantum effects give them exceptional physical, chemical, and optical capabilities. Metal nanoparticles have a wide range of applications in fields, such as catalysis, electronics, sensing, medicine, and environmental remediation. Different metal nanoparticles and their application are listed in the following.

- *Gold nanoparticles.* Used in biomedical applications for targeted drug delivery, biosensors, bioimaging, and electronics [39].
- *Silver nanoparticles.* Used for antimicrobial application, wound healing, textiles, and water purification [38].

- *Iron oxide nanoparticles.* Applied in magnetic resonance imaging (MRI) contrast agents, targeted drug delivery systems, cancer therapy, and tumor tomography [40].

#### 15.2.2.2 Metal Oxide Nanoparticles

*Metal oxide nanoparticles* are nanoscale particles composed primarily of metal atoms bonded to oxygen atoms. These nanoparticles exhibit unique properties due to their small size, high surface-area-to-volume ratio, and quantum effects. Some metal oxide nanoparticles are discussed here.

- *Titanium dioxide nanoparticles.* Used in sunscreen for UV protection, antifogging mirrors, and different photocatalysis applications [40].
- *Zinc oxide nanoparticles.* Employed in rubber, paints, cosmetics, coatings, and antimicrobial agents [41].

#### 15.2.2.3 Quantum Dots

*Quantum dots* are semiconductor nanoparticles with unique optical and electronic properties. They are widely used in sensors, LED screens, solar cells, lasers, drug delivery, and biological imaging [42]. These nanocrystals exhibit quantum confinement effects, which arise from their small size and confinement of charge carriers, leading to discrete energy levels and tunable optical properties.

#### 15.2.2.4 Silica Nanoparticles

*Silica nanoparticles* are nanoscale particles composed primarily of silicon dioxide ($SiO_2$) molecules arranged in a crystalline or amorphous structure. They offer versatile solutions for widely used applications in therapeutic drug delivery systems, catalysis, imaging agents, diagnostic sensors, and reinforcing agents in composites [43].

#### 15.2.2.5 Ceramic Nanoparticles

*Ceramic nanoparticles* are nanoscale particles composed primarily of ceramic materials, typically inorganic compounds composed of metals and nonmetals bonded together through ionic or covalent bonds. Due to their unique characteristics, ceramic nanoparticles are used in scratchproof eyeglasses, crack-resistant paints, antigraffiti coatings, transparent sunscreens, stain-repellent fabrics, self-cleaning window catalysis, and as reinforcements in materials [44].

Because of their distinct size-dependent features, inorganic nanoparticles are the subject of substantial research and application across multiple domains. Their uses include environmental remediation, electronics, medicine, and catalysis. Researchers are still looking into the safety and environmental impact of nanoparticles. Nanocomposites still have a long way to go. Research on nanomaterials, production techniques, and obstacles to wider industrial adoption are all ongoing goals. Eco-friendly nanoparticles and bio-based polymers are emerging to shape the next generation of nanocomposites as sustainability becomes a major concern. These materials are paving the way for a future where the limits of material performance are continuously pushed to new heights thanks to their unmatched qualities and diverse applications.

## 15.3 POLYMER COMPOSITES WITH FILLER COATING

Filler coating is a state-of-the-art technique used in polymer composites that aims to optimize the interaction between the reinforcing filler components and the polymer matrix. This method applies a thin film coating to the filler particles to solve dispersion, compatibility, and general integration problems inside the composite material. The objective is to produce a synergistic effect that enhances mechanical, thermal, and electrical qualities while maintaining the inherent flexibility and adaptability of polymer composites. When the filler particles are coated with an appropriate material, they become more dispersible, preventing clumping and ensuring a homogenous mixture. This is particularly crucial because it may be challenging to disperse fillers that are at the nanoscale, such as nanoparticles or nanotubes, evenly. Interfacial adhesion is increased and mechanical properties are improved when coated fillers and the polymer matrix are more compatible. The purpose of filler coating techniques is to enhance the compatibility of filler particles with the polymer matrix by altering their surface in composite materials. This modification may significantly influence the final composite's performance and characteristics. Treatments of either the chemical or physical kind, or both, may be used in this modification. Improving coating fillers' interactions with the polymer matrix is the main objective since it promotes improved adhesion, dispersion, and overall composite performance.

By chemically treating the filler surface with coupling agents, surfactants, or other chemicals, the process known as "chemical functionalization" can improve its compatibility with the polymer matrix. By using physical means, an alternate material, including metal nanoparticles, polymer coatings, or other nanoparticles, like silica, is applied to the filler surface in a thin layer. Coated fillers improve the mechanical properties of the composite and reduce the likelihood of agglomeration by fortifying their bonds with the polymer matrix. Fillers can be dispersed more evenly inside the polymer matrix by using coating techniques, resulting in products with homogenous and consistent properties. Filler coatings allow for the customization of composite properties, like mechanical strength, electrical conductivity, and thermal conductivity. These filler coating techniques offer a range of ways to alter the properties of polymer composites for specific applications, such as electronics, automotive, aerospace, and energy storage.

### 15.3.1 Better Dispersion and Compatibility

To achieve homogeneous filler dispersion inside the polymer matrix, filler coating is essential. When a suitable substance is applied to the filler particles, they become more dispersible, preventing clumping and ensuring a uniform mixture. This is particularly crucial because it may be challenging to disperse fillers that are at the nanoscale, such as nanoparticles or nanotubes, evenly. Better compatibility between the coated fillers and the polymer matrix increases interfacial adhesion and improves mechanical characteristics. Chen et al. illustrated that mixing different toughness agents with carbon nanotubes significantly improved the dispersion mechanism of carbon nanotubes [45].

### 15.3.2 Customizing Mechanical Properties

The mechanical performance of polymer composites is significantly influenced by the nature of the filler–polymer interface. Filler coating modifies the mechanical characteristics by maximizing the interaction between the coated fillers and the polymer matrix. By acting as a bridge, the coated surface ensures that the filler and matrix attach firmly and transfer loads efficiently. As a result, the composite material's tensile strength, modulus, and impact resistance are all increased, making it appropriate for various structural applications. The addition of different toughness agents with carbon nanotubes effectively increased the strength and toughness of the printed parts, as reported by Chen et al. [45].

### 15.3.3 Enhanced Thermal Conductivity and Stability

A crucial component of many applications is thermal management, particularly in heat dissipation or temperature swings. The addition of filler coating to polymer composites results in improved heat conductivity. Effective heat transfer inside the composite is facilitated by coating the fillers with materials that conduct heat, including certain metal oxides. Guo et al. inferred the $Al_2O_3$ (f-$Al_2O_3$)–deposited graphene as fillers improve the thermal conductivity and the electrical insulation property of nanofibrillated cellulose (NFC) due to the bridging effect of f-$Al_2O_3$ among graphene, increasing the number of heat conduction paths [46]. Additionally, the coated fillers frequently enhance thermal stability, extending the use of the composites in hot conditions.

### 15.3.4 Improving Electrical Conductivity

Filler coating plays a crucial role in the applications of electrical conductivity, including electronic devices or conductive coatings. Covering the fillers with conductive substances, like metals or conductive polymers, makes the composite more conducive to electron transmission. Because of the increased electrical conductivity that follows, the composite may now meet the demands of a wide range of electrical and electronic applications. Sha et al. introduced a hybrid method of carbon fiber–reinforced polymers (CFRPs) with the matrix with silver nanowires (AgNWs) to improve the electrical conductivity of CFRP by functionalizing carbon fibers with vertical graphene (VG) [47].

### 15.3.5 Compatibility with Different Polymers

Filler coatings can be modulated for optimal compatibility with different polymer matrices. Because of their adaptability, composite materials can meet the requirements of specific applications. The properties of different polymers can be enhanced by coated fillers, expanding the possible applications. Nassar et al. investigated the filler–polymer compatibility in terms of physical, mechanical, chemical, and thermal by introducing olive pits (OP) fillers with recycled low-density polyethylene (rLDPE) matrix. An excellent combination of the properties was reported with varying OP fillers [47].

### 15.3.6 Barrier Properties and Functionalization

Polymer composites with filler coating have the potential to improve their barrier qualities and become impermeable to liquids or gases. One way to assist a protective barrier inside the composite is to coat the fillers with materials resistant to permeability. Daradmare et al. demonstrate a composite coating of electroexfoliated few-layer graphene (FLG) sheet and epoxy composite with 20 times improved corrosion resistance properties compared to bare epoxy coating [47]. Furthermore, depending on the application requirements, the coating process enables the functionalization of the filler surfaces, adding attributes like flame resistance or hydrophobicity.

### 15.3.7 Sustainable and Eco-Friendly Approaches

Developments in filler coating methods have led to an increasing focus on environmentally friendly and sustainable methods. Researchers are investigating ecologically friendly coating techniques and materials in line with the worldwide movement in materials science toward green and sustainable practices. Green fillers, which include post-agricultural waste powders, like cashew nut powder, coconut shell powder, rice husk powder, and date seed powder, as well as natural fibers, like bamboo, hemp, banana bark, pineapple leaf, waste tea, and palm ash, are the main focus of recent studies in the literature. Researchers are moving away from conventional nonbiodegradable fillers and toward these sustainable materials as reinforcements for epoxy resins. Green fillers are becoming increasingly popular due to their affordability, sustainability, and capacity to repurpose waste materials in a way that supports eco-friendly practices [48].

Although filler coating has many advantages, there are also certain drawbacks, such as attaining a precise and consistent coating, process scalability, and financial concerns. To solve these issues and accelerate the broad deployment of filler-coated polymer composites, ongoing research is concentrated on creating cutting-edge coating processes and environmentally acceptable coating ingredients. In polymer composites, filler coating represents a calculated strategy to maximize fillers' capabilities by bringing accuracy and control to their matrix integration. This method not only overcomes dispersion issues but also provides opportunities to modify the characteristics of polymer composites, enabling them to be tailored to a wide range of industrial applications. Filler coating, which combines performance, adaptability, and sustainability, is expected to play a major part in developing new composite materials as research continues.

## 15.4 HYBRID APPROACH: A SYNERGISTIC APPROACH

Integrating a hybrid approach, combined with nanocomposites and filler coating, profoundly impacts the electrical conductivity of polymer composites. Nanocomposites, such as carbon nanotubes, graphene, or quantum dots, bring unique characteristics to polymer matrices, including their high aspect ratio, broad surface area, and strong interfacial interactions [48]. But the agglomeration and homogeneous dispersibility within the polymer matrix are the problems that must be addressed. Conversely, fillers such as fibers or nanoparticles can have a covering put to their surface by filler coating. By promoting more dispersion, this layer inhibits agglomeration and improves the filler–matrix interaction. The intentional construction of electrical pathways inside the composite structure distinguishes this metamorphosis. By integrating cutting-edge filler coating processes with the distinct features of nanomaterials, these hybrid composites demonstrate exceptional multifunctionality, concurrently augmenting mechanical strength, thermal stability, and electrical conductivity. The hybrid approach, combining nanocomposites and filler coating, creates synergies by leveraging the unique strengths of both techniques [49].

*Synergy* refers to the interaction of elements whose combination produces a total effect greater than the sum of the effects of the individual elements. In other words, synergy is the joint action of agents that can each increase the effectiveness of the other when combined. When composites containing hybrid fillers are observed to exhibit better qualities than those obtained by merely mixing the components, this suggests a positive synergistic effect [49]. The combination of filler coating and nanocomposites produces synergistic effects that improve the polymer composite's overall performance. The coating resolves issues like aggregation and uneven dispersion that arise from incorporating nanocomposites. Concurrently, the intrinsic qualities of nanocomposite materials facilitate mechanical, thermal, and electrical improvements. The benefits of each methodology work in concert to enhance the overall performance of the polymer composite, resulting in a synergistic impact. The majority of research in this area characterizes the synergy of the reinforcing materials and looks at the emergence of the so-called hybrid effect. Composite materials benefit from a synergistic effect that improves their overall characteristics when filler coating methods and nanoparticles are combined. This synergy is produced by the interaction of several mechanisms and leads to enhanced materials with superior performance characteristics, as follows.

### 15.4.1 Improved Interfacial Bonding

The improved interfacial bonding between the filler coatings, the polymer matrix, and the nanomaterials is a key mechanism underlying the synergy. Interfacial modification can significantly improve the mechanical properties of polymer composites, including their modulus, impact resistance, and tensile strength. When interfaces are suitably modified to enable efficient load transfer between the reinforcement and the matrix, structural integrity is enhanced. By slowing down the rate of heat transfer across the interface, interfacial modification can improve the thermal stability of polymer composite materials. This is particularly crucial for applications requiring resilience to high temperatures. To improve electrical conductivity in conductive polymer composites, interfacial modification is necessary. Effective electron transport between the conductive filler and the polymer matrix can be ensured by altering the interface. Polymer composites can be made more resilient to environmental elements, including moisture, chemicals, and UV light, through the process of interfacial modification. The increased resilience of these materials broadens their potential uses. Better dispersion of reinforcing materials inside the polymer matrix is ensured by appropriate interfacial modification. This improved dispersion leads to more homogenous properties and reduces the risk of agglomeration. Feng et al. inferred the enhancement of the interfacial adhesion with the fabrication of carbon fiber (CF)–graphene oxide (GO)–carbon nanotubes (CNTs)–polyamide (PA) and CF-CNTs-GO-PA due to the interfacial interactions and mechanical interlocking effect of the composite [50]. By choosing the right modifiers to satisfy certain application needs, researchers can optimize certain qualities.

Poor interfacial adhesion between fillers and the polymer matrix can result in a range of challenges, of limited durability, including reduced mechanical properties, lower thermal stability, inefficient filler dispersion, decreased electrical conductivity, and difficulties in tailoring material properties. In conductive polymer composites, low interfacial adhesion can obstruct the electron flow between conductive fillers, reducing electrical conductivity. This limitation is detrimental in applications where electrical properties are crucial. Weak interfacial bonds can make the composite susceptible to environmental factors, such as moisture, chemicals, and UV radiation. This can lead to degradation, reduced lifespan, and decreased performance over time. Poor interfacial adhesion often results in uneven filler dispersion within the polymer matrix, and agglomeration of fillers can occur, leading to variations in material properties and reduced overall performance. Weak interfacial adhesion limits the ability to tailor the properties of polymer composites according to specific application requirements. Various strategies (e.g., surface modification of fillers, advanced coating techniques, compatibilization agents, in situ polymerization, etc.) have been devised to augment interfacial bonding in polymer composites, ensuring a robust connection between the filler materials and the matrix [51].

In order to promote robust chemical interactions at the interface, the coating materials are selected to be chemically compatible with both the nanomaterials and the polymer matrix. The introduction of nanomaterials, such as graphene or carbon nanotubes, into this environment strengthens the bonding even more due to their distinct chemical characteristics and large surface area.

### 15.4.2 Uniform Dispersion

Coated fillers are essential to disperse nanoparticles uniformly throughout the polymer matrix. Because of their high surface energy, nanomaterials are prone to clumping together, which can cause the composite to have uneven qualities and weak points. Coatings act as a barrier between nanomaterials, preventing them from clustering together. Szeluga et al. report the uniform dispersion behavior of the hybrid polymer/nanocomposite materials due to the reinforcing effect related to the particle structure and geometry [52]. This uniform dispersion makes sure they are dispersed equally throughout the matrix to maximize the reinforcing effects of the nanomaterials. Such homogeneity would be difficult to achieve without coatings, and the performance of the composite would suffer.

### 15.4.3 Tailored Surface Properties

Coatings can be designed to give nanoparticles particular surface qualities, including hydrophobic or hydrophilic qualities. Better compatibility with the polymer matrix and control over the composite's overall performance in various conditions are made possible by this customization, for example, coatings can be made to make the nanoparticles hydrophobic in situations where moisture resistance is essential, guaranteeing that the composite maintains its mechanical integrity even in humid environments [53]. This surface property tuning increases the composite's adaptability.

### 15.4.4 Enhanced Thermal Conductivity

The combination of coated fillers and nanoparticles can result in improved thermal conductivity in applications where thermal management is crucial, such as in the electronics or aerospace sectors. High-thermal-conductivity nanomaterials effectively transmit heat inside the composite. Zhao et al. strategically constructed a three-dimensional thermally conductive network with 2D boron nitride nanosheets (BNNSs) and 0D boron nitride microspheres (BNMSs) to enhance the thermal conductivity of epoxy composites [54]. Moreover, the coating materials might aid in promoting thermal stability by halting the degradation of the polymer matrix over time. Together, these elements create composites that effectively dissipate heat, preventing overheating and ensuring the long-term reliability of their component parts.

### 15.4.5 Barrier against Moisture/Contaminants

In order to keep gases, moisture, and other contaminants out of the composite material, coatings function as a barrier. The mechanical properties and dimensional stability of the composite material must be maintained by this barrier in dry or harsh environments. Without the protective layer, the composite's performance could progressively deteriorate. The combined use of nanomaterials and filler coating techniques results in composite materials with enhanced resistance to moisture and contaminants, homogenous dispersion, greater heat conductivity, and customized surface properties. When combined, these mechanisms offer state-of-the-art materials with remarkable performance characteristics suitable for a range of applications.

A composite material with enhanced mechanical strength, thermal conductivity, and electrical characteristics is demonstrated via the hybrid technique. This multifunctionality broadens the scope of applications, ranging from high-performance electrical gadgets to aerospace structural components, where a blend of exceptional qualities is crucial. Although there are many benefits to the hybrid approach, issues including consistency in production, cost-effectiveness, and scalability must be resolved. Persistent investigation endeavors to surmount these obstacles and enhance the hybrid methodology for extensive industrial implementation. The effective fusion of coated filler techniques with nanomaterials has produced impressive results in a number of sectors to demonstrate the tangible benefits and enhanced properties achieved through this synergistic approach.

## 15.5 CHALLENGES AND FUTURE PERSPECTIVES

The field of hybrid polymer composites has a number of challenges, especially when it comes to materials strengthened by nanoparticles and sophisticated filler coatings. Achieving uniform dispersion of nanoparticles within the polymer matrix and ensuring compatibility pose significant hurdles. The transition from lab-scale production to large-scale manufacturing without compromising quality and cost-effectiveness is a substantial challenge, requiring innovative solutions. Additionally, addressing the cost-efficiency of nanomaterials and advanced coatings is imperative for their widespread application, calling for strategies to reduce production costs. Environmental and safety concerns associated with nanomaterials must also be addressed to promote sustainable development. Establishing industry standards and regulations for hybrid materials is crucial to ensure their safe and consistent use. Wu et al. also report the different challenges and opportunities of biodegradable polymers, specifically in the packaging sector [53].

Researchers are expected to concentrate on developing multifunctional hybrid materials suited to specific uses. In line with international environmental goals, there will likely be an increase in demand for environmentally friendly and sustainable nanomaterials and coatings. Advanced manufacturing techniques, particularly innovations in additive manufacturing, will streamline the production of intricate hybrid structures [55]. Hybrid materials are poised to play a pivotal role in biomedical breakthroughs, revolutionizing drug delivery systems, tissue engineering, and personalized medicine [56]. Integrating artificial intelligence (AI) and machine learning is expected to optimize material design and predictive modeling for hybrid composites [57–60]. In summary, this comprehensive examination of challenges and future perspectives serves as a valuable guide for researchers, engineers, and industry professionals navigating the evolving landscape of hybrid polymer composites.

## 15.6 CONCLUSIONS

The hybrid nanocomposite and filler coating approach opens future research and innovation avenues. Continuous advancements in coating techniques, compatibility optimization, and sustainable practices will likely contribute to the evolution of this hybrid strategy. Hybrid nanocomposites provide not only better dispersion or compatibility with the polymer but also the improved performance for various applications. The hybrid approach of combining nanocomposites with filler coating in polymer composites represents a cutting-edge solution that maximizes the strengths of both techniques. This synergy results in a composite material with superior mechanical, thermal, and electrical properties, paving the way for a new generation of high-performance materials across diverse industries.

## ACKNOWLEDGMENTS

First and foremost, I am deeply thankful to Dr. Priya Goel, my supervisor, for their guidance, encouragement, and unwavering support throughout the writing process. Their expertise and insights have been instrumental in shaping the content of this chapter. Furthermore, I am grateful to GSFC University, Department of Chemical Sciences, for providing the necessary resources and facilities that facilitated the research and writing of this chapter. I would like to express my sincere gratitude to Mr. Debajyoti Palai and all co-authors who collaborated with me on this chapter to enrich the content and strengthen the overall quality of the chapter.

## REFERENCES

[1] R. Hsissou, R. Seghiri, Z. Benzekri, M. Hilali, M. Rafik, A. Elharfi, Polymer composite materials: A comprehensive review, Composite Structure 262 (2021) 113640. https://doi.org/10.1016/J.COMPSTRUCT.2021.113640.

[2] K. Bin Liew, C.F. Goh, S. Asghar, H.K. Syed, Overview of mechanical and physicochemical properties of polymer matrix composites, Encyclopedia of Materials: Composites 1 (2021) 565–576. https://doi.org/10.1016/B978-0-12-819724-0.00049-5.

[3] V.V. Kumar, G. Balaganesan, J.K.Y. Lee, R.E. Neisiany, S. Surendran, S. Ramakrishna, A review of recent advances in nanoengineered polymer composites, Polymers (Basel) 11 (2019). https://doi.org/10.3390/polym11040644.

[4] H.S. Gupta, S.K. Sethi, A. Verma, Applications of coating materials: A critical overview, in: Materials horizons: From nature to nanomaterials part F, Springer, Singapore, 2023: vol. 1086, pp. 81–109. https://doi.org/10.1007/978-981-99-3549-9_4/COVER.

[5] A. Imam, S.K. Sethi, M.M.H. Tusher, M. Afroja Khatun, H.S. Gupta, A. Verma, Application of molecular dynamics simulations in coatings and composites, in: Dynamic mechanical and creep-recovery behavior of polymer-based composites, Elsevier, 2024: pp. 419–449. https://doi.org/10.1016/B978-0-443-19009-4.00022-9.

[6] H.S. Gupta, P. Kumar, A.K. Sethi, A. Verma, S.K. Sethi, Software-based simulations of the creep recovery model of polymer composites, in: Dynamic mechanical and creep-recovery behavior of polymer-based composites, Elsevier, 2024: pp. 271–289. https://doi.org/10.1016/B978-0-443-19009-4.00016-3.

[7] S.K. Sethi, G. Manik, A combined theoretical and experimental investigation on the wettability of MWCNT filled PVAc-g-PDMS easy-clean coating, Progress in Organic Coatings 151 (2021) 106092. https://doi.org/10.1016/j.porgcoat.2020.106092.

[8] A. Verma, N. Jain, S.K. Sethi, Modeling and simulation of graphene-based composites, in: Innovations in graphene-based polymer composites, Elsevier, 2022: pp. 167–198. https://doi.org/10.1016/B978-0-12-823789-2.00001-7.

[9] S.K. Sethi, R. Gogoi, G. Manik, Plastics in self-cleaning applications, in: Reference module in materials science and materials engineering, Elsevier, 2021. https://doi.org/10.1016/B978-0-12-820352-1.00113-9.

[10] F.M.I.G., C.Pvt. Ltd., Advanced polymer composites market is anticipated to (2023) 4. https://www.globenewswire.com/en/news-release/2023/01/25/2595439/0/en/Advanced-Polymer-Composites-Market-is-Anticipated-to-Surpass-a-Valuation-of-US-19-73-Billion-by-the-year-2032-Advancing-at-a-CAGR-of-5-9-Future-Market-Insights-Inc.html (accessed January 25, 2024).

[11] A.P.C.M. Value, Advanced polymer composites market share & trends 2032|FMI (2022). https://www.futuremarketinsights.com/reports/advanced-polymer-composites-market (accessed January 25, 2024).
[12] S.S. Godara, A. Yadav, B. Goswami, R.S. Rana, Review on history and characterization of polymer composite materials, Materials Today: Proceedings 44 (2021) 2674–2677. https://doi.org/10.1016/J.MATPR.2020.12.680.
[13] Z. Benzait, L. Trabzon, A review of recent research on materials used in polymer–matrix composites for body armor application, Journal of Composite Materials 52 (2018) 3241–3263. https://doi.org/10.1177/0021998318764002.
[14] N.R.R. Anbusagar, K. Palanikumar, A. Ponshanmugakumar, Preparation and properties of nanopolymer advanced composites: A review, in: Polymer-based nanocomposites for energy and environmental applications: A volume in Woodhead Publishing series in composites science and engineering, Elsevier, 2018: pp. 27–73. https://doi.org/10.1016/B978-0-08-102262-7.00002-7.
[15] S. Malik, K. Muhammad, Y. Waheed, Nanotechnology: A revolution in modern industry, Molecules 28 (2023). https://doi.org/10.3390/molecules28020661.
[16] R. Gangopadhyay, A. De, Conducting polymer nanocomposites: A brief overview, Chemistry of Materials 12 (2000) 608–622. https://doi.org/10.1021/CM990537F/ASSET/IMAGES/MEDIUM/CM990537FN00001.GIF.
[17] R. Krishnamoorti, R.A. Vaia, Polymer nanocomposites, Journal of Polymer Science Part B: Polymer Physics 45 (2007) 3252–3256. https://doi.org/10.1002/POLB.21319.
[18] S. Fu, Z. Sun, P. Huang, Y. Li, N. Hu, Some basic aspects of polymer nanocomposites: A critical review, Nano Materials Science 1 (2019) 2–30. https://doi.org/10.1016/J.NANOMS.2019.02.006.
[19] A. Kausar, I. Ahmad, T. Zhao, O. Aldaghri, K.H. Ibnaouf, M.H. Eisa, Nanocomposite nanofibers of graphene: Fundamentals and systematic developments, Journal of Composites Science 7 (2023). https://doi.org/10.3390/jcs7080323.
[20] D.R. Paul, L.M. Robeson, Polymer nanotechnology: Nanocomposites, Polymer 49 (2008) 3187–3204. https://doi.org/10.1016/j.polymer.2008.04.017.
[21] R. Sengupta, M. Bhattacharya, S. Bandyopadhyay, A.K. Bhowmick, A review on the mechanical and electrical properties of graphite and modified graphite reinforced polymer composites, Progress in Polymer Science 36 (2011) 638–670. https://doi.org/10.1016/J.PROGPOLYMSCI.2010.11.003.
[22] M.S. Mauter, M. Elimelech, Environmental applications of carbon-based nanomaterials, Environmental Science and Technology 42 (2008) 5843–5859. https://doi.org/10.1021/es8006904.
[23] S. Parham, D.H.B. Wicaksono, S. Bagherbaigi, S.L. Lee, H. Nur, Antimicrobial treatment of different metal oxide nanoparticles: A critical review, Journal of the Chinese Chemical Society 63 (2016) 385–393. https://doi.org/10.1002/jccs.201500446.
[24] S.K. Soni, B. Thomas, V.R. Kar, A comprehensive review on CNTs and CNT-reinforced composites: Syntheses, characteristics and applications, Materials Today Communications 25 (2020). https://doi.org/10.1016/j.mtcomm.2020.101546.
[25] J. Ren, Q. Li, L. Yan, L. Jia, X. Huang, L. Zhao, Q. Ran, M. Fu, Enhanced thermal conductivity of epoxy composites by introducing graphene@boron nitride nanosheets hybrid nanoparticles, Materials & Design 191 (2020) 108663. https://doi.org/10.1016/J.MATDES.2020.108663.
[26] T. Shahnaz, G. Hayder, M.A. Shah, M.Z. Ramli, N. Ismail, C.K. Hua, N.M. Zahari, N.H. Mardi, F.E. Selamat, N. Kabilmiharbi, H.A. Aziz, Graphene-based nanoarchitecture as a potent cushioning/filler in polymer composites and their applications, Journal of Materials Research and Technology 28 (2024) 2671–2698. https://doi.org/10.1016/J.JMRT.2023.12.108.
[27] B.V. Basheer, J.J. George, S. Siengchin, J. Parameswaranpillai, Polymer grafted carbon nanotubes: Synthesis, properties, and applications: A review, Nano-Structures and Nano-Objects 22 (2020). https://doi.org/10.1016/j.nanoso.2020.100429.
[28] S.Y. Huang, T.I. Wu, C.M. Chou, V.K.S. Hsiao, Tailoring mechanical and electrical properties of polydimethylsiloxane nanocomposites With graphene and carbon nanotubes for wearable electronics, Polymers and Polymer Composites 31 (2023) 1–11. https://doi.org/10.1177/09673911231188296.
[29] V. Georgakilas, J.N. Tiwari, K.C. Kemp, J.A. Perman, A.B. Bourlinos, K.S. Kim, R. Zboril, Noncovalent functionalization of graphene and graphene oxide for energy materials, biosensing, catalytic, and biomedical applications, Chemical Reviews 116 (2016) 5464–5519. https://doi.org/10.1021/acs.chemrev.5b00620.
[30] S.R. Schaffazick, A.R. Pohlmann, T. Dalla-Costa, S.S. Guterres, Freeze-drying polymeric colloidal suspensions: Nanocapsules, nanospheres and nanodispersion: A comparative study, European Journal of Pharmaceutics and Biopharmaceutics 56 (2003) 501–505. https://doi.org/10.1016/S0939-6411(03)00139-5.

[31] A. Zielińska, B.P. Filipa Carreiró, A.M. Oliveira, A. Neves, D.N. Venkatesh, A. Durazzo, P. Eder, M. Lucarini, A.M. Silva, A. Santini, E.B. Souto, Polymeric nanoparticles: Production, characterization, toxicology and ecotoxicology, Molecules 25 (2020) 3731. https://doi.org/10.3390/molecules 25163731.

[32] H.M. Aliabadi, A. Lavasanifar, Polymeric micelles for drug delivery, Expert Opinion on Drug Delivery 3 (2006) 139–162. https://doi.org/10.1517/17425247.3.1.139.

[33] L. Peltonen, M. Singhal, J. Hirvonen, Chapter 1: Principles of nanosized drug delivery systems, in: Nanoengineered biomaterials for advanced drug delivery, 2020: pp. 3–25. https://doi.org/10.1016/B978-0-08-102985-5.00001-2.

[34] M. Mehta, T.A. Bui, X. Yang, Y. Aksoy, E.M. Goldys, W. Deng, Lipid-based nanoparticles for drug/gene delivery: An overview of the production techniques and difficulties encountered in their industrial development, ACS Materials Au 3 (2023) 600–619. https://doi.org/10.1021/acsmaterialsau.3c00032.

[35] A.M.J. Newton, S. Kaur, Solid lipid nanoparticles for skin and drug delivery: Methods of preparation and characterization techniques and applications, Nanoarchitectonics in Biomedicine (2019) 295–334. https://doi.org/10.1016/B978-0-12-816200-2.00015-3.

[36] P. Mittal, A. Saharan, R. Verma, F.M.A. Altalbawy, M.A. Alfaidi, G.E.S. Batiha, W. Akter, R.K. Gautam, M.S. Uddin, M.S. Rahman, Dendrimers: A new race of pharmaceutical nanocarriers, BioMed Research International (2021). https://doi.org/10.1155/2021/8844030.

[37] E. Abbasi, S.F. Aval, A. Akbarzadeh, M. Milani, H.T. Nasrabadi, Dendrimers: Synthesis, applications, and properties, Nanoscale Research Letters 9 (2014) 1–10.

[38] A. Mahapatro, D.K. Singh, Biodegradable nanoparticles are excellent vehicle for site directed in-vivo delivery of drugs and vaccines, Journal of Nanobiotechnology 9 (2011) 55. https://doi.org/10.1186/1477-3155-9-55.

[39] Á.I. López-Lorente, M. Valcárcel, Analytical nanoscience and nanotechnology, in: Comprehensive analytical chemistry, Elsevier, 2014: pp. 3–35. https://doi.org/10.1016/B978-0-444-63285-2.00001-8.

[40] A. Ali, H. Zafar, M. Zia, I. ul Haq, A.R. Phull, J.S. Ali, A. Hussain, Synthesis, characterization, applications, and challenges of iron oxide nanoparticles, Nanotechnology, Science and Applications 9 (2016) 49–67. https://doi.org/10.2147/NSA.S99986.

[41] S. Mehnath, A.K. Das, S.K. Verma, M. Jeyaraj, Biosynthesized/green-synthesized nanomaterials as potential vehicles for delivery of antibiotics/drugs, 1st ed., Elsevier B.V., 2021. https://doi.org/10.1016/bs.coac.2020.12.011.

[42] M.A. Cotta, Quantum dots and their applications: What lies ahead?, ACS Applied Nano Materials 3 (2020) 4920–4924. https://doi.org/10.1021/acsanm.0c01386.

[43] N. Sathishkumar, B.J. Toley, Paper-microfluidic signal-enhanced immunoassays, in: Progress in molecular biology and translational science, Academic Press, 2022: pp. 267–288. https://doi.org/10.1016/BS.PMBTS.2021.07.015.

[44] D. Singh, S. Singh, J. Sahu, S. Srivastava, M.R. Singh, Ceramic nanoparticles: Recompense, cellular uptake and toxicity concerns, Artificial Cells, Nanomedicine and Biotechnology 44 (2016) 401–409. https://doi.org/10.3109/21691401.2014.955106.

[45] J. Chen, R. An, W.S. Tey, Q. Zeng, L. Zhao, K. Zhou, In situ filler addition for homogeneous dispersion of carbon nanotubes in multi jet fusion-printed elastomer composites, Advanced Science 10 (2023) 1–11. https://doi.org/10.1002/advs.202300593.

[46] S. Guo, R. Zheng, J. Jiang, J. Yu, K. Dai, C. Yan, Enhanced thermal conductivity and retained electrical insulation of heat spreader by incorporating alumina-deposited graphene filler in nano-fibrillated cellulose, Composites Part B: Engineering 178 (2019) 107489. https://doi.org/10.1016/J.COMPOSITESB.2019.107489.

[47] Z. Sha, X. Cheng, M.S. Islam, P. Sangkarat, W. Chang, S.A. Brown, S. Wu, J. Zhang, Z. Han, S. Peng, C.H. Wang, Synergistically enhancing the electrical conductivity of carbon fibre reinforced polymers by vertical graphene and silver nanowires, Composites Part A: Applied Science and Manufacturing 168 (2023) 107463. https://doi.org/10.1016/J.COMPOSITESA.2023.107463.

[48] K. Formela, M.R. Saeb, Ł. Klapiszewski, N. Sienkiewicz, M. Dominic, J. Parameswaranpillai, Natural fillers as potential modifying agents for epoxy composition: A review, Polymers 14 (2022) 265. https://doi.org/10.3390/POLYM14020265.

[49] I. Charitos, G. Georgousis, P.A. Klonos, A. Kyritsis, D. Mouzakis, Y. Raptis, A. Kontos, E. Kontou, The synergistic effect on the thermomechanical and electrical properties of carbonaceous hybrid polymer nanocomposites, Polymer Testing 95 (2021) 107102. https://doi.org/10.1016/J.POLYMERTESTING.2021.107102.

[50] P. Feng, G. Song, X. Zhu, D. Lv, Y. Zhao, X. Yang, N. Li, L. Zhang, L. Ma, Enhanced interfacial adhesion of carbon fiber/epoxy composites by synergistic reinforcement with multiscale "rigid-flexible" structure at interphase, Composites Part B: Engineering 225 (2021) 109315. https://doi.org/10.1016/J.COMPOSITESB.2021.109315.
[51] S.Y. Mun, J. Ha, S. Lee, Y. Ju, H.M. Lim, D. Lee, Prediction of enhanced interfacial bonding strength for basalt fiber/epoxy composites by micromechanical and thermomechanical analyses, Composites Part A: Applied Science and Manufacturing 142 (2021) 106208. https://doi.org/10.1016/J.COMPOSITESA.2020.106208.
[52] U. Szeluga, B. Kumanek, B. Trzebicka, Synergy in hybrid polymer/nanocarbon composites. A review, Composites Part A: Applied Science and Manufacturing 73 (2015) 204–231. https://doi.org/10.1016/J.COMPOSITESA.2015.02.021.
[53] P.Y. Hung, K.T. Lau, Q. Guo, B. Jia, B. Fox, Tailoring specific properties of polymer-based composites by using graphene and its associated compounds, International Journal of Smart and Nano Materials 11 (2020) 173–189. https://doi.org/10.1080/19475411.2020.1786477.
[54] L. Zhao, L. Yan, C. Wei, Q. Li, X. Huang, Z. Wang, M. Fu, J. Ren, Synergistic enhanced thermal conductivity of epoxy composites with boron nitride nanosheets and microspheres, Journal of Physical Chemistry C 124 (2020) 12723–12733. https://doi.org/10.1021/ACS.JPCC.0C01377/SUPPL_FILE/JP0C01377_SI_001.PDF.
[55] F. Guo, S. Aryana, Y. Han, Y. Jiao, A review of the synthesis and applications of polymer-nanoclay composites, Applied Sciences (Switzerland) 8 (2018) 1–29. https://doi.org/10.3390/app8091696.
[56] U. Szeluga, B. Kumanek, B. Trzebicka, Synergy in hybrid polymer/nanocarbon composites: A review, Composites Part A: Applied Science and Manufacturing 73 (2015) 204–231. https://doi.org/10.1016/J.COMPOSITESA.2015.02.021.
[57] F. Kibrete, T. Trzepieciński, H.S. Gebremedhen, D.E. Woldemichael, Artificial intelligence in predicting mechanical properties of composite materials, Journal of Composites Science 7 (2023). https://doi.org/10.3390/jcs7090364.
[58] T.B. Martin, D.J. Audus, Emerging trends in machine learning: A polymer perspective, ACS Polymers Au 3 (2023) 239–258. https://doi.org/10.1021/acspolymersau.2c00053.
[59] S. Cassola, M. Duhovic, T. Schmidt, D. May, Machine learning for polymer composites process simulation: A review, Composites Part B: Engineering 246 (2022) 110208. https://doi.org/10.1016/J.COMPOSITESB.2022.110208.
[60] P. Pattnaik, A. Sharma, M. Choudhary, V. Singh, P. Agarwal, V. Kukshal, Role of machine learning in the field of Fiber reinforced polymer composites: A preliminary discussion, Materials Today: Proceedings 44 (2021) 4703–4708. https://doi.org/10.1016/J.MATPR.2020.11.026.

# 16 Porous Polymers and Composites *via* Electrospinning and Emulsion Templating

## *Fundamentals and Computational Approaches*

*Mukesh Bajya, Pramod M. Gurave, and Anilkumar Yadav*

## 16.1 INTRODUCTION

*Porosity*, defined as the space or cavity in the form of pores, holes, or channels, is nothing but the free volume in a given structure [1]. The International Union of Pure and Applied Chemistry (IUPAC) classifies *porous materials* as microporous, mesoporous, and macroporous, with pore size <2 nm, 2–50 nm, and >50 nm, respectively [2]. In some of the examples, materials with micrometer-sized pores, however, have also been considered as microporous, depending upon the field of the application. A number of porous materials are identified in nature, such as soil, rocks (volcanic, aquifers, and reservoirs), zeolites, sponge (an aquatic multicellular organism), beehive, wood, and biological tissues (lung and bones), where the uniqueness in their structure and properties originates from the pore architecture. Efforts to design and develop synthetic porous materials resembling the architecture and porosity of natural once have been made to take the advantage of porosity for various applications. Synthetic porous materials developed so far typically have organic, metallic, or ceramic (oxide-, nitride-, or carbide-based materials) origin [3]. The importance and demand of porous polymers in academia and industries may be visualized from the diagram showing the publication status of porous polymers (Figure 16.1). To check the publication status in terms of research papers in scientific journals and patents, a search in the SciFinder search engine was performed using the keywords *emulsion templating* and *electrospinning*. Interestingly, the number of research papers on electrospinning and emulsion templating published since 2010 to 2023 was 4,8000, respectively. Similarly, the number of patents on electrospinning and emulsion templating since 2010 accounted to about 21,000, respectively.

## 16.2 POROUS POLYMER PREPARATION TECHNIQUES

A number of conventional and advanced methods have been devised for fabrication of porous polymeric materials. Such materials are also referred to as scaffolds (3D porous structures), especially in the context of tissue engineering [4, 5]. In general, the techniques used for fabricating different porous polymers are classified as templating and nontemplating methods [6] .The templating methods are further divided in endo- and exotemplating. The endotemplating method involves the use of a sacrificial template, like porogen, which is eliminated after solidification of the matrix, forming a polymeric phase, to generate pores. Solvent casting, particulate leaching, microemulsion, and emulsion templating are some of the examples of such templating method. Exotemplating involves the infiltration of rigid porous solid template from the precursor material, which ultimately forms the porous structure after removal of the template. Activated-carbon route to create oxides with high

DOI: 10.1201/9781003476337-16

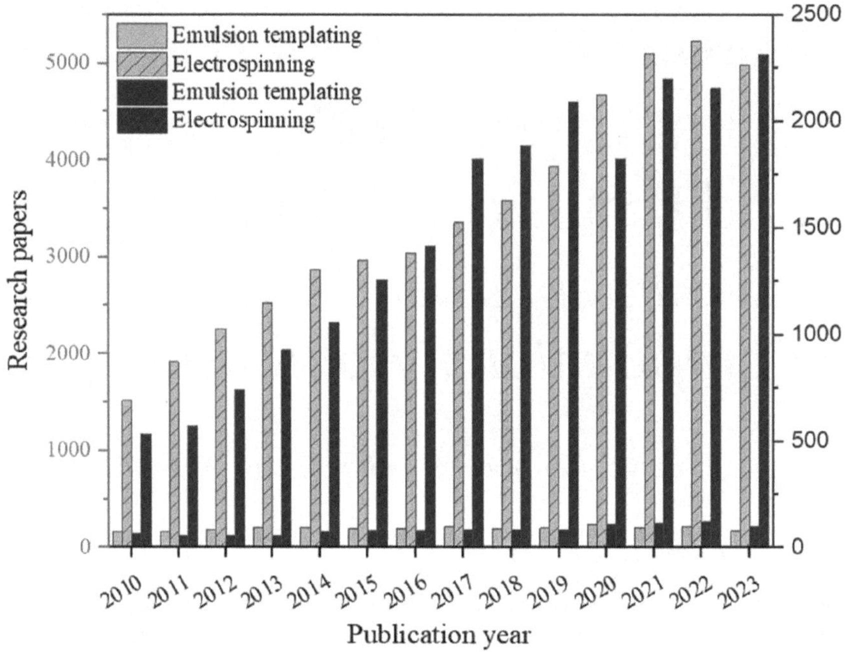


**FIGURE 16.1** Publication (research papers and patents) status of emulsion templating and electrospun nanofibers.

**Source: Statistical data obtained from SciFinder search in December 2023 with keywords *emulsion templating* and *electrospinning.***

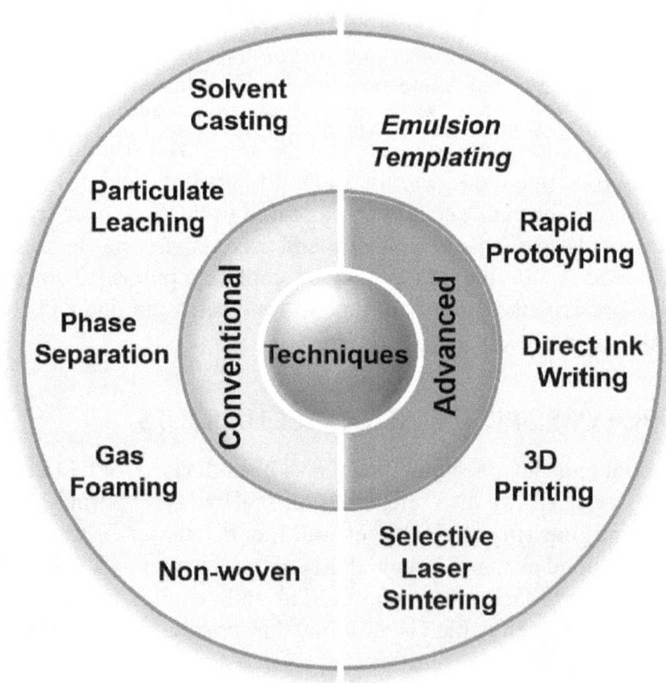


**FIGURE 16.2** Conventional and advanced techniques to design and develop porous polymers and composites thereof.

surface area, nanocasting procedures, and colloidal-crystal templating are examples of exotemplating. A nontemplating method, on the other hand, uses a polymer solution, where the occurrence of phase separation under specific conditions governs the formation of a solid matrix and fluidic porogen, and the removal of fluidic porogen generates porosity. Examples of nontemplating methods are thermally induced phase separation (solid–liquid, liquid–liquid, and polymerization-induced phase separation), cross-link-induced phase separation, gas blowing, and directional freezing.

The feasibility of melt or solution processing in various shapes and sizes and suitability for chemical modifications have made polymers a material of interest for developing different porous architectures. Porous polymers adorned with properties such as low specific density, high specific strength, controlled porosity with tunable pore size, and high surface area have thus been used in numerous household, industrial, and biomedical applications. In case of biomedical applications—specifically, tissue engineering—porous polymeric structures known as scaffolds are used to provide support and necessary nutrients for the regeneration of damaged tissues. Moreover, several synthetic and natural polymers are used to design and develop scaffolds (porous polymers) through different conventional and advanced scaffold fabrication techniques, which have been summarized in Figure 16.2 and in successive sections.

## 16.3 CONVENTIONAL TECHNIQUES

Conventional scaffolding techniques are relatively simple, inexpensive, user-friendly, and easy to operate. They do not require specialized equipment and are suitable for a wide range of polymers. Various conventional scaffolding techniques, such as particulate leaching, solvent casting, gas foaming, melt molding, membrane lamination, thermally induced phase separation, supercritical fluid method, and textile technologies (nonwoven materials), are discussed in the following.

### 16.3.1 Porogen Leaching

Porogen or particulate leaching is one of the popular techniques to fabricate porous polymeric scaffolds for tissue engineering applications. This method involves the mixing of particulate matters, such as salt, sugar, or wax, of required shape and size, in a polymer solution. The resulting mixture of porogen and polymer solution is casted into a mold, and a solid polymer matrix is obtained by solvent evaporation *via* room temperature (RT) drying, vacuum drying, and/or lyophilization. Porogen particles (salt and/or sugar) are extracted from the polymer matrix with repetitive washings in order to obtain a porous polymeric matrix to be useful as a scaffold. The porogen leaching method is distinguished as a simple method without any expensive equipment, and it is easy to operate. Control on the porosity level and shape and size of the pores in resulting scaffolds is achieved by varying the porogen loading and its shape and size. Non-uniform pore size distribution due to density difference between porogen and the polymer solution and difficulty in complete removal of porogen particles from the thick scaffolds are, however, the main limitations of this method. Moreover, skinning effect at the surface of scaffolds due to solvent evaporation inhibits the complete removal of porogen particles from inner polymer layers. Most of the scaffolds prepared by porogen leaching are thus porous thin films (0.5–2 mm). Layering and lamination of thin porous polymer films have been alternatively carried out to produce 3D scaffolds.

### 16.3.2 Solvent Casting

Solvent casting is another simple and inexpensive method to prepare porous polymeric scaffolds of high porosity without involvement of an expensive setup. It works on the principle of control evaporation of solvent from polymer solution for the generation of pores. Generally, two solvent casting routes are practiced for developing the porous polymeric scaffolds of solvent/non-solvent-induced phase separation in the form of a membrane. Use of toxic organic solvents, high processing time, and poor pore size distribution are the main drawbacks of this method. Use of high vacuum for the complete extraction of solvents and carrying out solvent casting with porogen leaching or phase separation is an improvement to overcome the limitations of this method.

### 16.3.3 Gas Foaming

Gas foaming is a relatively rapid method for preparing the scaffolds of desirable thickness through nucleation and growth of gas bubbles in a viscous polymer solution. *In situ* gas formation (dispersed phase) by a gas-forming agent—for example, sodium borohydride, hydrogen peroxide, sodium bicarbonate, and a mixture of ammonium bicarbonate and citric acid *via* chemical reactions at moderate temperature—transforms a polymer solution (continuous phase) into a foam. Stabilization of foam and solvent removal by lyophilization result in development of lightweight 3D porous matrix, highly suitable for various biomedical and industrial applications. Highly porous structures of polymers and ceramics have been described within gas foaming. Gas foaming through chemical reactions, and its combination with electrospinning or microfluidics, has also been described to create novel porous architectures. Due to their advantageous properties, such as excellent mass transfer, negligible surface tension, and variable density at different temperature and pressure, supercritical fluids have recently been considered the better choice of materials for obtaining a gas-saturated stage of a polymer. Gas foaming using supercritical fluid—generally, supercritical carbon dioxide ($scCO_2$; critical point, 31°C, and 73.8 bar)—is extensively used for creating porosity in amorphous and semi-crystalline polymers. The advantages of low viscosity, high diffusivity of $scCO_2$, and solvent-free operation at low temperature in dry conditions create a favorable environment for the incorporation of water, solvent, or heat-sensitive molecules, such as drugs, DNA, and proteins, in tissue engineering scaffolds. The requirement of an expensive infrastructure for liquefying the $CO_2$, critical process controls, and pore interconnectivity are the key challenges that need resolution before this process can be accepted at large scale.

### 16.3.4 Thermally Induced Phase Separation

Thermally induced phase separation (TIPS) has been used for producing several porous scaffolds with different morphological structures, and it has also been combined with other scaffolding methods, such as porogen leaching, to improve the scaffold's architecture. TIPS involves porosity generation through reducing the polymer solubility either by lowering the temperature or precipitating the polymer by freezing. There are two types of TIPS, liquid–liquid and solid–liquid phase separation. In liquid–liquid phase separation, lowering the temperature of the polymer solution results in the formation of rich and deprived polymer regions. The polymer-rich region solidifies, while the polymer-poor regions are extracted during the porosity generation. Solid–liquid phase separation progresses through freezing of polymer solution for developing porous polymers. The concentration of the polymer solution, rate of cooling, and temperature of phase separation are the factors that affect the porous morphology of resulting scaffolds. Polyurethane-based tissue engineering scaffolds for cardiovascular, bone, etc. has been extensively developed using TIPS.

### 16.3.5 Textile Technologies

Highly permeable, fiber-based scaffolds with especially designed structures and functionalities are possible to develop using textile materials, that is, nonwovens [5]. The preparation of nonwoven materials *via* fiber bonding processes, such as dry-laid, wet-laid, needle punching, thermal and chemical bonding, and fiber selection for nonwoven preparation, has been thoroughly documented. The nonwoven making processes are advantageous as they are simple, are environment-friendly, have high production rate, and offer the feasibility to fabricate precisely engineered structures. Moreover, nonwovens can be effectively designed with porous morphology analogues to natural extracellular matrix (ECM); thus, nonwoven structures are highly suitable as tissue engineering scaffolds. Nonwoven tissue engineering scaffolds based on collagen-enforced polylactic–glycolic acid, silk-fibroin, hyaluronan benzylic esters, and alginate-incorporated polyglycolic–polylactic copolymers have been reported.

### 16.3.6 Electrospinning

The fabrication of polymeric nanofibrous membrane using electrospinning was first reported in early 1930s and later patented in 1934 [7]. A number of electrospun nanofibrous membranes of natural and synthetic polymers have been demonstrated for usefulness in agriculture, sorption, filtration, separation [8], sensors, and biomedical fields. The four main components of electrospinning setup are a syringe pump, reservoir for polymer solution (generally a plastic syringe with metallic tip), high-voltage power supply, and grounded collector (electroconductive). Figure 16.3 depicts a typical electrospinning unit and scanning electron microscopic image (SEM) of electrospun nanofibrous membrane. In the span of the electrospinning process, three stages, viz. Taylor cone generation from droplet of polymer solution, jet stretching, and whipping action, constitute for the generation of fine fibers when a polymer solution with threshold viscosity comes under strong electric field. A high-potential difference is created between the metallic tip and grounded collector, and when the applied voltage overcomes the surface tension of the polymer solution, a fine jet of polymer solution emerges and deposits on the collector. While the polymer jet progresses toward the collector, the solvent evaporates, leaving behind a dry and randomly stacked fibrous mesh with fiber diameter ranging from nano- to micrometers. Parameters such as polymer solution concentration, polymer–solvent interactions, needle diameter, distance between tip and collector, applied voltage, etc. have significant effect on fiber diameter and overall performance of electrospun matrices.

Electrospinning is a relatively simple technique for fabricating nanofibrous scaffolds of synthetic and natural polymers; however, the use of toxic organic solvents, heterogeneity in pore size distribution, low production rate, poor mechanical properties, and complexity in producing the intricate shape structures are some of the major challenges associated with it. In addition to solution electrospinning, melt-electrospinning is also performed, which typically results in fibers with larger diameter compared to solution electrospinning. A typical EDX pattern of Zn-based, MOF-reinforced electrospun nanofibrous composite membrane is shown in Figure 16.4.

In respect to electrospun nanofibrous composites, solution and emulsion electrospinning approaches have been widely utilized. The basic aim in utilizing emulsion or dispersion electrospinning is to strengthen the nanofibrous membrane by producing a composite of polymer and nanoparticles; additionally, a one-step approach for composite formation succeeded in this technique [9, 10]. There are several ways to enhance the mechanical properties of fibrous membrane, such as coating, lamination, blending, cross-linking, and use of nanomaterials [11]. Each way has their own merits and limitations. Coating and lamination raise the problem of durability of the membrane, whereas the blending process is unsure about the compatibility of polymers and void formation. Cross-linking and the use of nanomaterials during the electrospinning process have been a widely used method to improve the serviceability of nanofibrous membrane. The use of nanomaterials such

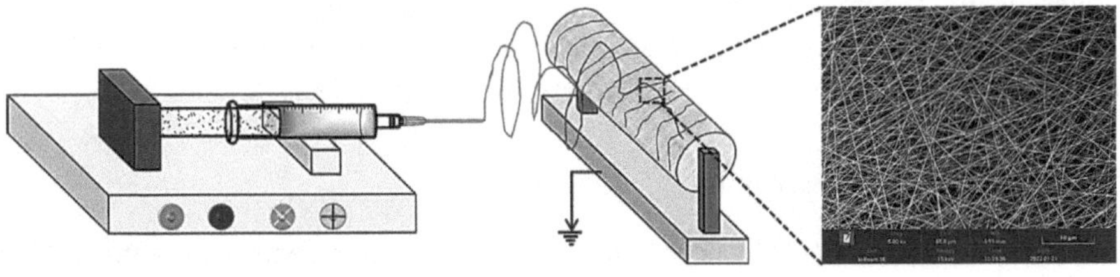

**FIGURE 16.3** Electrospinning unit and SEM image of electrospun nanofibrous porous membrane.

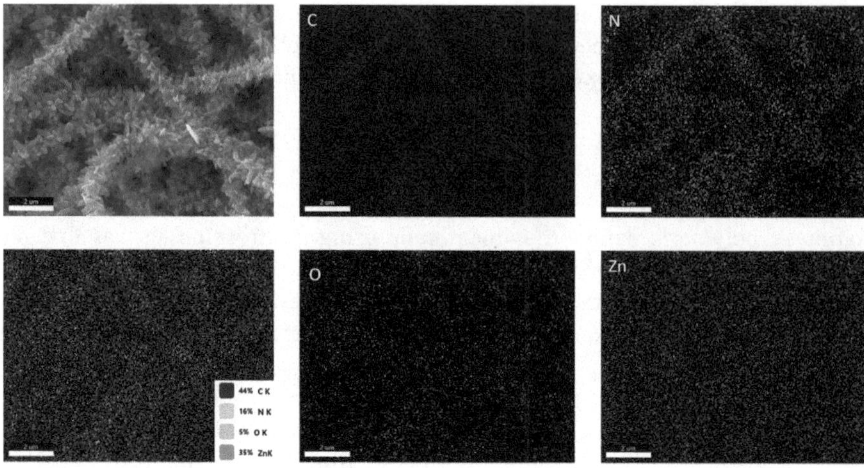


**FIGURE 16.4** EDX pattern of Zn-based, MOF-reinforced electrospun nanofibrous composite membrane.

as graphene oxides, silicone oxide, and metal organic frameworks in the microstructure of nanofiber enhances the various properties of the membrane, such as antimicrobial activity, mechanical and optical characteristics, and wettability modification of membrane surface. Recently, Gurave et. al. developed a unique technique named "near gel resin electrospinning" to produce robust yet flexible cross-linked nanofibrous membranes for water purification [12]. This type of electrospinning offers a one-step process for producing a functionalized cross-linked membrane with silica nanoparticles, with any need of preformed polymer and toxic solvent used as diluent for polymer in the spinning dope. The silica-composited nanofibrous membrane was utilized for the demulsification of various oil–water emulsions with efficient rate. The cross-linking and silica nanoparticles enhanced the wettability of electrospun membrane along with strength.

Nanomaterials provide great functionality to the electrospun membrane and enhance its relative surface-area-to-volume ratio. There are several reports available that discuss the importance of the integration of nanomaterials into the fine fibrillar microstructures of fibers. The incorporation of nanomaterials can be done *via* two methods, direct doping of particles during electrospinning and surface deposition on post-electrospinning of the membrane. The available functional groups on fiber surface as well surface of nanomaterials play a crucial role in deciding the successful integration into the fiber structure, antileaching property, and overall effectiveness of functionalized membranes. In these types, researchers have been used both methods depending on the property sought for. For instance, in applications where surface wettability—either hydrophobicity or hydrophilicity—needs to be improved, the nanomaterials should be present on the surface of the membrane. Primarily, modified silica; metal organic frameworks, such as ZIF-8, ZIF-67, and UiO-66 (pristine or modified); and candle soot, nanodiamonds, and carbon nanotubes have been used to modify the adhesion property of membrane toward water and oil media [13]. The bulk properties, such as antimicrobial, flame-retardancy, and strength enhancement, can be improved by incorporating nanomaterials into the fiber structure, and such membrane can be produced by electrospinning of polymer solution with well-dispersed nanomaterials. A glimpse of literatures studying electrospun nanofibrous and composite membranes is presented in Table 16.1.

## 16.4 ADVANCED TECHNIQUES

The use of organic solvents, lack of pore interconnectivity, poor control over porosity, pore size and shape, skinning effect, multistep processing, and difficulty in obtaining intricate 3D structures with interconnected pores and hierarchical pore structure are some of the serious limitations of

**TABLE 16.1**
**Nanofibrous Membranes and Composites Produced Using the Electrospinning Technique**

| Polymer System | Nanomaterial | Electrospinning Technique | Functionality |
|---|---|---|---|
| Polyvinyl alcohol, glutaraldehyde, phytic acid | Silica nanoparticles by Stober reaction | Solution electrospinning, followed by surface deposition of silica nanoparticles | Underwater superoleophobicity [14] |
| Polyacrylonitrile | ZnO | Coaxial electrospinning | UV protection [15] |
| Poly(ε-caprolactone) | Hydrophobic fumed silica nanoparticles | Pickering emulsion electrospinning | L929 fibroblast cell growth [16] |
| Cross-linked polyvinyl alcohol, cross-linked polystyrene | ZIF-8 MOF | Emulsion electrospinning | Anti-oil adhesion [17] |
| Poly(ε-caprolactone), lactose-modified chitosan | Silver nanoparticles | Solution electrospinning | Osteoblast adhesion and proliferation [18] |
| Polyimide | Laser-induced graphene | Electrospinning and laser ablation | Rapid and efficient interfacial water evaporation [19] |
| Poly(acrylonitrile), ferrous sulfate | $\alpha$-$Fe_2O_3$ nanoparticles | Electrospinning and carbonization | Acceleration of polysulfide conversion reactions in Li-S battery [20] |
| Polyvinyl alcohol, chitosan | Montmorillonite nanoclay | Single-dope electrospinning | Improved dye rejection [21] |
| Polyvinylidene fluoride-co-hexafluoropropylene, polydimethylsiloxane | Hydrophobically improved silica nanoparticles | Electrospinning and electrospraying | Superhydrophobicity (~170° water contact angle) [22] |
| Cellulose diacetate, tetraethyl orthosilicate | Vapor phase–based ZIF-8 | Sol-gel electrospinning and surface deposition | Environmental and biomedical applicability [23] |

conventional techniques. Such limitations of conventional scaffolding techniques have persuaded the need of rapid prototyping (RP) techniques which have also become popular as solid freeform fabrication (SFF) techniques. RP techniques can fabricate patient-specific tissue engineering scaffolds with ordered pore architecture using computer-aided designs (CAD) and computer-generated models. These techniques also offer the flexibility of processing materials like polymers, metals, and ceramics in the form of liquid, powder, filament, or a thin sheet in a layer-by-layer approach. Different RP techniques used to create polymeric scaffolds are briefly discussed in the next paragraphs.

### 16.4.1 3D Printing

3D printing is an additive manufacturing or RP technique which was invented at Massachusetts Institute of Technology in late 1980s. The importance of 3D printing is self-explanatory from its utility and the increasing number of publications (>5,000 in 2020) on 3D printing [24]. In 3D printing, layer-by-layer deposition, CAD, and software controls are integrated to construct the 3D objects of metals, ceramics, polymers, or composites. Objects are first developed by depositing the material in X–Y dimensions to form a base, and then building the desired 3D structure on this base through successive layer-by-layer material deposition in Z-direction. With the advent of 3D printing, it has been possible to develop 3D objects of desired shapes, and even with complex porous structures, to be useful in a number of advanced applications, including tissue engineering scaffolds. Advantages, such as the production of objects with requisite morphology and shape in a short time at a low cost, etc., are making 3D printing highly attractive. Its potential is further exploited in producing biomedical implants with aided functionalities.

### 16.4.2 Stereolithography

Patented in 1986, *stereolithography* is a rapid prototyping technique which develops 3D scaffolds with well-defined pore structure through a layer-by-layer solidification of photocurable resin *via* UV irradiation. In stereolithography, CAD files and software controls are applied to fabricate the intricate structures with high accuracy and complex pore morphology. For producing a required structure, a layer of photocurable resin is formed on a platform, which acts as a base for constructing the complete structure. Once the first layer is cured, the platform is lowered and the second layer is polymerized. The process is continued until the structure is complete, and once the desired 3D structure is obtained, it is further exposed to UV light for residual curing, and uncured resin, if any, is extracted. Reactive resins based on functionalized poly[(D,L-lactide)-co-(ε-caprolactone)], poly(trimethylene carbonate), and poly(propylene fumarate) macromers have been used to develop 3D porous scaffolds by stereolithography through photopolymerization. Generally, in stereolithography, photocured structures are rigid, and recent studies are focused on developing elastic and flexible scaffolds for soft tissue engineering applications. The association of stereolithography with magnetic resonance imaging or tomography techniques offers the advantage of fabricating patient-specific implants.

### 16.4.3 Selective Laser Sintering (SLS)

Developed in 1986 at University of Texas, Austin, the SLS technique fabricates the objects from a thin layer of small particles of ceramics, metals, or thermoplastic powders, where the coalescence of particles decreases their surface free energy and is the driving force for sintering of particles. Small particles are exposed to a high-intensity laser beam using computer controls and CAD file to build the 3D structure by combining the different sections, layer by layer. As the particles in the powder bed have relatively low packing density, the ultimate material has inherent porosity with interconnected morphology, which is highly suitable for tissue engineering scaffolds. In constructing the objects, for example, a porous scaffold, first, a 2D thin section, is formed, and the process is repeated to develop the overall structure. The exposure of powder bed to a high-intensity laser beam increases its temperature and results in the formation of a solid layer through melting and fusing of particles, and the same step is repeated until the structure is complete. SLS offers the advantages of fabrication of 3D intricate shape constructs with different nanofunctionalities using a powder material, whereas FDM and 3D printing typically require the feed material in the form of a filament, which is an additional processing step. Fabrication of hydroxyapatite-reinforced scaffolds of polylactides, poly(ε-caprolactone) (PCL), and poly(diether ketone) has been demonstrated using SLS. However, the need for a large quantity of feed material restricts the commercial viability, and the available particle size of many polymers is inappropriate for SLS.

### 16.4.4 Fused Deposition Modeling (FDM)

FDM is another rapid prototyping technique which extrudes the molten filament of thermoplastic materials and deposits it in layer-by-layer pattern to fabricate the 3D objects. Software controls and CAD files are used to deposit the extruded material in X–Y axes on a platform, and once the first layer is laid, the platform moves downward to deposit another layer of extruded material along the Z-axis. The same procedure repeats to build the 3D scaffolds with controlled porosity and pore size distribution by adjusting the rate of extrusion and deposition. FDM has been used to develop the 3D scaffolds of PCL, poly(L-lactide), and polypropylene reinforced with hydroxyapatite, tri-calcium phosphate, or alumina. High porosity, superior mechanical strength, nonnecessity of organic solvents, fabrication of multifaceted structures with ease of material handling, and fast production rate are the major advantages of FDM.

In summary, all the advanced and conventional techniques as mentioned earlier require a pre-synthesized polymer in the form of a fine powder, solution, or melt. Some of the techniques are relatively simple and inexpensive, but lack of control over the porosity and pore morphology and use of toxic organic solvents or high temperature are major issues associated with conventional techniques. The advanced techniques are beneficial in this respect to some extent; however, they are expensive and require specialized machines, software controls, and advanced infrastructure. Moreover, some of the techniques have constraints over achieving a lower pore size and the interconnectivity of pores. In this regard, emulsion templating offers the possibility of overcoming the challenges associated with conventional and advanced techniques.

## 16.5 EMULSION TEMPLATING

*Emulsions* are heterogeneous colloidal systems formulated by dispersing a liquid in another immiscible liquid with the help of a surface active agent (surfactant) or stabilizer. Thus, in an emulsion, one liquid is dispersed (internal) phase and the other one is continuous (external) phase. The size of the dispersed droplets of an emulsion categorizes the emulsions as macro- (>0.1 μm), mini- (50–1,000 nm), or nanoemulsions (20–200 nm). A variety of household and industrial applications consume emulsion-based products, for example, dairy, food, cosmetics, pharmaceutical, agriculture, paints, adhesives, coatings, lubrication, petroleum extraction, and material synthesis. The homogeneity of emulsion-based products is of utmost importance, which ultimately decides the product life. The stability of the emulsion depends upon several factors, such as dispersed phase volume, size and distribution of dispersed droplets, viscosity and density of dispersed and continuous phases, concentration, ionic nature, molecular weight and composition of emulsifier, rheological properties of emulsion, pH, temperature, additives, etc. Therefore, it is essential to prepare emulsions which have great resistance to destabilization and remain stable over a long period.

Water is often used as one of the phases of emulsion, thus creating either a *water-in-oil* (w/o) or an *oil-in-water* (o/w) emulsion, where the former phase is identified as the internal phase of the emulsion. Nonaqueous oil-in-oil emulsions (o/o) formulated with immiscible non-polar and polar organic liquids have also been reported in literature. O/o emulsions are advantageous as they offer the feasibility of working with emulsions involving high-temperature reactions using water-sensitive catalysts and/or monomers, thus expanding the span of material synthesis. However, o/o emulsions are more sensitive to destabilization as compared to their aqueous counterparts, and it has been observed that polymeric emulsifiers have better stabilizing action on o/o emulsions than the low-molecular-weight non-ionic emulsifiers. Further, based on the volume fraction of the dispersed phase ($\phi_d$), emulsions are distinguished as low- (<0.3), medium- (0.3–0.74), or high-internal-phase emulsion (HIPE, ≥0.74). HIPEs are characterized as highly concentrated gel emulsions due to their high viscosity, and few studies have been reported where emulsions with $\phi_d = 0.7$ are also considered as HIPEs.

In the perspective of material design and synthesis, specifically of polymer-based resources, emulsions are used as templates for preparing colloids (polymer latex), polymer composites, and porous polymers, using polymerizing monomers as the precursors in either or both of the phases of emulsion. For polymer latex preparation, in conventional emulsion polymerization, the monomer is polymerized in monomer-swollen micelles. In case of polymer composites, both dispersed and continuous phases of emulsion are solidified by the polymerization of monomers present in both the dispersed as well as continuous phases. The emulsion-templated porous polymers are synthesized by polymerizing the monomer present in the continuous phase of the emulsion and extracting the dispersed phase, which creates the porosity in the polymer matrix. In this regard, when HIPE is used as the emulsion template, the resulting porous polymer is known as polymerized HIPE, or polyHIPE.

PolyHIPEs are the exact replica of HIPEs, where the polymerization or solidification of the continuous phase and the removal of the dispersed phase during purification (washing and drying) develop porous polymers with pore size ranging from ten to hundreds of micron. A typical process of polyHIPE preparation and a SEM image and an EXD pattern of polyHIPE reinforced with silica nanoparticles are shown in Figure 16.5 and Figure 16.6, respectively. PolyHIPEs have a typical interconnected pore morphology due to the presence of interconnecting pores (or windows) on the pore surface. A change in the density of continuous phase due to volume contraction by conversion of monomer to polymer governs the formation of windows, thereby making the structure bicontinuous or biphasic. The selection of an emulsifier (or stabilizer) and its concentration is of prime importance for HIPE stability, where restoring the microstructure of HIPE during polymerization results in the development of low-density polyHIPE.

The use of conventional emulsifiers, for example, Span 80 (sorbitan monooleate), Tween 80 (polyoxyethylene sorbitan monooleate), and polymeric emulsifiers based on poly(ethylene oxide) and poly(propylene oxide), amphiphilic nanomaterials (Pickering stabilizers), and their synergistic mixtures, has been frequently reported for HIPE stabilization. Generally, a large amount of emulsifier (5–50 wt.% of continuous phase) is consumed with increasing amount of dispersed phase to prevent phase inversion and to retain the HIPE morphology up to the point of gelation of emulsion. The use of such a large quantity of emulsifiers, and their removal from the

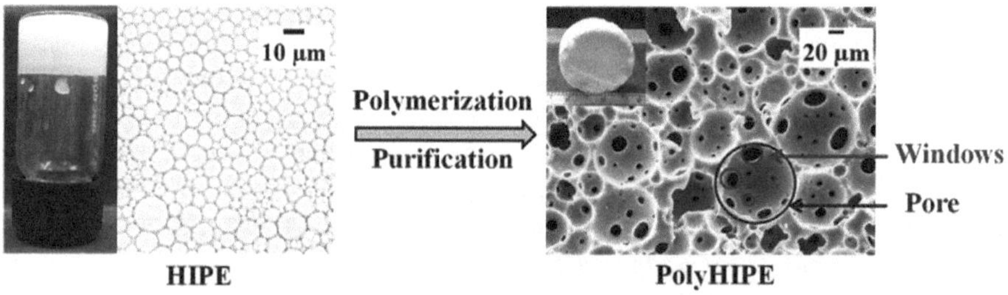


**FIGURE 16.5** Generation of porous polymer, that is, polyHIPE, through HIPE templating.

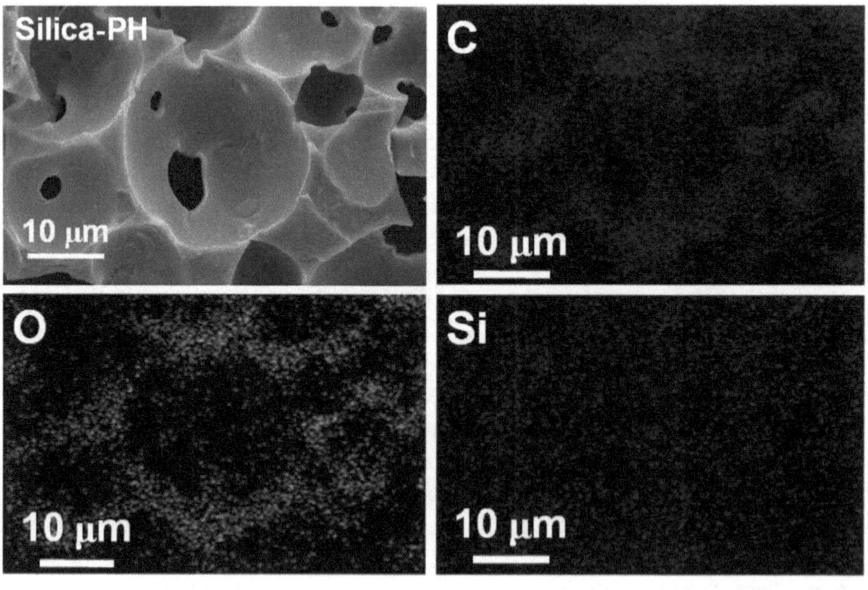


**FIGURE 16.6** SEM image and EDX pattern of silica nanoparticle reinforced polyHIPE composite.

final polyHIPEs, adds to the overall processing costs and causes other environmental issues. Reactive emulsifiers (which covalently bond with polyHIPEs), high-molecular-weight emulsifiers (difficult to leach out), and organic or inorganic Pickering stabilizers (cellulose nanocrystal, polymeric, Janus, silica, titania, and hydroxyapatite, particulate matters, etc.) are studied as potential alternatives to the conventional low-molecular-weight emulsifiers. Interestingly, the Pickering HIPE templates, in addition to emulsifier-free HIPE templates, offer the production of poly(Pickering HIPEs), that is, porous polyHIPE nanocomposite scaffolds derived from the Pickering HIPE templates through solvent evaporation or polymerization route. Hence, literatures expressing the development of porous composite scaffolds through Pickering emulsion templates are presented in Table 16.2–16.4. Few studies also demonstrate the development of polyHIPE composites through emulsifier-assisted Pickering HIPE templates, where the presence of the emulsifier in the continuous phase of HIPE tunes the HIPE morphology and microstructure of the resultant polyHIPE composites.

Free radical polymerization (FRP) of vinylic monomers using an oil-soluble initiator is the most employed method so far to derive polyHIPEs from w/o, o/w, and o/o HIPEs. Typically, the synthesis of hydrophobic polyHIPEs has been reported using w/o HIPEs of water-insoluble monomers, such as styrene, divinylbenzene, 4-vinylbenzylchloride, etc. Conversely, the polymerization of monomers, including acrylic acid, acrylamide, $scCO_2$-in-water, and $scCO_2$-in-ionic liquids, results in the development of hydrophilic polyHIPEs. Multiple emulsions and nonaqueous HIPE formulations having an oil (n-hexane, n-heptane, n-dodecane, n-tetradecane, and n-hexadecane), ionic liquids, and deep eutectic solvents (DES) have also been employed as the internal phase of HIPEs. Especially, in the last two decades, a variety of polymerization chemistry, including ATRP, AGET-ATRP, RAFT, step-growth, click-chemistry (thiol-ene), ring opening metathesis, and ring opening polymerization have been researched to expand the scope of polyHIPE synthesis not only to aqueous or nonaqueous HIPEs but also for their stabilization using various emulsifiers and Pickering stabilizers.

### 16.5.1 PolyHIPEs Derived within w/o HIPEs

The pioneering work in emulsion templating was initiated and patented by Unilever, which disclosed the synthesis of polystyrene foams within w/o HIPEs. Several fundamental and advanced studies have been reported on w/o HIPE templates primarily using vinylic monomers and focused on the effect of emulsion-processing parameters on overall performance and properties of the resulting polyHIPEs or polyHIPE nanocomposites. W/o HIPEs have been majorly devoted for preparing a variety of polyHIPEs as electrodes, hybrid foams, porous nanocomposites, hyper-cross-linked foams, elastic foams, stationary phase for chromatography columns, adsorbents, carbon foams, metal–organic frameworks, magnetic foams, and tissue engineering scaffolds. A brief summary of various polyHIPEs derived using w/o HIPEs is presented in Table 16.2.

### 16.5.2 PolyHIPEs-Derived Using o/w HIPEs

Generally, synthesis of polymeric porous materials using o/w HIPE templates is more challenging as compared to w/o emulsion templates due to the destabilization issues pertaining to o/w HIPEs. O/w HIPEs are more prone to destabilization, and the stabilization of o/w HIPEs requires better control during emulsification than conventional w/o HIPEs. Moreover, the use of a low-boiling solvent such as n-hexane causes environmental impacts, and high-boiling organic solvents like paraffin oil are difficult to remove during purification. The stabilization of o/w HIPEs has been reported using non-ionic emulsifiers with high HLB value (such as Pluronic F-68 (F68), Brij 98, Brij 58, Triton X-405) and/or Pickering stabilizers, such as egg yolk, chitosan–casein, yeast, starch–nanocrystals, CNC, gelatin, etc. A summary of polyHIPEs developed using o/w HIPE templates is presented in Table 16.3.

**TABLE 16.2**
**PolyHIPEs Derived Using w/o HIPE Templates**

| Continuous Phase | Dispersed Phase | Polymerization Mechanism and Conditions | Remarks |
|---|---|---|---|
| Styrene, divinyl benzene (DVB), Span 80 | Aqueous solution of potassium persulfate (KPS) and calcium chloride hexahydrate ($CaCl_2.6H_2O$) | FRP, 60°C, for 5 hr | Studied the formation of windows during polymerization of HIPE using cryo-SEM technique [25] |
| Styrene, DVB, Span 80, 2-ethylhexyl acrylate (EHA), or 2-ethylhexyl methacrylate | Aqueous solution of KPS and potassium persulfate and $CaCl_2.6H_2O$ | FRP, 65°C, for 48 hr | Preparation of flexible polyHIPEs using acrylate monomers [26] |
| Styrene, DVB, Span 80 | Aqueous solution KPS, $\phi_d = 0.8$ | FRP | Sulfonated polystyrene foams with acidic functional groups for absorbing water-soluble salts [27] |
| Styrene, vinylbenzyl chloride, DVB, Span 80, 2,2'-azobisisobutyronitrile (AIBN) | Aqueous solution of sodium sulfate | FRP, 60°C, for 24 hr | Grafting of polyHIPEs with 4-vinylpyridine increased heavy metal ion sorption [28] |
| Styrene, DVB, EHA, Span 80 | Aqueous solution of KPS, $\phi_d = 0.8$ | FRP, 60°C, for 6 hr | Electroplating and degradation of polyHIPE foams to produce nickel–polymer cathodes [29] |
| Styrene, vinylsilane, Span 80 and/or DVB, methacryloxypropyltrimethoxysilane (MPS) | Aqueous solution of KPS and calcium chloride dihydrate ($CaCl_2.2H_2O$) | FRP, 65°C, for 24 hr | Hybrid polystyrene foams with inorganic polysilsesquioxane network of superior mechanical and thermal properties [30] |
| Styrene, DVB, AIBN, functionalized titania nanoparticles | Aqueous solution of $CaCl_2.2H_2O$ | FRP, 70°C, for 24 hr | Synthesis of poly(Pickering HIPE) foams using functionalized titania nanoparticles [31] |
| Styrene or EHA, DVB, Span 80, or benzoyl peroxide | Aqueous solution of KPS, potassium sulphate, AM, and MBA | FRP, 65°C, for 24 hr | PolyHIPEs with dual characteristics, that is, hydrophilic-hydrophobic property [32] |
| Styrene, DVB, and Span 80 or polymeric emulsifier | Aqueous solution of KPS, MBA, and acrylic acid or N-isopropyl acrylamide | FRP, 65°C, for 24 hr | pH- or temperature-responsive polyHIPE materials [33] |
| 1,2-dichlorobenzene, styrene, DVB, Span 80 | Aqueous solution of KPS and $CaCl_2.6H_2O$, $\phi_d = 0.95$ | FRP, 60°C, for 24 hr | Conducting foams through chemical oxidative treatment with pyrrole [34] |
| Styrene, DVB, EHA, AIBN, Span 80 | Aqueous solution of acrylic acid and KPS | FRP, 60°C. for 24 hr | Carboxylic acid functionalization of polyHIPE favoring cell adhesion [35] |
| Dispersion of hydrophobically modified nanoclay in solution of PCL | Aqueous PVA | – | 3D-printed nanoclay-reinforced PCL composites with hierarchical porosity [36] |

**TABLE 16.3**
**PolyHIPEs Derived Using o/w HIPE Templates**

| Continuous Phase | Dispersed Phase | Polymerization Mechanism and Conditions | Remark |
|---|---|---|---|
| Solution of 4,4'-diazidostilbene-2,2' -disulfonic acid disodium salt, tripropargylamine and copper sulfate pentahydrate, Tween 60 in water/DMF (2/1) | Paraffin liquid, $\phi_d$ = 0.75–0.9 | 80°C for 24 hr | PolyHIPEs with tailor-made pore morphology and significant heavy metal sorption capacity (52 mg/g for Cu(II) ions) [37] |
| Aqueous solution of acrylic acid (AA), MBA, modified yeast (Pickering stabilizer), TMEDA, and APS | Canola oil of food grade | FRP, 70°C, for 24 hr | Poly(Pickering HIPEs) with excellent and sustainable sorption capacity for heavy metal ions [38] |
| Aqueous solution of AM, MBA, KPS, Ag-incorporated melamine-based microporous particles | Cyclohexane, $\phi_d = 0.8$ | FRP, 60°C, for 1 hr | Nanocomposite polyHIPEs demonstrating excellent catalytic reduction of 4-nitrophenol [39] |
| Aqueous dispersion of poly(urethane urea) or poly(vinyl ester) resin nanoparticles | Cyclohexane, $\phi_d = 0.7–0.9$ | Freeze-drying | Nanocomposite foams with tailor-made properties derived using Pickering HIPE templates [40] |
| Aqueous mixture of poly(N-isopropylacrylamide-co-methacrylic acid) microgel particles, titania nanoparticles | n-hexane, $\phi_d = 0.75$ | Freeze-drying and calcination | Nanocomposite polyHIPEs for photocatalytic degradation of Rhodamine B [41] |
| Dispersion of nHAp in aqueous solution of GMA-grafted dextran and APS | Methyl myristate or soybean oil, $\phi_d = 0.8$ | FRP, 60°C, for 24 hr | Macroporous nanocomposite hydrogels through inversion of Pickering HIPEs [42] |

### 16.5.3 PolyHIPEs Derived Using Nonaqueous HIPEs

Recently, nonaqueous o/o HIPEs comprising of polymerizable organic continuous phase and an organic liquid (aprotic organic solvents, DES, or ionic liquid) as the dispersed phase, are gaining attention to synthesize polyHIPE scaffolds. O/o HIPEs are less common and advantageous over conventional aqueous emulsions, where the presence of water is detrimental for the material synthesis involving the use of water-sensitive monomers, high temperature, or pressurized reactions. Similar to aqueous emulsions, Pickering particles and non-ionic polymers have also been studied as stabilizers for o/o HIPEs. A summary of o/o HIPE-templated porous polymers is presented in Table 16.4.

**TABLE 16.4**
**PolyHIPEs Derived Using Nonaqueous HIPE Templates**

| Continuous Phase | Dispersed Phase | Polymerization Mechanism and Conditions | Remark |
|---|---|---|---|
| MMA, EDGMA, AIBN, non-ionic emulsifier, nHAp (<200 nm) | DES of urea and choline chloride, $\phi_d = 0.8$ | FRP, 60°C, for 24 hr | Development of polyHIPE nanocomposites with *in vivo* biocompatibility [43] |
| Aqueous solution of deep-eutectic polymer of urea and 2,5-dihydroxy-1,4-benzoquinone, Triton X-405 | Cyclohexane or solution of chain extender in toluene and cyclohexane | Freeze-drying for 72 hr | Carbon monoliths with hierarchical porosity and specific surface area > 812 $m^2/g$ [44] |
| Hydrophobically modified fumed silica nanoparticles, polystyrene solution in xylene | Ethylene carbonate, $\phi_d = 0.75$ | – | Ethylene carbonate–filled polystyrene nanocomposites [45] |
| Solution of mannitol, poly(hexamethylene diisocyanate), and Pluronic P-123 in DMSO | Paraffin oil, $\phi_d =$ 0.75–0.85 | 60°C for 24 hr | Polyurethane-based polyHIPEs for oil spill reclamation *via* step-growth polymerization [46] |
| Dispersion of pristine cellulose particles (50 μm) in DMSO having pre-dissolved Pluronic F-127 and poly(hexamethylene diisocyanate), dibutyltin dilaurate | Liquid paraffin | 65°C for 24 hr | Cellulose-embedded polyurethane polyHIPE composites with affinity for different organic solvents [47] |
| ε-Caprolactone (CL), bis-ε-caprolactone-4-yl (BCY), Pluronic F-127 | n-Hexadecane | ROP, 120°C for 8 hr | Single-step synthesis of cross-linked PCL-based biodegradable polyHIPEs with high liquid absorptivity [48–50] |
| ε-Caprolactone (CL), bis-ε-caprolactone-4-yl (BCY), Pluronic F-127, nanographene oxide | n-Hexadecane | ROP, 120°C for 8 hr | Nanographene oxide–reinforced macroporous cross-linked PCL composite scaffolds [51] |
| ε-Caprolactone (CL), bis-ε-caprolactone-4-yl (BCY), hydrophobically modified fumed silica nanoparticles | n-Hexadecane | ROP, 120°C for 8 hr | Cross-linked PCL-based biodegradable macroporous nanocomposite scaffolds through polymerization of Pickering HIPE [52, 53] |

## 16.6 CHARACTERIZATIONS

Exploring the structure–property relationship is an important aspect for engineered materials to estimate the suitable end use and/or performance in an intended application. In this regard, the electrospun nanofibrous membranes and emulsion-templated porous polymers, neat or composites, have also been thoroughly investigated using various routine characterization techniques, summarized in Table 16.5, considering the new researchers pioneering in the field of porous polymer composites.

In addition to the characterizations listed in Table 16.5, several other properties, like density, porosity, swelling index, gel content, etc., are also determined through the gravimetric method. The density is determined as the mass-to-volume ratio, and porosity is determined using the equation porosity (%) = $(1 - \rho_b/\rho_a) \times 100$, where $\rho_a$ and $\rho_b$ are the density of nonporous and porous samples of identical dimensions. The swelling index of porous composites indicates the liquid sorption capacity of porous composites at a given temperature. Further, through the selling index measurements, the rate of liquid sorption kinetics may also be determined. The gel content analysis refers to the amount of insoluble polymeric content present in a given mass of porous polymeric composite, thus reflecting the extent of cross-linking of polymer chains. Thermal conductivity and sound insulation (acoustic property) of porous composites has also been investigated in a number of literatures. Besides the assessment of macromolecular structure of polymer composites, $^1$H-NMR has also been reported to determine the diffusion coefficient of water in emulsions as well.

For evaluating the mechanical properties of electrospun nanofibrous composites, a universal testing machine equipped with low-capacity load cells like 10 N or 50 N is used in a tensile mode. Similarly, for polyHIPE nanocomposites load cell like 500 N is majorly used to assess the compressibility (single or cyclic loading and unloading) and recovery from compression. The dynamic

**TABLE 16.5**
**Characterization Techniques for Porous Polymers Composites**

| Characterization | Inference |
|---|---|
| Optical microscopy | Droplet size distribution of emulsions, surface morphology of polyHIPE composites, or nanofibrous composite membranes |
| Rheology | Viscoelastic behavior of emulsions or polymer solutions to be used for preparation of porous polymer composites |
| SEM | Surface morphology of polyHIPE nanocomposites (pore size distribution) or nanofibrous composite membranes (fiber dimeter distribution) |
| Micro-CT | Pore size distribution in the bulk of polyHIPE composites |
| Transmission electron microscopy | Surface morphology of polyHIPE nanocomposites or nanofibrous composite membranes, and surface morphology and size of nanomaterials |
| FTIR | Functionalities of nanomaterials, macromolecular structure of polymer composite, functionality of polymer composites, polymer–nanomaterial interactions |
| Nuclear magnetic resonance spectroscopy (NMR) | $^1$H-NMR and/or $^{13}$C-NMR may be used to determine the macromolecular structure and polymerization kinetics during in situ synthesis of porous polymer composites |
| XRD | Crystallinity of polymer nanocomposite |
| Mechanical strength | Tensile strength and compressibility, recovery from compression |
| BET or mercury porosimetry | Surface area, pore volume, pore volume distribution |
| Differential scanning calorimetry | Thermal transitions (glass, transition, melting, and crystallization temperature) and crystallinity, polymer blend miscibility |
| Thermal gravimetric analysis | Thermal stability (temperature of safe use), extent of nanomaterial loading |
| Dynamic mechanical analysis | Thermal transitions below the melting temperature if polymer composites, polymer–nanomaterial interaction |
| Atomic force microscopy | Surface roughness |
| Contact angle | Hydrophobicity or oleophilicity of polymer composites |

mechanical analyzer is another important technique to assess the mechanical properties of the polyHIPE composites under dynamic loading over a distinct temperature range. In case of surface area determination, owing to the presence of macroporous morphology, polyHIPE composites possess very low specific surface area, whereas using the hyper-cross-linking approach, polyHIPEs with exceptionally high specific surface area have been reported. The biocompatibility of porous polymer composites specifically for biomedical application is determined through *in vitro* and/or *in vivo* cell growth experiments. *In vitro* cytocompatibility through MTT analysis is highly popular in this regard. The biodegradability of porous composites may be determined through an alkaline aqueous media, and also through the soil burial test.

## 16.7 COMPUTATIONAL AND MODELING ASPECTS

Computational and mathematical modeling approaches are a very important aspect for reducing the time and resources to achieve the desired functional properties of porous polymer composites. There are various computational and modeling approaches to demonstrate porous polymer composite based on the electrospinning and emulsion templating techniques. Over the past two decades, various techniques to derive the constitutive relationships and to model the porous nanocomposite based on the electrospinning and emulsion templating techniques have been adopted to determine the properties of the porous composite, such as porosity, strength, elasticity, and compressibility. Models such as analytical or mathematical, semiempirical, empirical, and numerical are based on various effects and phenomena associated with the porosity, mechanical properties, and process parameters. Many simplified modeling approaches have been developed to obtain realistic results; however, there is no universal modeling approaches which reproduce and represent all the phenomena at the same time. The analytical models are based on the general continuum mechanics, such as the conservation of momentum and energy [54, 55]. To develop the analytical model, a set of various structural and nonstructural parameters should be involved during the fabrication of the porous polymer composites as per the applications and performance. Structural parameters such as porosity and mechanical properties are utilized to formulate the equations as given in what follows, where $\rho$ is the density of dilated microsphere, $u_0$ is the velocity of the charged jet at $r = r_{cr}$, $R$ is the maximal radius of the microsphere, and $u_{min}$ is the minimal velocity.

$$\pi r_{cr}^2 \rho u_0 = \pi R^2 \rho u_{min} = Q \tag{16.1}$$

Equation (16.1), based on the equation of continuity, which indicates the density, radius, and velocity of electrospun fibers, is indirectly related to the microsphere, pores, and voltage applied during the electrospinning process. A higher applied voltage value indicates a higher jet speed value and more drastic electrospinning process happened, resulting in lower density, smaller size of microsphere and pores. Nonstructural parameters such as process parameters are also responsible in predicting the structural parameters of the porous composite. The drawback of this analytical modeling approach is the complexity due to the involvement of numerous variables. Several research groups who have worked in this respect have obtained success, yet there is a lot of scope to improve these models.

In contrast, empirical studies involve examining the fabric response and obtaining constitutive relationships and failure criterion based on the analysis of data obtained through experimental work. Techniques like curve fitting, nonlinear regression analysis of experimental data, and statistical distributions are put into use. The method is effective only for a small set of variables. Moreover, its accuracy depends on the accuracy of the data obtained. Numerical computational modeling approaches are based on techniques like finite element and finite volume methods. With the emergence of supercomputers, commercial packages like Material Studio Software, ABAQUS, DYNA3D, ANSYS, and LSDYNA for conducting analysis or simulation are becoming popular, owing to their lower cost against conventional expensive and destructive experiments [56, 57]. Most of the available literature focuses on experimental and theoretical analysis of the porous composite materials. The analysis becomes more complicated with an increase in the number of variables and components. Furthermore, multiscale modeling approaches are used to predict the real-time behaviors of porous polymeric composites. Finite volume and finite element methods are also modeled according to the length scale as per the product applications [58]. Based on the length scale, it is classified into micro-model, macro-model, and mixture-model. Kravchenko et al. [59] developed the micromechanics-based computational model of open-cell polyHIPE foams to study the compressive behavior, as shown in Figure 16.7a–f. The model is based on morphological parameters, which are determined by the scanning electron micrograph. These parameters are foam density, mean diameter of void, etc.

The computational model well predicts the compressive behavior of foam with the experimental data. It also provides the pathway for understanding the microstructure, the manufacturing process, and the role of the morphological parameter of polyHIPE for targeted mechanical properties. Bismarck et al. applied the finite element analysis for a polystyrene polyHIPE foam to assess the effect of hierarchical pores on mechanical properties through periodic compressions [60]. For the purpose, simulated 2D porous geometries were generated for polyHIPE foams with hierarchical and nonhierarchical pores. Improvements in Young's moduli for polyHIPEs with hierarchical pores were suggested by the modeling study and confirmed that hierarchical pores of polyHIPEs are analogous to lattice girders in a bridge. Furthermore, a distinct element method (type of simulation) was used to estimate the permeability of porous media resembling emulsion-templated macroporous polymers [61]. Results showed that the porosity and diameter of interconnected pores tuned the permeability of polyHIPEs; however, adjusting the porosity and diameter of interconnected pores, independently, was challenging. Silevrstein et al. used the modeling approach to quantitively describe the shape recovery behavior of stearyl acrylate– and stearyl methacrylate–based polyHIPE composites reinforced with modified fumed silica nanoparticles [62]. The shape recovery behavior of polyHIPE composites was described using Kelvin–Voigt units, while the temperature-dependent viscosity was described using the Williams–Landel–Ferry equation.

In respect to open cell foams, Shen et. al. developed the 3D Voronoi models to investigate the effect of small imperfections upon the simulated macroscopic behavior of linearly elastic foams. The dependence of the Young's modulus, Poisson's ratio, and bulk modulus of the foams on relative

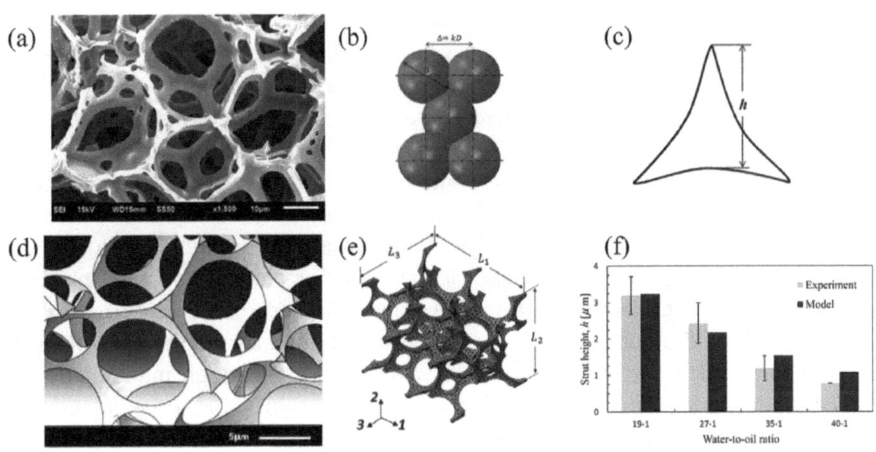


**FIGURE 16.7** Experimental and computational morphological parameters of polyHIPE: (a) SEM image, (b) hexagonal arrangement of dominant voids, (c) strut cross section height, (d) computational images, (e) foam microstructure from the computational model, and (f) experimental and computational results.

**Source: Derived from [59].**

density was evaluated through finite element analysis. The results showed the in the low-density regime, the Young's and bulk modulus of random Voronoi foams tuned by the microstructural imperfections of foams and the microbuckling of cell edges was accountable for the failure of the foams during compression [63]. For porous foams, other studies also demonstrated that the Young's modulus of the porous foams increased with increasing diameter of pores and decreasing porosity. In this context, attempt has been made to correlate the theoretical models and experimental results to investigate the effect of strut thickness, shape, and pore size distribution, impacting the mechanical properties of porous foams. Cyclic olefin copolymer–fumed silica nanocomposite foams prepared through the supercritical carbon dioxide at different pressures were assessed for the structural property relationship [64]. The thermomechanical properties of nanocomposite foams increased progressively with increasing content of fumed silica nanoparticles. Use of a theoretical model, the Gibson–Ashby, predicted the stiffening effect developing due to the interaction between the fumed silica nanoparticles and the polymer matrix, which was dominated by the formation of the networked structure of fumed silica nanoparticles rather than alignment of silica aggregates along the cell wall direction. Apart from the preceding discussion, computational modeling approaches, artificial neural networks, and statistical methods (such as design of experiments) have also been explored for the prediction and optimization of the properties of porous polymers and porous polymer composites [65].

In addition to polyHIPEs, modeling studies on diffusion in highly concentrated emulsions (HIPEs) suggest several physicochemical models based on various fundamental theories [66]. These models are primarily useful to predict the solute release profile (for example, drug) soluble in the dispersed phase of HIPE and encapsulated within the continuous phase. The models are beneficial to design stimuli-responsive delivery systems and in the preparation of formulations used for decontamination purposes. Additionally, the diffusion profile for predicting the stability of HIPE has also been explained. Several reports describing the stability, morphological characteristics, and rheological behavior of emulsions are also available, where experimental and theoretical modeling have been correlated [67–69].

Modeling of the electrospinning process producing nanofibrous membranes and nanofibrous composite membranes has also been studied widely and condensed by various researchers through review articles and book chapters [70–72]. The prime objective of these computational and modeling studies is to assess the cumulative effects of electrospinning parameters, the properties of polymer and solvent, the polymer–solvent interactions, and the presence of additives, such as nanomaterials and co-solvent, for the polymer in the electrospinning dope on the characteristics of nanofibers developed ultimately. One of such studies reviews the several mathematical models developed for the electrospun continuous nanofibers and nanoporous microspheres. For instance, one-dimensional steady model, Spivak–Dzenis model, Siro-electrospinning, vibration-electrospinning, etc. are discussed to improve electrospinability and overcome the instabilities arising during electrospinning [55]. Lan Xu established a mathematical model for a charged polymeric jet of the electrospinning process considering the various complex phenomena, for example, electric field, magnetic field, vibration force, and air drag [73]. The shape of the collector plays a key role in the orientation of deposited nanofibers [74]. Hence, the effect of especially designed collector with two parallel cylindrical rods of finite diameter was studied experimentally, and for the analytical description of the electric field applied across, the collector was described through mathematical modeling. The developed models computed the path of an electrified liquid jet, and the calculated values of critical voltage were verified experimentally as well. Response surface methodology (RSM) used to model and optimize the electrospinning parameters of *Bombyx mori* silk fibers was extended to obtain poly(vinyl alcohol) electrospun nanofibers with minimal diameter to be used for sound insulation application. The study involved the modeling and optimization of electrospinning parameters through the design of experiments, analysis of variance (ANOVA and the Taguchi orthogonal array, L27OA), and RSM [75]. A simulation of the whipping process using the numerical methods, coupled with polynomial equation, has been focused to study the solvent evaporation during electrospinning process [76]. The modeling reports available on electrospinning process encompass the theoretical model for electrospinning of fibers with nanopores on their surface [54] and melt electrospun PCL nanofibrous scaffolds [77, 78].

## 16.8 SUMMARY

The present chapter summarizes the different techniques to design and develop porous polymers and porous polymeric composites. The addition of nanomaterials in the matrix forming the polymeric phase leads to the generation of polymer–composite materials. Among several techniques described to develop porous polymers, electrospinning and emulsion templating are highly fascinating, as they generate the porous polymers and composites of controlled morphology, tailor-made properties, and reproducible performance. For instance, the dispersion of nanomaterials in an electrospinning dope or a continuous phase of HIPE, or simply through Pickering emulsions, porous polymeric composites of tailor-made physiochemical properties may be easily developed. Plausible applications of electrospun fibrous membranes, emulsion-templated polyHIPEs, and porous composite materials developed through these two techniques have also been delineated. Several techniques practiced for characterizing and understanding the structure–property relationship which ultimately affects the performance during the end use of electrospun nanofibrous membranes and emulsion-templated polyHIPEs have been summarized. The computational and modeling approaches to predict and analyze the experimental and theoretical results have been designated briefly so that the readers may be exposed to another paradigm in respect to porous polymers and composites generated using the techniques electrospinning and emulsion templating. Rather than encompassing the several complex mathematical equations of computational and modeling approaches, a glimpse of those studies has defined to spark the researcher toward those original literatures. It is expected that the readers will be definitely benefited with the content of the present chapter, and they can design and develop novel porous materials finding applications into commodity to advanced scientific applications.

## ACKNOWLEDGMENTS

The authors express a deep gratitude to Dr. B. R. Ambedkar National Institute of Technology Jalandhar and the Indian Institute of Technology–Delhi for providing the necessary resources and facilities to compose the chapter, focusing on porous polymer composites. The corresponding author is also grateful to the co-authors for their continuous support in articulating the present book chapter.

## REFERENCES

[1] Zhao, D.; Thallapally, P. K.; Petit, C.; Gascon, J. Advanced Porous Materials: Design, Synthesis, and Applications in Sustainability. *ACS Sustain. Chem. Eng.* **2019**, *7* (9), 7997–7998. https://doi.org/10.1021/acssuschemeng.9b02021.

[2] Sing, K. S. W. Reporting Physisorption Data for Gas/Solid Systems with Special Reference to the Determination of Surface Area and Porosity (Recommendations 1984). *Pure Appl. Chem.* **1985**, *57* (4), 603–619. https://doi.org/10.1351/pac198557040603.

[3] Yang, X. -Y.; Chen, L.-H.; Li, Y.; Rooke, J. C.; Sanchez, C.; Su, B.-L. Hierarchically Porous Materials: Synthesis Strategies and Structure Design. *Chem. Soc. Rev.* **2017**, *46* (2), 481–558. https://doi.org/10.1039/C6CS00829A.

[4] Hutmacher, D. W. Scaffolds in Tissue Engineering Bone and Cartilage. *Biomaterials* **2000**, *21* (24), 2529–2543. https://doi.org/10.1016/S0142–9612(00)00121–6.

[5] Woodruff, M. A.; Hutmacher, D. W. The Return of a Forgotten Polymer: Polycaprolactone in the 21st Century. *Prog. Polym. Sci.* **2010**, *35* (10), 1217–1256. https://doi.org/10.1016/j.progpolymsci.2010.04.002.

[6] Kimmins, S. D.; Cameron, N. R. Functional Porous Polymers by Emulsion Templating: Recent Advances. *Adv. Funct. Mater.* **2011**, *21* (2), 211–225. https://doi.org/10.1002/adfm.201001330.

[7] Avossa, J.; Herwig, G.; Toncelli, C.; Itel, F.; Rossi, R. M. Electrospinning Based on Benign Solvents: Current Definitions, Implications and Strategies. *Green Chem.* **2022**, *24* (6), 2347–2375. https://doi.org/10.1039/D1GC04252A.

[8] Gurave, P. M.; Rastgar, M.; Mizan, M. M. H.; Srivastava, R. K.; Sadrzadeh, M. Superhydrophilic Electrospun Polyamide-Imide Membranes for Efficient Oil/Water Separation Under Gravity. *ACS Appl. Eng. Mater.* **2023**, *1* (11), 3134–3146. https://doi.org/10.1021/acsaenm.3c00550.

[9] Ghosh, S.; Yadav, A.; Gurave, P. M.; Srivastava, R. K. Unique Fiber Morphologies From Emulsion Electrospinning: A Case Study of Poly(ε-Caprolactone) and Its Applications. *Colloids Interfaces* **2023**, *7* (1), 19. https://doi.org/10.3390/colloids7010019.

[10] Pal, J.; Wu, D.; Hakkarainen, M.; Srivastava, R. K. The Viscoelastic Interaction Between Dispersed and Continuous Phase of PCL/HA-PVA Oil-in-Water Emulsion Uncovers the Theoretical and Experimental Basis for Fiber Formation During Emulsion Electrospinning. *Eur. Polym. J.* **2017**, *96*, 44–54. https://doi.org/10.1016/j.eurpolymj.2017.09.004.

[11] Gurave, P. M.; Singh, S.; Yadav, A.; Nandan, B.; Srivastava, R. K. Electrospinning of a Near Gel Resin to Produce Cross-Linked Fibrous Matrices. *Langmuir* **2020**, *36* (9), 2419–2426. https://doi.org/10.1021/acs.langmuir.9b03870.

[12] Gurave, P. M.; Dubey, S.; Nandan, B.; Srivastava, R. K. Pickering Emulsion-Templated Nanocomposite Membranes for Excellent Demulsification and Oil–Water Separation. *ACS Appl. Mater. Interfaces* **2022**, *14* (48), 54233–54244. https://doi.org/10.1021/acsami.2c16483.

[13] Mo, J.; Hu, D.; Liu, X.; Sun, Y.; Li, X.; Wang, C.; Chai, B.; Ding, D.; Yan, J.; Sun, L. Eco-Friendly ZIF-67/Rice Straw-Derived Cellulose Acetate Electrospun Nanofiber Mats for Efficient $CO_2$ Capturing and Selectivity Removal of Methyl Orange Dye. *J. Environ. Chem. Eng.* **2022**, *10* (6), 108989. https://doi.org/10.1016/j.jece.2022.108989.

[14] Qing, W.; Li, X.; Wu, Y.; Shao, S.; Guo, H.; Yao, Z.; Chen, Y.; Zhang, W.; Tang, C. Y. In Situ Silica Growth for Superhydrophilic-Underwater Superoleophobic Silica/PVA Nanofibrous Membrane for Gravity-Driven Oil-in-Water Emulsion Separation. *J. Membr. Sci.* **2020**, *612*, 118476. https://doi.org/10.1016/j.memsci.2020.118476.

[15] Wang, S.; Chen, W.; Wang, L.; Yao, J.; Zhu, G.; Guo, B.; Militky, J.; Venkataraman, M.; Zhang, M. Multifunctional Nanofiber Membrane with Anti-Ultraviolet and Thermal Regulation Fabricated by Coaxial Electrospinning. *J. Ind. Eng. Chem.* **2022**, *108*, 449–455. https://doi.org/10.1016/j.jiec.2022.01.022.

[16] Samanta, A.; Takkar, S.; Kulshreshtha, R.; Nandan, B.; Srivastava, R. K. Facile Fabrication of Composite Electrospun Nanofibrous Matrices of Poly(ε-Caprolactone)–Silica Based Pickering Emulsion. *Langmuir* **2017**, *33* (32), 8062–8069. https://doi.org/10.1021/acs.langmuir.7b02119.

[17] Gurave, P. M.; Nandan, B.; Srivastava, R. K. Fish-Gill Inspired Multifunctional Nanofibrous Membrane for Efficient Demulsification and Pollutant Sorption. *J. Membr. Sci.* **2023**, *683*, 121850. https://doi.org/10.1016/j.memsci.2023.121850.

[18] Porrelli, D.; Mardirossian, M.; Musciacchio, L.; Pacor, M.; Berton, F.; Crosera, M.; Turco, G. Antibacterial Electrospun Polycaprolactone Membranes Coated with Polysaccharides and Silver Nanoparticles for Guided Bone and Tissue Regeneration. *ACS Appl. Mater. Interfaces* **2021**, *13* (15), 17255–17267. https://doi.org/10.1021/acsami.1c01016.

[19] Chen, Z.; Li, Q.; Chen, X. Porous Graphene/Polyimide Membrane with a Three-Dimensional Architecture for Rapid and Efficient Solar Desalination Via Interfacial Evaporation. *ACS Sustain. Chem. Eng.* **2020**, *8* (36), 13850–13858. https://doi.org/10.1021/acssuschemeng.0c05306.

[20] Raulo, A.; Gupta, A.; Srivastava, R.; Nandan, B. Excellent Electrochemical Performance of Lithium-Sulfur Batteries Via Self-Standing Cathode From Interwoven α-$Fe_2O_3$ Integrated Carbon Nanofiber Networks. *J. Electroanal. Chem.* **2021**, *880*, 114829. https://doi.org/10.1016/j.jelechem.2020.114829.

[21] Hosseini, S. A.; Vossoughi, M.; Mahmoodi, N. M.; Sadrzadeh, M. Clay-Based Electrospun Nanofibrous Membranes for Colored Wastewater Treatment. *Appl. Clay Sci.* **2019**, *168*, 77–86. https://doi.org/10.1016/j.clay.2018.11.003.

[22] Deka, B. J.; Lee, E.-J.; Guo, J.; Kharraz, J.; An, A. K. Electrospun Nanofiber Membranes Incorporating PDMS-Aerogel Superhydrophobic Coating with Enhanced Flux and Improved Antiwettability in Membrane Distillation. *Environ. Sci. Technol.* **2019**, *53* (9), 4948–4958. https://doi.org/10.1021/acs.est.8b07254.

[23] Rahmanian, V.; Ahmad Ebrahim, M. Z.; Razavi, S.; Abdelmigeed, M.; Barbieri, E.; Menegatti, S.; Parsons, G. N.; Li, F.; Pirzada, T.; Khan, S. A. Vapor Phase Synthesis of Metal–Organic Frameworks on a Nanofibrous Aerogel Creates Enhanced Functionality. *J. Mater. Chem. A* **2024**, *12* (1), 214–226. https://doi.org/10.1039/D3TA05299K.

[24] Wang, C.; Huang, W.; Zhou, Y.; He, L.; He, Z.; Chen, Z.; He, X.; Tian, S.; Liao, J.; Lu, B.; Wei, Y.; Wang, M. 3D Printing of Bone Tissue Engineering Scaffolds. *Bioact. Mater.* **2020**, *5* (1), 82–91. https://doi.org/10.1016/j.bioactmat.2020.01.004.

[25] Cameron, N. R.; Sherrington, D. C.; Albiston, L.; Gregory, D. P. Study of the Formation of the Open-Cellular Morphology of Poly(Styrene/Divinylbenzene) PolyHIPE Materials by Cryo-SEM. *Colloid Polym. Sci.* **1996**, *274* (6), 592–595. https://doi.org/10.1007/BF00655236.

[26] Cameron, N. R.; Sherrington, D. C. Preparation and Glass Transition Temperatures of Elastomeric PolyHIPE Materials. *J. Mater. Chem.* **1997**, *7* (11), 2209–2212. https://doi.org/10.1039/a702030i.

[27] Wakeman, R. J.; Bhumgara, Z. G.; Akay, G. Ion Exchange Modules Formed From Polyhipe Foam Precursors. *Chem. Eng. J.* **1998**, *70* (2), 133–141. https://doi.org/10.1016/S0923-0467(98)00088-8.

[28] Benicewicz, B. C.; Jarvinen, G. D.; Kathios, D. J.; Jorgensen, B. S. Open-Celled Polymeric Foam Monoliths for Heavy Metal Separations Study. *J. Radioanal. Nucl. Chem.* **1998**, *235* (1–2), 31–35. https://doi.org/10.1007/BF02385933.

[29] Brown, I. J.; Sotiropoulos, S. Electrodeposition of Ni From a High Internal Phase Emulsion (HIPE) Template. *Electrochimica Acta* **2001**, *46* (17), 2711–2720. https://doi.org/10.1016/S0013-4686(01)00481-9.

[30] Tai, H. Organic–Inorganic Networks in Foams From High Internal Phase Emulsion Polymerizations. *Polymer* **2001**, *42* (10), 4473–4482. https://doi.org/10.1016/S0032–3861(00)00820-X.

[31] Menner, A.; Ikem, V.; Salgueiro, M.; Shaffer, M. S. P.; Bismarck, A. High Internal Phase Emulsion Templates Solely Stabilised by Functionalised Titania Nanoparticles. *Chem. Commun.* **2007**, *41*, 4274. https://doi.org/10.1039/b708935j.

[32] Gitli, T.; Silverstein, M. S. Bicontinuous Hydrogel–Hydrophobic Polymer Systems Through Emulsion Templated Simultaneous Polymerizations. *Soft Matter* **2008**, *4* (12), 2475. https://doi.org/10.1039/b809346f.

[33] Kovačič, S.; Jeřabek, K.; Krajnc, P. Responsive Poly(Acrylic Acid) and Poly(*N*-Isopropylacrylamide) Monoliths by High Internal Phase Emulsion (HIPE) Templating. *Macromol. Chem. Phys.* **2011**, *212* (19), 2151–2158. https://doi.org/10.1002/macp.201100229.

[34] Karimian, H.; Moghbeli, M. R. Conducting Poly(Styrene-*Co*-Divinylbenzene)/Polypyrrole PolyHIPE Composite Foam Prepared by Chemical Oxidative Polymerization. *J. Appl. Polym. Sci.* **2013**, *127* (1), 804–811. https://doi.org/10.1002/app.37793.

[35] Hayward, A. S.; Sano, N.; Przyborski, S. A.; Cameron, N. R. Acrylic -Acid-Functionalized PolyHIPE Scaffolds for Use in 3D Cell Culture. *Macromol. Rapid Commun.* **2013**, *34* (23–24), 1844–1849. https://doi.org/10.1002/marc.201300709.

[36] Ghosh, S.; Yadav, A.; Rani, S.; Takkar, S.; Kulshreshtha, R.; Nandan, B.; Srivastava, R. K. 3D Printed Hierarchical Porous Poly(ε-Caprolactone) Scaffolds From Pickering High Internal Phase Emulsion Templating. *Langmuir* **2023**, *39* (5), 1927–1946. https://doi.org/10.1021/acs.langmuir.2c02936.

[37] Zhang, H.; Zhu, Y.; Chen, J.; Zhang, S. Preparation of polyHIPE via CuAAC "Click" Chemistry and Its Application as a Highly Efficient Adsorbent of Cu(II) Ions. *J. Polym. Sci. Part Polym. Chem.* **2017**, *55* (13), 2129–2135. https://doi.org/10.1002/pola.28548.

[38] Lu, T.; Zhu, Y.; Wang, W.; Qi, Y.; Wang, A. Interconnected Superporous Adsorbent Prepared Via Yeast-Based Pickering HIPEs for High-Efficiency Adsorption of Rb+, Cs+ and $Sr^{2+}$. *Chem. Eng. J.* **2019**, *361*, 1411–1422. https://doi.org/10.1016/j.cej.2018.11.006.

[39] Lee, J.; Chang, J. Y. A Hierarchically Porous Catalytic Monolith Prepared From a Pickering High Internal Phase Emulsion Stabilized by Microporous Organic Polymer Particles. *Chem. Eng. J.* **2020**, *381*, 122767. https://doi.org/10.1016/j.cej.2019.122767.

[40] Zhu, Y.; Zhang, R.; Zhang, S.; Chu, Y.; Chen, J. Macroporous Polymers with Aligned Microporous Walls From Pickering High Internal Phase Emulsions. *Langmuir* **2016**, *32* (24), 6083–6088. https://doi.org/10.1021/acs.langmuir.6b00794.

[41] Li, X.; Sun, G.; Li, Y.; Yu, J. C.; Wu, J.; Ma, G.-H.; Ngai, T. Porous $TiO_2$ Materials Through Pickering High-Internal Phase Emulsion Templating. *Langmuir* **2014**, *30* (10), 2676–2683. https://doi.org/10.1021/la404930h.

[42] Zhou, S.; Bismarck, A.; Steinke, J. H. G. Interconnected Macroporous Glycidyl Methacrylate-Grafted Dextran Hydrogels Synthesised From Hydroxyapatite Nanoparticle Stabilised High Internal Phase Emulsion Templates. *J. Mater. Chem.* **2012**, *22* (36), 18824. https://doi.org/10.1039/c2jm33294a.

[43] Carranza, A.; Romero -Perez, D.; Almanza-Reyes, H.; Bogdanchikova, N.; Juarez-Moreno, K.; Pojman, J. A.; Velasquillo, C.; Mota-Morales, J. D. Nonaqueous Synthesis of Macroporous Nanocomposites Using High Internal Phase Emulsion Stabilized by Nanohydroxyapatite. *Adv. Mater. Interfaces* **2017**, *4* (16), 1700094. https://doi.org/10.1002/admi.201700094.

[44] Kapilov-Buchman, K.; Portal, L.; Zhang, Y.; Fechler, N.; Antonietti, M.; Silverstein, M. S. Hierarchically Porous Carbons From an Emulsion-Templated, Urea-Based Deep Eutectic. *J. Mater. Chem. A* **2017**, *5* (31), 16376–16385. https://doi.org/10.1039/C7TA01958K.

[45] Cai, D.; Thijssen, J. H. T.; Clegg, P. S. Making Non-Aqueous High Internal Phase Pickering Emulsions: Influence of Added Polymer and Selective Drying. *ACS Appl. Mater. Interfaces* **2014**, *6* (12), 9214–9219. https://doi.org/10.1021/am501328r.

[46] Zhang, T.; Gui, H.; Xu, Z.; Zhao, Y. Hydrophobic Polyurethane polyHIPEs Templated From Mannitol Within Nonaqueous High Internal Phase Emulsions for Oil Spill Recovery. *J. Polym. Sci. Part Polym. Chem.* **2019**, *57* (12), 1315–1321. https://doi.org/10.1002/pola.29392.

[47] Zhang, T.; Zhao, Y.; Silverstein, M. S. Cellulose-Based, Highly Porous Polyurethanes Templated Within Non-Aqueous High Internal Phase Emulsions. *Cellulose* **2020**, *27* (7), 4007–4018. https://doi.org/10.1007/s10570-020-03059-z.

[48] Yadav, A.; Pal, J.; Nandan, B.; Srivastava, R. K. Macroporous Scaffolds of Cross-Linked Poly(ε-Caprolactone) via High Internal Phase Emulsion Templating. *Polymer* **2019**, *176*, 66–73. https://doi.org/10.1016/j.polymer.2019.05.034.

[49] Erdal, N. B.; Lando, G. A.; Yadav, A.; Srivastava, R. K.; Hakkarainen, M. Hydrolytic Degradation of Porous Crosslinked Poly(ε-Caprolactone) Synthesized by High Internal Phase Emulsion Templating. *Polymers* **2020**, *12* (8), 1849. https://doi.org/10.3390/polym12081849.

[50] Yadav, A.; Pal, S.; Kumar, D.; Nandan, B.; Srivastava, R. K. Polymer Crystallization Under Dual Confinement of High Internal Phase Emulsion Templated Crosslinked Polymer. *Colloids Surf. Physicochem. Eng. Asp.* **2020**, *600*, 124938. https://doi.org/10.1016/j.colsurfa.2020.124938.

[51] Yadav, A.; Erdal, N. B.; Hakkarainen, M.; Nandan, B.; Srivastava, R. K. Cellulose-Derived Nanographene Oxide Reinforced Macroporous Scaffolds of High Internal Phase Emulsion-Templated Cross-Linked Poly(ε-Caprolactone). *Biomacromolecules* **2020**, *21* (2), 589–596. https://doi.org/10.1021/acs.biomac.9b01330.

[52] Agrawal, M.; Yadav, A.; Nandan, B.; Srivastava, R. K. Facile Synthesis of Templated Macrocellular Nanocomposite Scaffold *Via* Emulsifier-Free HIPE-ROP. *Chem. Commun.* **2020**, *56* (83), 12604–12607. https://doi.org/10.1039/D0CC05331G.

[53] Agrawal, M.; Yadav, A.; Takkar, S.; Kulshreshtha, R.; Nandan, B.; Srivastava, R. K. Dual-Functionalized Pickering HIPE Templated Poly(ε-Caprolactone) Scaffold for Maxillofacial Implants. *Int. J. Pharm.* **2023**, *633*, 122611. https://doi.org/10.1016/j.ijpharm.2023.122611.

[54] Xu, L.; Liu, F.; Faraz, N. Theoretical Model for the Electrospinning Nanoporous Materials Process. *Comput. Math. Appl.* **2012**, *64* (5), 1017–1021. https://doi.org/10.1016/j.camwa.2012.03.019.

[55] He, J.; Xu, L.; Wu, Y.; Liu, Y. Mathematical Models for Continuous Electrospun Nanofibers and Electrospun Nanoporous Microspheres. *Polym. Int.* **2007**, *56* (11), 1323–1329. https://doi.org/10.1002/pi.2370.

[56] Sotomayor, O. E.; Tippur, H. V. Role of Cell Regularity and Relative Density on Elastoplastic Compression Response of 3-D Open-Cell Foam Core Sandwich Structure Generated Using Voronoi Diagrams. *Acta Mater.* **2014**, *78*, 301–313. https://doi.org/10.1016/j.actamat.2014.06.051.

[57] Kumar, V.; Bajya, M.; Majumdar, A.; Dubey, D. K. Design of Cost-Effective Hybrid Soft Armour Panels by Strategic Replacement of Backing Layers with a Non-Ballistic Material: A Parametric Study. *Int. J. Impact Eng.* **2023**, *180*, 104678. https://doi.org/10.1016/j.ijimpeng.2023.104678.

[58] Guo, Y.; Wang, X.; Shen, Y.; Dong, K.; Shen, L.; Alzalab, A. A. A. Research Progress, Models and Simulation of Electrospinning Technology: A Review. *J. Mater. Sci.* **2022**, *57* (1), 58–104. https://doi.org/10.1007/s10853-021-06575-w.

[59] Kravchenko, O. G.; Gedler, G.; Kravchenko, S. G.; Feke, D. L.; Manas-Zloczower, I. Modeling Compressive Behavior of Open-Cell Polymerized High Internal Phase Emulsions: Effects of Density and Morphology. *Soft Matter* **2018**, *14* (9), 1637–1646. https://doi.org/10.1039/C7SM02043K.

[60] Wong, L. L. C.; Baiz Villafranca, P. M.; Menner, A.; Bismarck, A. Hierarchical Polymerized High Internal Phase Emulsions Synthesized from Surfactant-Stabilized Emulsion Templates. *Langmuir* **2013**, *29* (20), 5952–5961. https://doi.org/10.1021/la3047643.

[61] Manley, S. S.; Graeber, N.; Grof, Z.; Menner, A.; Hewitt, G. F.; Stepanek, F.; Bismarck, A. New Insights into the Relationship Between Internal Phase Level of Emulsion Templates and Gas–Liquid Permeability of Interconnected Macroporous Polymers. *Soft Matter* **2009**, *5* (23), 4780. https://doi.org/10.1039/b900426b.

[62] Gurevitch, I.; Silverstein, M. S. Shape Memory Polymer Foams From Emulsion Templating. *Soft Matter* **2012**, *8* (40), 10378. https://doi.org/10.1039/c2sm26404h.

[63] Gan, Y. X.; Chen, C.; Shen, Y. P. Three-Dimensional Modeling of the Mechanical Property of Linearly Elastic Open Cell Foams. *Int. J. Solids Struct.* **2005**, *42* (26), 6628–6642. https://doi.org/10.1016/j.ijsolstr.2005.03.002.

[64] Pegoretti, A.; Dorigato, A.; Biani, A.; Slouf, M. Cyclic Olefin Copolymer–Silica Nanocomposites Foams. *J. Mater. Sci.* **2016**, *51* (8), 3907–3916. https://doi.org/10.1007/s10853-015-9710-9.

[65] Pervez, Md. N.; Yeo, W. S.; Mishu, Mst. M. R.; Talukder, Md. E.; Roy, H.; Islam, Md. S.; Zhao, Y.; Cai, Y.; Stylios, G. K.; Naddeo, V. Electrospun Nanofiber Membrane Diameter Prediction Using a Combined Response Surface Methodology and Machine Learning Approach. *Sci. Rep.* **2023**, *13* (1), 9679. https://doi.org/10.1038/s41598-023-36431-7.

[66] Calderó, G.; Patti, A.; Llinàs, M.; García-Celma, M. J. Diffusion in Highly Concentrated Emulsions. *Curr. Opin. Colloid Interface Sci.* **2012**, *17* (5), 255–260. https://doi.org/10.1016/j.cocis.2012.07.001.

[67] Vishal, B.; Ghosh, P. Nonlinear Viscoelastic Behavior of Aqueous Foam Under Large Amplitude Oscillatory Shear Flow. *Korea-Aust. Rheol. J.* **2018**, *30* (3), 147–159. https://doi.org/10.1007/s13367-018-0015-9.

[68] Lefebvre, G.; Riou, J.; Bastiat, G.; Roger, E.; Frombach, K.; Gimel, J.-C.; Saulnier, P.; Calvignac, B. Spontaneous Nano-Emulsification: Process Optimization and Modeling for the Prediction of the Nanoemulsion's Size and Polydispersity. *Int. J. Pharm.* **2017**, *534* (1–2), 220–228. https://doi.org/10.1016/j.ijpharm.2017.10.017.

[69] Ježková, M.; Jelínek, P.; Marelja, O.; Trunov, D.; Jarošová, M.; Slouka, Z.; Šoóš, M. The Preparation of Mono- and Multicomponent Nanoparticle Aggregates with Layer-by-Layer Structure Using Emulsion Templating Method in Microfluidics. *Chem. Eng. Sci.* **2022**, *247*, 117084. https://doi.org/10.1016/j.ces.2021.117084.

[70] Kulkarni, A.; Bambole, V. A.; Mahanwar, P. A. Electrospinning of Polymers, Their Modeling and Applications. *Polym. Plast. Technol. Eng.* **2010**, *49* (5), 427–441. https://doi.org/10.1080/03602550903414019.

[71] Mohammadzadehmoghadam, S.; Dong, Y.; Davies, I. J. Modeling Electrospun Nanofibers: An Overview From Theoretical, Empirical, and Numerical Approaches. *Int. J. Polym. Mater. Polym. Biomater.* **2016**, *65* (17), 901–915. https://doi.org/10.1080/00914037.2016.1180617.

[72] Akdere, M.; Schneiders, T. Modeling of the Electrospinning Process. In *Advances in Modeling and Simulation in Textile Engineering*; Elsevier, 2021; pp. 237–253. https://doi.org/10.1016/B978-0-12-822977-4.00015-7.

[73] Xu, L. A Mathematical Model for Electrospinning Process Under Coupled Field Forces. *Chaos Solitons Fractals* **2009**, *42* (3), 1463–1465. https://doi.org/10.1016/j.chaos.2009.03.054.

[74] Šimko, M.; Erhart, J.; Lukáš, D. A Mathematical Model of External Electrostatic Field of a Special Collector for Electrospinning of Nanofibers. *J. Electrost.* **2014**, *72* (2), 161–165. https://doi.org/10.1016/j.elstat.2014.01.003.

[75] Elkasaby, M.; Hegab, H. A.; Mohany, A.; Rizvi, G. M. Modeling and Optimization of Electrospinning of Polyvinyl Alcohol (PVA). *Adv. Polym. Technol.* **2018**, *37* (6), 2114–2122. https://doi.org/10.1002/adv.21869.

[76] Maghsoodlou, Sh.; Noroozi, B.; Haghi, A. K. A Simple Model for Solvent Evaporation in Electrospinning Process. *Nano* **2017**, *12* (3), 1750028. https://doi.org/10.1142/S179329201750028X.

[77] Ko, J.; Bhullar, S. K.; Mohtaram, N. K.; Willerth, S. M.; Jun, M. B. G. Using Mathematical Modeling to Control Topographical Properties of Poly (ε-Caprolactone) Melt Electrospun Scaffolds. *J. Micromech. Microeng.* **2014**, *24* (6), 065009. https://doi.org/10.1088/0960–1317/24/6/065009.

[78] Ko, J.; Mohtaram, N. K.; Lee, P. C.; Willerth, S. M.; Jun, M. B. G. Mathematical Model for Predicting Topographical Properties of Poly (ε-Caprolactone) Melt Electrospun Scaffolds Including the Effects of Temperature and Linear Transitional Speed. *J. Micromech. Microeng.* **2015**, *25* (4), 045018. https://doi.org/10.1088/0960–1317/25/4/045018.

# 17 Molecular Dynamics Simulation of Graphene-Based Composites

*Sanjay Krishna*

## 17.1 INTRODUCTION

In the realm of materials science, graphene has ascended as a transformative material, captivating researchers and engineers alike due to its exceptional mechanical, electrical, and thermal properties. This two-dimensional hexagonal lattice of carbon atoms has proven to be a versatile building block, giving rise to a new class of materials known as graphene-based composites. These composites, formed by integrating graphene with various matrix materials, exhibit extraordinary synergistic properties that surpass those of their individual components. From unprecedented tensile strength to remarkable electrical conductivity, the integration of graphene into composites holds immense promise for a wide range of applications, spanning electronics, energy storage, and structural materials.

However, the full realization of the potential of graphene-based composites necessitates a comprehensive understanding of their atomic-scale behavior and interactions. Experimental techniques provide valuable insights, but the atomic-level details often elude direct observation. Enter molecular dynamics (MD) simulation—a computational approach that has become instrumental in unraveling the mysteries of materials at the atomic and molecular scale. This chapter embarks on a journey into the world of MD simulations, exploring their application in deciphering the dynamic behaviors of graphene-based composites and bridging the gap between theoretical predictions and experimental observations.

Before delving into the intricacies of MD simulations, it is imperative to establish the foundational properties of graphene that make it an attractive candidate for composite materials. Graphene's remarkable mechanical strength, stemming from its sp2-hybridized carbon–carbon bonds, has been extensively studied and validated experimentally. Its exceptional electrical conductivity, thermal stability, and flexibility further contribute to its allure as a building block for advanced composites.

The two-dimensional nature of graphene, with its unique electronic structure and high surface area, opens up avenues for diverse applications. Its incorporation into composites promises improvements in mechanical strength, electrical conductivity, and thermal transport—properties crucial for the development of next-generation materials. By varying metal substrates, carbon sources, and growth conditions, like temperature and pressure, researchers have developed diverse protocols for graphene growth, resulting in varied growth patterns and sample structures [1–5]. Understanding these growth mechanisms requires a thorough knowledge of the underlying processes. However, capturing the atomic details of graphene growth, especially under high-temperature conditions, remains a significant challenge. Consequently, kinetic data often serves as the primary experimental basis for theorizing about growth mechanisms.

For geometry optimization and energy calculations, researchers commonly employ first-principles, semiempirical, and empirical theoretical methods. These methods, offering atomistic resolution, are instrumental in understanding growth mechanisms by comparing the stability of different species on metal surfaces [6–9]. While these results are enlightening, the equilibrium state property of energetics is challenging to directly correlate with the overall kinetics observed experimentally.


DOI: 10.1201/9781003476337-17

Bridging this gap between theoretical and experimental understanding necessitates explicit simulations of graphene growth. Due to the involved complexities, a range of multiscale simulation techniques is employed. These include molecular dynamics (MD), kinetic Monte Carlo (kMC), rate equations, kinetic Wulff construction, phase field, and computational fluid dynamics (CFD) [10–18]. For a comprehensive mechanistic understanding, maintaining atomistic resolution within these simulations is essential. Various computational techniques have been employed to understand the effect of graphene on polymer composites, which include continuum mechanics, quantum-based, molecular dynamics, molecular mechanics, density functional theory, and atomistic modeling [19–25].

Understanding these intrinsic characteristics of graphene forms the bedrock upon which the subsequent exploration of MD simulations is built. Molecular dynamics simulation, a computational technique rooted in Newtonian mechanics, provides a virtual laboratory for studying the dynamic evolution of atoms and molecules over time. At its core, MD simulation involves numerically solving the equations of motion for a system of interacting particles, allowing researchers to observe the evolution of a material's structure and properties in unprecedented detail. The versatility of MD simulations enables the exploration of a wide range of phenomena, from the dynamics of liquids to the behavior of complex biomolecular systems.

In the context of graphene-based composites, MD simulations serve as a powerful tool for unraveling the intricate interplay between graphene and the surrounding matrix materials. These simulations enable researchers to study the mechanical deformation, thermal behavior, and electronic interactions at the atomic level, some of which are challenging to analyze experimentally. The predictive capabilities of MD simulations make them particularly valuable for optimizing composite design and predicting performance under varying conditions.

Understanding the principles and methodologies that underpin MD simulations is crucial for extracting meaningful implications from the computational experiments. MD simulations involve discretizing time and space, with the integration of Newton's equations of motion providing a trajectory of atomic positions over time. The accuracy of these simulations depends on the force fields employed, which describe the interactions between atoms based on empirical or quantum mechanical principles.

This chapter sets out to provide a comprehensive exploration of molecular dynamics simulation as a tool for unraveling the atomic-scale dynamics of graphene-based composites. By combining a deep understanding of graphene's intrinsic properties with the principles and applications of MD simulations, researchers and academicians can gain unprecedented insights into the behavior of these advanced materials.

## 17.2 THEORETICAL BACKGROUND

### 17.2.1 Graphene Structure and Properties

*Graphene*, a monolayer of carbon atoms arranged in a hexagonal lattice, is a fundamental constituent of various carbon allotropes. Its unique structure is characterized by sp2-hybridized carbon atoms, forming a two-dimensional honeycomb lattice. This hexagonal arrangement imparts exceptional properties to graphene, making it a subject of intense research and technological interest.

The carbon–carbon bonds in graphene are strong covalent bonds, providing the material with remarkable mechanical strength. Each carbon atom in the lattice is bonded to three neighboring carbon atoms, resulting in a highly stable structure. The planar, two-dimensional nature of graphene allows for a high degree of delocalization of π-electrons across the hexagonal lattice, contributing to its outstanding electronic properties. Figure 17.1 shows a graphite structure modeled by stacking layers of graphene at a distance of 3.4 Å with each other [26].

One of the defining features of graphene is its extraordinary electronic properties. The electronic structure of graphene is characterized by a Dirac cone–shaped energy dispersion, a consequence of

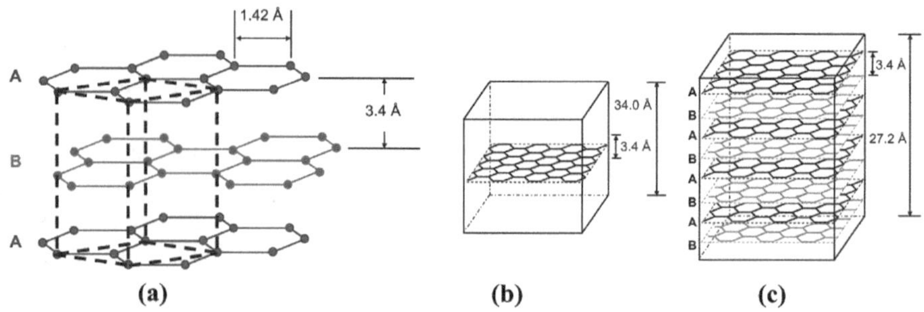


**FIGURE 17.1** Schematic representation of a (a) graphene structure, (b) atomistic model of graphene sheet, and (c) atomistic model of graphite flakes.

**Source: Based on [26].**

its unique hexagonal lattice. Near the Fermi level, the energy of charge carriers is proportional to the momentum, resulting in massless Dirac fermions. This unique electronic band structure leads to high electron mobility and exceptional electrical conductivity, making graphene a promising material for electronic applications. The π–electron cloud in graphene's hexagonal lattice facilitates rapid electron transport, and its behavior as massless Dirac fermions contributes to its remarkable electronic conductivity. The electronic properties of graphene, including its quantum Hall effect, have opened up new possibilities for electronic devices with unprecedented performance.

Graphene's mechanical properties are equally impressive. Despite being only one atom thick, graphene is incredibly strong. The intrinsic tensile strength of graphene exceeds 130 GPa (gigapascals), making it one of the strongest materials known. Additionally, graphene exhibits exceptional flexibility, allowing it to withstand substantial deformation without structural failure. This combination of strength and flexibility makes graphene an ideal candidate for reinforcement in composite materials. The remarkable mechanical properties of graphene are a result of its strong carbon–carbon bonds and the two-dimensional lattice structure. The planar configuration enables graphene to distribute mechanical stress effectively, contributing to its outstanding mechanical strength. Understanding these properties is crucial for utilizing graphene in various applications, from structural materials to flexible electronics.

Graphene also boasts remarkable thermal properties, with a thermal conductivity that surpasses many conventional materials. The absence of out-of-plane acoustic phonon modes and the two-dimensional nature of graphene contribute to its excellent thermal conductivity. Phonon transport in graphene is highly efficient, making it an attractive material for applications requiring effective thermal management. The high thermal conductivity of graphene, combined with its other exceptional properties, positions it as a key component in materials designed for heat dissipation. Understanding the thermal behavior of graphene is essential for developing advanced thermal interface materials and enhancing the efficiency of electronic devices.

In summary, graphene's unique structure and properties make it a material of immense scientific and technological significance. Its two-dimensional hexagonal lattice, coupled with outstanding electronic, mechanical, and thermal characteristics, opens up a myriad of possibilities for diverse applications. Whether in electronics, composites, or thermal management, the understanding of graphene's structure and properties is foundational for harnessing its full potential in various fields.

The incorporation of graphene into composites is driven by its high aspect ratio, large surface area, and unique electronic properties. Graphene can be integrated into different matrix materials, such as polymers, metals, ceramics, and more, depending on the desired properties of the composite. The interaction between graphene and the matrix material significantly influences the overall performance of the composite. Proper dispersion of graphene in the matrix and strong interfacial bonding are crucial for achieving optimal properties. Graphene-based composites find applications in various fields, including aerospace, electronics, energy storage, and structural materials. For example, in polymer composites, graphene serves as a reinforcing filler, improving mechanical strength, thermal conductivity, and electrical conductivity.

### 17.2.2 Molecular Dynamics Simulation

Understanding the behavior of graphene-based composites at the molecular level requires advanced simulation techniques. Molecular dynamics (MD) simulation is a powerful computational method used to study the motion and interactions of atoms and molecules over time. In the context of graphene-based composites, MD simulations enable the investigation of structural stability, mechanical properties, thermal behavior, and dynamic aspects at the atomic scale. The success of MD simulations relies on accurate force fields, which describe the potential energy and forces between atoms. For graphene-based composites, force fields must capture the interactions between graphene and the matrix material accurately. Various force fields, such as the CHARMM, AMBER, and GROMACS force fields, have been adapted and developed to simulate carbon-based materials and their interactions with other compounds.

MD simulations involve solving Newton's equations of motion iteratively to predict the trajectory of each atom in the system. These simulations reveal the dynamic evolution of graphene-based composites under different conditions, allowing researchers to explore the effects of temperature, pressure, and external forces on the material's behavior. The application of MD simulations to graphene-based composites has become increasingly prevalent in materials science research. By providing a molecular-level understanding of composite materials, MD simulations contribute to the design and optimization of graphene-enhanced materials for specific applications.

## 17.3 METHODOLOGY

### 17.3.1 Simulation Setup

Molecular dynamics (MD) simulations of graphene-based composites require a carefully configured simulation setup to accurately capture the behavior of atoms and molecules within the system. This section delves into the essential elements of simulation setup, encompassing parameters, initial configurations, and boundary conditions.

#### 17.3.1.1 Choice of Force Fields and Parameters

The foundation of any MD simulation lies in the selection of appropriate force fields and parameters. These force fields govern the interactions between atoms and molecules in the simulation, dictating the potential energy landscape and forces experienced by each particle. For graphene-based composites, selecting force fields that accurately describe the interactions between graphene and the matrix material is crucial.

Several force fields have been developed and adapted for simulating carbon-based materials and their interactions with other compounds. Examples include the CHARMM force field, AMBER force field, and GROMACS force field, among others. The choice of force field depends on the specific materials under investigation and the level of accuracy required for the simulation.

Force field parameters, such as bond lengths, bond angles, and dihedral angles, need to be carefully calibrated to reproduce the structural and dynamic properties of the materials accurately. Parameterization often involves fitting to experimental data or quantum mechanical calculations to ensure that the force field captures the underlying physics of the system.

#### 17.3.1.2 Initial Configurations

The initial configuration of the simulation system plays a critical role in determining the trajectory of the simulation. For graphene-based composites, the initial arrangement of graphene and the matrix material is of paramount importance. Depending on the research objectives, various initial configurations can be employed. When it comes to dispersion style of fillers, one common approach is to randomly disperse graphene sheets or nanoparticles within the matrix material. This random dispersion mimics the distribution of graphene in a composite material and allows researchers to investigate the effects of graphene on the overall properties of the composite. In some cases, researchers may choose to create ordered or layered structures by placing graphene sheets at specific positions within the matrix. These ordered configurations can help study the influence of the spatial arrangement of graphene on composite properties.

When examining the behavior of graphene at interfaces, such as the graphene–matrix interface in a composite, special attention must be paid to the initial configuration. Ensuring that the interface is well-defined and that graphene–matrix interactions are accurately represented is crucial for meaningful results. Figure 17.2 presents a schematic diagram that depicts the interactions at the interface between graphene and the polymer matrix [27]. When a nanocomposite undergoes tensile stress, one possibility is the breaking of graphene's carbon–carbon (C-C) bonds. However, according to a study by Lee et al., the strength of these graphene C-C bonds is around 130 GPa, which is significantly higher than the usual strength of polymers (under 100 MPa) and that of nanocomposites, which rarely exceeds 100 MPa [28]. Thus, the likelihood of C-C bond breakage in these circumstances is minimal. Moreover, in nanocomposites with nonfunctionalized carbonaceous reinforcements, the bonding between the reinforcement and the matrix typically involves weaker van der Waals forces, physical bonding, and mechanical interlocking. Van der Waals forces play a pivotal role in the interfacial interactions within carbon nanotube (CNT)–polymer nanocomposites. These forces are substantially weaker than covalent C-C bonds, making it more probable for graphene sheets to be pulled out of the matrix. Overcoming the van der Waals interactions and physical bonding to extract graphene requires energy, which in turn enhances the mechanical properties of the nanocomposites. However, this improvement might not fully reflect graphene's potential as a reinforcement material. Consequently, substantial research has focused on altering the surface of graphene through functionalization to strengthen these interfacial interactions.

#### 17.3.1.3 Boundary Conditions

Boundary conditions in MD simulations dictate how the simulation system interacts with its surroundings. Commonly used boundary conditions include periodic boundary conditions (PBC) and fixed boundary conditions. The choice of boundary conditions can significantly impact the simulation results. PBC is often employed in MD simulations to create an infinite, periodic replica of the simulation cell. This approach allows researchers to study bulk properties and mitigate finite-size effects. PBC is particularly useful when simulating large systems or when investigating phenomena

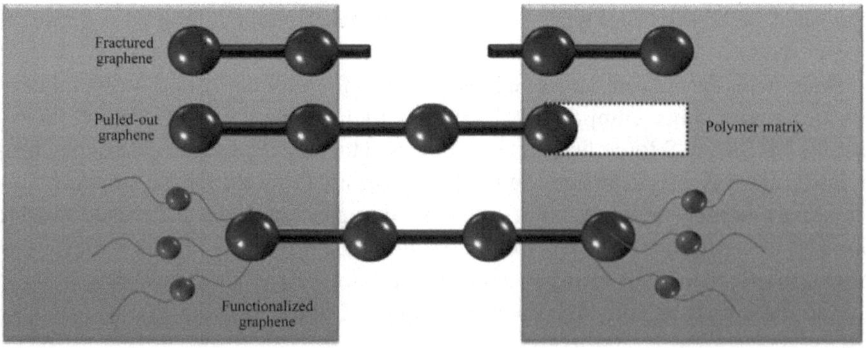


**FIGURE 17.2** Interfacial interactions for a graphene–polymer system.

**Source: Based on [27].**

that require a representative sample of the material. In some cases, fixed boundary conditions may be applied to restrict the movement of specific atoms or molecules in the simulation. Fixed boundary conditions can be useful when studying confined systems or when researchers want to investigate the behavior of specific regions of the material.

#### 17.3.1.4 Temperature and Pressure Control

Controlling temperature and pressure within the simulation cell is essential for reproducing realistic thermodynamic conditions. Temperature control methods, such as the Nosé–Hoover thermostat or the Langevin thermostat, can be employed to maintain the desired temperature. Pressure control methods, such as the Parrinello–Rahman barostat, are used to regulate pressure within the system. The choice of temperature and pressure control parameters should be based on the conditions relevant to the study, and care should be taken to ensure that these controls do not introduce artifacts into the simulation results.

#### 17.3.1.5 Time Step Considerations

The choice of time step in MD simulations determines the granularity of the simulation, with smaller time steps allowing for more accurate integration of the equations of motion but requiring longer simulation times. Time step selection should strike a balance between computational efficiency and accuracy, considering the dynamics of the system being simulated.

In conclusion, the setup of MD simulations for graphene-based composites involves careful choices related to force fields, parameters, initial configurations, boundary conditions, temperature, pressure, and time steps. These choices are essential for accurately capturing the behavior of atoms and molecules within the composite system and ensuring that simulation results provide meaningful insights into the material's properties and behavior.

### 17.3.2 Simulation Protocols

Molecular dynamics (MD) simulations of graphene-based composites require well-defined simulation protocols to ensure the accuracy and reliability of the results. Different simulation protocols, including the simulation environment, choice of force fields and parameters, as well as considerations for temperature and pressure control, have been discussed in this section.

#### 17.3.2.1 Simulation Environment

The choice of simulation environment plays a pivotal role in MD simulations. In the context of graphene-based composites, it is essential to create a realistic representation of the composite system. To achieve this, researchers typically build their simulation cells with periodic boundary conditions. This approach allows for the replication of the unit cell throughout space, mimicking an infinite material. Periodic boundary conditions are especially suitable for graphene composites, as they can account for the 2D nature of graphene layers and the interactions between graphene and other components.

The size of the simulation cell is a critical consideration. Ideally, it should be large enough to prevent interactions between periodic images of the composite, which could lead to artifacts. The choice of cell size depends on the specific system under investigation and the desired level of accuracy. Researchers often perform convergence tests to determine the appropriate simulation cell dimensions that minimize finite-size effects.

#### 17.3.2.2 Choice of Force Fields and Parameters

The accuracy of MD simulations heavily relies on the choice of force fields and parameters that describe the interatomic interactions within the composite system. For graphene-based composites, researchers commonly employ empirical force fields, like the CHARMM, AMBER, or OPLS force fields, which have been extensively validated for carbon-based materials.

The selection of the force field should take into account the nature of the composite constituents. In the case of graphene, the interaction parameters for carbon–carbon interactions should accurately reproduce the mechanical and electronic properties of graphene sheets. Additionally, the force field should be compatible with the other materials present in the composite, such as polymers, metals, or nanoparticles.

Parameters like bond lengths, bond angles, dihedral angles, and nonbonded interactions (van der Waals and electrostatic interactions) must be carefully calibrated to match experimental data and quantum mechanical calculations. Parameterization is often a time-consuming process that requires careful validation against experimental observations, but it is essential for the accuracy of MD simulations.

#### 17.3.2.3 Temperature and Pressure Control

Maintaining the appropriate temperature and pressure conditions during MD simulations is crucial for mimicking real-world scenarios. The choice of ensemble (e.g., NVE, NVT, NPT) depends on the research objectives and the physical conditions of interest.

- *NVE ensemble.* In the microcanonical ensemble (NVE), the number of particles ($N$), the volume ($V$), and the total energy ($E$) are conserved. NVE simulations are useful for studying the microstructural evolution of the composite without any exchange of heat or particles with the surroundings.
- *NVT ensemble.* The canonical ensemble (NVT) keeps the number of particles, volume, and temperature ($T$) constant. NVT simulations are widely used for equilibrium studies, where temperature control is essential to investigate thermodynamic properties.
- *NPT ensemble.* In the isothermal-isobaric ensemble (NPT), the number of particles, pressure ($P$), and temperature are maintained. NPT simulations are crucial when studying systems with variable volume or when investigating composite behavior under different pressure conditions.

Temperature control in MD simulations is typically achieved using thermostat algorithms, like the Berendsen thermostat, the Nosé–Hoover thermostat, or the Andersen thermostat. These algorithms control the kinetic energy of the particles to maintain the desired temperature. Pressure control, on the other hand, is realized through barostat algorithms, such as the Berendsen barostat, Parrinello–Rahman barostat, or Martyna–Tobias–Klein barostat. These algorithms adjust the volume of the simulation cell to maintain the desired pressure. The choice of thermostat and barostat algorithms should be made carefully, considering the computational cost and their suitability for the specific research objectives. Additionally, it is essential to ensure that the chosen algorithms do not introduce artifacts that could affect the results.

In summary, simulation protocols are a critical component of MD simulations of graphene-based composites. Creating a realistic simulation environment, selecting appropriate force fields and parameters, and implementing effective temperature and pressure control are essential for obtaining meaningful and reliable results. These protocols ensure that the simulations accurately capture the behavior of graphene-based composites and explicating their structural and mechanical properties.

## 17.4 MATERIAL ANALYSIS

### 17.4.1 Structural Analysis

One of the fundamental aspects of understanding graphene-based composites is a thorough analysis of their structural stability and the impact of graphene on the overall composite structure. Molecular dynamics simulations provide valuable insights into the atomic-level arrangement and stability of these composites. In this section, structural aspects of graphene-based composites and their findings have been discussed.

Effects of chemisorption on the interfacial bonding characteristics of graphene–polymer composites were studied by Lv et al. [29]. This study underlines the importance of chemical functionalization in developing strong interfacial bonding in graphene–polymer composites. The bonding energy and shear stress between graphene and polymer increased with the concentration of functionalized groups on the graphene surface. Specific chemical modifications of graphene significantly influence the strength of interfacial bonding. Attaching suitable chemical groups to graphene can greatly enhance load transfer between graphene and polymer in nanocomposites. Different chemical functionalizations on graphene (like -COOH, -OH, $-NH_2$) show varying degrees of enhancement in interfacial bonding energy and shear stress. Chemical functionalization can effectively increase the interfacial shear stress and bonding energy, enhancing the mechanical properties of the composites.

#### 17.4.1.1 Bond Lengths and Angles

The molecular dynamics simulations revealed important information about the bond lengths and angles within the graphene-based composites. As expected, the carbon–carbon (C-C) bonds in graphene sheets maintained their typical $sp^2$ hybridization, resulting in a bond length of approximately 0.142 nm, consistent with the values reported in the literature.

In the composite systems, variations are observed in bond lengths, particularly in the vicinity of the graphene sheets. The interaction between the graphene and the neighboring composite material had a noticeable influence on bond lengths. The C-C bonds adjacent to the graphene sheets experienced slight elongation or compression, depending on the specific arrangement and interactions. This phenomenon can be attributed to the transfer of charge and stress between the graphene and the surrounding material. Furthermore, the angles between the carbon–carbon bonds in graphene remained close to the ideal $sp^2$ bond angle of 120°. However, deviations from this angle were observed near the edges and defects of graphene sheets. These structural irregularities at the edges were expected and are consistent with previous studies. The interactions with the composite matrix also had subtle effects on the bond angles, with localized distortions detected in the immediate vicinity of the graphene sheets.

#### 17.4.1.2 Intermolecular Interactions

The intermolecular distances between the graphene sheets and the surrounding composite material are crucial for understanding the interaction and adhesion between the two components. In cases where strong interactions were observed between graphene and the surrounding material, such as in graphene–polymer composites, the intermolecular distance decreased significantly compared to the pristine graphene sheet. This decrease in distance indicated a strong adhesive interaction between graphene and the polymer chains. Such interactions are essential for enhancing the mechanical properties and load transfer capabilities of the composite.

On the other hand, in composites with weaker graphene–matrix interactions, such as graphene–metal composites, the intermolecular distances remained relatively larger. In some instances, a slight expansion of the intermolecular region occurs due to the thermal expansion of the composite material. These variations in intermolecular distances highlight the tunable nature of graphene-based composites, where the strength of interaction can be tailored to suit specific applications.

Interaction energy calculation helps in understanding the effect of surface modification, where this energy is calculated by deducting the total energy of the system with the energies of the polymer matrix and the filler. Figure 17.3 provides the structures to get the respective energies, calculating the interaction energy between epoxy and graphene [30]. The interaction energy was observed to be higher for the surface-modified one as compared to pure graphene, which led to improved thermal and mechanical properties of nanocomposites.

Miyazaki et al. examined the structural properties of graphene oxide–reinforced poly(styrene sulfonate) sodium salt (GPSS) through classical MD simulations [31]. The simulations showed the arrangement of the system comprising poly(styrene sulfonate) (PSS), water, and reduced graphene oxide (rGO) effectively equilibrated at standard temperature and pressure. The observed mass

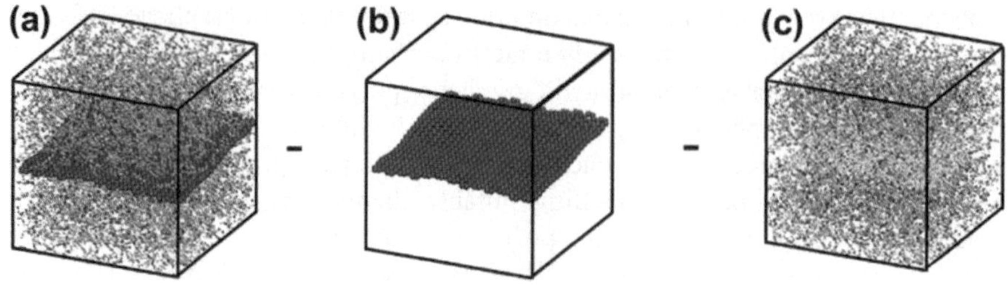


**FIGURE 17.3** Interaction energy calculation between polymer matrix and graphene.

**Source: Based on [30].**

density of this arrangement was approximately 1.4 g/cm³, notably higher than the 0.8 g/cm³ found in bulk PSS. Key findings include the identification of the radial distribution function (RDF) between oxygen atoms in sulfonate groups and the hydrogen in the hydroxyl and carboxyl groups. The simulations suggested that the primary interaction between PSS and rGO occurs through hydroxyl and carboxyl groups forming hydrogen bonds. Another significant observation was the extended conformation of PSS, with sulfonated styrene groups positioned adjacent to rGO. This orientation maximizes interactions between sulfite and OH groups while reducing electrostatic repulsion among neighboring sulfonate groups. This strong interaction was interpreted to restrict the vibrational movement of the PSS chain, a hypothesis supported by FTIR analysis. These findings aligned well with Raman analysis outcomes, indicating a tendency for PSS to be situated near rGO nanoplatelets without undergoing a chemical reaction. Overall, the study dealt with the interactions and structural dynamics within the GPSS system.

#### 17.4.1.3 Defects and Structural Integrity

Another critical aspect of structural analysis involves examining defects and their impact on the overall structural integrity of graphene-based composites. Defects can arise from various sources, including thermal fluctuations, mechanical stresses, and manufacturing processes.

Dynamics of polymer chains (polyethylene (PE), polystyrene (PS), and poly(methyl methacrylate) (PMMA)) near graphene layers were studied, revealing that polymer segments close to graphene exhibit slower relaxation compared to segments further away [32]. The study finds significant differences in the dynamics of the three polymers. PE showed much faster segmental dynamics compared to PS and PMMA. The presence of graphene layers led to a more complex dynamical behavior, with variations in relaxation times and distributions across different distances from the graphene surface. For PE, a well-ordered layered structure near the graphene layer significantly influences its dynamical behavior, showing slower dynamics in these regions. The study quantified these dynamic heterogeneities using parameters like the segmental relaxation time and KWW (Kohlrausch–Williams–Watts) stretch exponential functions.

Guryel et al. provided insights into the interactions between different polymers and graphene in nanocomposites [33]. Graphene was found to induce order in polyethylene (PE) and polyvinylidene fluoride (PVDF) systems by acting as a nucleation site for crystallization. This effect was less pronounced in polystyrene (PS), which remains disordered even in the presence of graphene. The polymer chains of PE and PVDF exhibit distinct orientations when interacting with graphene, with PE chains aligning parallel to the graphene surface and PVDF chains displaying a variety of orientations. The interaction strength between the graphene and the different polymers (PE, PS, PVDF) is influenced by the chemical structure and conformation of the polymer chains.

The immiscibility of graphene with linear aliphatic polyesters demanded a molecular study to look into the dynamic and conformational properties of the polymer chains at graphene's surface. This led to the study that supplemented existing empirical data by providing physical interpretations about the adsorption behavior of poly(ethylene succinate) (PES) chains on graphene nanosheets (GNS) [34]. It revealed that PES, a crystallizable semiflexible polymer, had undergone significant changes in conformation and dynamics upon interacting with GNS. Factors like GNS functionalization, temperature, and chain length significantly impacted the PES chain's conformational order and dynamics. On adsorption, PES adopted a folded configuration, facilitating structural ordering at the interface and acting as a nucleation stage for crystallization. The study also highlighted a crossover in chain dynamic mode, crucial for estimating nucleation stage rate.

During the simulations, formation and propagation of defects are monitored, such as vacancies, Stone–Wales defects, and edge reconstructions, in graphene sheets. Interestingly, the results indicated that the presence of the composite matrix could suppress the formation and migration of certain defects. The surrounding material acted as a stabilizing influence, reducing the mobility of defects in graphene. Furthermore, the structural integrity of the composites was evaluated under different loading conditions. Mechanical tests, including tensile and compressive loading, were conducted in silico to assess the response of the composites to external forces. The findings demonstrated that the presence of graphene enhanced the structural robustness of the composites, leading to improved mechanical properties.

In conclusion, the structural analysis of graphene-based composites through molecular dynamics simulations helped in understanding their atomic-level arrangement, bond lengths, intermolecular distances, and defect formation. These findings contribute to a better understanding of how graphene interacts with composite materials and offer opportunities for tailoring the structural properties of composites for specific applications.

### 17.4.2 Mechanical Properties

The mechanical properties of graphene-based composites play a pivotal role in determining their suitability for various applications, including aerospace materials, structural components, and advanced electronics. In this section, different aspects of mechanical properties have been discussed using molecular dynamics (MD) simulations to elucidate how graphene reinforcement affects the mechanical behavior of these composites.

#### 17.4.2.1 Tensile Strength and Young's Modulus

Tensile strength and Young's modulus are fundamental mechanical properties that define a material's resistance to deformation under an applied load. MD simulations have generally revealed that the incorporation of graphene into composite materials significantly enhances both of these properties.

Graphene's exceptional intrinsic tensile strength, with values exceeding 130 GPa, is well-documented in the literature. When graphene sheets are introduced into a composite, their high tensile strength acts as a reinforcement mechanism. A noticeable increase in tensile strength as compared to the pristine matrix material is observed, which can be attributed to the strong covalent bonds within the graphene lattice, which distribute the load efficiently and resist crack propagation. Young's modulus, a measure of a material's stiffness, also benefits from graphene reinforcement. The modulus of the composite is higher than that of the matrix material alone, signifying increased rigidity. This enhancement can be attributed to the stiff and two-dimensional nature of graphene, which resists deformation and aids in maintaining the composite's structural integrity under load.

When functionalized graphene sheets were reinforced into poly(methyl methacrylate) (PMMA) nanocomposites, mechanical properties were investigated to study the filler's effect [35]. The simulation methodology involved detailed atomistic simulations using the Dreiding all-atom force field, and treatment of mechanical response to deformation was performed for estimating elastic constants at room temperature. Specifically, at a 5.67 wt% loading of functionalized graphene, there was an approximately 74% improvement in Young's modulus, 19% in the bulk modulus, and 83% in the shear modulus. This study revealed the effect of ultralow concentrations of graphene sheets on the mechanical properties of the polymer, owing to the strong adsorption of PMMA on the fillers, as they contain pendant hydroxyl groups forming hydrogen bonds with PMMA's carbonyl groups.

Rissanau et al. conducted both qualitative and quantitative research to investigate how graphene layers affect the dynamic properties at the interfaces of graphene/PMMA, PS, and PE hybrid polymeric systems [36]. They discovered that polyethylene (PE) exhibits much faster terminal and segmental dynamics compared to PMMA and PS, with PMMA being the slowest. In another study, they explored hybrid nanostructured systems with varying film thicknesses between 2.60 and 13.35 nm at a constant temperature [37]. They observed that the density of PMMA polymer chains near the graphene layers is higher than in the bulk, up to distances of approximately 1.0–1.5 nm.

In other work, adding 1.0 wt% of graphene to PMMA led to an 80% increase in Young's modulus and a 20% increase in ultimate tensile strength (UTS) [38]. Jang et al. used MD simulation to examine the interfacial adhesion between a vinyl ester (VE) matrix and three-layers-thick graphite nanoplatelets [39]. Honarvar et al. also employed MD simulation, focusing on graphene with low and high aspect ratios embedded in SU8 resin [40]. It was concluded that composites reinforced with high-aspect-ratio graphene (single layer, 2.1 wt%) showed increases of 63%, 82%, and 202% in Young's, shear, and bulk modulus, respectively, compared to pure SU8. For composites with the same aspect ratio (double layer, 4.2 wt%), the enhancements were 81%, 83%, and 216%. However, composites containing low-aspect-ratio graphene exhibited more wrinkling, leading to lower mechanical properties.

#### 17.4.2.2 Fracture Analysis

Fracture toughness is a crucial parameter for assessing a material's ability to resist crack propagation and withstand catastrophic failure. A substantial improvement in fracture toughness when graphene was incorporated into the composites can be attributed to the crack-deflection and crack-bridging mechanisms facilitated by the graphene sheets. As cracks propagate within the composite, the graphene layers act as barriers, forcing the cracks to deviate from their path and dissipate energy, thereby increasing the material's resistance to fracture.

MD simulations were employed to explore how hydrogen functionalization affects the fracture strength of graphene and its nonbonded interactions within graphene–polymer nanocomposites [41]. The study found that while hydrogen functionalization is advantageous in certain nanomechanical applications, it can weaken graphene and deteriorate the interfacial properties of graphene–polymer composites. A notable decrease in graphene's fracture strength was observed with rising temperatures. Hydrogen passivation of carbon bonds led to reduced fracture strength, suggesting that first-principle simulations, which inherently involve hydrogen passivation, might underestimate graphene's fracture strength. An atomistic model was developed to evaluate the fracture strength of both pristine and hydrogen-functionalized graphene, considering the influence of temperature and strain rate. This model indicated that molecular dynamics simulations, typically conducted at higher strain rates, tend to overestimate graphene's strength at increased temperatures. This atomistic model serves as an effective computational tool for assessing graphene's design strength under various processing conditions.

Classical continuum concepts were applied to accurately model both the fracture strength of graphene and the interfacial properties of graphene–polymer composites. These insights are particularly valuable for the analysis and design of graphene-based nanoelectromechanical systems (NEMS) and composite materials, especially in scenarios where hydrogen adatoms and high temperatures are present. Additionally, the study offered computationally efficient continuum-based models for analyzing the fracture strength of graphene and the interfacial properties of advanced graphene-based nanocomposites.

#### 17.4.2.3 Temperature and Loading Rate Dependence

To further explore the mechanical behavior of graphene-based composites, simulations have been conducted at different temperatures and loading rates, enabling the assessment of the temperature and strain rate dependencies of the mechanical properties.

At elevated temperatures, a reduction in the tensile strength and Young's modulus of the composites was observed, owing to thermal vibrations, which disrupt the crystalline order of the graphene sheets and reduce their load-carrying capacity. However, even at elevated temperatures, the composites maintained a mechanical advantage over the matrix material alone, underlining their resilience to thermal stress. The loading rate dependence of the mechanical properties revealed interesting trends. At higher loading rates, the composites exhibited a stiffer response, with increased Young's modulus. This behavior can be attributed to the shorter timescales of high-rate loading, which limit the ability of defects to form and propagate within the graphene sheets. Consequently, the composites maintain their structural integrity and exhibit enhanced stiffness. These temperature and loading rate dependencies have significant implications for real-world applications of graphene-based composites. For instance, in aerospace materials, understanding how these composites perform under varying temperature and loading conditions is critical for ensuring their reliability and safety in extreme environments.

Many of the computational findings aligned with experimental observations and emphasized the potential of graphene-based composites for high-performance applications, where mechanical robustness is paramount. These findings not only advance the fundamental understanding of graphene-based composites but also provide valuable guidance for the design and engineering of materials with tailored mechanical properties for specific applications.

### 17.4.3 Thermal Properties

Understanding the thermal properties of graphene-based composites is crucial for a wide range of applications, including thermal management in electronics, aerospace materials, and energy storage systems. In this section, the thermal behavior of these materials has been discussed, with a focus on thermal conductivity, heat transfer mechanisms, and the influence of temperature on the composite structure.

#### 17.4.3.1 Thermal Conductivity

Thermal conductivity (TC) is a key property of materials that governs their ability to conduct and transfer heat. In the context of graphene-based composites, the inclusion of graphene sheets can significantly enhance thermal conductivity due to graphene's exceptional thermal properties. Enhancement of conductivity in the presence of reinforcement can be attributed to the efficient heat transfer pathways provided by graphene sheets. Graphene's high thermal conductivity allows for rapid phonon transport, which dominates the heat transfer mechanism in these composites.

Several studies have supported these findings. For instance, a study by Balandin et al. reported that graphene possesses an intrinsic thermal conductivity as high as 5,000 W/mK at room temperature [42]. When graphene is incorporated into composites, it serves as a thermal bridge, facilitating heat conduction between the composite's constituents. The thermal conductivity of the composite was found to be temperature-dependent. Gu et al. studied the thermal conductivity of graphene-based composite materials using reverse non-equilibrium molecular dynamics (RNEMD) method [43]. The simulation revealed that the addition of graphene (GR) to capric acid/ethylene-vinyl acetate (CA/EVA) composites affects their thermal conductivity. At a GR content of 1.8 wt%, the thermal conductivity increased, but beyond this concentration, a decrease in thermal conductivity was observed. This trend is contrary to the experimental findings where increased GR content consistently raised thermal conductivity. The study also found that the dispersion heterogeneity of GR in the composites differently influenced the experimental and simulation results. Excessive GR disrupted the microstructure of the composite, leading to reduced mean free path of phonons and stronger phonon scattering, thus restraining heat conduction.

Thermal conductivity was also studied for polyvinyl alcohol–based composites using the RNEMD [44]. The study quantifies the surface roughness of graphene and the distribution of functionalized graphene (FG) and polyvinyl alcohol (PVA) chains near the graphene surface. It's observed that PVA chains are more repelled by the carboxyl group than by hydroxyl and amino groups. Consequently, the hydroxyl group more effectively enhances interfacial thermal conductivity. However, FGs disrupt graphene's structure, lowering its intrinsic thermal conductivity. Using the effective medium approximation model and finite element method, the composite's overall thermal conductivity is calculated. FGs increase thermal conductivity in composites with shorter graphene lengths but decrease it in longer lengths, indicating a specific graphene length where functionalized and pristine graphene have equal thermal conductivities. As the length of graphene increases, the primary factor influencing overall thermal conductivity shifts from interfacial to intrinsic thermal conductivity. Additionally, the arrangement of graphene plays a more crucial role in determining overall thermal conductivity than the traditionally emphasized interfacial thermal conductivity.

Thermal conductances of interfaces between polymer and graphene/graphite were studied using steady-state non-equilibrium molecular dynamics (NEMD) [45]. It was discovered that long-wavelength phonons in graphene significantly contributed to thermal transport across these interfaces. Low-frequency modes, resulting from graphene's out-of-plane motions, enhanced the coupling between graphene and polymer spectra, thus facilitating thermal transport. Stronger interfacial interactions correlated with increased thermal conductance, an effect observable through the integrated autocorrelation function of net power exchange at the interface. Forming strong covalent bonds between graphene/graphite edges and polymers, or adjusting polymer density, improved interfacial thermal transport. These findings offered practical methods to enhance thermal conductivity in polymeric composites.

With respect to epoxy-based composites, conductivity was studied under different functionalizations of graphene [46]. It was found that embedding pristine graphene in an epoxy matrix reduced its TC by about 50% due to phonon energy damping. Functionalized graphene, however, showed a smaller reduction in TC (around 10%). Functionalization generally lowered graphene's TC significantly, primarily due to phonon scattering by functional groups. Interestingly, functionalization improved the interfacial thermal conductivity (ITC) between graphene and epoxy, especially with amino and TETA functional groups, through stronger bonds and better phonon vibrational matching. This enhanced ITC offset the lower TC of functionalized graphene in composites with lateral sizes below a critical few micrometers. Beyond this size, pristine graphene's higher intrinsic TC was more advantageous. The study suggests that the choice of functionalization and graphene size is crucial for optimizing graphene-based composites' TC.

Multiscale studies have also been carried out, where molecular dynamics (MD) and finite element (FE) methods were used to evaluate the thermal conductivity of graphene–epoxy nanocomposites [47]. The MD method investigated nanoscale thermal conduction in fillers and matrix, revealing a 30% reduction in the thermal conductivity of single-layer graphene in an epoxy matrix. Thermal boundary conductance between epoxy and graphene was also calculated. Finite element modeling then used these MD results to determine local thermal conductivity constants in nanocomposites. This method, validated against experimental data for various graphene concentrations, proved accurate for evaluating thermal conduction in polymer nanocomposites. The study also examined the impact of covalent bonds between graphene and epoxy atoms on nanocomposite properties, noting a slight reduction in effective thermal conductivity and significant effects on stiffness and failure reinforcement.

PMMA has shown promising results when it comes to mechanical response of graphene-based composites. Zhang et al. demonstrated that hydroxyl-functionalized graphene/PMMA interfaces, facilitated by hydrogen bonds, significantly improved thermal conductance, outperforming both pristine and methyl-functionalized graphene/PMMA interfaces [48]. The enhancement was attributed to three factors: strong hydrogen bonds enhancing interfacial association, improved coupling of low-frequency vibrational modes, and effective thermal channels created by hydrogen bond donors. The degree of functionalization influenced these effects, with hydroxyl functionalization leading to saturation and convergence of interfacial thermal conductance, while methyl functionalization presented a bell-shaped conductance curve. This suggested that targeting hydrogen-bonded interfaces led to improved interfacial heat transfer in nanocomposites. The study also notes that covalent functionalization can lower graphene's in-plane thermal conductivity while enhancing interfacial conductance, highlighting the potential of directional thermal conduction improvements in heterogeneous materials and structures for various applications. Further investigation is needed to understand the effects in nanocomposites with randomly oriented fillers, considering both interfacial and in-plane conductance.

Chang et al. worked on poly(p-phenylene sulfide) (PPS), simulating nanocomposites reinforced with defect graphene (DG) and graphene oxide (GO) using NEMD [49]. DG was dispersed in PPS at weight fractions below 10% but aggregated at higher fractions due to weaker interaction with PPS compared to their own van der Waals forces. In contrast, GO dispersed well in PPS due to similar solubility parameters. Thermal conductivity increased with the weight fractions of DG and GO, particularly beyond 20 wt%. The conductivity of PPS/GO is higher than for PPS/DG, attributed to aggregation of defect graphene at higher concentrations, owing to stronger interactions between graphene and the polymer molecules affecting GO's equilibrium structure.

#### 17.4.3.2 Heat Transfer Mechanisms

To gain a deeper understanding of the heat transfer mechanisms within graphene-based composites, the phonon dispersion spectra and thermal conductivity contributions were analyzed. *Phonons* are quantized vibrational modes that carry heat in crystalline materials, and their dispersion spectra expounded the mechanisms of heat transport.

The simulations revealed that the phonon dispersion spectra of the composite exhibited unique features compared to the pristine components. The presence of graphene sheets introduced additional phonon modes, especially in the high-frequency range. These modes contribute significantly to the overall thermal conductivity of the composite. Moreover, the analysis of phonon lifetimes demonstrated that the lifetimes of certain phonon modes in the composite were extended compared to those in the pure polymer or matrix material. This phenomenon can be attributed to the strong interaction between graphene sheets and phonons, which reduces phonon scattering and enhances heat conduction. In addition to phonon-mediated heat transfer, studies revealed that the interfacial thermal resistance between graphene sheets and the surrounding matrix played a crucial role in determining the overall thermal behavior of the composites. Effective thermal coupling between graphene and the matrix material is essential for efficient heat transfer.

Wang et al. discovered that covalent functionalization of graphene significantly enhances thermal transport across graphene–paraffin interfaces by reducing interfacial thermal resistance, which is linearly dependent on functional group coverage [50]. Butyl was identified as the most effective functional group for this purpose, surpassing methyl, phenyl, and formyl. In contrast, carboxyl, hydroxyl, and amine groups had negligible impact. The varying degrees of vibrational density of states (VDOS) overlap between different functionalized graphene and paraffin explain these differences in thermal resistance. Additionally, the number of graphene layers showed minimal effect on this interfacial transport. These findings establish the choosing of suitable functionalizations to improve graphene–polymer composite thermal interfaces, which could be crucial in developing high-performance graphene–polymer composite thermal interface materials. The authors also worked on thermal transport under noncovalent functionalization of graphene based on 1-pyrenebutylamine, 1-pyrenebutyl, and 1-pyrenebutyric acid [51]. It significantly reduced the interfacial thermal resistance, especially with increased functional molecule coverage. The functional molecules showed similar effectiveness in lowering this resistance. This reduction is attributed to improved vibrational density of states (VDOS) overlapping between graphene and paraffin. The effective medium theory revealed that such noncovalent functionalization enhances the thermal conductivity of graphene–paraffin composites while preserving mechanical properties, like Young's modulus and tensile strength. These findings demonstrated the development of graphene–polymer composite–based thermal interface materials (TIMs) for thermal management in electronics.

#### 17.4.3.3 Temperature Effects and Thermal Stability

Temperature can significantly influence the structural stability and properties of graphene-based composites. At elevated temperatures, a slight increase in the interlayer spacing between graphene sheets was observed. This expansion can be attributed to thermal expansion effects, which are typical of crystalline materials. However, it is important to note that the graphene sheets remained well-aligned and did not exhibit significant structural deformation or damage.

Furthermore, the simulations provided insights into the effects of temperature on the thermal conductivity of the composite. As expected, thermal conductivity decreased with increasing temperature due to enhanced phonon scattering and vibrational motion. However, the decrease was relatively gradual, indicating that graphene-based composites can maintain respectable thermal conductivity at elevated temperatures.

Rahmani et al. explored the thermal decomposition behavior of poly(ethylene oxide) (PEO) reinforced with pristine graphene (PG) and graphene oxide (GO), utilizing reactive molecular dynamics simulation [52]. The material system is particularly significant in the development of flexible solid-state polyelectrolyte films for wearable solar cells. The simulations were conducted in two modes: non-isothermal (dynamic gravimetric analysis) and isothermal decomposition. This dual approach provided a comprehensive understanding of the thermal stability under varying conditions. The inclusion of PG in the PEO system was observed to enhance the thermal stability of the polymer in both decomposition modes. Specifically, there was a delay in the onset temperature of decomposition in non-isothermal mode and a substantial increase (around 60%) in activation energy for decomposition in the isothermal mode. The improvement in thermal stability was more pronounced with higher PG concentrations. Contrastingly, introducing GO into the PEO system led to a deterioration in thermal stability. However, similar to the effect of PG, increasing the concentration of GO resulted in enhanced thermal stability, albeit at a lower level than that of the pristine PEO system. A detailed investigation into the confinement of PEO between PG and GO nanoplatelets revealed distinct effects on thermal stability. In the case of PG, the confinement improved the thermal stability of the polymer. This enhancement is attributed to the immobilization of polymer chains on the surfaces of PG sheets, a reduction in free volume, and a decrease in the occurrence of hot spots during decomposition. Conversely, the thermal stability of PEO confined between GO sheets showed significant deterioration compared to the bulk polymer. This decrease in stability is linked to accelerated polymer chain scission near GO sheets during temperature rise, prompted by the evolution of highly reactive hydroxyl radicals that attack the polymer chains. At higher concentrations of GO (and PG), the free radicals formed in the confined regions are restricted from moving to the bulk. This restriction results in improved non-isothermal and isothermal stability of the bulk PEO.

#### 17.4.3.4 Applications and Future Research

The findings gained from the thermal property simulations of graphene-based composites have significant implications for practical applications. These composites hold immense promise for thermal management in various industries, including electronics, aerospace, and energy storage systems. The exceptionally high thermal conductivity observed suggests that graphene-based composites can serve as efficient heat spreaders and thermal interface materials. These materials have the potential to enhance the thermal performance of electronic devices by dissipating heat more effectively, thereby preventing overheating and improving device reliability. Additionally, the temperature-dependent behavior of thermal properties can inform the design of graphene-based composites for specific applications. By tailoring the composite structure and graphene content, it may be possible to achieve the desired thermal performance across a wide range of operating temperatures.

While simulations have provided valuable information about the thermal properties of graphene-based composites, several avenues for future research remain unexplored, which include the influence of defects, anisotropic behavior, and dynamic behavior. These materials exhibit enhanced thermal conductivity due to the efficient heat transfer pathways provided by graphene sheets. Heat conduction is primarily phonon-mediated, with temperature-dependent behavior. The thermal expansion effects at elevated temperatures are relatively minor, highlighting the structural stability of these composites. These findings have significant implications for applications in thermal management and suggest avenues for future research to optimize the design and performance of graphene-based composites.

### 17.4.4 Dynamic Behavior

Understanding the dynamic behavior of graphene-based composites is crucial for predicting their performance in various applications. Molecular dynamics (MD) simulations provide a powerful tool to investigate the vibrational modes, frequencies, and dynamic responses of these materials.

#### 17.4.4.1 Vibrational Modes and Frequencies

One of the fundamental aspects of dynamic behavior in materials is the analysis of vibrational modes and frequencies. MD simulations allow us to examine how atoms within a composite material oscillate and interact with one another. For graphene-based composites, the vibrational modes can help understand the stability and mechanical properties of the material.

A variety of vibrational modes in graphene-based composites were analyzed, including stretching, bending, and torsional modes. These modes are influenced by the graphene sheets' arrangement and their interactions with the surrounding matrix material. The vibrational frequencies of these modes can be compared to experimental data to validate the accuracy of the simulations. It was found that the presence of graphene significantly affects the vibrational modes of the composites. The graphene sheets act as rigid layers that can influence the frequency and dispersion of vibrational modes in the adjacent matrix material. This effect can be attributed to the strong covalent bonds within the graphene lattice, which provide a stable framework for neighboring atoms.

#### 17.4.4.2 Effects of Temperature on Vibrations

To further investigate the dynamic behavior, simulations at different temperatures were caried out, ranging from cryogenic to elevated conditions. At low temperatures, there was reduction in the amplitude of vibrational modes, indicating decreased thermal energy in the system. Conversely, as the temperature increased, the vibrational amplitudes became more pronounced. This temperature dependence is consistent with the principles of statistical mechanics, where higher temperatures lead to increased atomic motion and, consequently, more vigorous vibrations. Graphene's unique thermal properties also played a role in influencing the composite's dynamic behavior. Graphene has exceptional thermal conductivity, which allowed it to dissipate heat efficiently within the composites. This property contributed to temperature homogeneity throughout the material, minimizing localized thermal effects on the vibrational modes.

#### 17.4.4.3 Role of Graphene and External Forces

The presence of graphene in composites has a profound impact on their dynamic stability. The strong covalent bonds in graphene sheets provide a stable scaffold that can enhance the overall structural integrity of the composite. Graphene layers acted as effective barriers against the propagation of defects and dislocations within the matrix material. Furthermore, the rigid and two-dimensional nature of graphene allowed it to distribute stress and strain more uniformly within the composite, resulting in improved dynamic stability. This effect is particularly significant in applications where mechanical vibrations or impacts are a concern, such as in the aerospace industry.

To assess the dynamic response of graphene-based composites to external forces, materials were subjected to mechanical perturbations. It was found that graphene-enhanced composites exhibited remarkable dynamic resilience. When subjected to mechanical stress or impact, the graphene layers absorbed and dispersed the energy efficiently, mitigating the risk of catastrophic failure. The dynamic response of these composites also revealed interesting nonlinear behavior. At high strain rates, the composites displayed stress–strain curves with non-elastic regions, indicative of plastic deformation. This behavior can be attributed to the interactions between the graphene sheets and the surrounding matrix, which undergo significant rearrangement under extreme mechanical loads.

#### 17.4.4.4 Tribological Properties

Tribological properties are studied using molecular dynamics (MD) simulations by modeling the atomic interactions of materials under tribological stress. These simulations provide a detailed understanding about the behavior of materials at the atomic level during friction and wear processes. MD simulations help in understanding how different materials, surface structures, and environmental conditions affect friction, wear, and lubrication. By simulating the movement of atoms and molecules in contact, researchers can predict the tribological performance of materials, analyze the mechanisms of wear and friction, and explore the effectiveness of lubricants. This method is particularly useful for studying new materials or surface treatments in a controlled virtual environment before conducting physical experiments.

Large-scale atomistic simulations were used to explore the tribological properties of polytetrafluoroethylene (PTFE) composites reinforced with carbon nanotube (CNT) and graphene (Gr) fillers [53]. The properties were studied using $Fe_2O_3$ sliders, where the rigid layer of the polymer substrate is fixed at the bottom and the slider is made to slide in the top layer at constant velocity under specific load. An atomic layer with free Newtonian dynamics is allowed to move freely. The friction process is then simulated through several steps, starting with Fe2O3 sliders moving toward the PTFE surface until a predefined normal force is reached. The $Fe_2O_3$ slider's rigid layer is subjected to constant normal pressure, followed by a relaxation phase. Finally, the $Fe_2O_3$ sliders are made to move laterally along the x-axis at a set speed. This methodology effectively simulates the friction process between the $Fe_2O_3$ slider and PTFE surface under controlled conditions. The composites demonstrated significantly improved wear resistance under high pressure compared to pure PTFE, with reductions in wear amounts of 76.2% for CNT and 85.7% for Gr. Both fillers were found to enhance deformation resistance within the composites and facilitated relative motions at the sliding interface by aligning along the shear direction. This resulted in thinner tribo-films compared to pure PTFE. Graphene fillers provided stronger reinforcements than CNT, attributed to stronger Gr/PTFE interfacial interactions and greater adsorption of PTFE chains on Gr surfaces. These findings contribute to understanding the reinforcement mechanism of low-dimensional fillers in PTFE composites.

## 17.5 APPLICATIONS AND FUTURE DIRECTIONS

Molecular dynamics simulations of graphene-based composites have emerged as a powerful tool not only for understanding the fundamental properties of these materials but also for their practical applications in various industries. This section has included the wide-ranging applications of graphene-based composites and discuss potential future directions that can further advance this field of research.

### 17.5.1 Engineering Applications

The aerospace industry is one of the sectors that can significantly benefit from graphene-based composites. These materials offer remarkable strength-to-weight ratios and exceptional thermal conductivity, making them suitable for the development of lightweight and high-performance aircraft components. Molecular dynamics simulations provide valuable insights into the mechanical properties and structural stability of graphene-based composites under extreme conditions, such as high temperatures and mechanical stress. Recent studies have demonstrated that graphene-reinforced composite materials can enhance the durability and longevity of aerospace components. Simulations have revealed the mechanisms by which graphene strengthens composite structures, leading to improved resistance to fatigue and mechanical failure.

Graphene-based composites have shown immense potential in the electronics and nanotechnology sectors. The unique electronic properties of graphene, such as high carrier mobility and electrical conductivity, make it an ideal candidate for use in transistors, sensors, and other electronic devices. Molecular dynamics simulations can aid in designing and optimizing the electronic properties of these composites at the atomic scale. Recent research has used simulations to investigate the behavior of graphene-based composites in nanoelectronic devices. By understanding the interactions between graphene and other materials in these devices, researchers have been able to fine-tune their performance, leading to more efficient and versatile electronics.

### 17.5.2 Optimization Strategies

One promising avenue for future research in molecular dynamics simulations of graphene-based composites lies in the functionalization and surface modifications of graphene sheets. Functional groups can be added to graphene to tailor its chemical and physical properties, allowing for the development of composites with specific functionalities. Simulations can predict the effects of functionalization on the overall behavior of composites, guiding experimental efforts. MD simulations have been employed to investigate the impact of functional groups on the mechanical and thermal properties of graphene-based composites. These simulations have revealed that judiciously chosen functionalization can enhance the adhesion between graphene and the matrix material, leading to improved load transfer and overall composite performance.

Multiscale modeling techniques combined with molecular dynamics simulations have become promising in understanding the material characteristics more accurately. Multiscale modeling can bridge the gap between atomistic and macroscopic scales, enabling researchers to predict the behavior of graphene-based composites at various levels of complexity, understanding the hierarchical structure and mechanical properties. These simulations can provide a holistic view of composite behavior, considering factors such as defects, interfacial interactions, and stress distribution at different scales.

Recent advancements in mesoscopic simulations have facilitated a comprehensive study of polypropylene (PP) composites reinforced with various forms of graphene [54]. This study primarily focused on employing a coarse-grained (CG) model to predict and analyze the thermal and mechanical properties of these composites. The developed CG model successfully replicated the properties of polypropylene (PP), including the glass transition temperature, Young's modulus, Poisson's ratio, and thermal conductivity. The model displayed remarkable accuracy, with the only noted discrepancy being an overestimation in the density of PP. In addition to PP, the mechanical and thermal properties of graphene were computed using CG potentials. The results obtained were in close alignment with existing literature, underscoring the reliability of the CG model. The study further extended its scope to include graphene oxide and reduced graphene oxide. Employing the same CG force field, the research provided valuable information about the degradation of mechanical properties as the oxidation level in graphene-based fillers increases. To enhance the accuracy of the CG model predictions and surpass traditional continuum methods, the modified Flory (MF) model with interphase was employed. This model incorporated properties calibrated based on mesoscopic model results, offering a detailed representation of filler–matrix interactions. The MF model with interphase displayed an improved capability in predicting the properties of the composites. The accuracy of this model was validated against experimental data, confirming the reliability of the CG model for predicting thermal and mechanical properties of graphene-reinforced polypropylene nanocomposites.

## 17.6 CONCLUSION

In conclusion, molecular dynamics simulations have opened up exciting possibilities for the applications and future directions of graphene-based composites. These simulations not only provide valuable insights into the behavior of these materials at the atomic level but also pave the way for their utilization in diverse industries, ranging from aerospace to electronics and biomedical applications. Furthermore, ongoing research in optimization strategies and emerging areas, such as environmental impact and multiscale modeling, holds the potential to revolutionize the field and lead to groundbreaking discoveries in materials science and engineering.

## ACKNOWLEDGMENTS

I would like to express my heartfelt gratitude to all those who contributed to the completion of this chapter. I extend my deepest appreciation to my colleagues and peers who generously shared their knowledge and expertise, helping shape the ideas and refine arguments. Their contributions have been instrumental in shaping the quality and depth of this chapter. I express my gratitude to the publishers and editorial team for their professionalism and assistance in bringing this chapter to fruition.

## REFERENCES

[1] X. Li, W. Cai, J. An, S. Kim, J. Nah, D. Yang, R. Piner, A. Velamakanni, I. Jung, E. Tutuc, S. K. Banerjee, L. Colombo, and R. S. Ruoff, 'Large-area synthesis of high-quality and uniform graphene films on copper foils', *Science*, vol. 324, no. 5932, 2009, doi: 10.1126/science.1171245.

[2] P. Sutter, J. T. Sadowski, and E. Sutter, 'Graphene on Pt(111): Growth and substrate interaction', *Phys Rev B Condens Matter Mater Phys*, vol. 80, no. 24, 2009, doi: 10.1103/PhysRevB.80.245411.

[3] Q. Yu, J. Lian, S. Siriponglert, H. Li, Y. P. Chen, and S. S. Pei, 'Graphene segregated on Ni surfaces and transferred to insulators', *Appl Phys Lett*, vol. 93, no. 11, 2008, doi: 10.1063/1.2982585.

[4] S. Bhaviripudi, X. Jia, M. S. Dresselhaus, and J. Kong, 'Role of kinetic factors in chemical vapor deposition synthesis of uniform large area graphene using copper catalyst', *Nano Lett*, vol. 10, no. 10, 2010, doi: 10.1021/nl102355e.

[5] H. Tetlow, J. P. de Boer, I. J. Ford, D. D. Vvedensky, J. Coraux, and L. Kantorovich, 'Growth of epitaxial graphene: Theory and experiment', *Phys Rep*, vol. 542, no. 3, 2014, doi: 10.1016/j.physrep.2014.03.003.

[6] H. Chen, W. Zhu, and Z. Zhang, 'Contrasting behavior of carbon nucleation in the initial stages of graphene epitaxial growth on stepped metal surfaces', *Phys Rev Lett*, vol. 104, no. 18, 2010, doi: 10.1103/PhysRevLett.104.186101.

[7] J. Gao, J. Yip, J. Zhao, B. I. Yakobson, and F. Ding, 'Graphene nucleation on transition metal surface: Structure transformation and role of the metal step edge', *J Am Chem Soc*, vol. 133, no. 13, 2011, doi: 10.1021/ja110927p.

[8] Q. Yuan, J. Gao, H. Shu, J. Zhao, X. Chen, and F. Ding, 'Magic carbon clusters in the chemical vapor deposition growth of graphene', *J Am Chem Soc*, vol. 134, no. 6, 2012, doi: 10.1021/ja2050875.

[9] P. Wu, W. Zhang, Z. Li, and J. Yang, 'Mechanisms of graphene growth on metal surfaces: Theoretical perspectives', *Small*, vol. 10, no. 11, 2014, doi: 10.1002/smll.201303680.

[10] Y. Wang, A. J. Page, Y. Nishimoto, H. J. Qian, K. Morokuma, and S. Irle, 'Template effect in the competition between Haeckelite and graphene growth on Ni(111): Quantum chemical molecular dynamics simulations', *J Am Chem Soc*, vol. 133, no. 46, 2011, doi: 10.1021/ja2064654.

[11] P. Wu, H. Jiang, W. Zhang, Z. Li, Z. Hou, and J. Yang, 'Lattice mismatch induced nonlinear growth of graphene', *J Am Chem Soc*, vol. 134, no. 13, 2012, doi: 10.1021/ja301791x.

[12] P. Wu, Y. Zhang, P. Cui, Z. Li, J. Yang, and Z. Zhang, 'Carbon dimers as the dominant feeding species in epitaxial growth and morphological phase transition of graphene on different cu substrates', *Phys Rev Lett*, vol. 114, no. 21, 2015, doi: 10.1103/PhysRevLett.114.216102.

[13] P. Li, Z. Li, and J. Yang, 'Dominant kinetic pathways of graphene growth in chemical vapor deposition: The role of hydrogen', *J Phys Chem C*, vol. 121, no. 46, 2017, doi: 10.1021/acs.jpcc.7b09622.

[14] L. Wang, X. Zhang, H. L. W. Chan, F. Yan, and F. Ding, 'Formation and healing of vacancies in graphene chemical vapor deposition (CVD) growth', *J Am Chem Soc*, vol. 135, no. 11, 2013, doi: 10.1021/ja312687a.

[15] V. I. Artyukhov, Y. Liu, and B. I. Yakobson, 'Equilibrium at the edge and atomistic mechanisms of graphene growth', *Proc Natl Acad Sci USA*, vol. 109, no. 38, 2012, doi: 10.1073/pnas.1207519109.

[16] V. I. Artyukhov, Y. Hao, R. S. Ruoff, and B. I. Yakobson, 'Breaking of symmetry in graphene growth on metal substrates', *Phys Rev Lett*, vol. 114, no. 11, 2015, doi: 10.1103/PhysRevLett.114.115502.
[17] E. Meca, J. Lowengrub, H. Kim, C. Mattevi, and V. B. Shenoy, 'Epitaxial graphene growth and shape dynamics on copper: Phase-field modeling and experiments', *Nano Lett*, vol. 13, no. 11, 2013, doi: 10.1021/nl4033928.
[18] G. Li, S. H. Huang, and Z. Li, 'Gas-phase dynamics in graphene growth by chemical vapour deposition', *Phys Chem Chem Phys*, vol. 17, no. 35, 2015, doi: 10.1039/c5cp02301g.
[19] B. Mortazavi and T. Rabczuk, 'Multiscale modeling of heat conduction in graphene laminates', *Carbon NY*, vol. 85, 2015, doi: 10.1016/j.carbon.2014.12.046.
[20] F. Liu, P. Ming, and J. Li, 'Ab initio calculation of ideal strength and phonon instability of graphene under tension', *Phys Rev B Condens Matter Mater Phys*, vol. 76, no. 6, 2007, doi: 10.1103/PhysRevB.76.064120.
[21] Y. Liu and Z. Xu, 'Multimodal and self-healable interfaces enable strong and tough graphene-derived materials', *J Mech Phys Solids*, vol. 70, no. 1, 2014, doi: 10.1016/j.jmps.2014.05.006.
[22] Z. Ni, H. Bu, M. Zou, H. Yi, K. Bi, and Y. Chen, 'Anisotropic mechanical properties of graphene sheets from molecular dynamics', *Phys B Condens Matter*, vol. 405, no. 5, 2010, doi: 10.1016/j.physb.2009.11.071.
[23] T. C. Theodosiou and D. A. Saravanos, 'Numerical simulation of graphene fracture using molecular mechanics based nonlinear finite elements', *Comput Mater Sci*, vol. 82, 2014, doi: 10.1016/j.commatsci.2013.09.032.
[24] Q. Lu, W. Gao, and R. Huang, 'Atomistic simulation and continuum modeling of graphene nanoribbons under uniaxial tension', *Model Simul Mat Sci Eng*, vol. 19, no. 5, 2011, doi: 10.1088/0965-0393/19/5/054006.
[25] Y. G. Yanovsky, E. A. Nikitina, Y. N. Karnet, and S. M. Nikitin, 'Quantum mechanics study of the mechanism of deformation and fracture of graphene', *Phys Mesomech*, vol. 12, nos. 5–6, 2009, doi: 10.1016/j.physme.2009.12.007.
[26] J. L. Tsai and J. F. Tu, 'Characterizing mechanical properties of graphite using molecular dynamics simulation', *Mater Des*, vol. 31, no. 1, 2010, doi: 10.1016/j.matdes.2009.06.032.
[27] R. Atif and F. Inam, 'Modeling and simulation of graphene based polymer nanocomposites: Advances in the last decade', *Graphene*, vol. 5, no. 2, 2016, doi: 10.4236/graphene.2016.52011.
[28] C. Lee, X. Wei, J. W. Kysar, and J. Hone, 'Measurement of the elastic properties and intrinsic strength of monolayer graphene', *Science*, vol. 321, no. 5887, 2008, doi: 10.1126/science.1157996.
[29] C. Lv, Q. Xue, D. Xia, M. Ma, J. Xie, and H. Chen, 'Effect of chemisorption on the interfacial bonding characteristics of graphene-polymer composites', *J Phys Chem C*, vol. 114, no. 14, pp. 6588–6594, 2010, doi: 10.1021/jp100110n.
[30] S. C. Shiu and J. L. Tsai, 'Characterizing thermal and mechanical properties of graphene/epoxy nanocomposites', *Compos B Eng*, vol. 56, 2014, doi: 10.1016/j.compositesb.2013.09.007.
[31] C. M. Miyazaki, M. A. E. Maria, D. D. Borges, C. Woellner, G. Brunetto, A. F. Fonseca, C. Constantino, M. Pereira-da-Silva, A. Siervo, D. Galvão, and A. Riul, 'Synthesis, characterization and computational simulation of graphene nanoplatelets stabilized in poly(styrene sulfonate) sodium salt', *arXiv Mater Sc*, 2017, https://api.semanticscholar.org/CorpusID:98937936.
[32] A. N. Rissanou and V. Harmandaris, 'Dynamics of various polymer–graphene interfacial systems through atomistic molecular dynamics simulations', *Soft Matter*, vol. 10, no. 16, pp. 2876–2888, 2014, doi: 10.1039/C3SM52688G.
[33] S. Güryel, M. Walker, P. Geerlings, F. De Proft, and M. R. Wilson, 'Molecular dynamics simulations of the structure and the morphology of graphene/polymer nanocomposites', *Phys Chem Chem Phys*, vol. 19, no. 20, pp. 12959–12969, 2017, doi: 10.1039/C7CP01552F.
[34] P. Kelich and A. Asadinezhad, 'Adsorption of poly(ethylene succinate) chain onto graphene nanosheets: A molecular simulation', *J Mol Graph Model*, vol. 69, pp. 26–38, 2016, doi: 10.1016/j.jmgm.2016.08.003.
[35] E. N. Skountzos, A. Anastassiou, V. G. Mavrantzas, and D. N. Theodorou, 'Determination of the mechanical properties of a poly(methyl methacrylate) nanocomposite with functionalized graphene sheets through detailed atomistic simulations', *Macromolecules*, vol. 47, no. 22, pp. 8072–8088, 2014, doi: 10.1021/ma5017693.
[36] A. N. Rissanou and V. Harmandaris, 'Dynamics of various polymer-graphene interfacial systems through atomistic molecular dynamics simulations', *Soft Matter*, vol. 10, no. 16, 2014, doi: 10.1039/c3sm52688g.
[37] A. N. Rissanou and V. Harmandaris, 'Structure and dynamics of poly(methyl methacrylate)/graphene systems through atomistic molecular dynamics simulations', *J Nanopart Res*, vol. 15, no. 5, 2013, doi: 10.1007/s11051-013-1589-2.

[38] T. Ramanathan, A. A. Abdala, S. Stankovich, D. A. Dikin, M. Herrera-Alonso, R. D. Piner, D. H. Adamson, H. C. Schniepp, X. Chen, R. S. Ruoff, S. T. Nguyen, I. A. Aksay, R. K. Prud'Homme, and L. C. Brinson, 'Functionalized graphene sheets for polymer nanocomposites', *Nat Nanotechnol*, vol. 3, no. 6, 2008, doi: 10.1038/nnano.2008.96.

[39] C. Jang, T. E. Lacy, S. R. Gwaltney, H. Toghiani, and C. U. Pittman, 'Interfacial shear strength of cured vinyl ester resin-graphite nanoplatelet from molecular dynamics simulations', *Polymer (Guildf)*, vol. 54, no. 13, 2013, doi: 10.1016/j.polymer.2013.04.035.

[40] F. Mohammadzadeh Honarvar, B. Pourabbas, M. Salami Hosseini, M. Kharazi, and H. Erfan-Niya, 'Molecular dynamics simulation: The effect of graphene on the mechanical properties of epoxy based photoresist: SU8', *Sci Iran*, vol. 25, no. 3F, 2018, doi: 10.24200/sci.2018.5697.1432.

[41] M. A. N. Dewapriya, R. K. N. D. Rajapakse, and N. Nigam, 'Influence of hydrogen functionalization on the fracture strength of graphene and the interfacial properties of graphene-polymer nanocomposite', *Carbon NY*, vol. 93, 2015, doi: 10.1016/j.carbon.2015.05.101.

[42] A. A. Balandin, S. Ghosh, W. Bao, I. Calizo, D. Teweldebrhan, F. Miao, and C. N. Lau, 'Superior thermal conductivity of single-layer graphene', *Nano Lett*, vol. 8, no. 3, pp. 902–907, 2008, doi: 10.1021/nl0731872.

[43] Y. Gu, T. Lu, Y. Zhou, J. Yang, C. Xie, and K. Zhang, 'Experimental research and molecular dynamics simulation on thermal properties of capric acid/ethylene-vinyl/graphene composite phase change materials', *J Ind Eng Chem*, vol. 99, pp. 256–263, 2021, doi: 10.1016/j.jiec.2021.04.039.

[44] S. Yang, W. Zhang, R. Ma, H. Li, Y. Lu, X. Zhao, L. Zhang, and Y. Gao, 'Manipulating the thermal conductivity of the graphene/poly(vinyl alcohol) composite via surface functionalization: A multiscale simulation', *Langmuir*, vol. 39, no. 28, pp. 9703–9714, 2023, doi: 10.1021/acs.langmuir.3c00677.

[45] T. Luo and J. R. Lloyd, 'Enhancement of thermal energy transport across graphene/graphite and polymer interfaces: A molecular dynamics study', *Adv Funct Mater*, vol. 22, no. 12, pp. 2495–2502, 2012, doi: 10.1002/adfm.201103048.

[46] X. Shen, Z. Wang, Y. Wu, X. Liu, and J.-K. Kim, 'Effect of functionalization on thermal conductivities of graphene/epoxy composites', *Carbon NY*, vol. 108, pp. 412–422, 2016, doi: 10.1016/j.carbon.2016.07.042.

[47] B. Mortazavi, O. Benzerara, H. Meyer, J. Bardon, and S. Ahzi, 'Combined molecular dynamics-finite element multiscale modeling of thermal conduction in graphene epoxy nanocomposites', *Carbon NY*, vol. 60, pp. 356–365, 2013, doi: 10.1016/j.carbon.2013.04.048.

[48] L. Zhang, Z. Bai, and L. Liu, 'Exceptional thermal conductance across hydrogen-bonded graphene/polymer interfaces', *Adv Mater Interfaces*, vol. 3, no. 13, p. 1600211, 2016, doi: 10.1002/admi.201600211.

[49] C.-Y. Chang, S.-P. Ju, J.-W. Chang, S.-C. Huang, and H.-W. Yang, 'The thermal conductivity and mechanical properties of poly(p-phenylene sulfide)/oxided-graphene and poly(p-phenylene sulfide)/defect-graphene nano-composites by molecular dynamics simulation', *RSC Adv*, vol. 4, no. 50, pp. 26074–26080, 2014, doi: 10.1039/C4RA02907K.

[50] Y. Wang, H. F. Zhan, Y. Xiang, C. Yang, C. M. Wang, and Y. Y. Zhang, 'Effect of covalent functionalization on thermal transport across graphene–polymer interfaces', *J Phys Chem C*, vol. 119, no. 22, pp. 12731–12738, 2015, doi: 10.1021/acs.jpcc.5b02920.

[51] Y. Wang, C. Yang, Y.-W. Mai, and Y. Zhang, 'Effect of non-covalent functionalisation on thermal and mechanical properties of graphene-polymer nanocomposites', *Carbon NY*, vol. 102, pp. 311–318, 2016, doi: 10.1016/j.carbon.2016.02.069.

[52] F. Rahmani, M. Mahdavi, S. Nouranian, and A. Al-Ostaz, 'Confinement effects on the thermal stability of poly(ethylene oxide)/graphene nanocomposites: A reactive molecular dynamics simulation study', *J Polym Sci B Polym Phys*, vol. 55, no. 13, pp. 1026–1035, 2017, doi: 10.1002/polb.24355.

[53] Q. Xu, J. Zhang, Y.-Z. Hu, and T.-B. Ma, 'Tribological behavior of poly(tetrafluoroethylene) and its composites reinforced by carbon nanotubes and graphene sheets: Molecular dynamics simulation', *Phys Status Solidi R*, vol. 16, no. 3, p. 2100298, 2022, doi: 10.1002/pssr.202100298.

[54] A. Muhammad, R. Srivastava, N. Koutroumanis, D. Semitekolos, E. Chiavazzo, P.-N. Pappas, C. Galiotis, P. Asinari, C. A. Charitidis, and M. Fasano, 'Mesoscopic modeling and experimental validation of thermal and mechanical properties of polypropylene nanocomposites reinforced by graphene-based fillers', *Macromolecules*, vol. 56, no. 24, pp. 9969–9982, 2023, doi: 10.1021/acs.macromol.3c01529.

# 18 Functional Nanomaterials
## *Synthesis, Characterization, and Versatile Application*

*Harsh Satiya, Annu Sharma, Sampat Bhati, and Kirtiraj K. Gaikwad*

## 18.1 INTRODUCTION

### 18.1.1 Background

The discovery of nanomaterials has brought about a profound change in the field of materials research. By fusing the remarkable qualities of nanomaterials with the innate strengths of polymer matrices, these intriguing materials close the gap between the microscopic and macroscopic realms. When particles known as nanofillers—which have at least one dimension in the nanoscale, or between 1 and 100 nm—are added to a polymer matrix, the resulting nanomaterials have significantly better mechanical, electrical, thermal, and functional properties than their pure counterparts [1–3]. To make nanomaterials better than ordinary composites, just very little amounts of a nanofiller need to be dispersed in the composites. Numerous fillers or reinforcements, including nanoclays, silica, carbon nanomaterials, nanotubes, and nanofibers, have been employed in polymer composites [4]. Fillers are used for a variety of goals, including material modification and cosmetic additions, whereas reinforcement is used to strengthen structures or behaviors. Often, the terms *fillers* and *reinforcements* are used interchangeably, though they vary from each other slightly. *Fillers* frequently address transitory or superficial needs, whereas *reinforcement* seeks long-term gains in strength and usefulness, notably in materials and building contexts. This chapter delves into the fascinating world of nanomaterials, unveiling their synthesis methods, characterization techniques, and diverse applications while shedding light on their unique merits and potential to revolutionize various fields [5–8].

### 18.1.2 Objectives of the Chapter

The goals of this chapter on polymeric nanomaterials are to present the idea and highlight the variety of uses for which they can be applied. It examines several nanofillers and describes their special qualities, such as nanoparticles and nanoclays. The significance of characterization techniques such as microscopy is explored, along with the synthesis processes and difficulties in preparation and processing. The performance and qualities of polymeric nanomaterials are a major area of focus. It is explained how the addition of nanofillers improves mechanical, thermal, and other properties, and the relationship between nanofiller concentration and composite performance is explored [3, 9, 10]. Successful case studies and real-world applications from a variety of industries are provided. There is discussion of the difficulties, potential paths, and environmental factors, including sustainability and toxicity. The chapter ends with a succinct summary of the major discoveries and some insights into the potential applications of polymeric nanomaterials in the future.

DOI: 10.1201/9781003476337-18

## 18.2 SYNTHESIS METHODS

Crafting these exceptional materials demands a deep understanding of the intricate relationship between synthesis methods and the resulting properties of the nanomaterials. Numerous techniques are employed to fabricate nanomaterials, each with advantages and limitations. Here, we explore some of the most prominent synthesis approaches. The various methods used for nanomaterials synthesis are broadly classified into top-down and bottom-up approaches, depending on the type of starting material, and into solid, liquid, and gas phase synthesis, depending upon the method adopted [11].

### 18.2.1 Top-Down Approaches

The top-down approach for nanomaterial synthesis involves the reduction of bulk materials or structures to the nanoscale, resulting in the creation of nanosized particles or structures [11]. This method allows for the production of nanomaterials through more extensive mechanical or physical manipulation of materials. Here's an in-depth look at the top-down approaches commonly employed in nanomaterial synthesis:

#### 18.2.1.1 Planetary Ball Milling

Ball milling is one of the most straightforward and efficient techniques to prepare nanomaterials. The process/technique uses a specified quantity of powder/material loaded into a milling vail and the grinding media (balls) [12, 13]. The coarse-grained materials are broken down into smaller

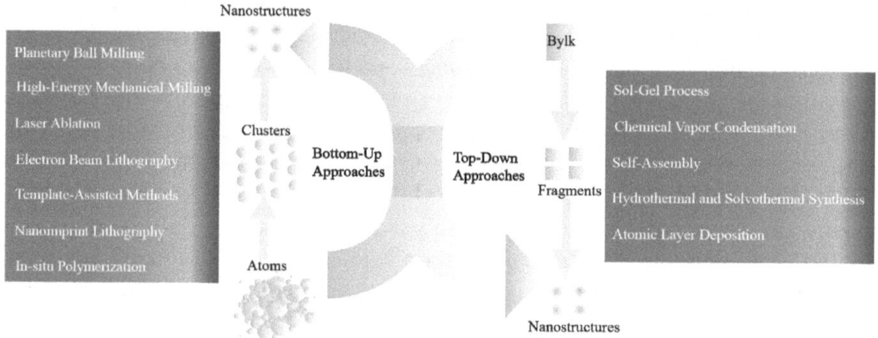


**FIGURE 18.1** Classification of different routes for synthesis of nanomaterials.

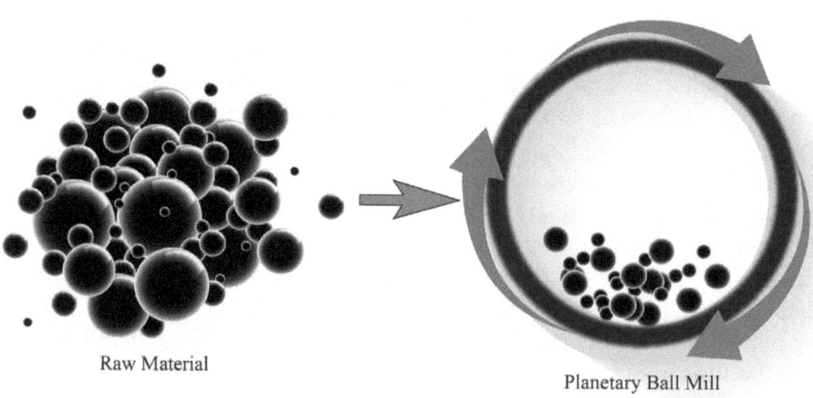


**FIGURE 18.2** Ball milling process.

particles with the help of mechanical energy due to the interaction of the material with the grinding media and by the friction between the grinding media and the walls of the vial. The material of grinding media depends on the type of material being milled. Some commonly used grinding media materials are stainless steel, hardened steel, tungsten carbide, etc. The ball mill is classified into various types depending upon the movement of the grinding media or milling balls. Some of them are vibration mills, planetary mills, and magneto-ball mills. The commonly used method is the planetary mill, as it is the most effective [13]. Specific parameters that affect the final result are ball material, ball size, ball-to-powder ratio, vial volume, milling rotational speed, milling atmosphere, and milling time [12]. The main advantage of this method is that it is simple and cost-effective and can be applied to various materials.

#### 18.2.1.2 High-Energy Mechanical Milling

High-energy mechanical or high-energy ball milling is similar to ball milling but is performed at higher speeds with specialized equipment. It involves working at high temperatures and pressure [14]. The method was first developed by John Benjamin in 1970 and uses the grinding media's kinetic energy to break the material's chemical bonds, resulting in the formation of smaller nanoparticles with greater surface area. Extensive plastic deformation and cold welding of powders occur during milling. Solid-state reactions can occur during high-energy mechanical milling that could lead to the formation of metastable phases. This method is more efficient than traditional ball milling as it also uses centrifugal and gravitational forces to crush the material [13]. This method can synthesize amorphous alloy powders, alloy powders with other metastable phases, intermetallic and different compounds, metal–ceramic composites, nanomaterial powders, and nanocrystalline powders. The method poses enhanced efficiency in reducing particle size and is also suitable for hard materials, which can be a limitation in the case of conventional ball milling.

#### 18.2.1.3 Laser Ablation

*Laser ablation* is the method by which solid targets can be realized into fabricating nanostructures of various materials with different morphologies, like nanoparticles, nanocubes, nanorods,

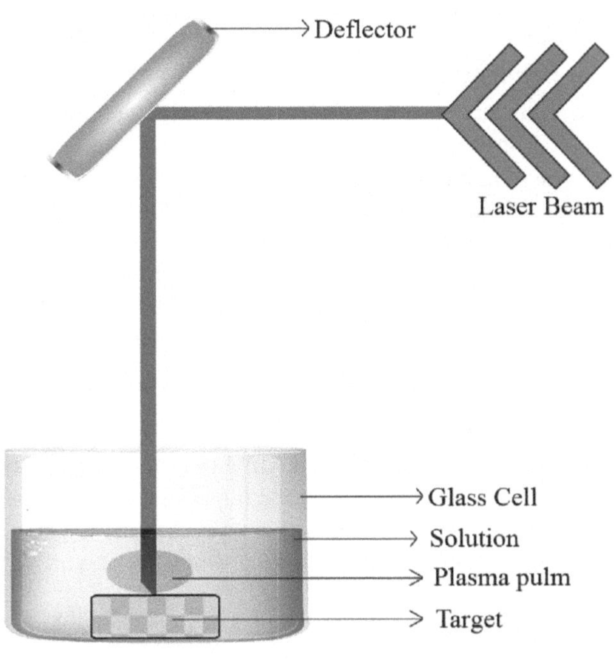


**FIGURE 18.3** Laser ablation.

nanomaterials, etc. In this method, laser pulses are focused on a target material. The incident pulsed laser penetrates the material's surface and removes electrons from the bulk material, which oscillate within the electromagnetic field and collide with the bulk atoms, transferring energy to the lattice. The penetration depth depends on the laser wavelength and the target material's refractive index. The irradiated surface then gets heated and vaporized, converting the material into plasma. A rapid plasma plume expansion is due to a significant pressure difference between the produced laser and the initial plasma seed. The plasma species then nucleate and grow into desirable nanostructures either on a substrate or in an excellent liquid medium under suitable condensation conditions [15, 16]. The main advantage of this method is that it allows for the synthesis of highly pure nanoparticles with precise control over particle size.

#### 18.2.1.4 Electron Beam Lithography

A process that transfers a pattern from one medium to another is called *lithography*. The material is selectively removed from a substrate, and patterns are created [2, 17, 18]. In electron beam lithography (EBL), an electron source is used, which has the advantage of having high diffraction-limited resolution and, hence, transfers patterns with the nanoscale feature. EBL can be of different types, with direct writing and projection printing methods being the most popular [17, 18]. This method is suitable for creating nanoscale surface features that can be coated with polymers. The technique poses an advantage in terms of high precision and provides excellent control over geometry.

#### 18.2.1.5 Template-Assisted Methods

Another method used for nanomaterial synthesis is the template-assisted method. This method uses a template substrate for the synthesis [19, 20]. Initially, the template is prepared using methods like precipitation, sol-gel, or chemical vapor deposition to synthesize the target production under the template function. After the synthesis, the template is removed, rendering a self-supported nanostructure. The mechanism behind the method implies a change in the morphology of particles by controlling the crystal's nucleation and growth [19–21]. The method can be classified into two types, depending on the kind of template: complex template method and soft template method [22]. The method allows control over size, shape, and orientation with broad flexibility for materials.

#### 18.2.1.6 Nanoimprint Lithography

Nanoimprint lithography is an essential top-down lithographic technique that creates nanometer-scale patterns. It is a simple, low-cost method that uses mechanical deformation of imprint resist and is based on compression molding of a plastic film [23, 24]. At first, a stamp of the desired shape is prepared using the standard fabrication method. Then, this stamp is pressed into the material from which the nanostructure will be formed. After curing, the stamp is removed, and the desired nanostructure is obtained. Usually, a polymeric substance, a curable monomer, heat, or UV light, is applied to the resist material. This technique is more effective than conventional lithography, which depends on light diffraction and beam scattering [23, 24]. The primary benefits of nanoimprint lithography lie in its high-throughput and large-scale production possibilities.

#### 18.2.1.7 In Situ Polymerization

Using this technique, the monomer is polymerized while the nanofiller is dispersed throughout the reaction medium. It provides better compatibility and interfacial adhesion between the matrix and filler because of its simultaneous growth, resulting in extremely homogenous nanomaterials [25]. This group includes methods such as sol-gel processing, emulsion polymerization, and free radical polymerization.

Every one of these top-down strategies has benefits and drawbacks. The required nanomaterial qualities, the kind of starting materials, the need for scalability, and the planned applications all influence the process choice.

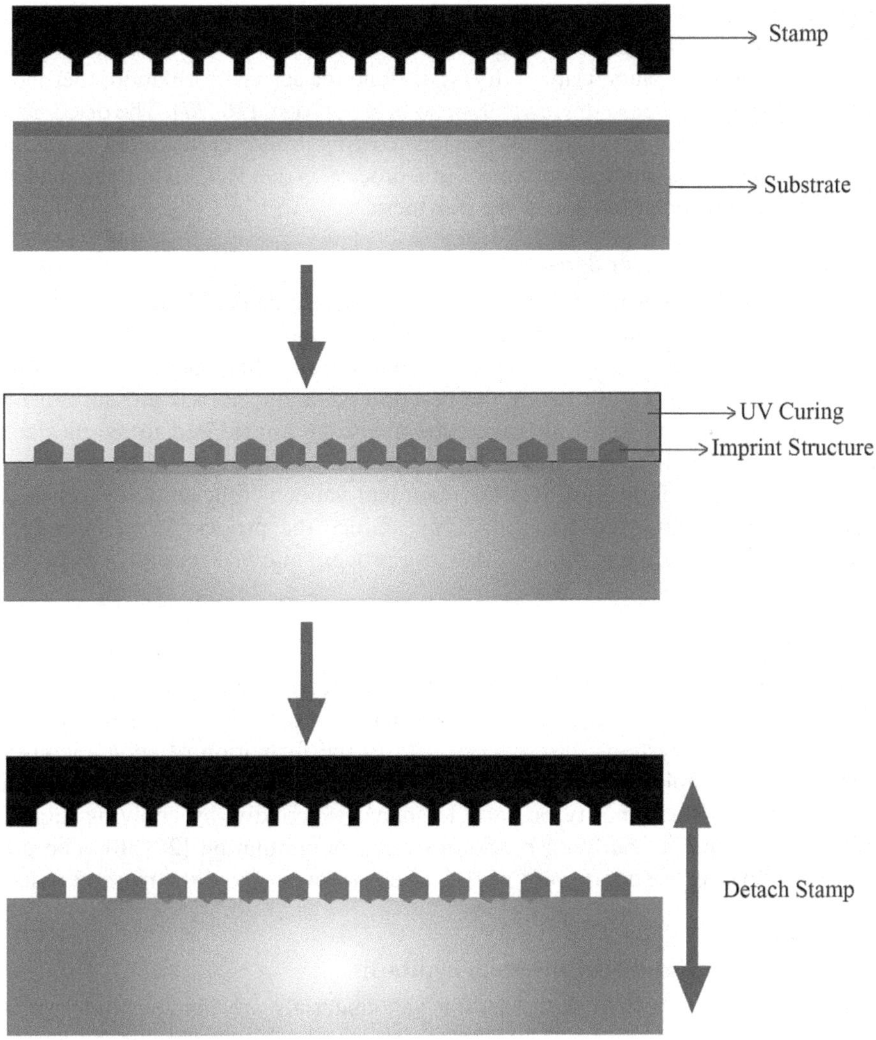


**FIGURE 18.4** Nanoimprint lithography.

### 18.2.2 Bottom-Up Approaches

The synthesis of nanomaterial materials using bottom-up methods entails assembling or growing nanoscale building blocks into a bigger structure. These techniques begin at the molecular or nanoscale level, providing fine control over the size, composition, and structure of the nanomaterials. An outline of a few popular bottom-up methods is provided in what follows:

#### 18.2.2.1 Sol-Gel Process

A wet chemical technique for synthesizing nanomaterials is the sol-gel process. Nitrides, sulfates, and chlorides are employed as precursors in this procedure. There are two categories of solvents that are utilized in this process: aqueous sol-gel and nonaqueous sol-gel. Whereas the nonaqueous sol-gel method uses an organic solvent, the aqueous sol-gel method uses water as a solvent medium [26, 27]. A sequence of chemical events, including hydrolysis, gelation, drying, and thermal treatment, is involved in this approach. First, the molecular precursors are dissolved in the solvent (hydration), and then they are heated (gelation) via polymerization to create gel. The final product is then obtained by drying the gel and doing proper heat treatment. The thermal treatment depends on the desired final product. Supercritical drying is done for aerogel formation, thermal drying is done to obtain xerogel, and freeze drying is done to bring cryogel [26, 27]. The dried gel is used in various industries for various purposes, like surface coating, building installation, and production of unique clothing [26]. The main advantage of this process is that it is versatile and allows for the incorporation of multiple materials, including polymers.

#### 18.2.2.2 Chemical Vapor Condensation

Chemical vapor condensation (CVC) is a material processing technique in which a substrate is exposed to one or more volatile precursors at high temperatures. The process involves the introduction of gaseous precursors into a reaction chamber and then condensed to form nanoparticles. Conditions like temperature inside the reaction chamber, partial pressure of monomers, vapor pressure of bulk solid, and residence time are to be considered to synthesize nanomaterials through this method. Hence, depending upon the modification in the process, it is also known as chemical vapor deposition (CVD), chemical vapor synthesis (CVS), chemical vapor reaction, and chemical vapor precipitation (CVP). Since the process is vacuum-deposited, it produces high-quality, ultrafine particles. Also, the process allows for control over composition and structure.

#### 18.2.2.3 Self-Assembly

Self-organization occurs when smaller subunits spontaneously combine to form larger, well-organized patterns [28]. The interactions between particles for achieving thermodynamic equilibrium and reducing the system's free energy lead to the formation of spontaneous assembly. Self-assembly can be completed in two ways. Firstly, the intrinsic properties are manipulated by changing interactions' directionality or particle shapes. Secondly, by applying and combining several effects to manipulate building blocks, external manipulation [28–30]. The process has the advantage of allowing the creation of well-defined structures, and the method is scalable and efficient [30].

#### 18.2.2.4 Hydrothermal and Solvothermal Synthesis

In hydrothermal and solvothermal methods, the nanomaterials are obtained through a heterogenous reaction carried out in a solvent at high temperature and high pressure in a sealed vessel around a critical point. The difference between the two methods is that the solvent in the hydrothermal method is an aqueous medium, and in the solvothermal method, the solvent is a nonaqueous medium. Both processes are done in closed systems. These methods can produce nanomaterials with various structures, like wires, rods, sheets, and spheres. These methods are often used for oxides and hydroxides.

#### 18.2.2.5 Atomic Layer Deposition (ALD)

Atomic layer deposition is a stepwise growth process in which the substrate surface is exposed to each precursor alternately. Typically, the ALD process consists of four steps: the precursor pulse, purge, reactant, and purge. This four-step process is repeated to grow a desired material with precise control over the thickness [31, 32]. All the surface reactions can be made sure to be completed and saturated before the start of the subsequent response by adjusting the duration of each exposure. This makes the process self-limiting and allows for conformal coatings, which is another advantage [32].

These bottom-up approaches offer a high degree of control over the properties of nanomaterials, making them suitable for various applications, including electronics, catalysis, sensors, and biomedical devices. Researchers often combine bottom-up and top-down methods to leverage the strengths of both approaches in creating advanced nanomaterials.

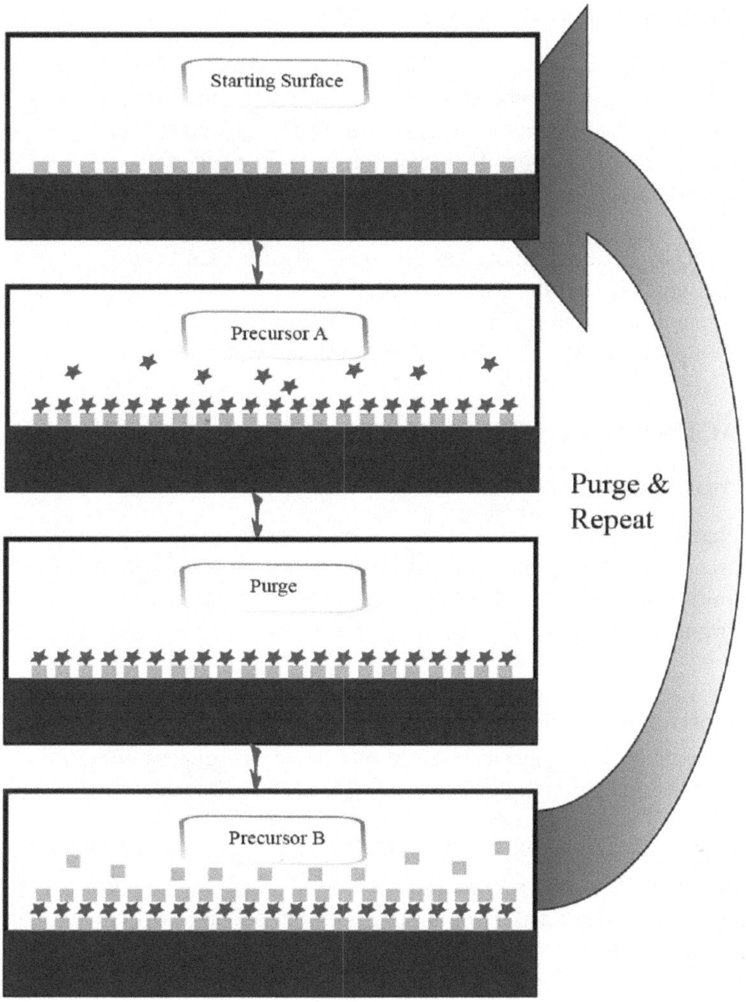


**FIGURE 18.5** Atomic layer deposition.

### 18.2.3 Solid Phase Synthesis

As the name suggests, these methods involve the reaction between the raw materials (precursors) in the solid state. The different solid phase synthesis methods are ball milling, atomic layer deposition, spark plasma sintering, and others. Ball milling is the most commonly used technique as it is relatively inexpensive [13]. Solid phase synthesis routes pose limitations like slow reaction rate, large particle size distribution, complex agglomerate formation, and impurity contamination.

### 18.2.4 Liquid Phase Synthesis

The precursors used in this synthesis route are in a liquid state, where the chemical reaction for the product formation occurs. The various liquid phase synthesis methods used are sol-gel [26], hydrothermal, solvothermal, spray pyrolysis, co-precipitation, and reverse co-precipitation. Among these methods, sol-gel and spray pyrolysis methods are generally used. Liquid phase synthesis methods provide the final product with better purity than solid phase synthesis methods. Also, the reaction rate is faster in this synthesis route than in the solid-state route.

### 18.2.5 Gas Phase Synthesis Method

This synthesis route involves converting the precursors/raw materials to the gaseous state for the reaction. The different gas phase synthesis methods used are flame synthesis, inert gas condensation, gas atomization, ion beam deposition, and others. The gas phase synthesis route is the fastest of the three routes available and typically yields impurity-free products.

Other synthesis methods, such as electrospinning, spray pyrolysis, dip-coating, and emulsion polymerization 3D printing [33, 34], are also used to synthesize nanomaterials.

These nanomaterials obtained using various methods are then used to make nanocomposites. For making nanocomposites, desired nanomaterial and matrix material are chosen depending on the application. Then, the nanomaterial and matrix materials are combined using various methods: melt mixing, in situ polymerization, ball milling, chemical vapor deposition, and various others.

## 18.3 CHARACTERIZATION TECHNIQUES

### 18.3.1 Morphological Characterization

Morphological characterization is a crucial aspect of understanding the structure and properties of polymeric nanomaterials. This process involves analyzing nanoparticles' or nanofillers' arrangement and distribution within the nanoscale polymer matrix. Several techniques are commonly employed for morphological characterization. Electron microscopy is one of the most prominently used techniques.

#### 18.3.1.1 Transmission Electron Microscopy

In a TEM, when an electron beam strikes a sample, the microscope monitors the changes in electron beam scattering within the sample. Because TEM offers information on structural features and chemical composition at several length scales, including atomic dimensions, it has become a crucial tool for nanomaterial characterization and interpretation [35–37]. Microstructure can be observed with TEM imaging; the diffraction pattern provides information on the phase of the material as well as crystallographic arrangement and orientation. The chemical composition of the nanomaterial can be deduced from the energy spectrum. A high-resolution photograph of the sample is also available. Because TEM can reach the nanometric range, it is helpful for morphological studies.

For TEM sample preparation, each has its own advantages and limitations. The most suitable method depends on the specific characteristics of your polymer nanomaterial, such as [38]:

1. *Conductivity.* Conductive samples are more accessible to prepare, while nonconductive ones require additional steps to prevent charging under the electron beam.
2. *Morphology.* The size and shape of the nanoparticles and polymer matrix influence the sectioning technique.
3. *Desired information.* Different methods offer varying levels of resolution and information about the sample's structure.

Some common methods include [39, 40]:

1. *Cryomicrotomy.* This technique involves freezing the sample under liquid nitrogen and then sectioning it with a microtome at cryogenic temperatures. This preserves the delicate nanostructure and minimizes artifacts.
2. *Ultramicrotomy.* Similar to cryomicrotomy but performed at room temperature. Suitable for more robust samples.
3. *Focused ion beam (FIB) milling.* This technique uses a focused beam of gallium ions to mill the sample into a thin lamella. Offers high precision and can target specific areas of interest.

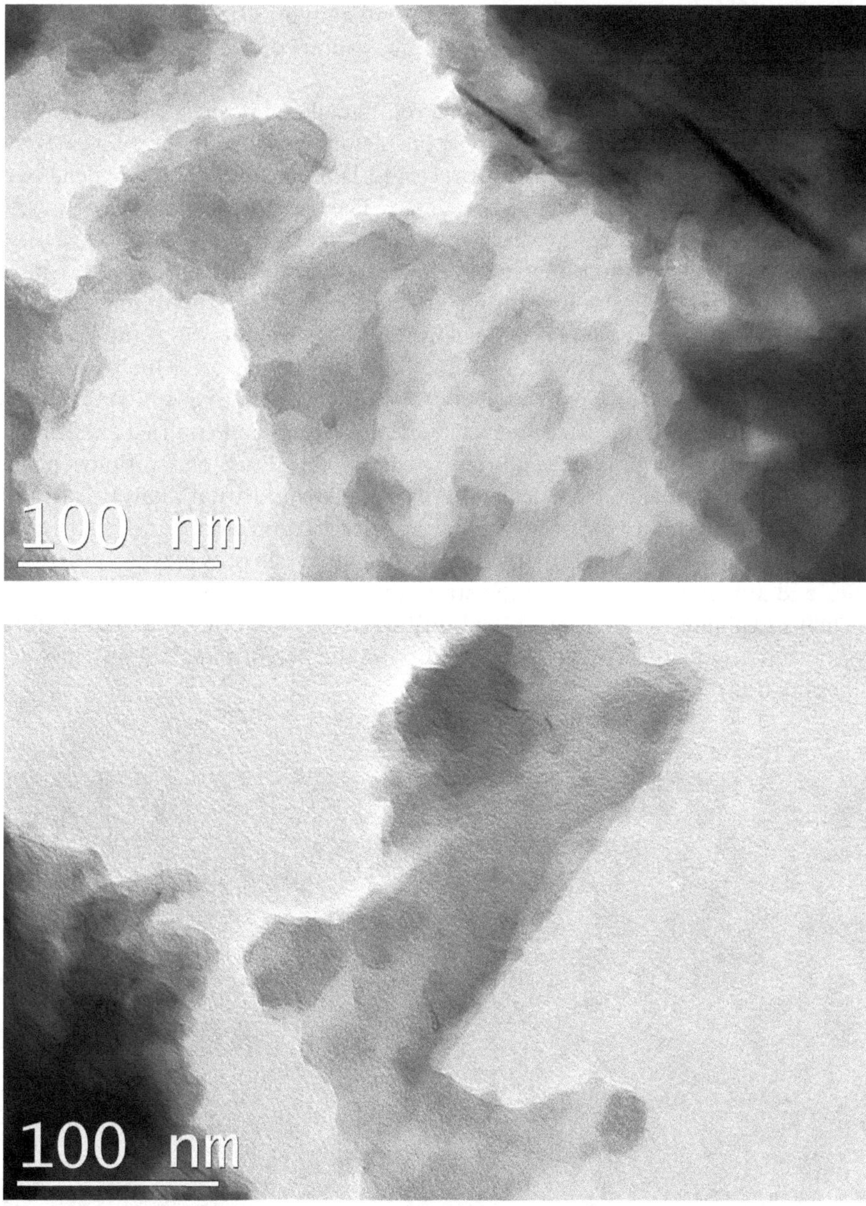


**FIGURE 18.6** SEQ Fig_3. * ARABIC 1 TEM image of graphene-reinforced composite.

The polymeric sample is prepared for TEM by cryomicrotomy (Figure 18.6a). In this, the graphene is evenly distributed in the entire composite. It is not concentrated in any specific region; instead, it is homogeneously distributed.

#### 18.3.1.2 Scanning Electron Microscopy

In an electron microscope, the sample surface is exposed to electron beams and scanned in a raster scan pattern. In this process, the electrons only interact with the sample surface, providing information about it alone [38]. Aside from nanomaterials, SEM can be used to scan biological specimens and computer microchips, and it has industrial uses in semiconductor inspection, production lines, and more [39]. They can also be helpful research instruments in sectors including biological science, medical and forensic science, and metallurgy [40].

Figure 18.6 shows the SEM image of graphene-reinforced composites. The SEM image clearly indicates the presence of graphene. Also, the type of fracture is visible in the picture. It is clear that the grain size is uneven, and that there is no structured laminar layer. The graphene particles are clearly unstable and randomly orientated, as observed. The epoxy resin chain is equally dispersed between graphene layers, increasing the strength of composites.

SEM study methods such as contrast enhancement, high magnification, elemental analysis, fracture surface investigation, and qualitative evaluation combined with one another offer an extensive knowledge of the interfacial adhesion between polymer and reinforcement in composite materials. Fine details are revealed by the high magnification, which makes it easier to see characteristics that could indicate adhesion or failure mechanisms. Compositional variations can be highlighted with the use of backscattered imaging and contrast enhancement. This improves the visibility of the polymer and reinforcing material, which facilitates phase separation and evaluation of interfacial properties. Elemental composition analysis at the interface helps evaluate chemical interactions by revealing information about the distribution of the polymer and reinforcing material. The fracture surface analysis provides information on adhesion qualities and aids in understanding the strength and quality of the interaction. Qualitative assessment helps evaluate the continuity and uniformity of the interface. This information is valuable for optimizing material design and predicting the performance of polymer composites in various applications.

**FIGURE 18.7** SEQ Fig_3. * ARABIC 2 SEM image of graphene-reinforced composite.

#### 18.3.1.3 Atomic Force Microscopy

AFM uses electrically conductive surfaces for measurement. The technology in question is a scanning probe microscopy imaging tool that utilizes a micromachined cantilever with a sharp tip at one end to detect cantilever tip deflection caused by attraction and repulsion between atoms on the measured surface as well as electrostatic forces [38]. The feedback parameter for AFM image surfaces is the force applied between the AFM probe and the sample. A tip placed on a cantilever spring is used to scan the material in order to create an AFM topographic image. The image is obtained by plotting the height position vs. its position on the sample, and a feedback loop maintains a constant force between the tip and the sample while studying by altering pixels with the aid of the scanner's height. Variations in various modes of operation are caused by the nature of tip motion and tip–sample interaction [40]. There is a defined distance between the tip and the sample as a result of either attractive or repulsive interactions. While maintaining a constant deflection, the tip is raster-scanned across the surface of the sample in contact mode (also known as static mode). The amplitude, frequency, or maximum force at the contact point indicate feedback, and the tip oscillates in tapping, noncontact, and peak force tapping (dynamic modes).

### 18.3.2 Structural Characterization

The structure of nanomaterials can be characterized using various techniques, like XRD, FTIR, and Raman spectroscopy, which provide information about the crystal structure and crystallinity of the nanofillers and their interaction with the polymer matrix. It allows for the identification of phases, lattice parameters, and the degree of interfacing between the components.

#### 18.3.2.1 X-Ray Diffraction

X-ray diffraction analysis is a popular technique for characterizing crystalline materials. The material under research uniquely diffracts X-rays in this kind of analysis, and the diffraction patterns produced contribute to a variety of relevant structural properties [3, 41]. Peak position, intensity, and form are critical in determining lattice parameters, space group, chemical composition, crystal structure, texture, qualitative and quantitative phase analyses, crystallite size, and so on. This approach provides structural information on bulk, thin film crystalline, or polycrystalline materials. The crystalline forms seen in nanomaterials can be recognized explicitly using XRD, which can also be used to measure structural properties, such as stress state, particle size, orientation, and defect structure [3, 42]. The XRD apparatus includes a source, sample pieces, and primary and secondary optics. The source's fundamental goal is to inject cathode rays or electrons from the cathode ray tube by increasing the filament's temperature and the electrical power or voltage to accelerate the electrons produced toward the specimen. When the electrons strike the specimen, they are scattered in the direction of the detector. The detector converts these signals or scattered electrons in the X-ray spectra. XRD is a completely safe and inert procedure.

The percentage of crystallinity can be determined from the XRD pattern by using the following equation [43, 44],

$$\%\,Crystallinity = \frac{I_c}{I_c + KI_a} 100 \tag{18.1}$$

Where $I_c$ is the intensity of crystalline peak, $I_a$ is the intensity of amorphous peak, and $K$ is the constant known as wave vector, whose value is taken as 0.9. Hence, the preceding equation gets modified as:

$$\%Crystallinity = \frac{I_c}{I_c + 0.9I_a} 100 \tag{18.2}$$

The direction and magnitude of the scattered X-rays are determined by the K-value, sometimes referred to as the wave vector, which is an essential parameter in X-ray diffraction. $K = 2\pi/\lambda$ is a mathematical equation where λ is the incident X-ray wavelength [43, 44]. Because it directly affects the scattering angle and, in turn, the diffraction pattern, the K-value is important. X-rays diffract at different angles, depending on the K-value, due to different crystallographic planes within the material. Researchers can learn important details about the crystal structure, such as lattice spacing and orientation, by examining the diffraction pattern. Consequently, it is crucial to comprehend and manage the K-value in order to properly interpret XRD data and derive important insights regarding the crystalline nature of materials.

#### 18.3.2.2 Fourier Transform Infrared Spectroscopy (FTIR)

FTIR analyzes the vibrational fingerprint of the nanomaterial, revealing the presence of functional groups, changes in bonding, and interactions between the filler and the matrix. Infrared spectroscopy is an important instrument for characterizing matter's structure at the molecular level [45]. The unique, resonant vibrational modes of various branches or body portions of a molecule can be used to reveal its constituents and bonding arrangement. The reverse procedure is also used since each molecule has a unique infrared spectrograph, similar to a fingerprint, that can reveal its identification. FTIR spectroscopy analysis is a method based on the principle of infrared spectroscopy that has been extended to the analysis of nanoscaled objects [45].

Based on a database of known vibrational frequencies, FTIR may not be able to show the chemical interactions between the nanoparticles and the functional molecules, making it challenging to characterize functionalized nanoparticles. This is due to the fact that the functionalized molecules and the nanomaterial typically generate unique chemical bonds. As a result, a method for demonstrating the resonant vibrational information between the grafted molecules and the nanosized particles must be developed.

#### 18.3.2.3 Raman Spectroscopy

A popular molecular spectroscopy technique for learning about polymeric systems from their vibrational characteristics is Raman scattering. An important benefit of Raman scattering is that it can analyze thick polymer materials, whereas infrared transmission spectroscopy can only look at very thin layers.

The characteristic of the vibrational modes of the sample that cause the scattering is the shift in frequency between the incident and scattered photons [46, 47]. The molecule's surroundings, symmetry, and chemical structure can all be determined by knowing the frequency of these modes. In fact, the selection principles derived from the symmetry characteristics of the molecule (crystal) under consideration control the vibrations' activity. Therefore, in infrared absorption spectroscopy, only vibrations that result in a change in the molecule's dipole moment will be detected, and in Raman scattering, only vibrations that cause a change in the polarizability of the molecule will be detected [46]. As a result, Raman spectroscopy is one of the most important techniques used to analyze composite properties.

### 18.3.3 Thermal Analysis

The creation and design of any product are based on its thermophysical properties; hence, it is crucial to precisely measure them with known precision. A collection of methods known as thermal analysis tracks how a sample's chemical or physical characteristics change over time when it is subjected to a temperature program. When compared to unfilled polymeric matrices, there was a noticeable improvement in the thermal stability of the nanotube-filled ones.

#### 18.3.3.1 Differential Scanning Calorimetry (DSC)

DSC has been extensively used to study a variety of processes, including glass transition ($T_g$), melting, crystallization, and curing, that arise during the thermal heating of polymer nanomaterials. Thermoanalytical DSC measures temperature and heat flux associated with a material's thermal transitions [9]. Using this thermoanalytical approach, the difference in the amount of heat needed to raise the temperature of a sample and a reference is assessed. The sample and reference pan are subjected to the same heating history, and both have a well-defined heat capacity over the temperature range being scanned. The fundamental idea behind DSC is straightforward: it calculates how much energy a sample absorbs or releases in order to keep its temperature constant relative to the reference pan. In addition to glass transition, melting and crystallization temperatures, fusion heat, and oxidation induction times, solidification can be determined by means of physical changes [48]. To study the α-transition in polymers and their composites, one of the most popular methods is the DSC method. A connection exists between the α-transition and the relaxation of dipoles associated with the main chains' Brownian motion at the glassy-to-rubbery-state transition. The DSC analysis of the curing process and its kinetics can give insight into the genuine mechanism of cure reaction and its effect on the degree of cross-linking and, as a consequence, on mechanical qualities [7, 9, 49]. Moreover, DSC has been used to measure the thermal conductivity of polymer nanomaterials.

The crystallinity of a sample can be determined using the following equation in the context of DSC [50]:

$$\text{Crystallinity } (\%) = \frac{\Delta H_{melt}}{\Delta H_{fusion}} \times 100 \tag{18.3}$$

Where $\Delta H_{melt}$ is the enthalpy change associated with the melting of the crystalline regions and $\Delta H_{fusion}$ is the enthalpy change of fusion for a fully crystalline sample of the same material.

The ratio of the enthalpy change during sample melting to the enthalpy change of fusion for a totally crystalline sample is what this equation uses to define the crystallinity as a percentage.

When examining phase transitions in materials, like melting and crystallization, DSC is especially helpful. Through the examination of heat transfer during these shifts, scientists can learn more about the behavior and thermal characteristics of numerous materials, including as metals, polymers, and medications. The degree of structural order in the material and the effectiveness of crystalline areas in storing thermal energy are shown by the crystallinity that is computed from DSC data.

#### 18.3.3.2 Thermogravimetric Analysis (TGA)

TGA is a useful tool for measuring the thermodynamic stability of polymer nanomaterials. A thermal analysis method called thermogravimetric analysis (TGA) calculates how a polymer's mass changes in response to temperature or time in a controlled setting [10, 51, 52]. From TGA measurements of the mass change of the nanomaterial as a function of temperature, it is possible to infer the presence of filler, a thermally stable nanomaterial, and decomposition patterns. To describe a material's physical and chemical changes with temperature, TGA tracks the material's mass change. Thermal phenomena such as desorption, absorption, sublimation, vaporization, oxidation, reduction, and decomposition can all be identified using TGA [49]. The temperature of the furnace is monitored by a thermocouple using a millivoltmeter when the sample is placed within a regulated furnace in a TGA apparatus. Constant mass determination is made possible by the balance. In addition to examining the kinetics of the physicochemical processes taking place in the sample, it provides the chance to characterize the breakdown and thermal stability of materials under various conditions.

A complimentary method that is frequently used in conjunction with thermogravimetric analysis (TGA) is differential thermogravimetry (DTG). The weight change rate of the sample with respect to temperature is represented by DTG, which is obtained from TGA data. By taking the derivative of the TGA curve, one can obtain the DTG curve, which offers important information about the thermal events that are taking place in a sample. The location and strength of peaks in the DTG curve, which represent particular thermal events, like oxidation, phase transitions, or breakdown, might provide information on the types of processes involved. The capacity of DTG to improve the interpretability of complicated TGA data accounts for its significance in TGA. While TGA provides data on weight growth or loss, DTG offers a more lucid picture of the particular temperature ranges linked to certain processes. Understanding the thermal behavior of the material under study is facilitated by the ability to identify and distinguish between overlapping thermal events provided by peaks in the DTG curve. To sum up, DTG is a useful technique in thermal analysis, especially when researching the processes of material transformation and breakdown. It facilitates the identification of particular temperature ranges linked to various occurrences, offering a more thorough and accurate examination of a sample's thermal behavior.

### 18.3.4 Mechanical and Rheological Characterization

The strength, stiffness, and toughness of the nanomaterial are evaluated by a range of mechanical tests, including tensile, flexural, and impact tests, in comparison to their pristine counterparts [1, 10, 14]. A material's tensile strength indicates how much axial or tensile load it can bear before breaking. The tensile strength of nanocomposites can be greatly affected by the addition of nanofillers. Increased tensile strength results from greater load transfer between the matrix and the filler, which is facilitated by the higher aspect ratio and increased surface area of nanofillers. To maximize tensile characteristics, the nanofillers must be evenly distributed throughout the matrix.

The ability of a material to tolerate bending forces is measured by its flexural strength, also referred to as its modulus of rupture or bending strength. Through matrix material reinforcement, nanofillers can improve flexural strength. Improved resistance against bending stresses is a result of the stiffer matrix and better interfacial adhesion between the nanofillers. To achieve optimal flexural strength, nanofillers must be properly dispersed and aligned inside the matrix. The capacity of a material to absorb energy following a sudden impact or shock loading is measured by its impact strength. Well-dispersed nanofillers can enhance the impact resistance of nanocomposites. More efficiently than the matrix alone, the impact energy is absorbed and dispersed by the nanofillers, which serve as energy-dissipating sites. Improving the impact strength of nanocomposites primarily involves achieving a homogenous distribution of nanofillers and encouraging strong interfacial bonding.

It is commonly acknowledged that in nanomaterials, there is an interphase between the matrix material and the nanofillers. Numerous researchers have examined the interphase behavior of polymer nanomaterials in nanomaterial settings. This interphase is a transition region that spans nanometers to micrometers, over which the mechanical and physical properties change from the properties of filler to the properties of the matrix [3, 34, 42, 9, 53–55].

The study of rheology examines the behavior of a fluid or soft solid under the influence of external forces. The rheological properties of nanofluids are a crucial and important area of research that affects our understanding and use of them [3, 34]. One of the most important properties of the intended usage of a nanofluid is its rheological behavior. The non-Newtonian fluid behavior and shear rate–dependent viscosity must be carefully considered in the engineering design of the processing and applications of nanofluids [3, 54]. The rheological behavior of these materials is characterized by a number of important factors [56, 57].

1. *Storage modulus (G')*. This gauges a material's capacity to hold energy during deformation and depicts its elastic behavior. It has to do with a material's capacity to regain its original shape following the application of stress.
2. *Loss modulus (G")*. This gauges a material's viscosity and quantifies how much energy it can release during deformation. It has to do with how resistant the material is to flow.
3. *Viscosity complexity (η*)*. This is a complicated number that combines elastic and viscous components. $\eta^* = G'/\omega + iG''$ is how it is expressed, with ω denoting the angular frequency.
4. *Plateau*. A *plateau* is a location in the frequency sweep where the storage modulus or loss modulus is largely constant, according to rheology. This frequently marks the change from one rheological behavior to another.
5. *Cole–Cole plot*. The Cole–Cole plot illustrates complex viscosity in a complex plane graphically. It is a semicircular plot that aids in illustrating a material's viscoelastic

behavior. The rheological properties of the material can be inferred from the plot's form and the curve's location.

6. *Shear rate ($\dot{\gamma}$).* The pace at which neighboring layers of a material move in relation to one another is known as the shear rate ($\dot{\gamma}$). It is an essential variable in figuring out how materials react to forces that are applied.
7. *Yield stress.* This is the stress at which a substance starts to behave more like a fluid and less like a solid. Only when the applied stress above this crucial value can materials with a yield stress undergo plastic deformation.
8. *Thixotropy.* Under continuous shear stress, this is a time-dependent decrease in viscosity. It is frequently seen in materials whose structure degrades with time when shear is applied.
9. *Creep and recovery.* Under a continuous load, a material will gradually deteriorate; when the strain is released, the material will recover and take on its previous shape.

In many industries, such as material development, food processing, medicines, and many more, where the flow and deformation characteristics of substances play a crucial role, an understanding of these rheological parameters is vital.

## 18.4 APPLICATIONS

Various industries have found nanomaterials useful due to their unique properties, which incorporate nanoscale fillers into a matrix material. Some notable applications of nanomaterials are shown in Figure 18.8.

### 18.4.1 Automotive Industry

The automobile sector is always working to create and use components with superior mechanical and tribological qualities that are also lightweight. Nanomaterials are a novel family of materials with superior mechanical and thermal properties. They are appropriate for use in the automotive

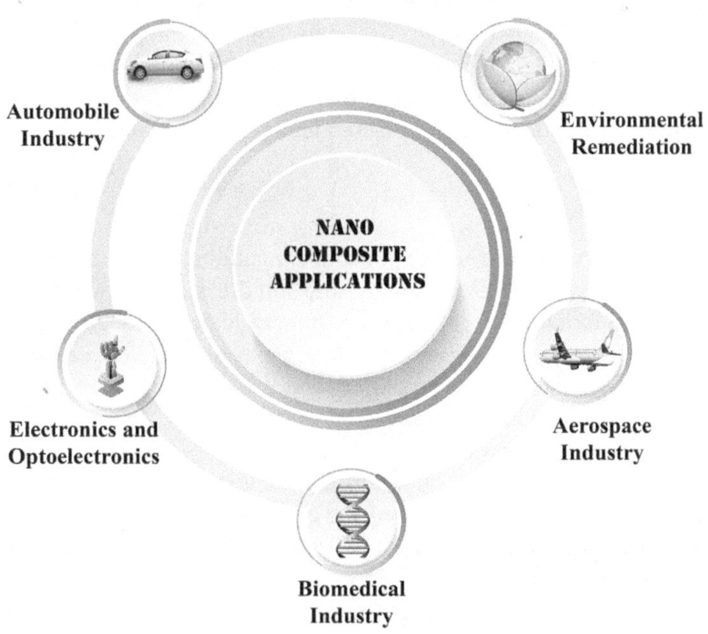


**FIGURE 18.8** SEQ Fig_4. * ARABIC 1 Application of nanomaterials.

industry because of their excellent performance [58, 59]. In the automotive industry, polymer nanocomposites are essential, especially for components that need to be lightweight and have improved mechanical qualities. These cutting-edge materials improve strength and fuel efficiency when used in structural components like bumpers, body panels, and frames. Nanocomposites provide increased durability to interior parts, such as dashboards and seats. Nanocomposites are used in tires' sidewalls and tread to improve traction and wear resistance. These materials are used in fuel systems' tanks and lines to provide excellent barrier qualities and increase safety. To further promote economy, safety, and sustainability throughout the vehicle, nanocomposites are also used in electrical components, coatings, battery enclosures, brake components, and air intake systems.

Significant advancements in a number of industries in recent years have inspired academics to focus on developing novel structural materials for improved performance in engineering applications. Other forms of reinforcement, such nanofibers, nanowhiskers, or nanoplatelets, have gradually supplanted nanoparticles as the reinforcing element in metal matrix composites. Reinforcing the oxides, carbides, nitrides, and borides is the primary division. The nanoparticle types SiC, TiC, WC, TaC, $TiB_2$, AlN, and $Al_2O_3$ are the most often utilized ones [60–62]. The growing automobile manufacturing and a growing need for lightweight materials are the primary drivers of the development in demand for nanomaterials. It is envisaged that the application of the structural and operating parts will boost the rate because polymer composites operate as an ideal substitute for metal parts.

### 18.4.2 Electronics and Optoelectronics

Organic light-emitting diodes and organic photovoltaic cells are two examples of optoelectronic and electrical devices employing organic materials as active elements that have garnered a lot of attention lately due to their potential to replace their inorganic counterparts in a number of applications [63, 64]. Graphene nanomaterials have emerged as the most effective nanomaterial for optoelectronic and electrical applications recently. In fact, optoelectronic components utilizing traditional semiconductors (silicon, gallium arsenide) require sophisticated processing that results in high production costs, especially for large area applications, despite a well-developed and mastered technology offering excellent stability and good performance devices [36, 64]. On the other hand, organic devices are more appealing for a number of particular applications because they make use of organic materials' advantages, which include easy processing, light weight, vast area, flexibility, and low cost. Numerous applications, like solar cells and small-size organic display screens, have already made it to the consumer market, demonstrating that the materials—despite their lower efficiency when compared to inorganic semiconductors—are ideally suited for industrial manufacturing.

### 18.4.3 Biomedical Applications

Polymer composites have a wide range of applications, including biomedicine and energy storage. In particular, they are actively being used as moist and soft dressings for wound healing, as vehicles for the controlled encapsulation and release of pharmaceuticals, as tough and adaptable biointerfaces for cellular and tissue engineering, and as nanoscale biosensors for diagnostics. In the field of biomedicine, polymer nanomaterials are a result of the wide range of uses and the highly diverse functional specifications associated with each of these uses [6, 35]. As a promising class of hybrid materials made of both synthetic and natural biodegradable polymers as well as organic and inorganic fillers, bionanomaterials have emerged. Different chemistries and compositions can lead to applications from tissue engineering to load-bearing composites for bone restoration. In biomedical applications, a wide range of polymers are useful. These include synthetic polymers like poly(lactic acid) (PLA), poly(ε-caprolactone) (PCL), poly(glycolic acid) (PGA), and poly(hydroxyl butyrate), as well as natural polymers like polysaccharides (starch, alginate, chitin/chitosan) and proteins (soy). Structural implants and scaffolds for tissue engineering are among the uses for the biodegradable polymer composite materials, which have lately become a new class of bioactive materials [65, 66].

With the stiffness, strength, and bioactive properties of polymeric nanocomposites, these composites combine the flexibility of polymers. Conventional (micron-size) bioactive glass particles have been used as fillers (or coatings) in the majority of composites research conducted to date.

### 18.4.4 Aerospace Applications

The aircraft industry uses nanomaterials for a variety of purposes—by employing nanomaterials to create materials that are both strong and lightweight for use in spacecraft and airplanes. An optimistic picture of the nanotechnology sector is painted by the use of nanomaterials in a number of aerospace industry subsystems, particularly due to the self-healing qualities of nanomaterial polymers. Nanomaterials are a perfect technology for aircraft applications because of their continual technological advancements and interwoven sustainability consequences [67]. For superior aerospace qualities, nanomaterial systems have also employed high-strength elements like titanium. High hardness, a crucial goal for many structural aerospace components, was achieved by using graphene oxide–reinforced nanopowders as a matrix system. By reducing waste creation in manufacturing procedures, nanomaterials can be included into intricate aircraft designs. Structures and fuselage that are lightweight and require little maintenance can be designed using this. Because of the nanomaterial's low grain growth at extremely high temperatures, it was applied as an overlay coating to the turbine blades of a jet engine and used in a variety of aerospace applications. For aircraft purposes, a robust nanomaterial covering was created. A corrosive environment such as space is the perfect fit for nanomaterials because of their unique combinations. The matrix, microelectronics, and impact studies are all optimized by the reinforcement [67, 68].

### 18.4.5 Environmental Remediation

One of the biggest issues facing modern society is, without a doubt, environmental degradation. To remove pollutants from the air, water, and soil, new methods are continuously being researched. These nanoparticles are employed in the removal of many environmental pollutants, including halogenated herbicides, heavy metals, dyes, organophosphorus chemicals, and chlorinated organic compounds. Each material has its own advantages and challenges connected to its application; we offer here an overall perspective of several nanoparticles that have been utilized in the context of environmental restoration [69].

Nanomaterials serve a critical role in a variety of environmental applications, providing novel solutions to pressing issues. Nanocatalysts, for example, improve water treatment operations by efficiently reducing pollutants and impurities. They can also be used in air purification systems to efficiently capture and remove dangerous contaminants. Nanomaterial-based sensors detect environmental pollutants sensitively and selectively, allowing for real-time monitoring and enhanced resource management. In the field of renewable energy, nanomaterials help enhance solar cells and fuel cells. They improve the efficiency of energy conversion processes, hence encouraging the development of renewable energy sources. In addition, nanocomposites in lightweight materials improve fuel efficiency and reduce environmental impact in transportation. Nanotechnology aids in soil remediation by tackling contaminated locations. Nanoparticles can immobilize or break down contaminants, thereby increasing soil quality and aiding in ecosystem restoration. Furthermore, nanoparticles play an important role in the development of environmentally friendly packaging materials and water purification technologies, encouraging sustainable practices across a variety of industries [70].

While nanomaterials offer intriguing answers, their research and deployment must be guided by responsible and ethical concerns in order to reduce potential environmental and health dangers. Continued study leads to a greater knowledge of their long-term effects, promoting the long-term incorporation of nanotechnologies into environmental applications.

## 18.5 FUTURE PERSPECTIVES

### 18.5.1 Emerging Trends

Throughout the past three decades, research on polymer-based nanomaterials has drawn a lot of attention. Incorporating tiny concentrations of organic or inorganic nanofillers into the parent polymer improves its qualities further and gives the nanomaterial new properties that are nearly identical to its initial processing properties. Phase interfaces and significant alterations in mechanical, optical, and electrical characteristics, among other effects, are also brought about by the high surface-area-to-volume ratio of nanofillers and the molecular interaction between the polymer matrix and the nanofiller. Therefore, in a variety of industries, including biomedical, electronics, construction, pipelines, intelligent coatings, automotive and transportation, and energy storage and generation, polymer-based nanomaterials may be able to replace current materials.

### 18.5.2 Challenges and Opportunities

In order to successfully move from the macroscale to the nanoscale, the production of nanomaterials still needs to overcome a number of obstacles, despite the endless potential uses. For instance, to comprehend and optimize the structure–process–properties relationships, accomplish a more straightforward and efficient particle exfoliation and dispersion, and lower manufacturing costs by utilizing more compostable and biodegradable polymers, a strong interdisciplinary collaboration between scientists and engineers is still required. One of the primary issues with incorporating nanoparticles into composites is agglomeration [71]. The primary obstacle to the commercialization of nanomaterials is the inability of the nanoparticles to disperse uniformly due to agglomeration. It is particularly challenging to achieve homogeneous characteristics when agglomeration occurs during bulk production.

## 18.6 CONCLUSION

The study of nanomaterials has grown in significance in the creation of novel materials for cutting-edge applications due to the recent rapid development of nanotechnology. Due to their high degree of integrated association and versatility, nanomaterials are the ideal material to meet the increasing need for multifunctional materials. This is a multidisciplinary field that involves understanding both technological and scientific foundations to produce macroscopic engineered materials made from nanoscale structures. These materials are appropriate for addressing the new needs brought about by advancements in science and technology. Researchers have been working on nanomaterials to improve their qualities, in addition to other areas. In order to conduct this study, we collected credible research on the applications and usage of nanoparticles in polymeric composites. Fillers in composites come in various forms, including nanomaterials and nanoclays, which improve ballistic qualities and lessen brittleness. These fillers can also strengthen mechanical, electrical, optical, EMI shielding, magnetic, and chemical/corrosion resistance. Additionally, it can enhance composites usefulness and stability. Improved property composites find extensive applications in thermoelectric, photovoltaic, industrial, biomedical, structural, and coating fields. Moreover, it produces hydrophobic surfaces and lessens delamination. It is used in food packaging, films, oil and gas pipelines, vehicles, airplanes, and electrical devices, in addition to environmental protection.

## ACKNOWLEDGMENTS

All the authors wish to thank IIT Roorkee for providing an excellent research facility which facilitated in writing this book chapter.

## REFERENCES

[1] E.P.S. Tan, C.T. Lim, Mechanical characterization of nanofibers: A review, Compos Sci Technol 66 (2006) 1102–1111. https://doi.org/10.1016/j.compscitech.2005.10.003.

[2] V. Beedasy, P.J. Smith, Printed electronics as prepared by inkjet printing, Materials 13 (2020). https://doi.org/10.3390/ma13030704.

[3] N.R. Patra, Y.S. Negi, Thermal, structural, and rheological modifications in recycled polyethylene terephthalate for a sustainable alternative source for additive manufacturing, Polym Eng Sci 62 (2022) 2486–2497. https://doi.org/10.1002/pen.26022.

[4] M. Sathwane, M. Chhajed, C. Verma, K.K. Gaikwad, P.K. Maji, Synergistic effect of hybrid hydroxylated boron nitride and cellulose nanocrystals for enhancing the thermal, mechanical, and hydrophobic properties of composite film, Polym Compos 44 (2023) 492–504. https://doi.org/10.1002/pc.27112.

[5] F. Daver, E. Baez, R.A. Shanks, M. Brandt, Conductive polyolefin-rubber nanocomposites with carbon nanotubes, Compos Part A Appl Sci Manuf 80 (2016) 13–20. https://doi.org/10.1016/j.compositesa.2015.10.002.

[6] Y. Zare, I. Shabani, Polymer/metal nanocomposites for biomedical applications, Mater Sci Eng C 60 (2016) 195–203. https://doi.org/10.1016/j.msec.2015.11.023.

[7] C.E. Corcione, M. Frigione, Characterization of nanocomposites by thermal analysis, Materials 5 (2012) 2960–2980. https://doi.org/10.3390/ma5122960.

[8] T. Hassan, A. Salam, A. Khan, S.U. Khan, H. Khanzada, M. Wasim, M.Q. Khan, I.S. Kim, Functional nanocomposites and their potential applications: A review, J Polym Res 28 (2021). https://doi.org/10.1007/s10965-021-02408-1.

[9] J. Golebiewski, A. Galeski, Thermal stability of nanoclay polypropylene composites by simultaneous DSC and TGA, Compos Sci Technol 67 (2007) 3442–3447. https://doi.org/10.1016/j.compscitech.2007.03.007.

[10] Z. Weng, J. Wang, T. Senthil, L. Wu, Mechanical and thermal properties of ABS/montmorillonite nanocomposites for fused deposition modeling 3D printing, Mater Des 102 (2016) 276–283. https://doi.org/10.1016/j.matdes.2016.04.045.

[11] N. Baig, I. Kammakakam, W. Falath, I. Kammakakam, Nanomaterials: A review of synthesis methods, properties, recent progress, and challenges, Mater Adv 2 (2021) 1821–1871. https://doi.org/10.1039/d0ma00807a.

[12] P. Kuziora, M. Wyszyńska, M. Polanski, J. Bystrzycki, Why the ball to powder ratio (BPR) is insufficient for describing the mechanical ball milling process, Int J Hydrogen Energy 39 (2014) 9883–9887. https://doi.org/10.1016/j.ijhydene.2014.03.009.

[13] H.X. Kho, S. Bae, S. Bae, B.-W. Kim, J.S. Kim, Planetary ball mill process in aspect of milling energy, J Korean Powd Metall Inst 21 (2014) 155–164. https://doi.org/10.4150/kpmi.2014.21.2.155.

[14] D.L. Zhang, Processing of advanced materials using high-energy mechanical milling, Prog Mater Sci (2004) 537–560. https://doi.org/10.1016/S0079-6425(03)00034-3.

[15] H. Zeng, X.W. Du, S.C. Singh, S.A. Kulinich, S. Yang, J. He, W. Cai, Nanomaterials via laser ablation/irradiation in liquid: A review, Adv Funct Mater 22 (2012) 1333–1353. https://doi.org/10.1002/adfm.201102295.

[16] D. Tan, S. Zhou, J. Qiu, N. Khusro, Preparation of functional nanomaterials with femtosecond laser ablation in solution, J Photochem Photobiol C Photochem Rev 17 (2013) 50–68. https://doi.org/10.1016/j.jphotochemrev.2013.08.002.

[17] Y. Wang, J.A. Pan, H. Wu, D.V. Talapin, Direct wavelength-selective optical and electron-beam lithography of functional inorganic nanomaterials, ACS Nano 13 (2019) 13917–13931. https://doi.org/10.1021/acsnano.9b05491.

[18] A.A. Tseng, K. Chen, C.D. Chen, K.J. Ma, Electron beam lithography in nanoscale fabrication: Recent development, IEEE Trans Electron Packag Manuf 26 (2003) 141–149. https://doi.org/10.1109/TEPM.2003.817714.

[19] A. Kaur, B. Bajaj, A. Kaushik, A. Saini, D. Sud, A review on template assisted synthesis of multi-functional metal oxide nanostructures: Status and prospects, Mater Sci Eng B 286 (2022). https://doi.org/10.1016/j.mseb.2022.116005.

[20] V. Pavlenko, S. Khosravi H, S. Żółtowska, A.B. Haruna, M. Zahid, Z. Mansurov, Z. Supiyeva, A. Galal, K.I. Ozoemena, Q. Abbas, T. Jesionowski, A comprehensive review of template-assisted porous carbons: Modern preparation methods and advanced applications, Mater Sci Eng R Rep 149 (2022). https://doi.org/10.1016/j.mser.2022.100682.

[21] R.R. Poolakkandy, M.M. Menamparambath, Soft-template-assisted synthesis: A promising approach for the fabrication of transition metal oxides, Nanoscale Adv 2 (2020) 5015–5045. https://doi.org/10.1039/d0na00599a.

[22] Z. Qiao, J. Zhang, X. Hai, Y. Yan, W. Song, S. Bi, Recent advances in templated synthesis of metal nanoclusters and their applications in biosensing, bioimaging and theranostics, Biosens Bioelectron 176 (2021). https://doi.org/10.1016/j.bios.2020.112898.

[23] L.C. Glangchai, M. Caldorera-Moore, L. Shi, K. Roy, Nanoimprint lithography based fabrication of shape-specific, enzymatically-triggered smart nanoparticles, J Control Release 125 (2008) 263–272. https://doi.org/10.1016/j.jconrel.2007.10.021.

[24] B. Nowduri, A. Britz-Grell, M. Saumer, D. Decker, Nanoimprint lithography-based replication techniques for fabrication of metal and polymer biomimetic nanostructures for biosensor surface functionalization, Nanotechnology 34 (2023). https://doi.org/10.1088/ 1361–6528 /acb35b.

[25] V.V. Shinde, A.D. Celestine, L.E. Beckingham, B.S. Beckingham, Stereolithography 3 D printing of microcapsule catalyst-based self-healing composites, ACS Appl Polym Mater 2 (2020) 5048–5057. https://doi.org/10.1021/acsapm.0c00872.

[26] D. Bokov, A. Turki Jalil, S. Chupradit, W. Suksatan, M. Javed Ansari, I.H. Shewael, G.H. Valiev, E. Kianfar, Nanomaterial by sol-gel method: Synthesis and application, Adv Mater Sci Eng 2021 (2021). https://doi.org/10.1155/2021/5102014.

[27] M.F. Bertino, Rapid fabrication of hybrid aerogels and 3 D printed porous materials, J Solgel Sci Technol 86 (2018) 239–254. https://doi.org/10.1007/s10971-018-4649-3.

[28] A. Rao, S. Roy, V. Jain, P.P. Pillai, Nanoparticle self-assembly: From design principles to complex matter to functional materials, ACS Appl Mater Interfaces 15 (2023) 25248–25274. https://doi.org/10.1021/acsami.2c05378.

[29] F. Bai, K. Bian, B. Li, C. Karler, A. Bowman, H. Fan, Self-assembly of functional nanoscale materials, MRS Bull 45 (2020) 135–141. https://doi.org/10.1557/mrs.2020.21.

[30] B. Zhang, Principles, methods, formation mechanisms, and structures of nanomaterials prepared via self-assembly, in: Physical fundamentals of nanomaterials, Elsevier, 2018: pp. 177–210. https://doi.org/10.1016/b978-0-12-410417-4.00005-8.

[31] X. Meng, X. Wang, D. Geng, C. Ozgit-Akgun, N. Schneider, J.W. Elam, Atomic layer deposition for nanomaterial synthesis and functionalization in energy technology, Mater Horiz 4 (2017) 133–154. https://doi.org/10.1039/c6mh00521g.

[32] Z. Zhang, Y. Zhao, Z. Zhao, G. Huang, Y. Mei, Atomic layer deposition-derived nanomaterials: Oxides, transition metal dichalcogenides, and metal-organic frameworks, Chem Mater 32 (2020) 9056–9077. https://doi.org/10.1021/acs.chemmater.9b04414.

[33] F. Finaa, C.M. Madlaa, A. Goyanesb, J. Zhanga, S. Gaisforda, A.W. Basita, Fabricating 3 D printed orally disintegrating printlets using selective laser sintering, Int J Pharm (2018). https://doi.org/10.1016/j.ijpharm.2018.02.015.

[34] M. Bertolino, D. Battegazzore, R. Arrigo, A. Frache, Designing 3 D printable polypropylene: Material and process optimisation through rheology, Addit Manuf 40 (2021) 101944. https://doi.org/10.1016/j.addma.2021.101944.

[35] N.H. Solangi, R.R. Karri, N.M. Mubarak, S.A. Mazari, Mechanism of polymer composite-based nanomaterial for biomedical applications, Adv Ind Eng Polym Res (2023). https://doi.org/10.1016/j.aiepr.2023.09.002.

[36] A.Y. Yassin, Synthesized polymeric nanocomposites with enhanced optical and electrical properties based on gold nanoparticles for optoelectronic applications, J Mater Sci Mater Electron 34 (2023). https://doi.org/10.1007/s10854-022-09402-3.

[37] Z. Liu, Y. Wang, B. Wu, C. Cui, Y. Guo, C. Yan, A critical review of fused deposition modeling 3 D printing technology in manufacturing polylactic acid parts, Int J Adv Manuf Technol 102 (2019) 2877–2889. https://doi.org/10.1007/s00170-019-03332-x.

[38] A. Mayeen, L.K. Shaji, A.K. Nair, N. Kalarikkal, Morphological characterization of nanomaterials, in: Characterization of nanomaterials: Advances and key technologies, Elsevier, 2018: pp. 335–364. https://doi.org/10.1016/B978-0-08-101973-3.00012-2.

[39] S. Kumar, I. Singh, S.S.R. Koloor, D. Kumar, M.Y. Yahya, On laminated object manufactured FDM-printed ABS/TPU multimaterial specimens: An insight into mechanical and morphological characteristics, Polymers (Basel) 14 (2022). https://doi.org/10.3390/polym14194066.

[40] M. D'Arienzo, R. Scotti, B. Di Credico, M. Redaelli, Synthesis and characterization of morphology-controlled $TiO_2$ nanocrystals opportunities and challenges for their application in photocatalytic materials, in: Studies in surface science and catalysis, Elsevier Inc., 2017: pp. 477–540. https://doi.org/10.1016/B978-0-12-805090-3.00013-9.

[41] A.M. Jordan, K. Kim, D. Soetrisno, J. Hannah, F.S. Bates, S.A. Jaffer, O. Lhost, C.W. Macosko, Role of crystallization on polyolefin interfaces: An improved outlook for polyolefin blends, Macromolecules 51 (2018) 2506–2516. https://doi.org/10.1021/acs.macromol.8b00206.

[42] V. Selvakumar, K. Palanikumar, K. Palanivelu, Studies on mechanical characterization of polypropylene/$Na^+$-MMT nanocomposites, J Miner Mater Charact Eng 9 (2010) 671–681. https://doi.org/10.4236/jmmce.2010.980482010.

[43] İ. Uzun, Methods of determining the degree of crystallinity of polymers with X-ray diffraction: A review, J Polym Res 30 (2023). https://doi.org/10.1007/s10965-023-03744-0.

[44] S. Kavesh, Meaning and measurement of crystallinity in polymers: A review, Polym Eng Sci 9 (1969) 452. https://doi.org/10.1002/pen.760090612.

[45] C. Baudot, C.M. Tan, J.C. Kong, FTIR spectroscopy as a tool for nano-material characterization, Infrared Phys Technol 53 (2010) 434–438. https://doi.org/10.1016/j.infrared.2010.09.002.

[46] L. Bokobza, M. Couzi, J.L. Bruneel, Raman spectroscopy of polymer-carbon nanomaterial composites, Rubber Chem Technol 90 (2017) 37–59. https://doi.org/10.5254/rct.16.83759.

[47] B. Singh, R. Kumar, J.S. Chohan, Polymer matrix composites in 3 D printing: A state of art review, Mater Today Proc 33 (2019) 1562–1567. https://doi.org/10.1016/j.matpr.2020.04.335.

[48] L. Wang, W.M. Gramlich, D.J. Gardner, Y. Han, M. Tajvidi, Spray-dried cellulose nanofibril-reinforced polypropylene composites for extrusion-based additive manufacturing: Nonisothermal crystallization kinetics and thermal expansion, J Compos Sci 2 (2018). https://doi.org/10.3390/jcs2010007.

[49] M. Kodal, N. Karakaya, A.A. Wis, G. Ozkoc, Thermal properties (DSC, TMA, TGA, DTA) of rubber nanocomposites containing carbon nanofillers, in: Carbon-based nanofillers and their rubber nanocomposites: Fundamentals and applications, Elsevier, 2019: pp. 325–366. https://doi.org/10.1016/B978-0-12-817342-8.00011-1.

[50] R. Chatterjee, M. Mondal, S. Acharya, J. Chanda, D. Basu, P. Ghosh, R. Mukhopadhyay, A. Bandyopadhyay, Material extrusion of SEEPS blended isotactic PP for possible application as automotive bumper: Performance analysis through finite element simulation, J Manuf Process 111 (2024) 166–179. https://doi.org/10.1016/j.jmapro.2024.01.025.

[51] I. Blanco, A brief review of the applications of selected thermal analysis methods to 3D printing, Thermo 2 (2022) 74–83. https://doi.org/10.3390/thermo2010006.

[52] A. Alam, Y. Zhang, H.C. Kuan, S.H. Lee, J. Ma, Polymer composite hydrogels containing carbon nanomaterials: Morphology and mechanical and functional performance, Prog Polym Sci 77 (2018) 1–18. https://doi.org/10.1016/j.progpolymsci.2017.09.001.

[53] H. Hu, L. Onyebueke, A. Abatan, Characterizing and modeling mechanical properties of nanocomposites-review and evaluation, J Miner Mater Charact Eng 9 (2010) 275–319. https://doi.org/10.4236/jmmce.2010.94022.

[54] S. Thankappan, J. Abraham, S.C. George, S. Thomas, Rheological characterization of nanocomposites, in: Characterization of nanomaterials: Advances and key technologies, Elsevier, 2018: pp. 167–189. https://doi.org/10.1016/B978-0-08-101973-3.00006-7.

[55] Z.M. Ariff, A. Ariffin, N. Azura, A. Rahim, Rheological behaviour of polypropylene through extrusion and capillary rheometry coating view project plastic product failure associated with polymer flow behaviour in complex geometries view project, 2014. https://www.researchgate.net/publication/236737926.

[56] C.J. Phillips, T.R.H. Davies, Determining rheological parameters of debris flow material, Geomorphology, 1991. https://doi.org/10.1016/0169-555X(91)90022-3.

[57] G.B. Thurston, Rheological parameters for the viscosity viscoelasticity and thixotropy of blood, Biorheology 16(3) (1979) 149–162. https://doi.org/10.3233/BIR-1979-16303.

[58] S. Salifu, D. Desai, O. Ogunbiyi, K. Mwale, Recent development in the additive manufacturing of polymer-based composites for automotive structures: A review, Int J Adv Manuf Technol 119 (2022) 6877–6891. https://doi.org/10.1007/s00170-021-08569-z.

[59] F.T. Jahromi, M. Nikzad, K. Prasad, J. Norén, M. Isaksson, A. Arian, I. Sbarski, Additive manufacturing of polypropylene micro and nano composites through fused filament fabrication for automotive repair applications, Polym Adv Technol 34 (2023) 1059–1074. https://doi.org/10.1002/pat.5952.

[60] S. Veličković, B. Stojanović, L. Ivanović, S. Miladinović, S. Milojević, Application of nanocomposites in the automotive industry, Mob Veh Mech 45 (2018). https://doi.org/10.24874/mvm.2019.45.03.05.

[61] F.T. Jahromi, M. Nikzad, K. Prasad, J. Norén, M. Isaksson, A. Arian, I. Sbarski, Additive manufacturing of polypropylene micro and nano composites through fused filament fabrication for automotive repair applications, Polym Adv Technol (2022). https://doi.org/10.1002/pat.5952.

[62] J.W. Lee, S.J. Park, Y.H. Kim, Numerical prediction of thermal stress-strain behavior on the wire-directed energy deposition additive manufacturing for automotive component, Mod Phys Lett B 36 (2022). https://doi.org/10.1142/S0217984922420040.

[63] S. Sagadevan, M.M. Shahid, Z. Yiqiang, W.C. Oh, T. Soga, J. Anita Lett, S.F. Alshahateet, I. Fatimah, A. Waqar, S. Paiman, M.R. Johan, Functionalized graphene-based nanocomposites for smart optoelectronic applications, Nanotechnol Rev 10 (2021) 605–635. https://doi.org/10.1515/ntrev-2021–0043.

[64] T.P. Nguyen, Polymer-based nanocomposites for organic optoelectronic devices: A review, Surf Coat Technol 206 (2011) 742–752. https://doi.org/10.1016/j.surfcoat.2011.07.010.

[65] A.R. Boccaccini, M. Erol, W.J. Stark, D. Mohn, Z. Hong, J.F. Mano, Polymer/bioactive glass nanocomposites for biomedical applications: A review, Compos Sci Technol 70 (2010) 1764–1776. https://doi.org/10.1016/j.compscitech.2010.06.002.

[66] Y. Bozkurt, E. Karayel, 3 D printing technology; methods, biomedical applications, future opportunities and trends, J Mater Rese Technol 14 (2021) 1430–1450. https://doi.org/10.1016/j.jmrt.2021.07.050.

[67] G.J. Schiller, Additive manufacturing for aerospace, in: IEEE aerospace conference proceedings, IEEE Computer Society, 2015. https://doi.org/10.1109/AERO.2015.7118958.

[68] A. Bhat, S. Budholiya, S.A. Raj, M.T.H. Sultan, D. Hui, A.U.M. Shah, S.N.A. Safri, Review on nanocomposites based on aerospace applications, Nanotechnol Rev 10 (2021) 237–253. https://doi.org/10.1515/ntrev-2021–0018.

[69] F.D. Guerra, M.F. Attia, D.C. Whitehead, F. Alexis, Nanotechnology for environmental remediation: Materials and applications, Molecules 23 (2018). https://doi.org/10.3390/molecules23071760.

[70] A.K. Maurya, R. Gogoi, G. Manik, Polymer-based nanocomposites for removal of pollutants from different environment using catalytic degradation, in: A.E. Shalan, A.S. Hamdy Makhlouf, S. Lanceros-Méndez (Eds.), Advances in nanocomposite materials for environmental and energy harvesting applications, Springer International Publishing, Cham, 2022: pp. 331–368. https://doi.org/10.1007/978-3-030-94319-6_11.

[71] P. He, B. Derby, Inkjet printing ultra-large graphene oxide flakes, 2 D Mater 4 (2017). https://doi.org/10.1088/ 2053–1583 /aa629e.

# Index

## M



## N



## O



## P



## Q



## R



## S



## T



## U



## V



## W



## X



## Y



## Z

FSC
www.fsc.org
MIX
Papier | Fördert gute Waldnutzung
FSC® C083411

Kolibri 360 GmbH
Ferdinand-Jühlke-Straße 7
99095 Erfurt, Deutschland
produktsicherheit@kolibri360.de